Polymeric Materials

Polymeric Materials

Surfaces, Interfaces and Bioapplications

Special Issue Editors

Marta Fernández-García
Alexandra Muñoz-Bonilla
Coro Echeverría
Agueda Sonseca
Marina P. Arrieta

MDPI • Basel • Beijing • Wuhan • Barcelona • Belgrade

MDPI

Special Issue Editors

Marta Fernández-García
Instituto de Ciencia y Tecnología
de Polímeros (ICTP-CSIC)
Spain

Alexandra Muñoz-Bonilla
Instituto de Ciencia y Tecnología
de Polímeros (ICTP-CSIC)
Spain

Coro Echeverría
Instituto de Ciencia y Tecnología
de Polímeros (ICTP-CSIC)
Spain

Agueda Sonseca
Instituto de Ciencia y Tecnología
de Polímeros (ICTP-CSIC)
Spain

Marina P. Arrieta
Universidad Complutense de
Madrid (UCM)
Spain

Editorial Office
MDPI
St. Alban-Anlage 66
4052 Basel, Switzerland

This is a reprint of articles from the Special Issue published online in the open access journal *Materials* (ISSN 1996-1944) from 2018 to 2019 (available at: https://www.mdpi.com/journal/materials/special_issues/polymeric_bioapplications).

For citation purposes, cite each article independently as indicated on the article page online and as indicated below:

LastName, A.A.; LastName, B.B.; LastName, C.C. Article Title. *Journal Name* **Year**, *Article Number*, Page Range.

ISBN 978-3-03897-962-3 (Pbk)
ISBN 978-3-03897-963-0 (PDF)

Contents

About the Special Issue Editors

Marta Fernández-García is a research scientist at the Institute of Polymer Science and Technology. She belongs to the Spanish National Research Council (ICTP-CSIC) and is a leader of the Macromolecular Engineering Group (MacroEng). She is a co-author of ca. 160 articles and several book chapters and a co-editor of a book (Royal Society of Chemistry). She has supervised eight PhD theses and some minor theses. She has served as an international advisory board member of *European Polymer Journal* (Elsevier) and now is an editor member of *International Journal of Molecular Sciences*: Material Science (MDPI) and *International Journal of Polymer Science* (Hindawi). Her research interests include block copolymers, glycopolymers, antimicrobial and antifouling polymers, sustainable materials, and recycling. She is member of the Interdisciplinary Platform for Sustainable Plastics towards a Circular Economy (SusPlast-CSIC).

Alexandra Muñoz-Bonilla completed her PhD in 2006 at the Institute of Polymer Science and Technology (ICTP-CSIC), in the area of controlled radical polymerization. During her PhD studies, she also held a one-year position at the University of Warwick, UK, as a Marie Curie visiting research student. Afterwards, she carried out postdoctoral research at the University of Bordeaux (France) in the field of polymer surfaces and interfaces and held a second postdoctoral position at the Eindhoven University of Technology (The Netherlands) in the field of colloidal systems. Currently, she is a tenured scientist at the ICTP-CSIC, having published nearly 90 articles and several book chapters and co-edited a RSC book. Now, she serves as editor for *Coatings* (MDPI) and *Advances in Polymer Technology* (Hindawi/Wiley). Her main research interests are the development of antimicrobial materials and bio-based and biodegradable polymeric systems, also participating in SusPlast-CSIC advancement.

Coro Echeverria received her PhD in applied chemistry and polymeric materials in 2011 at the University of the Basque Country (EHU, Spain). Her main research area is in the field of polymeric materials, its characterization, and applications, and her scientific interest is the study and development of multifunctional stimuli-responsive polymeric systems. During her postdoctoral period, 2013–2017, she worked at the New University of Lisbon (UNL) and CENIMAT on the project "Cellulose in Motion", which focused on the development of bio-based actuators from cellulosic liquid crystalline systems with an emphasis on the structure–properties relationship and flow behavior. Since 2018, she has been a Juan de la Cierva researcher at ICTP-CSIC. Her research is devoted to the development of multifunctional stimuli-responsive bio-based polymer systems with antimicrobial properties, aligned with the objectives of SusPlast-CSIC.

Agueda Sonseca obtained her PhD in 2015 at the Polytechnic University of Valencia, supported by a FPU predoctoral fellowship in the field of biodegradable nanocomposites with shape memory properties and possible biomedical applications. Afterwards, she was hired as a postdoctoral researcher in the Division of Biomaterials and Microbiological Technologies of the West Pomeranian University of Technology (Poland), where she was involved in the development of new biodegradable elastomeric templates for heart tissue engineering. Since September 2017, she has been a part of MacroEng group at ICTP-CSIC, being financially supported by several postdoctoral contracts obtained in national and autonomic competitive calls. She is also member of SusPlast-CSIC, and her research interests are mainly oriented towards the development of multifunctional bio-based, biodegradable, and biocompatible polymers with well-defined structures by enzymatic polymerization.

Marina P. Arrieta currently works at the Organic Chemistry Department of the Complutense University of Madrid (UCM, Spain). She holds a BS in biochemistry from the National University of Córdoba (Argentina), a MS in food technology from the Catholic University of Cordoba (Argentina), a MS in polymer science and technology from UNED (Spain), and an international PhD in science, technology, and food management from the Polytechnic University of Valencia (Spain), awarded with the extraordinary PhD-Thesis award. She has been an AECID, Santiago Grisolía, Juan de la Cierva, and UCM fellow. She has experience with the synthesis, processing (melt-blending, extrusion, injection molding, electrospinning, etc.), and characterization of bio-based and biodegradable polymers and their nanocomposites with active and multifunctional properties for sustainable food packaging or agricultural applications. She has published 42 papers with more than 1300 cites (h-index: 18) and several book chapters in the field of bio-based and biodegradable polymers.

materials MDPI

Editorial

Polymeric Materials: Surfaces, Interfaces and Bioapplications

Alexandra Muñoz-Bonilla [1,2] , Coro Echeverría [1,2] , Águeda Sonseca [1,2] , Marina P. Arrieta [3] and Marta Fernández-García [1,2,*]

[1] Instituto de Ciencia y Tecnología de Polímeros (ICTP-CSIC), C/Juan de la Cierva 3, 28006 Madrid, Spain; sbonilla@ictp.csic.es (A.M.-B.); cecheverria@ictp.csic.es (C.E.); agueda@ictp.csic.es (A.S.)
[2] Interdisciplinary Platform for Sustainable Plastics towards a Circular Economy, SUSPLAST, 28006 Madrid, Spain
[3] Facultad de Ciencias Químicas, Universidad Complutense de Madrid (UCM), Av. Complutense s/n, Ciudad Universitaria, 28040 Madrid, Spain; marina.arrieta@gmail.com
* Correspondence: martafg@ictp.csic.es; Tel.: +34-912587530

Received: 16 April 2019; Accepted: 21 April 2019; Published: 22 April 2019

Abstract: This special issue "Polymeric Materials: Surfaces, Interfaces and Bioapplications" was proposed to cover all the aspects related to recent innovations on surfaces, interfaces and bioapplications of polymeric materials. The collected articles show the advances in polymeric materials, which have tremendous applications in agricultural films, food packaging, dental restoration, antimicrobial systems and tissue engineering. We hope that readers will be able to enjoy highly relevant topics that are related to polymers. Therefore, we hope to prove that plastics can be a solution and not a problem.

Keywords: surface modification/functionalization; surface segregation; micro- and nanopatterned films; blends and (nano)composites; coatings; surface wettability; stimuli-responsive materials/smart surfaces; bioapplications

---

Polymeric materials have moved from making the progress of the twentieth century to becoming the materials of the future to be reviled and persecuted by problems that were mainly generated by the ignorance of citizens, businesses and governments. These problems have resulted in the planet being contaminated and the resulting consequences. A world without plastics is hardly imaginable and for this reason, the European Community is proposing some goals related to the production, use and recyclability of plastics: (1) 60% reuse and recycling of all plastic packaging by 2030; and (2) 100% reuse, recycling and/or recovery of all plastic packaging in the whole EU by 2040 [1]. Recently, Devasahayam et al. [2] pointed out the advantages of recycling polymers in mineral and metallurgical processing. For example, plastics in e-wastes can be used as fuels and reductants in recovering valuable metals. In another example, the epoxy resins can be used as a binder/reductant or fuel source, which offers high compression strength under ambient conditions. This far exceeds the heat induration strength and provides savings in terms of costs, energy and emissions during the iron ore pelletization. Moreover, several review have focused on solid plastic waste recycling, discussing both mechanical and chemical recycling [3,4]. Of all types of waste, the largest amount of waste produced is packaging waste (near 40%), which has short life times. Therefore, EU has placed a limitation on single-use plastic to decrease this ratio. Another alternative to the recycling process and reducing the production of plastics from non-renewable resources is the use of biodegradable and/or bio-based polymers, respectively.

One of the reviews in this special issue (SI) focuses on the use of natural and bio-based polymers as antimicrobial systems and their potential mainly in biomedical and food applications, but also in water purification and coating technology [5]. However, natural and bio-based materials frequently have lower performance than traditional synthetic polymers. Therefore, it is necessary to make modifications or adjustments during the processing steps in order to modulate their final performances. In one of the SI articles, Samper et al. [6] analyzed the influence of small amounts of biodegradable polymers, such as poly(lactic acid), polyhydroxybutyrate and thermoplastic starch in the recycled polypropylene. It is shown that the recycling of polypropylene blended with these bio-based and biodegradable polymers is hardly affected when it is used at a proportion higher than 5 wt %. In this sense, the review of Luzi et al. [7] presents the blending of bio-based and/or biodegradable polymers with traditional synthetic polymers for packaging applications with an optional use of bio-based nanofillers. This nicely highlighted how these bio-based materials enhance the gas/water/light barrier properties and the compostability and migration performance of blends. Moreover, they also discuss the effect of incorporating bio-based nanofillers on the overall behavior of nanocomposite systems that is constituted of synthetic polymers, which is combined with biodegradable and/or bio-based plastics.

The use of natural polymers is also presented in another article wherein alginate crosslinking by $CaCl_2$ is obtained to create modified-release drug delivery systems with mucoadhesive properties [8]. The authors present the production of microparticles by the spray drying technique, which enables us to obtain microparticles with a low moisture content, high drug loading, a high production yield and a prolonged release of soluble drugs. Peng's group [9] reported the use of chitosan with wood auto-hydrolysates that are obtained in the pulping process by hydrothermal extraction, which contains a considerable amount of hemicelluloses and slight lignin, in order to form films by the casting method. These films possess a higher tensile strength, better thermal stability, higher transmittances, lower water vapor permeability and superior oxygen barrier properties compared to those without chitosan due to the crosslinking interaction between the components, which occurs due to the Millard reaction. In another article, Ma et al. [10] used fibers from waste corn stalks as reinforcing materials in friction composites. They found that the incorporation of corn stalk fibers had a positive effect on the friction coefficients and wear rates of friction composites. The results revealed that the satisfactory wear resistance performances of these materials are associated with their worn surface morphologies and the formation of secondary contact plateaus.

Moreover, another polysaccharide, chitosan, is applied for scaffold preparation in tissue engineering. In more detail, Francolini et al. [11] analyzed the conjugation of chitosan with graphene oxide. Depending on its oxidation degree, the resulting scaffolds present improved or reduced mechanical performance and best or worst cytocompatibility as tested in human primary dermal fibroblasts. Another review of Foster's group [12] meticulously displays the problem of disc degeneration, which affects a great part of population, by describing the anatomy of the spine, the functions and biological aspects of the intervertebral discs. They point out that although there are numerous studies focusing on tissue engineering for disc degeneration, more progress needs to be made.

Focusing on some actual problems, dental restoration failures remain a major challenge in dentistry. In another review, Xu's group [13] provided information on the development and properties of innovative antibacterial dental polymeric composites, antibacterial bonding agents, bioactive root caries composites, adhesives and antibacterial and protein-repellent endodontic sealers. These polymeric materials substantially inhibit biofilm growth and greatly reduce acid production and polysaccharide synthesis of biofilms. Following with antimicrobial polymeric materials, Lienkamp's group [14] describes the development of amphiphilic copolymers of oxanorbornene monomer bearing N-*tert*-butyloxycarbonyl protected cationic groups with an oxanorbornene-functionalized poly(ethylene glycol) macromonomer. After this, these comb-like copolymers are surface-attached to polymer hydrogels, giving rise to a material that is simultaneously antimicrobial and protein-repellent. In another article, Ji et al. [15] analyzed the antifouling behavior directly in the natural seawater of different carbon nanotubes-modified polydimethylsiloxane nanocomposites by using the multidimensional scale analyses method.

Palza et al. [16] carefully reviewed the development of polymeric materials with electroactivity, such as intrinsically electric conductive polymers, percolated electric conductive composites and ionic conductive hydrogels. They evaluated their use in the electrical stimulation of cells, drug delivery, artificial muscles and as antimicrobial materials.

On the other hand, the breath figures approach is presented as an efficient method to obtain highly ordered porous materials with potential applications in cell culture and antimicrobial coatings, respectively [17,18]. These articles discuss the influence of the chemical nature of polymers, the solvent or the humidity in the preparation process on the final properties (porous size, surface energy, etc.). Another approach is presented in the article of Lavieja et al. [19], where the use of a green laser in the range of nanosecond pulses was an effective method to obtain superhydrophobic and superhydrophilic surfaces on a white commercial acrylonitrile-butadiene-styrene copolymer and therefore, to control its wettability. The last article deals with the surface modification method to produce gradient wrinkles using a gradient light field. Li et al. [20] described the easy control of the gradient wavelength of wrinkles by modulating the distance between the lamp and the substrate.

Finally, we would like to thank all authors for contributing to this collection in "Polymeric Materials: Surfaces, Interfaces and Bioapplications".

Funding: This research was funded by MINECO, Project MAT2016-78437-R, the Agencia Estatal de Investigación (AEI, Spain) and Fondo Europeo de Desarrollo Regional (FEDER, EU).

Acknowledgments: C.E. and Á.S acknowledge the Juan de la Cierva contracts (IJCI-2015-26432 and FJCI-2015-24405, respectively) from the Spanish Ministry of Science, Innovation and Universities. M.P.A. thanks Universidad Complutense de Madrid for her postdoctoral contract (Ayuda posdoctoral de formación en docencia e investigación en los Departamentos de la UCM).

Conflicts of Interest: The authors declare no conflict of interest.

References

1. Europe, P. An analysis of european plastics production, demand and waste data. In *Plastics—The facts 2018*; PlasticsEurope: Brussels, Belgium, 2018.

2. Devasahayam, S.; Raman, R.K.S.; Chennakesavulu, K.; Bhattacharya, S. Plastics-villain or hero? Polymers and recycled polymers in mineral and metallurgical processing—A review. *Materials* **2019**, *12*, 655. [CrossRef] [PubMed]

3. Singh, N.; Hui, D.; Singh, R.; Ahuja, I.P.S.; Feo, L.; Fraternali, F. Recycling of plastic solid waste: A state of art review and future applications. *Compos. Part B Eng.* **2017**, *115*, 409–422. [CrossRef]

4. Ragaert, K.; Delva, L.; Van Geem, K. Mechanical and chemical recycling of solid plastic waste. *Waste Manag.* **2017**, *69*, 24–58. [CrossRef] [PubMed]

5. Muñoz-Bonilla, A.; Echeverria, C.; Sonseca, Á.; Arrieta, M.P.; Fernández-García, M. Bio-based polymers with antimicrobial properties towards sustainable development. *Materials* **2019**, *12*, 641. [CrossRef] [PubMed]

6. Samper, M.D.; Bertomeu, D.; Arrieta, M.P.; Ferri, J.M.; López-Martínez, J. Interference of biodegradable plastics in the polypropylene recycling process. *Materials* **2018**, *11*, 1886. [CrossRef] [PubMed]

7. Luzi, F.; Torre, L.; Kenny, J.M.; Puglia, D. Bio- and fossil-based polymeric blends and nanocomposites for packaging: Structure–property relationship. *Materials* **2019**, *12*, 471. [CrossRef] [PubMed]

8. Szekalska, M.; Sosnowska, K.; Czajkowska-Kośnik, A.; Winnicka, K. Calcium chloride modified alginate microparticles formulated by the spray drying process: A strategy to prolong the release of freely soluble drugs. *Materials* **2018**, *11*, 1522. [CrossRef] [PubMed]

9. Xu, J.-D.; Niu, Y.-S.; Yue, P.-P.; Fu, Y.-J.; Bian, J.; Li, M.-F.; Peng, F.; Sun, R.-C. Composite film based on pulping industry waste and chitosan for food packaging. *Materials* **2018**, *11*, 2264. [CrossRef] [PubMed]

10. Ma, Y.; Wu, S.; Zhuang, J.; Tong, J.; Xiao, Y.; Qi, H. The evaluation of physio-mechanical and tribological characterization of friction composites reinforced by waste corn stalk. *Materials* **2018**, *11*, 901. [CrossRef] [PubMed]

11. Francolini, I.; Perugini, E.; Silvestro, I.; Lopreiato, M.; Scotto d'Abusco, A.; Valentini, F.; Placidi, E.; Arciprete, F.; Martinelli, A.; Piozzi, A. Graphene oxide oxygen content affects physical and biological properties of scaffolds based on chitosan/graphene oxide conjugates. *Materials* **2019**, *12*, 1142. [CrossRef] [PubMed]

12. Frost, B.A.; Camarero-Espinosa, S.; Foster, E.J. Materials for the spine: Anatomy, problems and solutions. *Materials* **2019**, *12*, 253. [CrossRef] [PubMed]

13. Zhang, K.; Baras, B.; Lynch, C.D.; Weir, M.D.; Melo, M.A.S.; Li, Y.; Reynolds, M.A.; Bai, Y.; Wang, L.; Wang, S.; et al. Developing a new generation of therapeutic dental polymers to inhibit oral biofilms and protect teeth. *Materials* **2018**, *11*, 1747. [CrossRef] [PubMed]

14. Kurowska, M.; Widyaya, V.T.; Al-Ahmad, A.; Lienkamp, K. Surface-attached poly(oxanorbornene) hydrogels with antimicrobial and protein-repellent moieties: The quest for simultaneous dual activity. *Materials* **2018**, *11*, 1411. [CrossRef] [PubMed]

15. Ji, Y.; Sun, Y.; Lang, Y.; Wang, L.; Liu, B.; Zhang, Z. Effect of cnt/pdms nanocomposites on the dynamics of pioneer bacterial communities in the natural biofilms of seawater. *Materials* **2018**, *11*, 902. [CrossRef] [PubMed]

16. Palza, H.; Zapata, P.A.; Angulo-Pineda, C. Electroactive smart polymers for biomedical applications. *Materials* **2019**, *12*, 277. [CrossRef] [PubMed]

17. Muñoz-Bonilla, A.; Cuervo-Rodríguez, R.; López-Fabal, F.; Gómez-Garcés, J.L.; Fernández-García, M. Antimicrobial porous surfaces prepared by breath figures approach. *Materials* **2018**, *11*, 1266. [CrossRef] [PubMed]

18. Ruiz-Rubio, L.; Pérez-Álvarez, L.; Sanchez-Bodón, J.; Arrighi, V.; Vilas-Vilela, J.L. The effect of the isomeric chlorine substitutions on the honeycomb-patterned films of poly(x-chlorostyrene)s/polystyrene blends and copolymers via static breath figure technique. *Materials* **2019**, *12*, 167. [CrossRef] [PubMed]

19. Lavieja, C.; Oriol, L.; Peña, J.-I. Creation of superhydrophobic and superhydrophilic surfaces on abs employing a nanosecond laser. *Materials* **2018**, *11*, 2547. [CrossRef] [PubMed]

20. Li, H.; Sheng, B.; Wu, H.; Huang, Y.; Zhang, D.; Zhuang, S. Ring wrinkle patterns with continuously changing wavelength produced using a controlled-gradient light field. *Materials* **2018**, *11*, 1571. [CrossRef] [PubMed]

Review

![materials logo] **MDPI**

Bio-Based Polymers with Antimicrobial Properties towards Sustainable Development

Alexandra Muñoz-Bonilla [1], Coro Echeverria [1] , Águeda Sonseca [1] , Marina P. Arrieta [2] and Marta Fernández-García [1,*]

[1] Instituto de Ciencia y Tecnología de Polímeros (ICTP-CSIC), C/Juan de la Cierva 3, 28006 Madrid, Spain; sbonilla@ictp.csic.es (A.M.-B.); cecheverria@ictp.csic.es (C.E.); agueda@ictp.csic.es (Á.S.)
[2] Facultad de Ciencias Químicas, Universidad Complutense de Madrid (UCM), Av. Complutense s/n, Ciudad Universitaria, 28040 Madrid, Spain; marina.arrieta@gmail.com
* Correspondence: martafg@ictp.csic.es; Tel.: +34-912587530

Received: 28 January 2019; Accepted: 15 February 2019; Published: 20 February 2019

Abstract: This article concisely reviews the most recent contributions to the development of sustainable bio-based polymers with antimicrobial properties. This is because some of the main problems that humanity faces, nowadays and in the future, are climate change and bacterial multi-resistance. Therefore, scientists are trying to provide solutions to these problems. In an attempt to organize these antimicrobial sustainable materials, we have classified them into the main families; i.e., polysaccharides, proteins/polypeptides, polyesters, and polyurethanes. The review then summarizes the most recent antimicrobial aspects of these sustainable materials with antimicrobial performance considering their main potential applications in the biomedical field and in the food industry. Furthermore, their use in other fields, such as water purification and coating technology, is also described. Finally, some concluding remarks will point out the promise of this theme.

Keywords: bio-based polymers; antimicrobial; biodegradable; sustainable; eco-friendly

1. Introduction

Nowadays, plastics have gone from being outstanding materials that make life easier for us to being a serious concern for our ecological system. The European Council has pointed out the need to reduce our dependency on fuel and gas imports and to create sustainable energy, that is, achieve sustainable development by 2030. The 17 goals that cover this sustainable development include food security, health, sustainable consumption and production, the sustainable management of natural resources, clean oceans, and climate change [1]. Bio-based polymers have emerged as a potent solution for replacing petroleum-based polymeric materials and reducing the dependence on the depleting crude oil reserve. Besides this, many of the existing bio-based polymers can be biodegradable; in particular, natural bio-based polymers, such as polysaccharides and proteins, but also several synthetic biopolymers, such as poly(lactic acid). Biodegradability is also an important and desired property in many applications, including food packaging and agricultural applications, and contributes to sustainability as it reduces the waste impact of oil-based polymers. Nowadays, although the bio-plastics market represents only about 1% of the 335 million tons of plastic that the world produces annually [2,3], their production is continuously growing [4]. In some of the uses of biopolymers, additional properties are also needed; for instance, antimicrobial properties are desired in food packaging and biomedical devices, wherein microbial contamination can cause serious problems for public health and safety.

On account of this background, in this article we intend to show the capacities of bio-based polymers to be antimicrobial materials, centered on both natural and synthetic polymers.

There are extensive and excellent reviews about antimicrobial polymeric materials [5–12] in which the methodologies of encapsulation and blending with antimicrobial organic and inorganic compounds as well as their possible mechanisms of action are discussed [9]. However, most of them are mainly focused on fossil-oil derivatives. On the other hand, there are also many reviews about bio-based polymers [13–20]; however, only a few are related to antimicrobial activity [21,22]. Therefore, this review does not intend to gather all of the works performed to date but give hints on the subject and make the general public aware of the great possibilities of sustainable polymeric materials.

First, we will mention polysaccharides, which are the most abundant and exploited family. Following the natural systems, the proteins with antimicrobial activity will be described. Then, synthetic systems based on natural products will be analyzed; specifically, polyesters and polyurethanes. Since the literature regarding natural and bio-based antimicrobial polymeric materials is significantly wide, we focus the analysis mainly on the research performed in the field during recent years. It is not our purpose to do an extensive review; instead, we will highlight some of these interesting materials. Finally, we will conclude with some reflections on this hot topic.

2. Polysaccharides

Polysaccharides are the macromolecules that belong to the components of life, together with proteins and nucleic acids. They determine the functionality and specificity of species. Their functionalities divide them into structural, storage, and gel-forming polysaccharides. Due to their abundance and excellent properties, such as biodegradability, they are unique materials to develop interesting antimicrobial bio-based materials.

2.1. Chitosan

Chitosan (CS) is a linear polysaccharide with inherent antimicrobial activity that is derived from naturally occurring chitin, which is, after cellulose, the most common biopolymer on earth. It is sourced mainly from crustacean shellfish and certain fungi. Chitosan is a partially or completely *N*-deacetylated derivative of chitin, chemically composed of *N*-acetylglucosamine and glucosamine units joined through β(1−4)glycosidic linkages, and has primary amino groups that provide a positive charge under acidic pH (pK$_a$ about 6.3) and decent antimicrobial properties against a wide range of micro-organisms (Figure 1) [23,24].

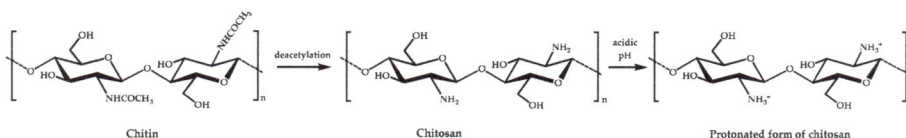

Figure 1. The chemical structure of chitin and chitosan and the protonated form of chitosan.

Although the exact mechanism of action is still not completely understood, the most accepted mechanism is based on electrostatic interactions between positively charged chitosan and the negatively charged micro-organism membrane [25]. Nevertheless, other modes of action, such as interactions with DNA or the formation of complexes with metal ions, seem to be involved [26]. This antimicrobial activity is strongly affected by its structural characteristics, such as molecular weight or degree of deacetylation, and by environmental conditions, such as pH, temperature, or ionic strength [27]. Compared with other antimicrobial polymers, chitosan offers several advantages, as it has a natural origin, is biodegradable, biocompatible, and nontoxic for mammalian cells, and has been approved by the U.S. Federal Drug Administration (FDA) and the E.U. as safe (GRAS, Generally Recognized As Safe) for tissue engineering, drug delivery, wound dressing, dietary use, and plant protection applications. Besides this, chitosan has excellent film-forming ability and good mechanical and barrier properties; thus, it has great potential in food packaging [21]. However, its biocidal activity and solubility are reduced in neutral pH conditions [28], which limit its use in many applications. Therefore, chemical modifications of chitosan, typically either at amino (the secondary C2 NH$_2$ group) or hydroxyl groups (the primary C6 OH and secondary C3 OH groups), aim to produce derivatives with enhanced properties to widen its applications [29,30]. A huge number of studies have been carried out on the preparation of antimicrobial chitosan derivatives mainly via quaternization and carboxylation. However, all of these modifications propose to improve its solubility and antimicrobial activity while also maintaining its original biodegradability and biosafety. Next, the most common and recent functional groups and derivatives used to improve its antimicrobial activity without affecting its inherent properties are discussed viz. by chemical modification and blending with organic and inorganic antimicrobial agents.

2.1.1. Chitosan Modification

Probably the most common method for introducing a permanent positive charge into chitosan chains is by the formation of quaternary ammonium groups by either direct quaternization of the primary amino group at the C2 position or by incorporating such groups at any of the reactive moieties (hydroxyl and amino groups). For instance, in a recent study, chitosan derivatives with triple quaternary ammonium groups were synthesized via Schiff-based reactions. Although the resulting samples with a high positive charge exhibit significantly enhanced antifungal activity, the preparation method required multiple steps [31].

In another study, chitosan derivatives were prepared by reaction with different quaternary ammonium salts containing a bromide end-group capable of reacting with the amino or hydroxyl groups of chitosan [32]. The ammonium salts benzalkonium bromide, pyridinium bromide, and triethyl ammonium bromide were previously obtained by a quaternization reaction between 1,4-dibromobutane and the respective tertiary amines. These chitosan derivatives with quaternary ammonium groups showed much lower minimum inhibitory concentration (MIC) values against Gram-negative *Escherichia coli* and Gram-positive *Staphylococcus aureus* bacteria than neat chitosan. Also, in the case of *S. aureus*, the type of substitution influences the activity, with better properties for the pyridinium derivative. Although an important improvement of the activity is generally obtained [33], these chemical modifications often lead to unselective reactions at the amine, the hydroxyl, or both, as occurred in the last example. For instance, the *N*-methylation with methyl iodide typically provokes partial *O*-methylation [34]. Similarly, chitosan derivatives only modified at the OH positions are exceptional. Besides this, it is difficult to obtain a high degree of substitution in most of the cases, in particular with long alkyl chains, as these syntheses normally need to be carried out in acidic conditions or heterogeneous media [35].

Recent studies have been directed at obtaining better selectivity and a high degree of substitution by using several protecting groups. Sahariah et al. [36,37] have developed an efficient method for the selective modification of chitosan with up to 100% substitution of the amino groups. They prepared protected di-tertbutyldimethylsilyl (TBDMS) chitosan and introduced quaternary ammonium groups with different alkyl chain lengths by reductive amination. All of the prepared derivatives showed bactericidal properties and good selectivity when tested with human red blood cells (RBCs). It was also shown that the activity was influenced by the length of the alkyl chain and by the tested micro-organisms; derivatives with a short alkyl chain presented high activity against *S. aureus*, while longer alkyl chains were more active against Gram-negative *E. coli* and *Enterococcus faecalis* bacteria [36]. These derivatives also demonstrated effectiveness towards *S. aureus* biofilms, especially those with short alkyl chains [37].

In another recent work, the quaternary ammonium groups were introduced exclusively at the hydroxyl groups by previous protection of the −NH$_2$ groups via a Schiff-based condensation reaction with benzaldehyde [38]. By this way, it is possible to prepare positively charged chitosan derivatives with free primary amino groups, which is important as these amino groups have a key role in the biological activity of chitosan, such as in its antioxidant activity. The obtained *O*-quaternized chitosans showed an improved water solubility and antibacterial activity against Gram-positive bacteria. Remarkably, the cytotoxicity for the AT2 cell line was significantly lower than that of the free quaternary ammonium salts.

A different strategy was followed in a recent work, in which quaternized chitosan samples with alkyl chains were prepared by an external acid-free method [39]. In this approach, a quaternary ammonium molecule containing carboxylic acid was synthesized, which acted as a reactant for attachment onto the amino groups of chitosan via 1-ethyl-3-(3-dimethylaminopropyl) carbodiimide/*N*-hydroxysuccinimide (EDC/NHS) chemistry and also acted as an acid for the dissolution of chitosan. The antimicrobial activity was tested against planktonic Gram-positive *Staphylococcus epidermidis* and *E. coli* bacteria, and *Candida albicans* fungi, by evaluating the MIC and minimum bactericidal and fungicidal concentration (MBC and MFC, respectively) and against biofilms. Although the chitosan derivative showed growth inhibition and biocidal effects, the results against Gram-negative bacteria were modest.

The preparation of carboxyalkyl chitosan derivatives, especially carboxymethyl chitosan (CMC) polymers, is also a common strategy to improve the water solubility of chitosan and enhance the antimicrobial activity over the whole range of pH [40]. Typically, carboxymethyl chitosan is synthesized through carboxymethylation of the primary amino and alcohol groups, leading to *N*-CMC and *O*-CMC chitosan, respectively, as well as *N,N*-CMC and *N,O*-CMC derivatives. The incorporation of the carboxymethyl group at the reactive positions of chitosan (amine and hydroxyl) can be controlled by the reaction conditions, such as temperature or concentration [41–43]. In general, the antimicrobial activity of *O*-CMC is greater in comparison with the rest of the derivatives, as it presents a higher number of free amino groups [41], although other parameters, such as the degree of deacetylation, the degree of substitution, the molecular weight, and the pH of the medium, can affect its antimicrobial capacity [44].

In addition, carboxyalkyl chitosan is commonly modified by the introduction of other functional groups in order to improve the activity. For instance, several thiosemicarbazone *O*-carboxymethyl CS derivatives have been prepared towards a condensation reaction of thiosemicarbazide *O*-carboxymethyl CS with *o*-hydroxybenzaldehyde, *p*-methoxybenzaldehyde, and *p*- chlorobenzaldehyde [45]. The antimicrobial activity of the prepared derivatives was tested against Gram-positive *Bacillus subtilis* and *S. aureus*, Gram-negative *E. coli* bacteria, and *Aspergillus fumigatus*, *Geotrichum candidum*, and *C. albicans* fungi using the inhibition zone method. The microbiological results showed that both the antibacterial and antifungal activities of the thiosemicarbazone *O*-carboxymethyl chitosan derivatives were better than those of the original *O*-carboxymethyl chitosan, especially the chloro-derivative.

In recent years, the effectiveness of carboxyalkyl chitosan against biofilm formation has also been studied [46,47]. A carboxymethyl chitosan with an *O*-carboxymethylation degree of ~90% was tested against Gram-positive and Gram-negative bacterial biofilm formation, and the results indicated that CMC provoked a reduction of 74.6% at 2.500 mg/mL and 81.6% at 0.156 mg/mL, respectively [46]. CMC also demonstrated the capability to prevent bacterial biofilm formation in dynamic conditions. Although the mechanism of action was not fully understood, it seems that the presence of CMC induces the flocculation of bacteria by surface charge neutralization that prevents initial bacterial adherence and cell–cell interaction. This group has also demonstrated the efficacy of CMC on the inhibition of a fungal biofilm of *Candida tropicalis*, *Candida parapsilosis*, *Candida krusei*, and *Candida glabrata* [47].

Chitosan and chitosan derivatives have also been conjugated with several cationic amino acids and antimicrobial peptides with the purpose of improving their activity. The incorporation of amino acids, such as arginine [25,48], typically leads to derivatives with excellent antimicrobial properties and high water solubility. Arginine has a guanidine group with a pKa of ~12.5; thus, it is positively charged over almost the whole range of pH. Arginine can be easily attached onto chitosan by using, for example, an EDC/NHS coupling reaction between the amino group of chitosan and the carboxylic acid of arginine [48]. Guanidine molecules have been also attached onto chitosan [49–51]; however, their selective introduction requires more complex strategies, such as the use of protecting groups [52].

Although the coupling of amino acids onto CS has been extensively explored, in the last few years the incorporation of antimicrobial peptides has attracted more attention. Antimicrobial peptides (AMPs), both host defense peptides and their synthetic analogues, are promising candidates as antimicrobial agents due to their high efficiency and low probability to induce bacterial resistance [53–55]. Furthermore, the preparation of these cationic peptide-polysaccharides might enhance the selectivity towards bacteria in comparison with mammalian cells, as their structure mimics the peptidoglycans found in the bacterial membrane [56]. Cationic chitosan-*graft*-polylysine and chitosan-*graft*-poly(lysine-*ran*-phenylalanine) have been prepared by N-carboxyanhydride (NCA) ring-opening copolymerization of α-amino acids initiated from the amino groups of chitosan. The resulting derivatives showed outstanding broad spectrum antimicrobial properties against Gram-negative *Pseudomonas aeruginosa* and *E. coli* and Gram-positive *S. aureus* bacteria (MIC values between 5 and 20 µg/mL) and the fungi *C. albicans* and *Fusarium solani* (MIC values between 0.2 and 0.9 µM,) while maintaining very high selectivity over human RBCs. In another strategy, the antimicrobial peptide poly(lysine$_{11}$-*stat*-phenylalanine$_{10}$), prepared by NCA ring-opening copolymerization, was modified by hexamethylene diisocyanate and then statistically grafted onto the acid-functionalized chitosan [57]. In this approach, the residual −COOH groups had to be further esterified, as an important decrease in the antibacterial activity was observed that partially counteracted the positive charge.

In more recent studies, click chemistry has been used to couple AMPs onto chitosan. For instance, a potent antimicrobial peptide, Dhvar-5 (sequence LLLFLLKKRKKRKY), with an N-terminal propargylglycine was attached onto azide-functionalized CS via Cu(I)-catalyzed azide-alkyne cycloaddition (CuAAC) [58]. Similarly, an anoplin peptide was grafted onto chitosan by using a CuAAC coupling reaction [59]. Azide moieties were anchored onto the amine groups of chitosan by using *tert*-butyldimethylsilyl (TBDMS) protection, whereas the anoplin peptides were synthesized bearing either N– or C–terminal alkyne groups (Figure 2).

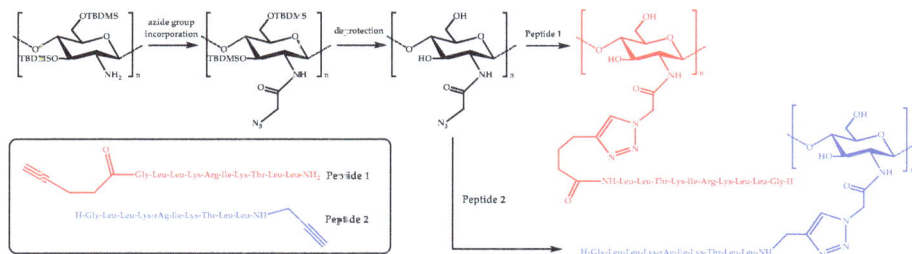

Figure 2. A schematic representation of the preparation of anoplin-chitosan derivatives by CuAAC click chemistry.

Then, several conjugates were obtained by varying the content of the attached peptide. Some of the resulting derivatives exhibited enhanced antimicrobial activity against *S. aureus*, *E. faecalis*, *E. coli*, and *F. aeruginosa* bacteria compared to anoplin or chitosan. In particular, the conjugates were very effective against *E. coli*, with MIC values as low as 4 µg/mL. More importantly, the hemotoxicity was significantly reduced in the case of the anoplin-chitosan derivatives.

Other click chemistry reactions, such as thiol-ene click chemistry, have been employed to prepare peptide–chitosan conjugates. The cationic peptide ε-poly(L-lysine) was attached onto chitosan in order to prepare broad-spectrum antimicrobial compounds, as the ε-poly(L-lysine) is effective against bacteria but presents poor activity against fungi [60]. In this coupling reaction, the chitosan was first functionalized with methacrylate groups, and the terminal amino group of the peptide was thiolated with homocysteine thiolactone hydrochloride. After the click reaction, the obtained cationic peptide-polysaccharides demonstrated both antibacterial and antifungal activities. In addition, the conjugates showed low hemolytic activity, good in-vitro biocompatibility when tested with bone mesenchymal stem cells, and scant evidence of in vivo toxicity.

In another recent example, cysteine-terminated HHC10 (KRWWKWIRW) AMP was grafted onto the C2 (amino) or C6 (hydroxyl) reactive centers of CS by thiol-maleimide click conjugation [61] (Figure 3). Remarkably, the peptide-polysaccharide with free amino groups of chitosan backbone (CSO-HHC) displayed higher antibacterial activity than the corresponding conjugate with the modified amino groups (CSN-HHC) due to its capacity for protonation, which increases its water solubility and the positive charge. Likewise, both conjugates showed lower hemolytic activity and cytotoxicity than the free peptide due to the effect of chitosan.

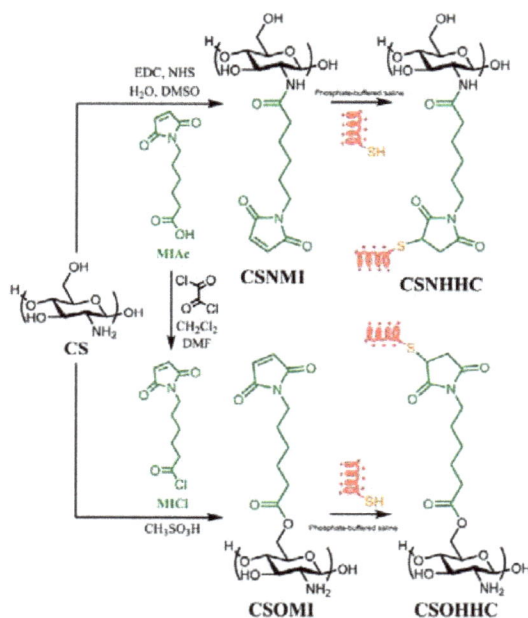

Figure 3. Preparation of the peptido-polysaccharides by conjugation of the HHC10 antimicrobial peptide (AMP) to the C-2 (amine) or C-6 (hydroxyl) positions of chitosan. Adapted from [62].

2.1.2. Chitosan Mixed with Antimicrobial Organic Compounds

Essential oils (EOs) are a product of aromatic plants, which contain multiple substances, including terpenes and aromatic and aliphatic compounds, such as esters, ethers, aldehydes, ketones, lactones, and phenols [62,63]. They have received great attention during the last few years due to their antioxidant and antimicrobial activities, in particular in the food industry [64]. However, the direct use of essential oils for food preservation is often limited due to their cost, poor solubility, toxicity, and aroma, which may impact on the sensory perception of foods. In this sense, essential oils can be incorporated into films, coatings, or capsules in a reduced dose that maintains their efficacy. Much effort has been made in the development of chitosan materials with essential oils, as chitosan has great potential as an active ingredient as well as in bio-based packaging films and edible films [65]. In general, the incorporation of essential oils improves the effectiveness of chitosan against fungi and food-borne bacteria. Essential oils can be incorporated either directly in the formulation of chitosan films [66] or previously encapsulated [67]. Likewise, the preparation of micro- and nanocapsules of chitosan derivatives as food additives has been extensively explored [68] as well as the covalent attachment of some components, such as gallic acid, onto chitosan [69,70]. For instance, in a recent publication, a rosemary essential oil was incorporated onto chitosan-montmorillonite nanocomposite films in different amounts (0.5%, 1%, and 2% v/v) [71]. Films containing the essential oil exhibited antimicrobial activity on the contact surface for both Gram-positive and Gram-negative bacteria tested by the inhibition zone method, whereas the chitosan films did not present any activity. Results in growth media obtained by the colony forming units (CFU) method, indicated, however, that the presence of the essential oil did not affect the antibacterial activity, and the montmorillonite decreased the activity as it can interact with chitosan and phenolic compounds of the rosemary essential oil. Thus, in the preparation of composite films for a food packaging application, the components and additives that are added to improve the mechanical and physico-chemical properties could affect their antimicrobial properties, so an optimal design is normally required. In another example, nanoemulsions of carvacrol were incorporated onto carboxymethyl chitosan films that were previously obtained by electrospray from CMC microgels [72]. The resulting composite films showed good antibacterial activity against *S. aureus* and *E. coli* and also the capability to prolong the shelf-life of wheat bread.

In the last few years, the nanoencapsulation of essential oils has attracted more attention compared to microencapsulation as smaller particles improve the solubility and dispersibility of the compounds. A clove essential oil was encapsulated by chitosan nanoparticles via the emulsion ionic gelation technique [73]. The resulting loaded nanoparticles demonstrated enhanced fungal activity against *Aspergillus niger* in comparison with empty chitosan nanoparticles and free oil. Similar results were found for a rosemary essential oil nanoencapsulated in chitosan/γ-polyglutamic acid nanoparticles, with a significant increase in the antibacterial activity against *B. subtilis* [74]. Likewise, an essential oil of Cyperus articulates was loaded into chitosan nanoparticles by an oil-in-water mixture and ionic gelation method [75]. These loaded particles also showed lower MIC values against *S. aureus* and *E. coli* compared to free oil and unloaded chitosan nanoparticles. However, these nanoparticles exhibited a higher cytotoxicity effect against MDA-MB-231 cells, probably due to the slow release of oil components encapsulated in the chitosan nanoparticles.

2.1.3. Chitosan with Metallic Nanoparticles

An important and highly explored strategy for improving the antimicrobial activity of chitosan is the incorporation of metal or metal-oxide nanoparticles (NPs), including Ag, Cu, ZnO, and TiO$_2$ NPs, and the preparation of nanocomposites. Among all existing nanoparticles, silver nanoparticles (AgNPs) have attracted much attention due to their potent antimicrobial activity. Several approaches have been used to prepare chitosan/AgNP nanocomposites, including physical and chemical strategies, in which the main objective is to reduce agglomeration, which is considered to be an important factor that affects the antimicrobial efficacy in the nanocomposites. Common methods imply the in situ preparation of AgNPs by the chemical reduction of silver salts; however, the used reducing agents may exhibit toxicity and also could interact with the functional groups of chitosan. Thus, more environmental friendly methods are becoming a priority nowadays [76]. In this sense, it was demonstrated that chitosan can act as both a reducing and stabilizing agent in the synthesis of AgNPs [77]. For instance, AgNPs stabilized with chitosan were synthesized at a large scale by a green method using autoclave, in which chitosan functions as a reducing agent as well as a stabilizer [78]. It was shown that, while chitosan only can prevent the growth of *S. aureus*, AgNPs stabilized with chitosan are also able to inhibit the growth of *E. coli* bacteria. Moreover, the inhibition zone from a disk diffusion test increased with the presence of the AgNPs. This was due to the additional modes of action of AgNPs by disrupting the cell wall of bacteria via several pathways and also by the release of Ag$^+$ ions, which can interact with bacterial DNA and proteins [79]. They also showed that the stabilization of chitosan reduces the cytotoxicity of AgNPs against L-929 fibroblast cells, which might be due to limited contact between the NPs and the cells and the controlled release of Ag$^+$.

Nevertheless, there is a serious concern related to the possible toxicity of AgNPs for the human body; for instance, when they are used in food packaging [80,81]. A possible solution might be the immobilization of silver nanoparticles to limit the leakage and diffusion of AgNPs. For example, laponite, which is a synthetic clay with a nano-sized and layered structure, was used to immobilize silver nanoparticles in chitosan films [82]. In this approach, quaternized chitosan was used as a reducing agent for the synthesis of AgNPs embedded in laponite, and subsequently the modified laponite was mixed with chitosan to prepare films by the casting solvent evaporation technique. Remarkably, the resulting films only released about 5.6% of AgNPs, which was much lower than films without laponite (about 29.1%). In addition, the films showed low toxicity and good antimicrobial activity against *E. coli* and *S. aureus* bacteria and *A. niger* and *Penicillium citrinum* fungi, and were capable of extending the shelf-life of fresh litchi.

In addition to AgNPs, other nanoparticles, such as Cu and CuO NPs [83,84], ZnO NPs [85,86], and TiO$_2$ NPs [87,88], have been used to improve the antimicrobial activity of chitosan. In the case of TiO$_2$ nanoparticles, it is accepted that their antimicrobial activity is based on photocatalytic processes under UV-light irradiation that generate reactive oxygen species (ROS) [89,90]. However, recent studies have also demonstrated biocidal properties of TiO$_2$ nanocomposites under visible light [88,91], which might improve their applicability since there is a low proportion of UV light in the total solar irradiance. Zhang et al. [88] prepared chitosan-TiO$_2$ composites with efficient antimicrobial activities under visible light. They incorporated TiO$_2$ nanoparticles into chitosan and evaluated the antimicrobial behavior against food-borne pathogenic microbes, including *E. coli*, *S. aureus*, *C. albicans*, and *A. niger*, under visible light irradiation (a 20 W daylight lamp). The films exerted high antimicrobial activity against the tested strains with 100% sterilization in 12 h. The good performance obtained under visible light irradiation was attributed to the decreased transmittance found in the visible light region, which enabled the films to have a photocatalytic antimicrobial effect.

2.2. Cellulose

Cellulose is a linear syndiotactic and semi-rigid homopolymer consisting of D-anhydro glucopyranose units (AGU), where each unit has three hydroxyl (OH) groups at the C2, C3, and C6 positions (see Figure 4). Cellulose is a semi-crystalline and high-molecular-weight homopolymer of β-D-glucopyranose units linked by β-1,4-linkages, where the repeat unit is a dimer of glucose: cellobiose. Cellulose and its derivatives are one of the most abundant natural biopolymers, and much progress has been made towards their study, characterization, and applications. Most cellulose derivatives are commercially available. Some are water-soluble, biodegradable, electro-neutral, and biocompatible. They are used in many industrial applications, such as packaging and textile production; however, in recent years, cellulose-based materials have been investigated regarding new advanced applications, such as sensors, liquid crystal polymers, soft-actuators, and biomaterials [92–94]. The increased interest in this natural polymer, obtained from renewable biomass feedstock, responds to the urgent need for the replacement of synthetic polymers to reduce the actual global dependence on fossil fuel sources, making possible the development of sustainable and ecofriendly functional materials.

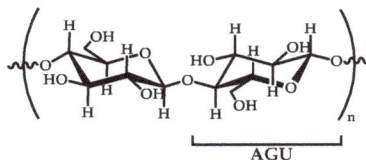

Figure 4. A schematic representation of cellobiose that shows the repeated anhydroglucopyranose units (AGUs).

As stated in the introduction, there is also an actual challenge involving the development of antimicrobial materials, which includes the development of biopolymers (natural polymers) and bio-based polymers with antimicrobial properties. Indeed, there is a deep concern regarding the increased resistance of microbes (bacteria, fungi) against actual antimicrobial agents. Micro-organisms are everywhere and they require only moisture, a source of carbon, and mild temperatures to multiply and prosper. Unfortunately, cellulose and its derivatives are an excellent medium that can serve as a supplier of moisture and even the growth of micro-organisms. So, modifications need to be done to impart antimicrobial properties to this natural polymer [95]. To do so, and also in the case of polysaccharides, three main approaches have been followed: cellulose and cellulose derivative modification (functionalization or grafting); blending with cationic molecules, essential oils, or antimicrobial polymers; and the incorporation of antimicrobial metal nanoparticles (silver, gold, etc.). Special recognition of nanocellulose is given below.

2.2.1. Cellulose Modification

The first strategy followed for the development of sustainable antimicrobial cellulosic materials considers the chemical modification/functionalization or grafting onto cellulose derivative surfaces [96] so that non-leachable materials are obtained. This strategy is often applied for the modification of nanocellulose, or uses microfibrillated cellulose as starting material, as we will describe in the following section. However, based on the literature from the last three years, research related to cellulose derivatives is scarce. As an example of this approach, Wu et al. reported the preparation of nisin-grafted cellulose membranes [97]. The authors first oxidized native cellulose using sulfuric acid, and then bonded nisin amino groups onto aldehyde groups of the oxidized cellulose. The obtained nisin-grafted cellulose membranes were then tested against Gram-variable *Alicyclobacillus acidoterrestris* bacteria, which are not pathogenic to humans but are implicated in the spoilage of fruits and cause a bad taste and flavor. The performed antimicrobial test confirmed the antimicrobial activity of the cellulosic material. Besides this, the authors determined that the antimicrobial efficacy increased as the oxidation time of the native cellulose increased.

Li et al. [98], using as starting material fully bleached eucalyptus kraft pulp fibers, took advantage of the layer-by-layer (LbL) technique to modify these fibers' surfaces with chitosan and lignin (LS), which present both antimicrobial and antioxidant properties. The electrostatic LbL technique is a simple and versatile polymer surface modification method that builds a nanostructured multilayer onto a solid substrate surface with the desired composition and properties [99]. For the preparation of multilayer deposition onto cellulosic fibers, the authors immersed the fibers into CS for a short period of time and rinsed the surface with water to remove the excess. The same procedure was followed for the deposition of the LS layer. By repeating those steps up to four times, the authors fabricated a multilayer of alternant CS and LS layers over cellulose fibers. They evaluated their antimicrobial activity by measuring their MIC against *E. coli* by the standard broth microdilution method. For the test, fibers modified with different numbers of layers, and fibers where the outermost layer was either CS or LS, were selected. The obtained results revealed that as the content of the bilayer increased the growth inhibition degree also increased, but always when the CS was located in the outermost layer.

2.2.2. Cellulose Mixed with Antimicrobial Organic Compounds

Another strategy described in the literature is the incorporation of essential oils into cellulose-based matrices imparting the antimicrobial property that cellulosic material lacks [100]. For instance, Heredia-Herrero et al. developed an antimicrobial plastic film based on the cellulose derivative ethyl cellulose (EC) [101]. This derivative comes from the substitution of some hydroxyl groups of the cellulose backbone with ethyl ether groups. EC is soluble in organic solvents but insoluble in water, non-toxic, versatile, and edible. From EC, it is possible to form tough films that not very flexible; hence, the addition of plasticizers is needed. Common plasticizers, being low-molecular-weight components, have the problem of migration from the polymer matrix, which affects the material's performance on the desired application. However, what is even more important and relevant to this contribution is that certain plasticizers also lead to serious environmental pollution in addition to affecting human health [102]. The strategy followed by Heredia-Guerrero et al. was the combination of EC with acetoxy-polydimethylsiloxane (PDMS), which interact to form hydrogen bonds, and with a clove essential oil. The EC-based film's antimicrobial activity was tested against, *E. coli, P. aeruginosa*, and *S. aureus* bacteria. The obtained results revealed that biofilm formation by *E. coli* was significantly inhibited due to the presence of the essential oil, with a film inhibition of 44% and 57% after 24 h and 48 h, respectively. In the case of *S. aureus*, significant inhibition occurred after 48 h, with 62% of biofilm inhibition [101]. In conclusion, the authors provided the cellulosic films with antimicrobial properties as well as improved their flexibility so that they could have potential application in food packaging.

Following the strategy of incorporating an essential oil into cellulosic matrices, Liakos et al. created antimicrobial cellulosic nanofibers by an electrospinning technique. For that, the authors used cellulose acetate (CA) as a biopolymer matrix to encapsulate different EOs within the CA fibers: cinnamon, lemongrass, peppermint, rosemary, and oregano [103,104]. CA is a biodegradable compound formed from the acetylation of cellulose. This biopolymer is amorphous, odorless, non-toxic, and water-vapor permeable, and shows excellent optical properties besides a high resistance to heat and chemicals [105]. For the encapsulation of the essential oils, the authors first dissolved a cellulosic polymer in acetone and then added the corresponding essential oil to the CA/acetone solution. The obtained CA/EO electrospun fibers showed a diameter size that ranged from 1 to 3 μm approximately. The antimicrobial properties of the CA/EO fibers were evaluated against *E. coli* bacteria and *C. albicans* fungi. The results revealed that cellulosic fibers containing 6.2% and 25% EOs were able to inhibit the growth of *E. coli* bacteria. The cellulosic material owes its enhanced effectiveness to the nanostructured morphology that is provided by the used technique (electrospinning). Cellulosic electrospun fibers have a high exposed surface area compared to cellulosic flat films, allowing the micro-organisms to more easily penetrate inside, so that they can better sense the presence of the antimicrobial agent. Despite this advantageous morphology, the antifungal activity of the CA/EO fibers was not effective. The authors concluded that the lack of activity was due to the size of *C. albicans* fungi being four times larger than *E. coli* bacteria, such that they were not able to penetrate inside the cellulosic material and thus make contact with the encapsulated essential oil [103].

2.2.3. Cellulose Containing Antimicrobial Metal Nanoparticles

Tran et al. developed a novel method to prepare biocompatible antimicrobial composites from cellulose and keratin with silver nanoparticles [106]. The idea for the study came from the need to fix silver nanoparticles into a matrix so that nanoparticle agglomeration or coagulation could be hindered. Following this idea, they took advantage of a previous methodology used for the green synthesis of a cellulose and keratin antimicrobial composite [107], in which ionic liquids (ILs) were used as green solvents. In this case, they introduced silver salt into a cellulose-keratine-IL solution that was further reduced to obtain a biopolymer-based composite containing silver in either its ionic (Ag^+) or metallic (Ag^0) form. The antibacterial property of the obtained material was tested against *E. coli, S. aureus, E. faecalis*, and *P. aeruginosa*. To evaluate the antimicrobial activity, bacteria were grown in the presence of the composites with ionic or metallic Ag, and further measured by CFU counting compared to those for the cellulose/keratin composite and the control. As determined from the experiments, both composites exhibited excellent antibacterial activity against most of the studied bacteria; however, those with metallic silver nanoparticles showed slightly better performance compared to ionic silver. The interesting properties of the cellulose/keratin composite, together with the antimicrobial activity derived from the presence of silver nanoparticles, make this sustainable biopolymer-based material useful as a potential dressing for chronic wounds treatment.

As mentioned, micro-organisms are everywhere; for instance, in paper and the paper products that are widely used in our everyday life, from bank notes to newspapers, books, and packaging paper. The distribution of this kind of material can contribute to the contamination and spreading of infectious diseases. Taking this into consideration, Islam et al. integrated antimicrobial activity into cellulose paper by the incorporation of silver nanoparticles using a mussel-inspired strategy [108]. To do so, the authors functionalized cellulose paper with dopamine molecules. This functionalized paper was then immersed in an ammoniacal silver nitrate solution; at this stage, dopamine catechol groups reduced the silver salt and subsequently held the produced nanoparticles via strong adhesion. Finally the authors demonstrated the successful antimicrobial activity of the AgNP-decorated cellulose paper against some highly virulent fish and shrimp pathogenic bacterial strains, such as Gram-negative *Proteus mirabilis*, *Vibrio parahemolyticus*, *E. faecalis*, and *Serratia marcescens*.

Dairi et al. have recently developed a cellulose-acetate-based film with antimicrobial and antioxidant properties for packaging applications [109]. The strategy consisted of the use of CA and AgNPs prepared following a biogenic synthesis mediated by plants. This is a novel ecofriendly process to obtain AgNPs in which there is no need for high temperatures, high pressures, and the production of toxic chemicals [110–112]. In this particular case, the process consisted of the synthesis of AgNPs into a gelatin-modified montmorillonite organoclay (OM) using a *Curcuma longa* tuber aqueous extract. The final material consisted of plasticized CA films that were obtained by a solvent casting method from a solution of CA, thymol, and modified nanoparticles. The antimicrobial property of the obtained film was tested against *E. coli*, *P. aeruginosa*, *Salmonella enterica*, and *S. aureus* bacteria as well as *A. niger* and *Aspergillus flavus*. The antimicrobial test for the films was performed by an agar diffusion disc against micro-organisms. The authors found that the CA films presented a low bacterial inhibition zone indicative of a moderate antimicrobial activity, which is directly related to the low content of Ag within the CA film. The activity against *E. coli* bacteria was moderate even when this bacteria strain was found to be most sensitive to AgNPs. In addition, the authors concluded that the presence of the organoclay may contribute to control over the silver release for a long-lasting antimicrobial effect.

Although silver nanoparticles present antimicrobial properties, both ionic and metallic silver nanoparticles were found to be toxic above a certain concentration as mentioned above [106]. Being so, Tran et al. [113] developed a cellulosic-based biopolymer antimicrobial composite using gold nanoparticles as antimicrobial agents instead of silver. It is well-known that gold nanoparticles exhibit high antimicrobial activity against Gram-positive and Gram-negative bacteria alongside their antiviral function [114]. As the authors indicated in their study, most of the work done in this regard used synthetic polymers as carriers or matrixes for the growth or encapsulation of Au nanoparticles. Tran et al. focused their work on the use of cellulose as a matrix. They took advantage of the methodology used for the green synthesis of the cellulose and keratin antimicrobial composite mentioned earlier [106,107]. However in this case, the authors used two different ionic liquids as solvents of both the biopolymer matrix and the chloroauric acid to obtain a cellulose/keratin/Au NP composite with antimicrobial activity. The composite was tested against methicillin-resistant *S. aureus* (MRSA) and vancomycin-resistant *Enterococcus* (VRE). The assays demonstrated that the biopolymer composite is able to inhibit 97% and 98% of the VRE and MRSA bacteria, respectively, being the Au NPs responsible for the antibacterial effect. As toxicity is the drawback for silver nanoparticles, the authors also evaluated their biocompatibility. They evaluated the cytotoxicity of the composites using human fibroblasts. Interestingly, the results revealed that the cellulose/keratin/Au NPs composites were not cytotoxic. The authors demonstrated for the first time that any possible cytotoxicity that the gold nanoparticles may have had was removed when they were incorporated into the cellulose/keratin biopolymer matrix.

2.2.4. Nano-Cellulose-Based Materials (Nanocrystalline, Nanofibrillated, and Bacterial Cellulose) with Antimicrobial Activity

Research on cellulose-based materials has increased intensively; however, cellulose has some limitations related to its functionalities. In this sense, there is a growing interest regarding new nanocellulose materials, such as nanocrystalline cellulose [115–119], microfibrillar/nanofibrillar cellulose, and bacterial cellulose [120–124]. The three-dimensional hierarchical structures that compose nanocellulose open up new opportunities for new fields and applications [125]. However, as occurred with cellulose, nanocellulose-based materials lack antimicrobial properties, so it is necessary to provide them with this activity.

Nanofibrillated cellulose, obtained by mechanical disintegration from native plant fibers, is a cellulosic derivative used as a novel packaging material [126]. However, its major drawback is its vulnerability to microbe attacks, such as from cellulose-consuming fungi, for instance [127]. It is known that most of the bacterial cell walls are negatively charged; therefore, and as mentioned above, an interesting option to develop intrinsic antimicrobial materials is the use of quaternary ammonium compounds, molecules, or polymers [128]. Indeed, it has been demonstrated that these compounds could interact electrostatically with the negatively charged bacteria cell wall, causing a disruption of the membrane and posterior death [6,129,130]. With this in mind, Littunen et al. [127] proposed the chemical incorporation of quaternary ammonium compounds into nanofibrillated cellulose (NFC) to impart an antimicrobial property. For that, the authors developed two types of cationized and nanofibrillated cellulose via redox-initiated graft copolymerization with a [2-(methacryloyloxy)ethyl]trimethylammonium chloride (DMQ) monomer to obtain nanofibrillated cellulose grafted poly[2-(methacryloyloxy)ethyl]trimethylammonium chloride (NFC-PDMQ), and by etherification with a quaternary ammonium compound (NFCQ). They evaluated the antimicrobial activity of unmodified NFC and both NFCQ and NFC-PDMQ against three potential human pathogens: *Micrococcus luteus*, *E. coli* bacteria, and *Candida oleophila* yeast. As expected, unmodified NFC did not show pathogen growth inhibition. As the authors stated, the cationized sample NFCQ showed a strong broad-spectrum antimicrobial effect at a high concentration (2000 µg/mL). In contrast, the polymer-grafted NFC-PDMQ showed moderate antibacterial activity but a strong antifungal response. In addition, NFCQ was notably more efficient against the Gram-negative than the Gram-positive bacteria, but NFC-PDMQ exhibited consistent activity. A cytotoxicity test was also performed for both systems and confirmed the lack of toxicity.

Fernandes et al. [131], inspired by the intrinsic antimicrobial property of chitosan that is imparted by the amino groups along the polymer chain, chemically modified cellulosic fibrils' surface by grafting aminoalkyl groups. In particular, the authors chose bacterial cellulose nanofibrils as the matrix for the modification. Bacterial cellulose (BC), a high-purity cellulose that is produced mainly from the *Gluconacetobacter* genus, presents physical and mechanical properties that, together with its biocompatibility, make it interesting for biomedical applications. For the surface modification, the authors used a silane chemical grafting approach to produce BC-NH$_2$ nanofibrils. After confirming the surface modification of the bacterial cellulose, they evaluated the antimicrobial activity against *S. aureus* and *E. coli* bacteria using non-functionalized BC as a reference. Aminoalkyl-functionalized BC membranes showed a significant reduction in bacterial viability after 24 h.

Nanocellulose (NC) particles have been also used as reinforcing agents to improve the mechanical and viscoelastic properties of biomaterials, since it is known that the major drawback of bio-based polymeric materials is their poor mechanical, thermal, and barrier properties compared to synthetic polymers [132]. Besides this, the chemical versatility of nanocellulose allows for its modification/functionalization so that antimicrobial agents, such as nanoparticles, can be anchored [133]. As a result, nanocellulosic particles could act as antimicrobial agents in addition to reinforcing the bio-based material [118,119,125]. For instance, Spagnol et al. [134] developed silver-functionalized cellulose nanoparticles without using organic solvents that were further incorporated into a polymer matrix. To obtain nanocrystal (NC)/AgNPs, they first obtained cellulose NCs by acid hydrolysis using HCl; the hydrolysis lasted for different periods of time so that NCs with different dimensions were obtained. In the next step, the authors functionalized the NCs' surface with succinic anhydride (NCSA) to incorporate carboxylic groups. Then, the carboxylic groups were deprotonated by adding NCSA to a sodium bicarbonate solution to act as anchoring groups for AgNPs. In the last step, the deprotonated NCSA solution and the AgNO$_3$ solution were mixed together and further purified so that NCs functionalized with AgNPs were successfully obtained. The antimicrobial activity of this NC/AgNPs system was evaluated by determining their MIC against *S. aureus*, *B. subtilis*, and *E. coli* bacteria and *C. albicans* fungi by the standard broth microdilution method. The effectiveness was dependent on the morphology of the obtained NC/AgNPs. In fact, the best results were obtained for NC/AgNPs samples with smaller sizes (between 6 and 18 nm), since larger nanoparticles had difficulty penetrating into the micro-organism cells.

In a recent study, Lizundia et al. [135] designed innovative antimicrobial bio-based films composed of cellulose nanocrystals and metallic (silver, zinc oxide, and titanium dioxide) nanoparticles that show antimicrobial activity. As mentioned above, AgNPs present antimicrobial activity but with the limitation of toxicity above a certain concentration. Zinc-oxide nanoparticles have effective antibacterial activity and good catalytic, electrical, photochemical, and optical properties [136]. For the film preparation, Lizundia et al. first synthesized an NC by sulfuric acid hydrolysis of microcrystalline cellulose, giving rise to nanorods approximately 10 nm in diameter and 170 nm in length. Next, they dispersed the respective nanoparticles (rod-like ZnO, spherical Ag_2O, and TiO_2) in water through sonication prior to their incorporation into the aqueous NC suspension. In the final step, the prepared NC/nanoparticles dispersions were solvent-cast to form the films by evaporation-induced self-assembly. The properties of the obtained nanocellulosic films were evaluated in terms of the effect of NC dimension, shape, and chemistry in the final composite. Interestingly, the authors confirmed that the method used for the films' preparation was derived from the formation of liquid crystal phases with a chiral nematic (cholesteric) structure. However, the incorporation of NPs affects this cholesteric structure, as reflected by the different optical properties observed by UV. The authors evaluated the antimicrobial activity against *E. coli* and *S. aureus* bacteria and the cytotoxicity on planktonic cell cultures after being in contact with the different NC-based films, and on the cells adherent to the surface of materials to determine the number of live cells. As revealed from the results regarding *E. coli*, at 3 h of incubation time, the surviving fraction of cells adherent to the surfaces showed a significant reduction in NC/Ag_2O followed by NC/ZnO and, to a minor extent, on NC/TiO_2 films. A similar trend was observed for planktonic cells. In the case of *S. aureus* bacteria, the surviving fraction of cells adherent to NC-based films was significantly diminished on NC/Ag_2O, followed again by NC/TiO_2 and NC/ZnO. The difference was observed after the direct exposition of *S. aureus* cells to the NC-based film when no significant reduction was observed for NC/TiO_2 films, whereas for NC/Ag_2O and NC/ZnO films an important decrease in cell survivability was detected. After 24 h of incubation, the surviving fraction of adherent cells decreased significantly for all NC films, but differences were observed for the planktonic cells, in which relevant results were only observed with the NC/Ag_2O and NC/TiO_2 films. In summary, the direct exposition of *E. coli* cells to NC-based films containing Ag_2O or ZnO nanoparticles was effective at both shorter and longer incubation times. Similar results were obtained in the case of direct exposition of *S. aureus*, although its effectiveness is more significant on adherent cells at 3 h of incubation. The activity of NC-based films was less effective against planktonic cells at either 3 h or 24 h of incubation. These successful results indicated that the incorporation of nanoparticles could provide NC films with antimicrobial activity. The authors suggested their use as biomaterials and, in particular, as sustainable biomaterials for wound-healing applications.

2.3. Starch

Another interesting polysaccharide is starch, which is formed by a large number of glucose units linked by glycosidic bonds. Starch is a highly hydrophilic polymer that consists of linear amylose and highly branched amylopectin. It can be obtained from different botanical sources, such as potatoes, wheat, maize (corn), rice, and cassava. Starch has numerous applications in the food area because it is abundant, cheap, biodegradable, and edible [137]. However, its mechanical performance is poor; therefore, to overcome this limitation, it is usually is blended with another biopolymer, such as chitosan. In spite of this, starch does not have inherent antimicrobial properties, so these properties also need to be conferred on it.

Starch has been chemically modified to introduce cationic groups by etherification, graft copolymerization, or a combination of both. Yang and coworkers [138–141] have extensively used this type of modification to obtain flocculants for water treatment, as starch is a low-cost and effective system. These authors also tested starch's antimicrobial activity, which has scarcely been explored in this field. They showed that *E. coli* and *S. aureus* bacteria were almost unviable after flocculation. Cationic starch was also used in combination with starch and sodium alginate to obtain polyelectrolyte films with an antimicrobial character [142]. These films have inhibitory effects on *E. coli* and *S. aureus* that are greater against Gram-positive than against Gram-negative bacteria.

Another modification of starch performed by Guo's group was the introduction of 1,2,3-triazole via click chemistry, which reached high yields and degrees of substitution (see Figure 5) [143]. The resulting derivatives, 6-hydroxymethyltriazole-6-deoxy starch (HMTS), 6-bromomethyltriazole-6-deoxy starch (BMTS), 6-chloromethyltriazole-6-deoxy starch (CMTS), and 6-carboxyltriazole-6-deoxy starch (CBTS), were able to inhibit the growth of *E. coli* and *S. aureus* bacteria. The best system was CBTS, followed by CMTS, BMTS, and HMTS.

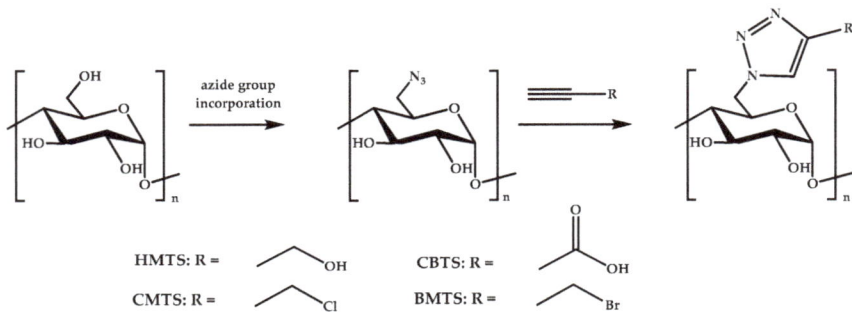

Figure 5. The synthesis of starch derivatives via click chemistry.

This group has also incorporated quaternized phosphonium salts into starch [144]. In this work, the derivatives were tested against the common plant-threatening fungi *Watermelon fusarium*, *Phomopsis asparagi*, *Colletotrichum lagenarium*, and *Fusarium oxysporum*. The most active derivatives were those with phenyl and cyclohexyl groups. The cytotoxicity of starch derivatives was also examined against HEK-293T cells using an MTT assay. These systems presented low cytotoxicity. The cytotoxicity was higher in those systems having alkyl groups.

Indeed, one of the most common approaches to the provision of antimicrobial activity is the incorporation of antibiotics into the formulation. Microparticles formed by a polyelectrolyte complex or self-aggregation are the preferred carriers for drug administration [145–147]. Nevertheless, its release has to be in a controlled manner and without a toxicity effect.

As mentioned for chitosan, at present, many studies are focused on the incorporation of natural compounds [148,149]. Pattanayaiying et al. [150] have evaluated the effect of the combination of nisin (a small antimicrobial peptide approved by the European food safety authority that is used as a food preservative [134,135]) and lauric arginate® (ethyl lauroyl arginate: a derivative of lauric acid, L-arginine, and ethanol) (LAE) in a thermoplastic starch/poly(butylene adipate terephthalate) film coated with gelatin against Gram-negative *Vibrio parahaemolyticus* and *Salmonella typhimurium* bacteria. This combination has a synergic effect in comparison with LAE alone, as occurred in pullulan, another polysaccharide film [151]. Recently, pouches of polyamide/low-density polyethylene were coated with blends of oxidized starch with gelatin containing LAE [152]. The authors found that the incorporation of LAE extends the shelf-life of chicken breast fillets without affecting the meat's oxidation. Nevertheless, the release of these natural products is not always in a controlled manner. Campos-Requena et al. [153] have developed thermoplastic starch/layered silicate (TPS/LS) bionanocomposite films for the controlled release of carvacrol. This is possible due to the formation of intercalated/exfoliated structures that can tune the migration of antimicrobial carvacrol [154], which results in the increase of its half-life.

Other biocomposite films have been obtained by a combination of pea starch and guar gum containing catechins from blueberry ash and macadamia as a natural extract, as well as epigallocatechin-3-gallate from green tea [155]. These films were tested against Gram-positive *Staphylococcus lugdunensis*, *S. epidermidis*, *B. subtilis*, and *E. faecalis* bacteria, Gram-negative *Pseudomonas fluorescence*, *Klebsiella pneumoniae*, *Enterobacter aerogenes*, *S. typhimurium*, and *E. coli* bacteria, and *C. albicans*, *A. niger*, *Geotrichum candidum*, *Penicillium italicum*, *Penicillium digitatum*, *Rhizopus* sp., and *Mucor* sp. fungi. These films were able to prevent the growth of food pathogenic and spoilage micro-organisms; therefore, they can be used as edible films.

The group of Chiralt [156–160] has also intensively worked toward the incorporation of antimicrobial essential oils in starch-based materials for their use as preservative coatings or packaging systems. Moreover, they have also introduced proteins into starch [161], such as lactoferrin and lysozyme, as efficient antioxidant/antimicrobial systems [162,163]. In this sense, the co-encapsulation of herb extracts and lysozyme into such polysaccharides as starch, chitosan, and alginate has been demonstrated to produce more stability and durability during storage [164]. Besides this, these particles were more effective against Gram-positive *B. subtilis* and *Micrococcus luteus* bacteria and Gram-negative *E. coli* and *Serratia marcescens* bacteria. Starch was also modified with octenyl succinic anhydride for microencapsulation by electrospray, in combination with gum Arabic and nutmeg oleoresin [165]. They exhibit excellent antioxidant activity and a high retention of phenolic and flavonoid content after 60 days of storage as well as antimicrobial activity against *E. coli* and Gram-positive *Bacillus cereus* bacteria. This modification of starch was also used to stabilize emulsions of nisin and thymol (2-isopropyl-5-methylphenol) (an isomer of carvacrol) cantaloupe juice [166]. The addition of modified starch to the juice increases its capacity to retain nisin and thymol over the storage period and to inhibit the growth of Gram-positive *Listeria monocytogenes* and *Salmonella enterica* serovar Typhimurium.

Starch has been also blended with antimicrobial polymers to provide a bioactive character. Chitin nanowhiskers were added (0.5–5%) to starch, and films obtained by solvent casting were tested against *L. monocytogenes* and *E. coli* to analyze their antibacterial properties [167]. These films showed more effectiveness against Gram-positive than against Gram-negative bacteria. Moreover, they exhibited improved thermal properties and mechanical strength in comparison to native maize starch.

Additionally, starch is also extensively used as a reducing and capping agent for the synthesis of metal and metal-oxide nanoparticles, as in the case of chitosan [168,169]. Taking advantage of this ability, antimicrobial chitosan–starch–silver-nanoparticle-coated [170] cellulose papers were obtained. In a first step, starch–silver nanoparticles were synthesized, and then blended with chitosan in solution at different compositions. Afterwards, the mixture was poured onto papers and the antimicrobial activities were tested against the *E. coli* DH5α and *S. aureus* bacterial strains and the *Penicillium expansum* fungal strain. The results showed that the chitosan–starch–AgNP papers were effective against these microorganisms in comparison with papers coated with chitosan or starch–AgNPs alone, which do not present antimicrobial properties.

In another approach, starch-based flexible coating papers with excellent hydrophobicity and antimicrobial activities were obtained [171]. These were prepared with ZnO NPs, in which carboxymethyl cellulose (CMCe) and chitosan were added to improve the compatibility between particles and matrix. The antimicrobial activity was improved with the addition of guanidine-based starch in different amounts (see Figure 6 for an illustration). Moreover, migration tests were performed in three food simulants (deionized water, 10% alcohol solution, and 3% acetic acid), according to the E.U. No. 10/2011 standard (see Figure 7). It seems clear that the migration of ZnO NPs is much higher in films than in coated papers, and, although there is migration in all of the simulants, it is within the overall migration limits prescribed by legislation.

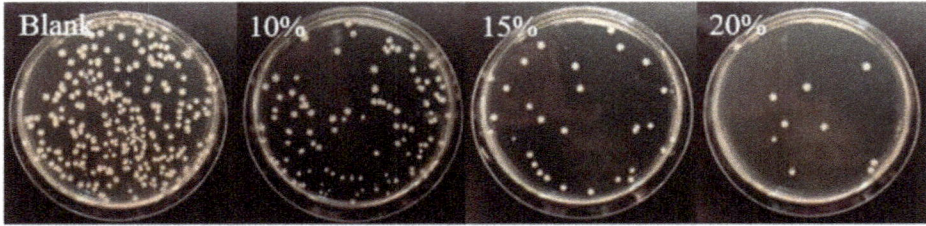

Figure 6. Images of the antimicrobial activities against *Escherichia coli* bacteria of coated papers with different amounts of guanidine-based starch using a shaking flask method. Reproduced from [171].

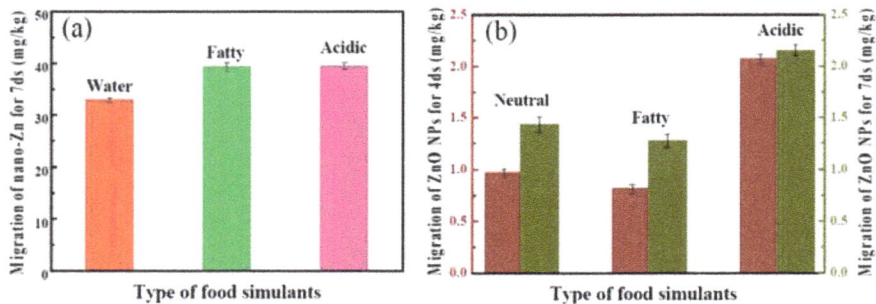

Figure 7. (**a**) The migration of ZnO nanoparticles (NPs) from films into different food simulants at 40 °C for 7 days; and (**b**) from coated papers for 4 and 7 days. Reproduced from [171].

In this sense, the group of Xiao has widely used guanidine-based systems to confer potent antimicrobial properties on cellulosic materials for their use as sanitary papers, filters, or food packaging papers [10,172–174].

Copper nanoparticles have also been incorporated into starch-based hydrogels to obtain antimicrobial systems [175]. These nanoparticles were synthesized in a starch medium followed by silica coating, which enhances their stability. The antibacterial activity of hydrogels with different amounts of NPs was evaluated against *E. coli* and *S. aureus,* and was maintained for at least four cycles of use. In addition, their dermal toxicity was studied, showing slight irritancy. Therefore, these hydrogels can be suitable wound-dressing materials.

Starch–graphene (G) hydrogels were obtained by Diels-Alder crosslinking reactions between furan-modified starch bismaleimide in the presence of graphene layers. These were incorporated as a conductive nanofiller to the mixture using Salvia extracts as dispersion stabilizers. The resulting hydrogels were tested against *E. coli* and *S. aureus* bacteria [176]. They present activity with low concentrations of extract, which confirms that the addition of G sheets also influences the antimicrobial efficiency. Moreover, these materials exhibit improved mechanical and conductivity performance.

2.4. Other Polysaccharides

In addition to the most abundant polysaccharides (starch, chitosan, and cellulose), other carbohydrate polymers, such as alginate, pectin, and κ-carrageenan, have been employed to prepare antimicrobial biopolymeric materials with high potential in a large variety of applications, especially in the food and biomedical fields [137].

Alginate is a linear anionic polysaccharide extracted from marine algae containing β-D-mannuronate and α-L-guluronate residues linked by (1,4)-glycosidic bonds. This biopolymer has found a variety of applications in biomedical science and the food industry due to its biocompatibility and gelation capability [177,178]. Several strategies have been followed in the last few years to confer an antimicrobial character on alginate-based materials. In recent investigations, sodium alginate/poly(ethylene glycol) hydrogels with antimicrobial activity were prepared by grafting the cysteine-terminated antimicrobial peptide HHC10–CYS, at different proportions, into the structure through a thiol-ene click reaction [179] Microbiological studies of the hydrogel against *E. coli* bacteria revealed that the activity increased with the content of peptide in the hydrogel. In addition to the strong antibacterial activity, the hydrogel showed good cytocompatibility. Nevertheless, most of the works related to alginate-based materials with an antibacterial character use approaches that are mainly centered on the incorporation of antimicrobial agents into the alginate material without any chemical reaction. In fact, there is a huge number of publications on the preparation of nano- and microcapsules of alginate for the encapsulation of antimicrobial components, such as essential oils [180,181], nisin [182,183], ZnO NPs [184], and AgNPs [185]. In this respect, a common strategy is the preparation of capsules by the formation of complexes between anionic alginate and cationic polysaccharides, such as chitosan [186], or cationic peptides, such as nisin [187], with inherent antimicrobial properties. Besides this, antimicrobial films based on alginate have also been prepared by incorporating the antimicrobial agents [142,188–190].

Pectin is another important anionic polysaccharide rich in galacturonic acids, with a potential use in many fields, especially the food industry. Pectin is found in the cell wall of most plants; however, apple and citrus peels are almost exclusively used for the commercial production of pectin. Likewise, it presents an ability to form gels and has good gas permeability properties [137]. Typically, in pectin-based materials, the pectin is crosslinked and blended with other components to improve their physical properties and water stability. An interesting strategy to impart antimicrobial activity to pectin is the use of ions as a crosslinking agent with an antimicrobial character, such as Zn ions [191]. Similarly to alginate-based materials, pectin has been employed to prepare capsules for loading antimicrobial agents, including nisin [192] and antibiotics [193]. In the last few years, there has also been interest in the preparation of antimicrobial films based on pectin by including such agents as essential oils [194–197] and AgNPs [198].

Carrageenan has been also studied as a anionic polysaccharide material for potential applications in packaging [199]. Carrageenan is a linear sulfated polysaccharide composed of D-galactose and L-anhydrogalactose obtained from marine red algae. Among all of the carrageenan types, the κ-carrageenan type is used the most due to its good properties. A number of studies have been published in recent years related to the development of antimicrobial films based on κ-carrageenan by addition of classical AgNPs [200,201], ZnO NPs [202], CuO NPs [203], essential oils [204], and clays [205]. For instance, carrageenan-based hydrogels and dry films with antimicrobial properties were prepared by their combination with CuO and ZnO NPs [203]. Several samples were prepared, containing 1% ZnO, 1% CuO, or 0.5% ZnO/0.5% CuO, and, in general, both the mechanical and antimicrobial properties were improved with the incorporation of nanoparticles. The films showed strong antibacterial activity against *E. coli* and *L. monocytogenes*; however, an insignificant difference in the activity was observed between the different types of incorporated NPs. Equally as described in starch and cellulose, chitin nanofibrils have been used to reinforce carrageenan and to impart antibacterial properties [206]. The tensile strength and modulus of carrageenan film increased significantly with up to 5 wt% of chitin nanofibers. With respect to the antibacterial activity, the films showed high activity against *L. monocytogenes* depending on the content of chitin, but insignificant activity against *E. coli*.

3. Proteins/Polypeptides

Within this challenge of finding sustainable bio-based materials, natural macromolecules as proteins are playing an important role due to their versatility and it being possible to modify them enzymatically, chemically, and physically so that the desired properties can be obtained for each specific application. In general, proteins are used as additives in polymeric matrices (nisin and corn zein [207,208], soy protein [209], and wheat gluten [210]), and some examples of this can be found in the different sections of this work. However, protein-based materials have also emerged in applications as diverse as packaging [211] and biomedicine [212]. In this section, we will briefly describe the most recent work regarding sustainable antimicrobial protein-based materials for some of the most used proteins, including caseinates, keratin, and collagen.

3.1. Caseinates

Among animal proteins, caseinates are considered to be attractive for use in the food-processing industry; i.e., food packaging and culinary applications, since they show numerous advantageous properties, such as their natural origin, edible character, water solubility, and ability to act as emulsifiers [213,214]. Caseinates show an ability to form networks, plasticity, and elasticity, which lead to the formation of transparent films with good performance as a barrier against oxygen, carbon dioxide, and aroma compounds [213–215]. Sodium caseinate is more frequently used than the other caseinates, such as calcium caseinate or potassium caseinate, and, importantly, caseinates are frequently plasticized with glycerol to obtain the required flexibility for the formation and manufacture of film [213,216,217]. Moreover, caseinate-based films have attracted interest as carriers of antimicrobial substances in food-related applications [217]. The main advantage of introducing antimicrobial agents into caseinate films, as in the other described systems, is the ability to slow the diffusion of the agents through the film, allowing for its availability at a desired concentration. Therefore, smaller amounts of antimicrobial additives are needed to achieve a targeted shelf-life extension, compared with the direct addition of the antimicrobial additives onto the food surface strategy, where they quickly diffuse away from the surface, and are rapidly diluted or react with food components [216].

Noori et al. [218] have recently developed a nanoemulsion-based edible coating with strong antimicrobial activity against psychrophilic bacteria to extend the shelf-life of chicken fillets. For that, ginger (*Zingiber officinale*) EO was added to sodium caseinate matrices and the obtained films showed comparable results to the antibiotic gentamicin. Arrieta et al. added 10 wt% of carvacrol into edible matrices of both sodium and calcium caseinate, and further studied the obtained plasticized films against *S. aureus* and *E. coli* bacteria by the agar diffusion method [214,217]. Although edible films of sodium caseinate and calcium caseinate with carvacrol showed antibacterial effectiveness against both *S. aureus* and *E. coli* bacteria, they showed a higher diffusion of carvacrol through an agar gel inoculated with *S. aureus* than that with *E. coli*, resulting in a higher inhibition zone around the edible film area [217]. Moreover, the *E. coli* inhibition zone was larger for sodium caseinate films than for calcium caseinate films (Figure 8). This behavior was related to the fact that divalent calcium cations in calcium caseinate promote crosslinking with protein chains [214], retaining carvacrol more efficiently due to the more tortuous structure, which releases the active agent more slowly [217].

Imran et al. [219] developed sodium caseinate films that incorporated nisin, one of the most-used bacteriocins for food conservation, with high antilisterial and antistaphylococcal activity. Meanwhile, Calderón-Aguirre et al. [216] introduced nisin as well as antimicrobial substances produced by *Streptococcus infantarius* into glycerol-plasticized sodium caseinate films, and observed that caseinate films containing bacteriocins produced by *S. infantarius* showed higher antilisterial effectiveness in long-term refrigeration storage (around 2 months) than nisin-incorporated ones.

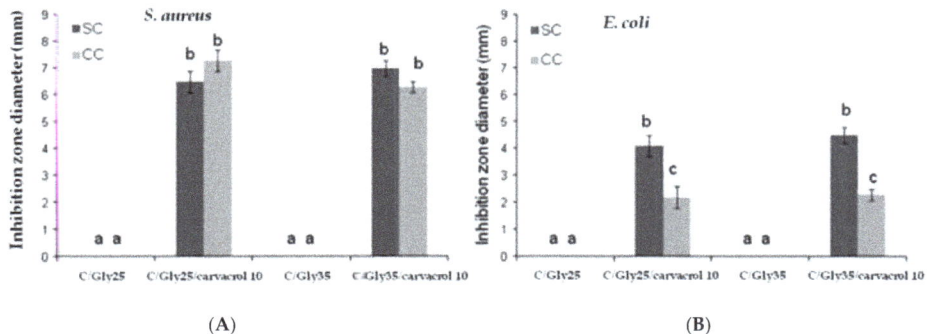

Figure 8. The inhibition zone (mm) observed in caseinate-based films plasticized with glycerol and loaded with carvacrol against (**A**) *S. aureus and* (**B**) *E. coli* bacteria. a–c Different letters on the bars indicate significant differences between formulations ($p < 0.05$). Reproduced from [217] with permission from Elsevier.

3.2. Keratin

Keratin is a protein found in mammalian hair, fur, wool, skin, hoofs, claws, and horns and in feathers of birds. This is an ancient material used for textile applications due to the early domestication of sheep and the use of the produced wool for such purposes. However, the need for non-contaminant sources for the design of sustainable bio-plastics has put the focus on keratin. Keratin extracted from such agricultural waste products as poor quality wool and chicken feathers has been used to produce films, fibers (electrospun fibers), and hydrogels and shown potential application as a scaffold for tissue engineering and tissue dressings and even as drug delivery systems [212,220–222]. Nevertheless, as is the case for most of the bio-based materials shown in this work, keratin is not antimicrobial by itself, so its functionalization or combination with antimicrobial agents is indispensable [223].

Having this in mind, Yu et al. [224] proposed the immobilization of quaternary ammonium moieties on a keratin-based substrate, thus turning keratin into an antimicrobial material for biomedical applications. The methodology consisted of the generation of thiols in wool keratin fibers (reduction of disulfide bonds using tris(2-carboxyethyl)phosphine hydrochloride) and then their reaction with the acrylate monomer [2-(acryloyloxy)ethyl]trimethylammonium chloride (2-AE) through click chemistry. In this way, a quaternary ammonium moiety was grafted onto reduced keratin fibers. The antimicrobial property of the obtained material and also of the untreated material was evaluated against *E. coli* bacteria using the agar diffusion plate test. Interestingly, the percentage of bacteria reduction obtained with the modified keratin was 94%, whereas for the untreated material there was no antibacterial effect. In this case, since the antimicrobial compound (the quaternary ammonium moiety) is covalently bonded to keratin, the antimicrobial activity is sustained in time and no leaking can occur. This interesting approach makes keratin-based materials applicable, for instance, as medical textiles.

Nayak et al. used a different approach by blending keratin with different polysaccharides (alginate, agar, and gellan) to obtain therapeutic porous dermal patches [225]. To impart antimicrobial activity, the obtained patches were coated with AgNPs. In particular, the antimicrobial activity of a keratin/agar patch was evaluated through the disk diffusion test against *S. aureus* and Gram-negative *Pseudomonas putida* bacteria and *A. niger* and *C. albicans* fungi pathogens. The results indicated the good antimicrobial activity of the patches against the tested pathogens, with a remarkable effect against *S. aureus* Thus, the broad antimicrobial activity of the keratin-based obtained patches was confirmed.

3.3. Collagen

Collagen is the main structural protein found in the extracellular matrix of various connective animal tissues. The amino acids that compose collagen are wound together in a triple-helix, giving rise to elongated fibrils. The main role of collagen is both structural and functional, since it contributes to some processes of tissue repair [226]. In addition, by partial hydrolysis of collagen and destabilization of the triple-helix, it is possible to obtain the natural polymer called gelatin, which has also attracted much industrial interest. In sum, collagen possesses such properties as biocompatibility, biodegradability, and non-toxicity, which makes it suitable for applications involving wound healing and tissue regeneration. However, as is the case with the aforementioned bio-based polymers, this protein lacks the antimicrobial activity that is rather important for those kinds of applications. So, the modification/functionalization of collagen or its combination with antimicrobial agents is required.

Having this in mind, Balaure et al. [227] recently developed a collagen dressing containing orange essential oil functionalized ZnO nanoparticles (*d* = 20 nm) inserted into a three-dimensional (3D) matrix. As was described above, ZnO nanoparticles present antimicrobial activity, and their incorporation into collagen to impart biocidal properties is a strategy that has also been followed by other authors [228,229]. In this particular case, for the preparation of the dressing, suitable amounts of previously synthesized ZnO nanoparticles, collagen, and glutaraldehyde solution (crosslinker) were added so that collagen–ZnO gels were formed. For the antibacterial activity, collagen–ZnO gels (with three different ZnO contents) were placed in petri dishes and inoculated with *S. aureus* and *E. coli* bacteria strains. For the sake of comparison, antibiotic disks were used as a control. After 24 h of incubation, the diameters of the inhibition zone were measured and compared with the control disks. The results revealed that the collagen–ZnO wound dressings presented a remarkable antimicrobial activity against the *S. aureus* strain, as the growth inhibition zones were 11.5 mm and comparable to the diameter of the inhibition zones obtained for the control antibiotics. When analyzing antimicrobial activity against *E. coli*, the influence of the ZnO content was evidenced. Besides the antimicrobial activity, the developed collagen dressings showed great regenerative capacity. This combination of outstanding properties makes this bio-based sustainable material a potential candidate for wound healing applications.

Similarly, You et al. developed a metallic silver nanoparticles–collagen/chitosan hybrid scaffold to be used as a dressing for burn wounds [230]. Ideal dressings are required to present antimicrobial activity among other factors, such as keeping moisture in the wound, removing exudates, or even providing drugs that contribute to the healing. However, in the case of burn wounds, where a large amount of fluid is lost and bacterial infections occur, it can become necessary to have dressings with antimicrobial activity. For the scaffold's preparation, bovine type-I collagen and chitosan were dissolved together in an acetic acid solution. In the next step, commercial AgNPs were added to the collagen–chitosan, poured in a plate and kept at 4 °C, frozen at −25 °C, and finally lyophilized to obtain the scaffolds. The antibacterial activity was evaluated against *S. aureus* and *E. coli* via the disk diffusion method, at which the zone of growth inhibition was measured after 24 h of incubation. An antimicrobial assay was performed for collagen and the hybrid collagen-based scaffolds, and, as expected, more activity was observed in the hybrid scaffold compared to the neat sample without AgNPs. Furthermore, the antimicrobial activity was improved as the amount of AgNPs in the scaffold increased. These bio-based hybrid scaffolds, besides their proven antimicrobial activity, exhibit anti-inflammatory properties that are also derived from the presence of AgNPs. As a conclusion, the authors postulated that the hybrid collagen-based dermal scaffolds could have application as a new antimicrobial dressing designed using non-toxic components.

Michalska-Sionkowska et al. [231] developed a collagen-based antimicrobial dressing that incorporated thymol as an essential oil to provide antimicrobial activity to the final material. As previously highlighted, it has been demonstrated that thymol shows antimicrobial activity against both Gram-positive and Gram-negative bacteria strains as well as against fungi and yeast [232,233]. For the preparation of the films, collagen obtained from rat tendons was dissolved in acetic acid. Then, different amounts of thymol were added to the collagen solution. For a better miscibility of the thymol in the collagen, a nonionic surfactant was also added to the solution. Finally, the solutions were poured, and, by the solvent evaporation method, collagen–thymol films were obtained. Since the main objective of the work was to develop an antimicrobial material against biofilm formation, the obtained films were tested by the agar diffusion method against *E. coli*, *P. aeruginosa*, *S. aureus*, *B. subtilis*, *E. aerogenes*, and *C. albicans* strains. The antimicrobial test revealed that the most sensitive micro-organism was *S. aureus* bacteria. This effect was even more significant with the increase in thymol content in collagen films. The growth of *E. coli* bacteria was also inhibited, although the dose needed was higher compared to that of *S. aureus*. Similar results were observed for *C. albicans*, *B. subtilis*, and *E. aerogenes*, but no inhibition was observed against *P. aeruginosa*. The biofilm formation for collagen–thymol and collagen control films was also evaluated against *S. aureus*. Through an SEM observation, it was concluded that the number of micro-colonies on the surface of the collagen-thymol film was smaller compared to that of the control film without the EO. Thus, collagen-based films were also effective in inhibiting biofilm formation.

4. Polyesters

Bio-based and biodegradable polyesters, such as poly(lactic acid) (PLA) and poly(hydroxyalkanoates) (PHAs), are currently the main drivers of the growth in the bioplastics market [2], as they have the potential to replace traditional polymers, such as poly(ethylene terephthalate) (PET).

As was already mentioned, antimicrobial systems based on biocompatible and biodegradable polymeric matrices have attracted interest for both biomedical and food-related applications. The main objective in developing antimicrobial polymeric systems is to lessen and subsequently prevent the micro-organisms from growing [234]. In this context, several strategies have been developed to provide materials with antimicrobial performance based on biopolyesters, most of which have been focused on adding additives with antimicrobial activity, including natural compounds (i.e., EOs), peptides (i.e., nisin), chelating agents, antibiotics, other polymers (i.e., chitosan), enzymes, and metals (zinc oxide, silver, etc.) [235]. Despite the wide range of antimicrobial agents that have been used during the last two decades for the development of biopolyester-based antibacterial materials, it seems that the route to introduce them is even more important than the agent itself in obtaining an effective antimicrobial activity at the surface of the material.

4.1. Poly(lactic acid) (PLA)

Among all sustainable polymers, PLA is currently the most promising one to replace petroleum-derived polymers. Moreover, PLA is biocompatible and biodegradable, and this is why it is widely used in the biomedical field as well as in a wide range of commodity applications in other industrial sectors, such as the agricultural and food packaging sectors [236,237]. PLA is chemically synthesized starting with simple sugars obtained from agro-resources, such as corn, potatoes, cane, and beet, and fermented to the lactic acid monomer [236,237]. The most common route to produce PLA at the industrial level is the ring-opening polymerization (ROP) of the cyclic lactide dimer by condensation with metal catalysts (e.g., tin octoate) with the elimination of water at a high temperature (but less than 200 °C) [238,239] (Figure 9). Another important advantage that has allowed us to rapidly introduce PLA into the market is that PLA can be processed with the same processing technology that is already used at the industrial level for traditional petroleum-based thermoplastics to obtain films, molded pieces, and fibers, including melt blending, extrusion, injection molding, thermoforming, film forming, and electrospinning [236,240,241].

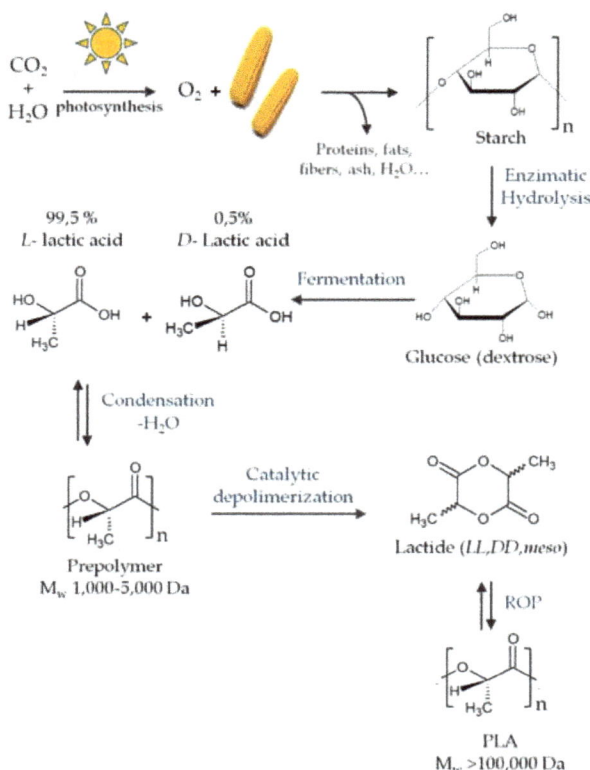

Figure 9. A schematic representation of high-molecular-weight PLA industrial production. Reproduced from [242].

A laminated chitin PLA–PLA composite has been recently prepared by the hot-press method. Chitin powder was dispersed into a PLA matrix and further processed by a hot press to obtain the inner layer, where a neat PLA layer was also obtained by the hot press method [243]. It was observed that the incorporation of 5 wt% of chitin in the chitin–PLA layer in direct contact with a Mueller–Hinton agar medium, inoculated with *E. coli* bacteria, was enough to produce a clear inhibition zone. Besides this, antibacterial PLA-based compounds have been recently developed by adding chitosan nanoparticles in 0.5, 1, and 2 wt% into a PLA matrix by melt extrusion followed by a solvent casting process to obtain nanocomposite films. The obtained nanocomposites showed higher antibacterial effectiveness with increasing content of nanochitosan as well as higher antibacterial effectiveness against *L. monocytogenes* than against *E. coli* bacteria [243].

Imran et al. [219] introduced nisin into several bio-based and biodegradable matrices, including sodium caseinate, chitosan, and PLA. The antibacterial effectiveness was tested against *L. monocytogenes* and *S. aureus*. Only the PLA–nisin disc films showed effectiveness against both bacteria due to the PLA's hydrophobic nature, which lead to a higher nisin retention ability.

EOs have been widely used as PLA antimicrobial additives for the development of active PLA-based materials with antibacterial activity, which have mainly been studied by the agar disk diffusion method to simulate the food-wrapping conditions [244–246]. Plasticized PLA composite films were developed by loading bimetallic silver–copper (Ag–Cu) nanoparticles and cinnamon essential oil into a polymer matrix via melt blending [247]. The obtained nanocomposites were used to pack contaminated chicken samples with Gram-negative *Campylobacter jejuni*, *L. monocytogenes*, and

S. Typhimurium, and the degree of bacteria survival after 21 days under refrigerated storage conditions was measured. The bacterial survival of food-borne pathogens decreased significantly from 6.65 to 3.87 log CFU/g for *L. monocytogenes*, from 5.40 to 2.59 log CFU/g for *C. jejuni*, and from 5.52 to 2.42 log CFU/g for *S. typhimurium*, resulting in interesting materials for antimicrobial food packaging systems. However, when EOs are directly incorporated into thermoplastic polymeric matrices that should be processed at a high temperature, high amounts of EOs are lost during processing due to their high volatility. For instance, plasticized poly(lactic acid)–poly(hydroxyalcanoates) (PLA-PHA)-based blends were directly incorporated with 10 wt% of carvacrol and processed by melt extrusion [245,246]. Around 25% of the carvacrol was lost during thermal processing [245]. Nevertheless, due to the high amount added into the blend formulations as well as the high antibacterial activity of carvacrol, the films showed some antibacterial surface effectiveness against *S. aureus*, with the observation of a clear zone of growth inhibition around plasticized film samples, while no inhibition halo was observed in unplasticized samples. The increased mobility of the macromolecular chains, due to the plasticizer's presence, promotes the diffusion of carvacrol from the polymeric matrix into the agar medium in a radial way, improving the antimicrobial effectiveness. However, no inhibition halo was observed in the agar plates inoculated with *E. coli* bacteria, even with plasticizer [246].

In this context, novel processing strategies other than thermal processing to obtain antimicrobial PLA systems have recently been developed to avoid the loss of volatile additives and to increase the antibacterial performance of the materials at the surface.

Supercritical impregnation has recently been proposed as an effective route to the introduction of volatile active compounds, such as EOs, into PLA matrices [248–251]. For instance, it has been recently used to incorporate thymol into PLA to develop materials with a wide range of applications [248,250]. Villegas et al. recently developed PLA films impregnated with cinnamaldehyde by supercritical impregnation with very effective antibacterial activity against *S. aureus* and *E. coli* bacteria [249].

Likewise, the production of ultrafine electrospun PLA fibers in the form of nonwoven materials has shown great potential in several fields, such as drug delivery, tissue engineering, filtration, catalytic processes, sensor development, and packaging. In fact, the electrospinning process has been suggested to be an effective way to produce nonwoven materials with active surfaces [251–253]. This technique allows for processing of the active substance with the PLA matrix at room temperature, avoiding the thermal degradation of the active substance. Moreover, during the preparation of a PLA material through the electrospinning technique, the high ionic strength as well as the rapid evaporation of the solvent induce the localization of the active compounds predominantly on the surface of the fibers [251,252]. For instance, cinnamaldehyde was incorporated into electrospun PLA-based materials by Lopez de Dicastillo et al. [251] following two approaches: direct incorporation into an electrospun PLA solution and supercritical impregnation of the electrospun PLA material. The authors observed that the materials into which cinnamaldehyde had been incorporated by supercritical impregnation showed a higher surface availability of the active compound, since those materials showed a higher release rate in fatty food simulants when compared to electrospun materials into which cinnamaldehyde had been incorporated during the electrospinning process.

Antimicrobial natural extracts with antibacterial activity can also be obtained from algae. For instance, natural extracts with some antimicrobial performance against *E. coli* have been obtained from *Durvillaea antarctica* algae and incorporated into a PLA matrix encapsulated in electrospun poly(vinyl alcohol) (PVA) fibers [253].

Nanostructured and aluminum-doped ZnO coatings were sputter-deposited onto an extruded PLA film by Valerini et al. [254] to functionalize its surface with antimicrobial activity. The materials with uniform surface coverage showed antibacterial effectiveness against *E. coli*. The authors concluded that the strong antibacterial effectiveness against Gram-negative bacteria is due to a polycrystalline structure, with the presence of cubic aluminum-doped zinc oxide ($ZnAl_2O_4$), ZnO, and aluminum oxide phases.

Antimicrobial electrospun PLA fibers have also been developed by adding silica nanoparticles functionalized with ZnO. The PLA-based materials' antibacterial effectiveness against *E. coli* was found to be concentration-dependent and size-dependent. They required at least 0.8 wt% of functionalized zinc-oxide-doped silica NPs to produce a reduction in bacteria growth. Moreover, the authors obtained nanoparticles with an average diameter of 5–20 nm and observed that when the particle size was reduced, the surface reactivity of the nanoparticles was enhanced, leading to electrospun materials with better antimicrobial performance [255].

Another interesting technique for the modification of PLA film surfaces and the improvement of their functionalization is by plasma treatments [256–258]. For instance, Hu et al. [257] have recently coated a PLA film's surface with nisin after modifying the film's surface by means of cold plasma treatment (CPT). They studied the surface antimicrobial activity of the films against *L. monocytogenes* bacteria by the inhibition zone method. The authors found that the inhibition zone was only formed in the area where the incubated agar medium was in direct contact with the PLA–nisin-based films due to the poor water solubility of nisin. The antibacterial effectiveness of the PLA–nisin films was then determined quantitatively by means of the viable cell count method, and the total viable counts (TVCs) of *L. monocytogenes* decreased from 3.01 to 2.25 $\log_{10}$ (CFU/mL) by increasing the time of functionalization of the PLA's surface by CPT from 15 to 60 s. These results suggest that increasing the CPT time positively affects the nisin adsorption capacity of the PLA's surface as well as allowing for nisin to be released from the PLA's surface when it is in an appropriate medium. Zhang et al. [258] treated electrospun materials with oxygen plasma to obtain a material with a hydrophilic surface that can be further functionalized with antimicrobial substances. Catechol possesses an essential role as an adhesive between interfaces; particularly, the synthetic catechol derivative dopamine methacrylamide monomer (DMA) possesses an exceptional adhesive property, which was used by these authors for surface modification in the development of biomedical devices. The electrospun PLA materials were modified with graphene oxide with DMA and the obtained nanocomposite showed excellent biocompatibility and exhibited antimicrobial properties against *S. aureus* and *E. coli* bacteria.

Munteanu et al. [259] prepared a nanocoating by encapsulating argan and clove oils into chitosan by coaxial electrospinning and then used it to functionalize electrospun PLA materials previously treated with CPT. The authors concluded that the electrospun coaxial encapsulation of clove and argan oils into chitosan led to enhanced antimicrobial activity against *E. coli*, *S. typhymurium*, and *L. monocytogenes*.

4.2. Poly(hydroxyalcanoates) (PHAs)

PHAs are a family of isotactic, semi-crystalline, and high-molecular-weight thermoplastic polyesters (Figure 10), which are biologically synthesized by controlled bacterial fermentation of a wide variety of both Gram-negative bacteria (i.e., those belonging to the genera *Azobacter*, *Alcaligenes*, *Bacillus*, and *Pseudomonas*) [260] and Gram-positive bacteria (i.e., those belonging to the genera *Nocardia*, *Rhodococcus*, and *Streptomyces*) [261]. In response to nutrient limitation (e.g., phosphorus, nitrogen, trace elements, or oxygen) and in the presence of an abundant source of carbon (e.g., glucose or sucrose) or lipids (e.g., vegetable oil or glycerine) [238], bacteria can accumulate up to 60–80% of their weight in PHA [261–263]. Therefore, PHAs are bio-based, biocompatible, and biodegradable, and exhibit thermal and mechanical properties that are comparable to those of other common polymers, such polystyrene (PS) and polypropylene (PP). In fact, the PHA family has been under development during the last few decades, the PHAs market is now emerging very fast, and the production capacity is predicted to quadruple in a few years [2].

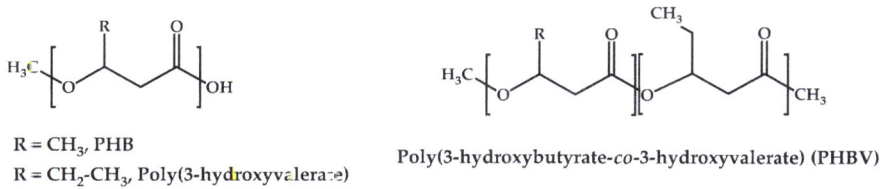

R = CH₃, PHB
R = CH₂-CH₃, Poly(3-hydroxyvalerate)

Poly(3-hydroxybutyrate-*co*-3-hydroxyvalerate) (PHBV)

Figure 10. The chemical structure of most typical PHAs.

Among the PHAs, poly(3-hydroxybutyrate) (PHB) is the most simple and common representative of PHAs. This is the reason why it is the most widely investigated for the development of PHA-based antimicrobial systems. From a processing point of view, the main drawback of PHB is its very low resistance to thermal degradation, since PHB has a melting temperature (around 170–180 °C) that is close to the degradation temperature (270 °C) [240]. Thus, although PHAs are still processed by melt compounding [264–267], the most recent studies on antimicrobial PHA systems are focused on the preparation of active PHA-based materials at room temperature [268–271].

PHB films with antibacterial properties have been developed by adding metal nanoparticles. In particular, PHB/AgNP nanocomposites are the most extensively developed group. PHB films have had AgNPs embedded into them by a two-step process consisting of helium plasma treatment followed by an immersion process in a silver nitrate solution [272]. Castro-Moyorga et al. [269] simultaneously biosynthesized AgNPs and PHB from the fermentation process of *Cupriavidus necator*. The obtained biopolymer and nanoparticles were used to produce bionanocomposite films with potential as active coatings for food packaging applications, since they showed antibacterial effectiveness against the food-borne pathogens *S. enterica* and *L. monocytogenes*.

Another PHA, poly(3-hydroxybutyrate-*co*-3-hydroxyvalerate) (PHBV), was loaded with less than 1 wt% of AgNPs and processed into an ultrafine electrospun fiber. It showed good in vitro cell compatibility and completely inhibited the proliferation of *S. aureus* as well as *K. pneumoniae* bacteria [270] (Figure 11). The amount of released AgNPs in distilled water increased with the incubation time, reaching a value of 0.537 ppm after 30 days (see Figure 12). The release is accelerated by the high surface area of nanofibrous scaffolds and the biodegradation of PHBV. Therefore, these materials may be good candidates for arthroplasty in regenerative medicine.

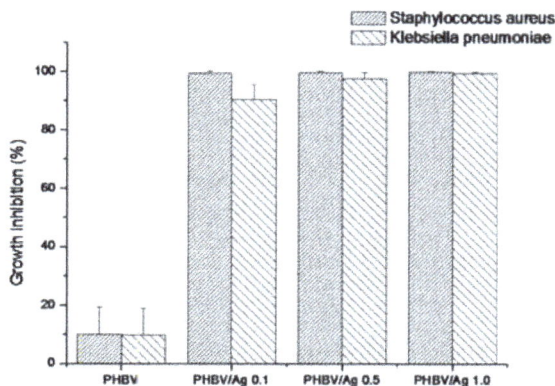

Figure 11. The growth inhibition of PHBV nanofibrous scaffolds with different amounts of silver against *S. aureus* and *Klebsiella pneumoniae*. Reprinted from [270].

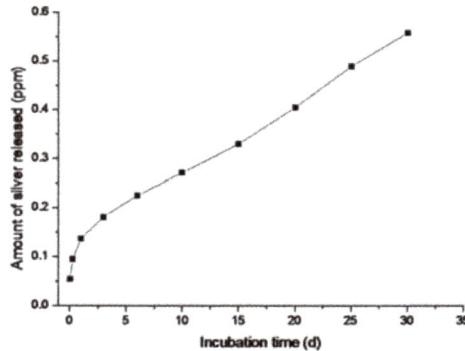

Figure 12. The amount of silver nanoparticles (AgNPs) released from PHBV/Ag 1.0 nanofibrous scaffolds as a function of the immersion time. Reprinted from [270].

Zinc-oxide nanoparticles were also recently used in combination with PHA matrices. For instance, PHB was melt compounded with bacterial cellulose nanofibers to obtain nanocomposites, which were further modified by plasma treatment and coated with ZnO nanoparticles dispersed in alcohol using an ultrasonic spraying device [273]. The obtained nanocomposites completely inhibited the growth of *S. aureus* bacteria.

Naveen et al. [274] have prepared PHB electrospun nanofibers loaded with kanamycin sulphate, a water-soluble aminoglycoside antibiotic, for the development of nanoscaffolds for cell growth and antimicrobial devices. The electrospun scaffolds promote cell attachment, while the hydrophilic antibiotic confers antimicrobial performance on the electrospun surface as revealed by the obtained good zone of inhibition against *S. aureus* bacteria. The results showed that the electrospinning process does not affect the antibacterial activity of the kanamycin sulphate drug.

Poly(hexamethylene guanidine hydrochloride) granular polyethylene derivatives are of particular interest as additives in PHB-based materials processed by melt blending technologies, since they possess high thermal resistance [275]. Walczak et al. developed melt-blended PHB films enriched with poly(hexamethylene guanidine hydrochloride) granular polyethylene wax in 0.6–1 wt% with respect to the polymeric matrix. The active films were able to avoid bacterial biofilm formation by *S. aureus* on their surface [275]. Thus, the resulting materials are very interesting for biomedical as well as for food-related applications.

4.3. Poly(butylene succinate) (PBS)

Poly(butylene succinate) is one of the most popular poly(alkylene dicarboxylate) polymers as it combines good properties with biodegradability, and can be produced by polycondensation of succinic acid and 1,4-butanediol (BDO) monomers, which are completely bio-based and obtained from refined biomass feedstock (from sugar-based feedstock by bacterial fermentation) [276–278]. In fact, many sources support the view that PBS is a material with the potential to replace polyolefins in the near future [279]. Nowadays, the easiest way and most-employed strategy to obtain succinic acid is from micro-organisms, such as fungi or bacteria, being the most intensively studied *Anaerobiospirillum succiniciproducens* [280] and *Actinobacillus succinogenes* [281] due to their ability to produce a relatively large amount of succinic acid. Interestingly, the use of glycerol as a carbon source substrate had been reported to lead to the best yields of succinic acid compared to other carbohydrates [280]. BDO can also be directly produced from biomass [277,282] or indirectly from bio-based succinic acid through its catalytic hydrogenation [283–285], being this latest route the most common. Figure 13 shows an innovative and promising pathway to obtain BDO in a one-step microbial fermentation process. Although, in most of the literature, PBS is produced using petroleum-based monomers, it is within the scope of the present work to include them, as nowadays both monomers that constitute PBS are bio-based, becoming more and more accessible, and slowly replacing the petroleum-based polymer. Therefore, PBS is expected to be completely included in the family of bio-based polymers in the near future as it is already produced by an industrial company [286,287].

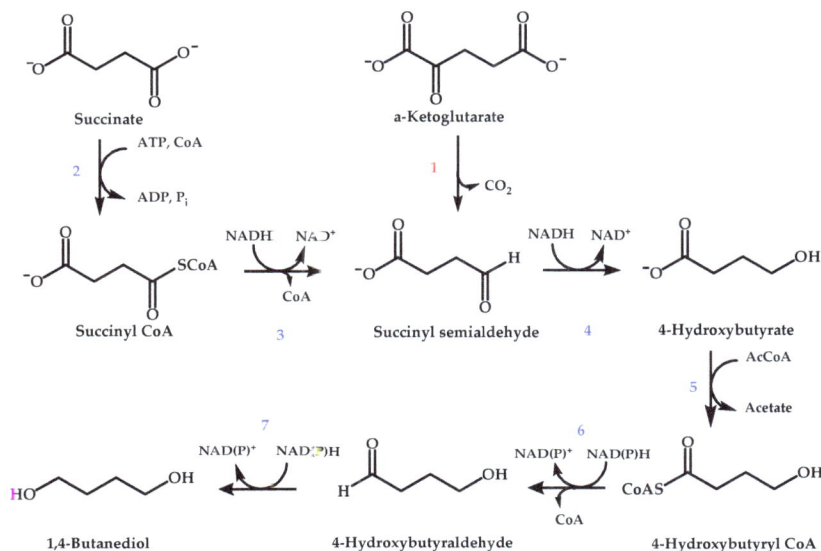

Figure 13. BDO biosynthetic pathways introduced into *E. coli* bacteria. Enzymes for each numbered step are as follows: (1) 2-oxoglutarate decarboxylase; (2) succinyl-coenzyme A (CoA) synthetase; (3) CoA-dependent succinate semialdehyde dehydrogenase; (4) 4-hydroxybutyrate dehydrogenase; (5) 4-hydroxybutyryl-CoA transferase; (6) 4-hydroxybutyryl-CoA reductase; (7) alcohol dehydrogenase. Steps 2 and 7 occur naturally in *E. coli*, whereas the others are encoded by introduced heterologous genes.

Typically, PBS is employed in medical equipment and in the food industry, where sterile conditions together with biodegradability are usually required and the safety and shelf-life of food products must be ensured/improved. In these regards, antimicrobial properties are added to the material by the addition of antimicrobial agents. Those additives can be included in the polyester matrix to achieve the desired activity through the migration of the agent, or can be bounded chemically to its surface through functionalization, as stated for the above-reviewed polymers.

The first strategy consists in bounding specific functional groups or molecules to the polymer backbone. This polyester, in general, possesses a lack of available functional groups for further functionalization; therefore, most of the works introduce cationic groups that will provide antimicrobial activity into PBS chains by melt polycondensation to achieve cationic copolymers. Bautista et al. [288] copolymerized either 2,2-(dihydroxymethyl)propyl-tributylphosphonium bromide (PPD), 2-(N,N,N-trimethylammonium)dimethyl-glutarate iodide (TMA-DMG-I), or 2-ammonium dimethyl glutarate hydrochloride (A-DMG-CL) with succinic acid and BDO to obtain pendant quaternary phosphonium/ammonium groups on the copolymer chains (Figure 14). In order to overcome the drawbacks of the high temperatures needed for polycondensation reactions that can decompose cationic compounds, the authors successfully applied an enzymatic catalyst that was able to avoid metal residues and drive the reaction of ammonium-containing PBS under mild conditions (vacuum; 80–115 °C; 40–1.6 × 10^{-3} mbar). However, although at lower temperatures than is usual, the introduction of phosphonium groups was only possible by conventional polycondensation with a titanium catalyst (190 °C; 2.9 × 10^{-3} mbar). The antibacterial activity of the PBS copolyesters containing ammonium or phosphonium side groups at different concentrations was explored against *E. coli* and *S. aureus*. The phosphonium derivatives showed a strong antimicrobial effect with only 15 mol% of this cationic compound, while in the case of ammonium, the best results were obtained when 50 mol% of ammonium groups in the PBS were present.

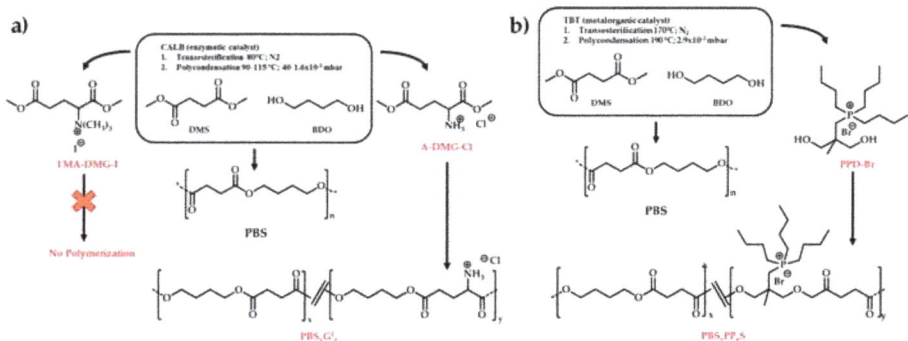

Figure 14. (**a**) An enzymatic synthesis route to a cationic PBS copolyester containing ammonium. (**b**) An organometallic catalyzed synthesis route to a cationic PBS copolyester containing phosphonium.

Another way to obtain specific functional antimicrobial groups on PBS is the one followed by Wang et al. [289]. The authors modified the PBS surface by O_2 or N_2 plasma immersion ion implantation (PIII), a versatile technique that introduces chemical groups onto the samples depending on the gas employed. The authors evaluated the changes that occurred on the surfaces (chemistry, hydrophilicity) and their effects on the behavior in the presence of osteoblasts and bacteria. When the PBS surface was treated with O_2, no difference was found with the control. However, when treated with N_2, an antibacterial effect of 91.41% and 90.34% against *S. aureus* and *E. coli*, respectively, were obtained (from the amounts of active bacteria). This fact was associated with the presence of $C=NH$ and $C-NH_2$ groups that not only enhance the antimicrobial properties but also promoted osteoblast proliferation, differentiation, and mineralization, thus representing potential materials for implants.

Most of the reviewed literature on antimicrobial PBS is based on the incorporation of such ions as Cu or Ag, or essential oils that slowly migrate towards a product or media, into a PBS matrix. Research that considers PBS containing covalently bound antimicrobial components remains at a really early stage. However, the promising results that have been achieved to date will probably attract the attention of many researchers in the future.

Generally, for the production of antimicrobial packaging, a common strategy is to use a bio-based EO and biopolymer (see Figure 15). Therefore, to produce antimicrobial PBS, Petchwattana et al. [290] incorporated thymol as an essential oil, which can be extracted from thyme (*Thymus vulgaris*), garlic (*Allium sativum*), and onions (*Allium cepa*) among other plants. As mentioned above, thymol possesses demonstrated antimicrobial efficiency for avoiding food spoilage and extending shelf-life [291]. Pure thymol has been demonstrated to possess antimicrobial efficacy against a broad range of micro-organisms, including Gram-positive *Listeria innocua* and *S. aureus* bacteria and *Saccharomyces cerevisiae* and *A. niger* fungi, with an MIC of 250 ppm for bacteria and mold and 125 ppm for yeast, retaining significant inhibition even under a microencapsulated condition [292]. In the same work, these authors also prepared PBS/thymol blown films, with antimicrobial food packaging applications as the target, containing 2, 4, 6, 8, and 10 wt% of thymol, and tested their antibacterial activity against *S. aureus* and *E. coli*. The release kinetic of the antimicrobial agent and its activity were evaluated. The MIC value was 10 and 6 wt% of thymol for *E. coli* and *S. aureus*, respectively. By incorporation of 10 wt% of thymol in PBS films, the antimicrobial agent release was effective over 15 days in all of the food simulants tested, while the maximum diffusivity was obtained in isooctane due to its identical polarity with thymol. In all of the tested systems, thymol migrated rapidly from the PBS matrix towards the food simulants, requiring 50–60 h to reach an equilibrium plateau in each case. Therefore, the authors claim that these materials are suitable for short-cycle food packaging applications, such as meat, vegetable, and fruit products.

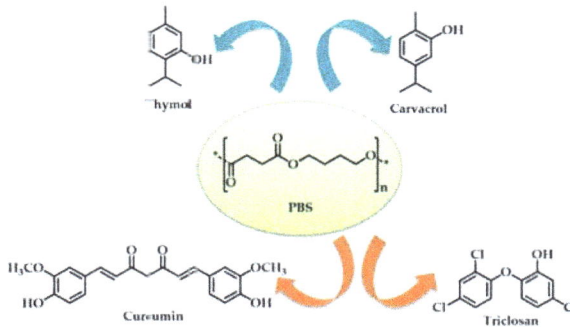

Figure 15. The chemical structure of PBS and agents employed to impart antimicrobial properties.

A similar study conducted by Wibunanawong et al. [293] was focused on the addition of carvacrol to a PBS matrix as an antimicrobial agent for the preparation of food packaging. The prepared materials showed clear zones of inhibition of *S. aureus* and *E. coli* growth at 4 and 10 wt% of carvacrol; however, no release studies were conducted.

Jie et al. used the extract from *Scutellaria* root (*S. baicalensis*), a herb traditionally employed in Chinese medicine [294], to achieve a dual effect of dyeing and antimicrobial activity in PBS matrices. Natural pigments were mixed at 1, 3, 5, 7, and 9 wt% with PBS to achieve dyed films. The materials were tested against *S. aureus* and *E. coli* bacteria, achieving antimicrobial properties at the highest pigment load (9 wt%) without notably affecting the crystallization and thermal stability of PBS [295].

Although EOs are considered to be potential antimicrobial agents for PBS matrices, some authors claim that their strong odors might affect their acceptance on the market, and hence have tested alternative fillers as inorganic particles. The antimicrobial activity of ZnO-modified PBS films was tested against representative food spoilage bacteria (*S. aureus* and *E. coli*), and it was observed that a minimum content of 6 wt% was required for their growth inhibition with a slight increase in the inhibition zone diameter with the ZnO content. Release of Zn^{2+} ions from PBS was measured in distilled water, 3% acetic acid, and 10% ethanol food simulants and was found to have a strong dependence on ZnO concentration. During the tested 15 days, in distilled water and ethanol, Zn^{2+} released slowly, reaching a maximum at values lower than 10 ppm, while a fast release (similar to the PBS/EO systems discussed above) was observed on samples immersed in acetic acid, reaching a plateau value of 15 ppm after 50 h [296]. An applied study of a PBS film filled with ZnO NPs (10–30 nm in size), used as packaging to preserve and prolong the shelf-life of fresh-cut apples, was reported by Naknaen [297]. The author, packed, sealed, and stored freshly cut apple slices in PBS plastic bags with different contents of ZnO NPs (0, 2, 4, and 6 wt%) at 10 °C. After a three-day period (over a total of 18 days), quality parameters, such as color, weight loss, total acidity, and concentration of sugar, were determined. Additionally, a microbiological analysis was conducted at the end of the study over homogenized, filtered, and diluted samples in peptone water. Samples were plated onto agar, incubated, and then plate counted, showing a lower total amount of bacteria for samples containing ZnO NPs. By increasing the content of antimicrobial agent, a decrease in the bacteria population was achieved, reaching equal values for the samples containing 4 and 6 wt% ZnO NPs. Taking into account the most common regulation that limits the maximum count of aerobic micro-organisms to 6 log CFU/g at the expiration date [298,299], the shelf-life of the fresh-cut apple slices was enhanced by 8 days when packed in PBS containing 4–6 wt% of ZnO NPs (from 9 to 18 days).

In the case of an application in the medical field, biodegradable polyesters are usually processed in the shape of fibers to find application as scaffolds. Tang et al. [300] coated PBS scaffolds, obtained using the salt leaching method, with copper-doped nano laponite (cnLAP). This material was to be applied in bone tissue engineering; therefore, the authors used nLAP to promote the osteogenic differentiation of human mesenchymal stem cells. Additionally, copper ions were expected to also promote bone regeneration and to inhibit infections. In fact, as is shown in Figures 16 and 17, after the scaffolds were cultured for 24 h, only a reduction of bacteria (*E. coli* and *S. aureus*) was found in the coated samples containing copper (cnLBC), with values around 90% at the same time as no cytotoxicity occurred. Certainly, the incorporation of laponite improves the adhesion, proliferation, and differentiation of bone mesenchymal stem cells, which were ascribed to the release of Mg, Si, and Li ions from the coating on the scaffolds into the media. All of these facts make this system a good candidate for bone regeneration.

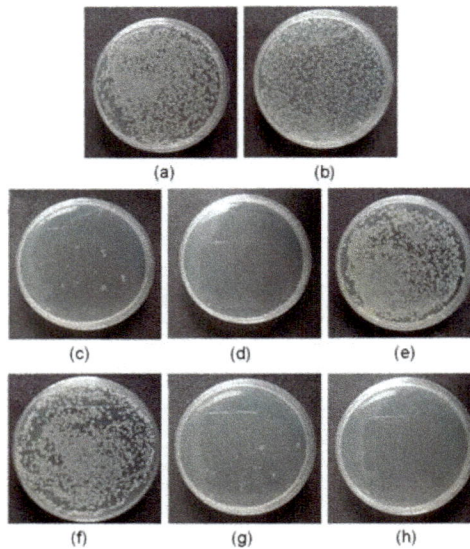

Figure 16. Photographs of *E. coli* (**a–d**) and *S. aureus* (**e–h**) colonies' growth after incubation for 24 h with PBS (**a,e**), nano laponite (nLAP)-coated PBS (nLBC) (**b,f**), and Cu–nLAP-coated PBS (**c,g**) scaffolds with vancomycin as a positive control (**d,h**). Reproduced from [300].

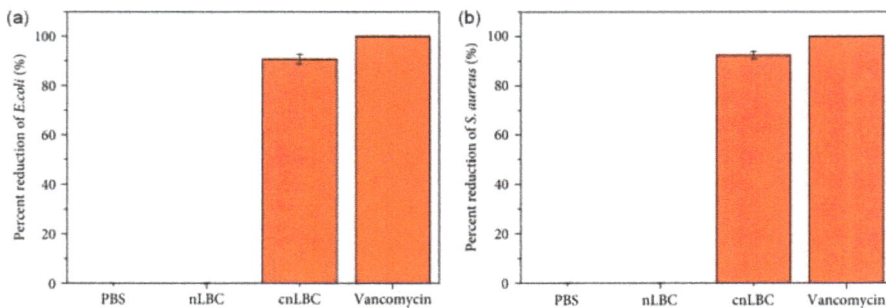

Figure 17. The percent reduction of *E. coli* (**a**) and *S. aureus* (**b**) bacteria on PBS, nLBC, and cnLBC scaffolds and vancomycin for 24 h. Reproduced from [300].

Tian et al. [301] introduced poly(vinyl pyrrolidone) capped with silver nanoparticles (PVP-capped AgNPs) into PBS electrospun materials to impart antimicrobial properties. PVP is often employed to stabilize AgNPs [302–304], and PVP-capped AgNPs have been reported to have better in vivo antimicrobial activity and less toxicity to mammalian cells [305] than other capped AgNPs [306]. Spherical capped AgNPs were successfully distributed and incorporated into the PBS electrospun fibers. The release of silver ions from the scaffolds was investigated in aqueous solution by an inductively coupled plasma spectrophotometer. After 2 weeks, the authors still detected silver ions released from the scaffolds due to the hydrophobic character of the PBS that hindered the permeation of water into the fibers and the diffusion of AgNPs from the fibers. Consequently, an antimicrobial capacity was confirmed against *S. aureus* and *E. coli*, obtaining an ability to inhibit bacterial growth in the long-term (more than 2 weeks).

Llorens et al. went further and developed a drug delivery scaffold constituted by electrospun poly(ethylene glycol) (PEG) and PBS blends. By coaxial electrospinning, different core–shell distributions were obtained, having either PEG or PBS in the outer or inner part (PEG-PBS and PBS-PEG core–shell distributions) in order to be compared with scaffolds obtained using a conventional setup (fibers obtained from a mixed solution of PEG and PBS). PEG and PBS solutions were loaded, respectively, with triclosan (polychlorophenoxy phenol at 1 w/v%), which is an antimicrobial and antifungal agent [307], and curcumin (0.5 w/v%), which is a natural phenol that seems to have beneficial effects on the treatment of several diseases [308,309]. The release profiles were studied for all of the scaffolds and revealed a high dependence on the media's hydrophobicity and the structure of the fibers. The authors achieved a different release of curcumin and triclosan and claimed that the solubility of PEG in aqueous media led to a fast release of the antibacterial compound, while the non-aqueous solubility of the curcumin-loaded-PBS component will permit a sustained anticancer effect with time [310]. Additionally, antibacterial tests were performed against *E. coli* and *M. luteus*, determining the bacterial adhesion and growth onto triclosan-loaded scaffolds. All drug-loaded scaffolds prevented bacterial colonization effectively, while coaxial samples were found to be more susceptible to bacteria colonization (growth inhibition measurements) than their uniaxial electrospun counterparts. Thus, 60–40% of inhibition was obtained from the coaxial samples, without a significant influence on the core–shell structure, while 90–75% of inhibition resulted from the uniaxial electrospun scaffolds.

5. Polyurethanes Based on Renewable Oils

Natural oils and fats from vegetable oils (VOs) are the most important renewable industrial feedstock for sustainable chemistry and, indeed, in polymer science [311]. VOs are triacylglycerols formed from glycerol and three fatty acids, being the most common the saturated capric (C10), lauric (C12), myristic (C14), palmitic (C16), and stearic (C18) acids and unsaturated oleic (C18, with one carbon–carbon double bond) and linoleic (C18, with two or three carbon–carbon double bonds) acids. In general, VOs do not contain hydroxyl groups; therefore, they are often modified chemically to introduce hydroxyl groups into their structures. Interestingly, polymer scientists have found plant-oil-based polyols to be attractive thanks to their unique chemical structure that offers numerous options for modification. VOs are cheap, non-toxic, biodegradable, and, moreover, their polar character confers enhanced antibacterial properties to the resultant polymeric systems. Most utilized polyols from VOs are derived from soybean, sunflower, and cottonseed and, together with fatty acids, have been used for many years in the production of polyols employed in polyurethane and, to a lesser extent, in polyester synthesis [312,313].

VOs have been employed for the synthesis of cationic polyurethane (PU) coatings with antimicrobial properties for application either in the biomedical or food packaging fields. Usually, cationic PUs are prepared by incorporating a tertiary amine diol or polyol treated with an acid that is able to bind microbes and disrupt their cell structure thanks to amino groups. Xia et al. [314] prepared cationic soybean-oil-based waterbone PU dispersions and coatings from five amino polyols

(Figure 18), and examined the effect of their structure and hydroxyl functionality over the antimicrobial, mechanical, and thermal properties.

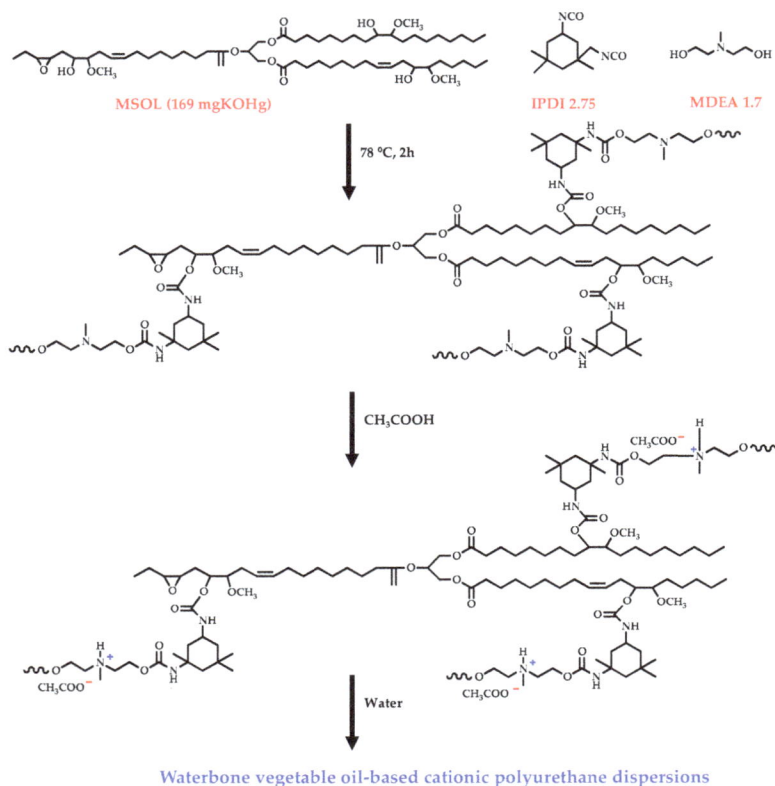

Figure 18. The synthesis route of the cationic soybean-oil-based polyurethane dispersions (PUDs). MSOL, Methoxylated soybean-oil polyol; IPDI, Isophorone diisocyanates; MDEA, *N*-methyldiethanolamine.

N-methyldiethanolamine (MDEA)- and *N*-ethyldiethanolamine (EDEA)-containing PUs provided the best antibacterial activity, which the authors associated with the smaller/shorter side chains attached to the nitrogen atoms that allow for better penetration of materials into cells. They observed a good antibacterial activity towards *L. monocytogenes* and *S. typhimurium* and against a Gram-negative structural mutant of *Salmonella minnesota* (R613) lacking a full outer membrane layer. Additionally, although the MDEA with a triethanolamine (TEA) residues structure had lower antibacterial activity than MDEA and EDEA (lower ammonium ion content), it provided the best balance of antimicrobial, thermal, and mechanical properties from all tested amino polyols thanks to its relatively high crosslink density. In a similar study, the same group [315] varied the molar ratio between hydroxyl and isocianate groups in the PU by varying the MDEA content. All formulations showed inhibitory activity against *S. typhimurium*, *L. monocytogenes*, and MRSA. Clearly, an increase in the ratio of ammonium cations enhanced the antibacterial properties. Additionally, they varied the crosslinking density of the materials by using methoxylated soybean polyols (MSOLs) with different numbers of hydroxyls. Lower crosslinking densities (a lower functionality of MSOLs) showed increased antimicrobial activities even though the concentration of quaternary ammonium was slightly lower than in their higher crosslinked counterparts. It seemed that a lower crosslinking density promoted the physical interaction with the target bacteria as the molecular mobility of the chains was higher, which helped to increase the antibacterial activity of the material.

Similarly, Liang et al. [316] employed castor oil, as a natural antimicrobial agent, and MDEA, as an ionic chain extender, to synthesize waterborne PUs. Samples were tested against *L. monocytogenes* and *V. parahaemolyticus* bacteria, being more effective by increasing either the MDEA content or the reduction of polyol functionality (a lower crosslinking density, higher mobility, and better interaction with bacteria).

Bakhshi et al. [317] functionalized intermediate tertiary amine soybean-oil-based polyols (TAPs) with ammonium salts by using either methyl iodine or benzyl chloride as alkylating agents instead of using an acid. These bio-based polyols containing quaternary ammonium salts were incorporated into polyurethanes using different diisocyanate monomers to obtain biocompatible and bactericidal coatings. When methyl iodine was used, the materials showed significant bacterial reduction (83–95%) against *E. coli* and *S. aureus* bacteria due to the higher amount of active groups in comparison to their benzyl-chloride-alkylated counterparts.

Thiol-ene (TEC) and thiol-yne (TYC) couplings are the most commonly employed chemistries for the preparation of plant-oil-derived polyols (hydroxyl building blocks) (Figure 19) [318,319]. By applying thiol-yne coupling chemistry to alkyne-derivatized fatty acids from naturally occurring oleic and 10-undecenoic acids (mainly obtained from sunflower oil saponification and castor oil pyrolysis), bio-based methyl-ester-containing polyols for PU technology were obtained (Figure 20).

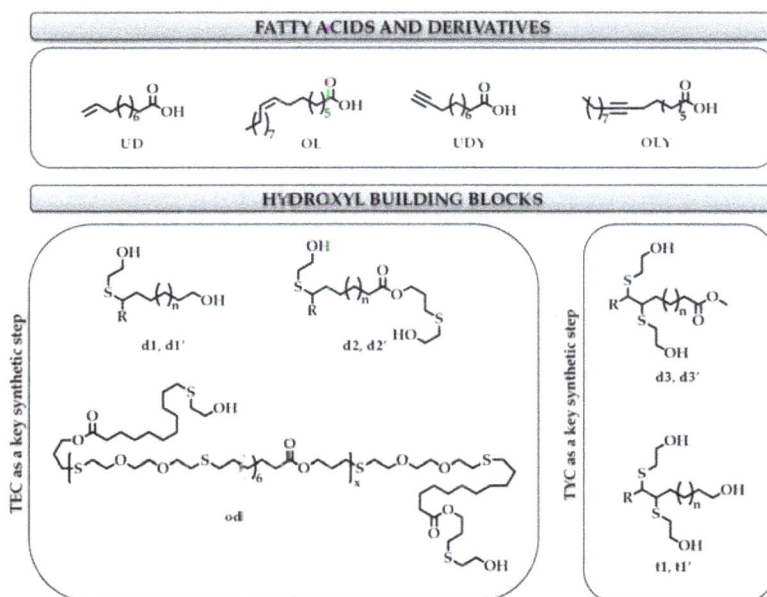

Figure 19. The chemical structures of 10-undecenoic (UD) acid and oleic (OL) acid, their alkyne derivatives (UDY and OLY), and the hydroxyl building blocks synthesized from them using either thiol-ene (TEC) or thiol-yne (TYC) as a key synthetic step (d, od, and t refer to diol, oligomeric diol, and triol, respectively, whereas 1, 2, and 3 refer to R = H and n = 6, and 1′, 2′, and 3′ to R = –(CH$_2$)$_7$–CH$_3$ and n = 5).

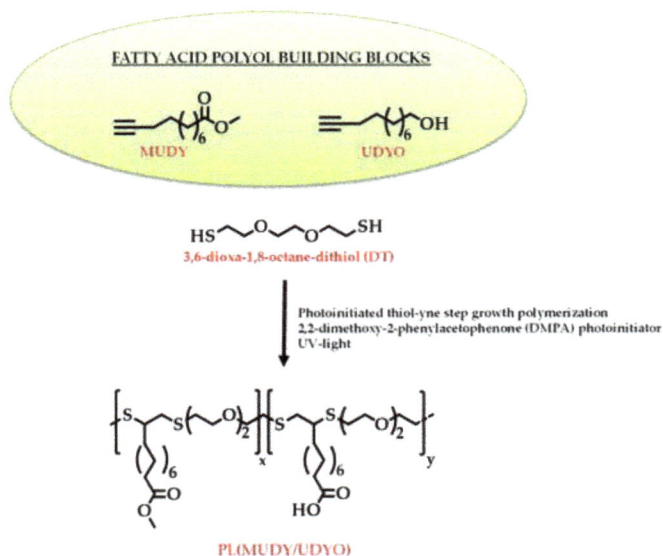

Figure 20. The chemical structures of reagents and the synthesis route employed to obtain methyl-ester-containing polyols from copolymerized methyl 10-undecynoate (MUDY) and 10-undecynyl alcohol (UDYO) fatty-acid-derived polyols. Varying MUDY/UDYO ratios allowed the researchers to obtain polyols with different hydroxyl contents for further reactions with diisocyanate to obtain polyurethanes (PUs).

The synthesized polyols were used to prepare several PU formulations with enhanced surface hydrophilicity and antimicrobial properties after aminolysis with poly(propylene glycol) monoamine (Jeffamine® M-600) and complexation with iodine. The antibacterial activity of these systems was tested against *P. aeruginosa*, *S. aureus*, and *C. albicans* [320], where effectiveness was only observed for Gram-positive bacteria and fungi but not for Gram-negative bacteria.

Algae oil and several di-acids from renewable sources (dimer acid, itaconic acid, maleic acid, and phthalic anhydride) were used to synthesize alkyd and polyesteramide polyols for PU coatings with anticorrosive and antibacterial properties [321]. Bactericidal properties were tested against *E. coli* and *S. aureus* by turbidimetry, showing that the uncoated polyols and the algae oil fatty amide (AOFA) and monoglyceride (MG) coatings had more bacterial attachment and less inhibition efficiency compared to algae-oil-modified PU coatings, since algae-oil-modified PU coatings have a higher percentage of oleic acid, which inhibits the bacterial growth.

In general, bio-based PU does not show antimicrobial activity and, in other polymeric families, the incorporation of uniformly dispersed metal or metal-oxide nanoparticles has been reported to result in nanomaterials with excellent antimicrobial activity [322]. These nanocomposites enhance the durability and efficacy of the antimicrobial effect through a controlled release of the fillers. The antimicrobial properties of *Mesua Ferrea L.* seed-oil-based hyperbranched and linear PU nanocomposites containing AgNPs were studied by Deka et al. [323]. The hyperbranched PU (HBPU) nanocomposite showed a better bactericidal effect over its linear counterpart (LPU) towards *S. aureus* and *E. coli* bacteria and against *C. albicans* yeast. The antibacterial activity was dose-dependent and, particularly at a high loading, the efficiency of the system was comparable to standard antibiotic and antifungal agents.

Moreover, Das et al. [324] reported the antibacterial activity of smart HBPU/Fe$_3$O$_4$ nanocomposites based on sunflower oil and with superparamagnetic-like behavior and shape-recovery effects. The authors concluded that the hyperbranched structure of the sunflower-oil-based HBPU matrix, as occurred with AgNPs, prevented agglomeration and led to a better dispersion of Fe$_3$O$_4$

NPs, resulting in better antibacterial performance against *S. aureus* and *K. pneumonia* with respect to bare Fe_3O_4. Both Fe_3O_4 nanomaterials and $HBPU/Fe_3O_4$ nanocomposites exhibited good antibacterial activity against infectious and biofilm-forming microbes.

Additionally, the same group demonstrated that the incorporation of multiwall carbon nanotubes (MWCNTs) decorated with Fe_3O_4 enhanced the antimicrobial capacity in comparison to non-nanohybrid HBPU nanocomposites filled either with Fe_3O_4 or MWCNTs [325]. Self-healable castor-oil-based polyurethane containing sulfur-nanoparticle-decorated reduced graphene oxide (SrGO), combining the potential of both antimicrobial agents (rGO and sulfur nanoparticles), was obtained by Thakur et al. [326] and further studied by Wu et al. [327]. A synergistic effect was obtained on the nanohybrid particles of SrGO, and, although a high dose of nanocomposite was needed, the HBPU–SrGO material showed an inhibitory effect on both Gram-positive and Gram-negative bacteria, enhancing the inhibitory effect of the neat matrix. Recently, Duarah et al. developed a bio-based hyperbranched PU from a starch-modified polyol filled with carbon dots and AgNPs (HPU/CD-Ag) as a material for rapid self-expandable stents. Importantly, HPU/CD-Ag nanohybrid membranes prevented biofilm formation against *E. coli* and *S. aureus* and bacterial adherence against *P. aeruginosa*, assessing the highest synergetic antibacterial activity in comparison to HPU/AgNP and HPU/CD nanocomposites [328].

A linseed-oil-based polyol was employed for PU synthesis, and nanocomposite films with 0.5–10 wt% of Biocera A$^{\circledR}$ commercial particles (composed of silver, zinc, magnesium, calcium phosphate, alumina, and silica, with a particle size of 3–4 μm) were obtained by the casting-evaporation technique. Interestingly, systems with and without commercial particles showed antibacterial performance against *E. coli*, *P. aeruginosa*, *S. aureus*, and *B. subtilis* and were not cytotoxic against the Murine fibroblast NIH 3T3 cell line. They suggested that these systems would be useful for wound dressing applications [329].

Boron-incorporated linseed oil polyols were also employed to obtain semi-inorganic vegetable-oil-based PUs [330]. Both polyols and their obtained PUs showed high antibacterial activity against *S. aureus*; however, although the polyols were completely inactive against *E. coli*, the PUs were found to be mildly to moderately active. Sharmin et al. [331] linked CuO NPs to a linseed oil polyol through an esterification reaction with copper (II) acetate (CuAc) in a one-pot, solvent-less process to obtain good antibacterial activity against *E. coli* and *S. aureus*, probably through membrane disruption and cell death. The authors proved as advantageous the small size of the particles, which enhanced the size of the contact surface area and improved the antibacterial action as the metal content increased. These materials may be useful for making a self-sterilizing biofilm that resists PU coatings and paints.

VOs and their derivatives have demonstrated huge potential as renewable feedstocks, receiving increasing interest from the research community for developing polyols and PUs from under-used sources of plant oils. However, there are still some challenges related to cost barriers, mainly due to the fact that they come from edible sources; so, their usage affects the cost of foodstuffs. Therefore, future efforts have to be concentrated towards the employment of non-edible seed oils, such as algae oils, for the preparation of bio-based polyols. Additionally, heterogeneity in terms of VO structure also represents a challenge and, additionally, new advances in bio-based isocyanates are expected to be made in the near future.

6. Concluding Remarks

There is a necessity to find new routes or alternatives to petroleum-based materials. The circular economy has to be reached in every corner of life, and, at the same time, we must be able to combat bacterial resistance to antibiotics. Thus, biopolymers are outstanding candidates to be modified or combined with an antimicrobial substance to obtain antimicrobial systems with application in several fields and in good alignment with the circular economy. This article describes some of the recent steps taken to reach these goals. However, more research and investment are needed to achieve fully sustainable materials with antimicrobial activity and effective substitutes for the existing ones. In this sense, nanotechnology as a method of reinforcement, nanoencapsulation, or nanostructuration has shown that it can help us to achieve our goals.

Author Contributions: Writing (original draft preparation), A.M.-B., M.P.A., C.E., Á.S. and M.F.-G.; Writing (review and editing), A.M.-B., M.P.A., C.E., Á.S. and M.F.-G.; Funding acquisition, M.F.-G.

Funding: This research was funded by MINECO, Project MAT2016-78437-R, the Agencia Estatal de Investigación (AEI, Spain), and Fondo Europeo de Desarrollo Regional (FEDER, EU).

Acknowledgments: C.E. and Á.S acknowledge the Juan de la Cierva contracts (IJCI-2015-26432 and FJCI-2015-24405, respectively) from the Spanish Ministry of Science, Innovation, and Universities. M.P.A. thanks Universidad Complutense de Madrid for her postdoctoral contract (Ayuda posdoctoral de formación en docencia e investigación en los Departamentos de la UCM).

Conflicts of Interest: The authors declare no conflict of interest.

References

1. Available online: http://europa.eu/rapid/press-release_MEMO-15-5709_en.htm (accessed on 3 January 2019).
2. European Bioplastics. Available online: https://www.European-bioplastics.Org/market/ (accessed on 15 January 2019).
3. Plastics Europe. An analysis of european plastics production, demand and waste data. In *Plastics—The Facts 2017*; Plastics Europe: Brussels, Belgium, 2017.
4. Ramesh, P.B.; Kevin, O.C.; Ramakrishna, S. Current progress on bio-based polymers and their future trends. *Prog. Biomater.* **2013**, *2*, 8.
5. Charnley, M.; Textor, M.; Acikgoz, C. Designed polymer structures with antifouling-antimicrobial properties. *React. Funct. Polym.* **2011**, *71*, 329–334. [CrossRef]
6. Muñoz-Bonilla, A.; Fernández-García, M. Polymeric materials with antimicrobial activity. *Prog. Polym. Sci.* **2012**, *37*, 281–339. [CrossRef]
7. Siedenbiedel, F.; Tiller, J.C. Antimicrobial polymers in solution and on surfaces: Overview and functional principles. *Polymers* **2012**, *4*, 46–71. [CrossRef]
8. Palza, H. Antimicrobial polymers with metal nanoparticles. *Int. J. Mol. Sci.* **2015**, *16*, 2099–2116. [CrossRef] [PubMed]
9. Muñoz-Bonilla, A.; Fernández-García, M. The roadmap of antimicrobial polymeric materials in macromolecular nanotechnology. *Eur. Polym. J.* **2015**, *65*, 46–62. [CrossRef]
10. Xue, Y.; Xiao, H.; Zhang, Y. Antimicrobial polymeric materials with quaternary ammonium and phosphonium salts. *Int. J. Mol. Sci.* **2015**, *16*, 3626–3655. [CrossRef]
11. Alvarez-Paino, M.; Munoz-Bonilla, A.; Fernandez-Garcia, M. Antimicrobial polymers in the nano-world. *Nanomaterials* **2017**, *7*, 48. [CrossRef]
12. Muñoz-Bonilla, A.; Fernández-García, M. Poly(ionic liquid)s as antimicrobial materials. *Eur. Polym. J.* **2018**, *105*, 135–149. [CrossRef]
13. Siqueira, G.; Bras, J.; Dufresne, A. Cellulosic bionanocomposites: A review of preparation, properties and applications. *Polymers* **2010**, *2*, 728–765. [CrossRef]
14. Raquez, J.M.; Deléglise, M.; Lacrampe, M.F.; Krawczak, P. Thermosetting (bio)materials derived from renewable resources: A critical review. *Prog. Polym. Sci.* **2010**, *35*, 487–509. [CrossRef]
15. Chen, G.-Q.; Patel, M.K. Plastics derived from biological sources: Present and future: A technical and environmental review. *Chem. Rev.* **2012**, *112*, 2082–2099. [CrossRef] [PubMed]
16. Rhim, J.-W.; Park, H.-M.; Ha, C.-S. Bio-nanocomposites for food packaging applications. *Prog. Polym. Sci.* **2013**, *38*, 1629–1652. [CrossRef]
17. Miller, S.A. Sustainable polymers: Opportunities for the next decade. *ACS Macro Lett.* **2013**, *2*, 550–554. [CrossRef]
18. Mosiewicki, M.A.; Aranguren, M.I. A short review on novel biocomposites based on plant oil precursors. *Eur. Polym. J.* **2013**, *49*, 1243–1256. [CrossRef]
19. Bugnicourt, E.; Cinelli, P.; Lazzeri, A.; Alvarez, V. Polyhydroxyalkanoate (pha): Review of synthesis, characteristics, processing and potential applications in packaging. *Express Polym. Lett.* **2014**, *8*, 791–808. [CrossRef]

20. Zhu, Y.; Romain, C.; Williams, C.K. Sustainable polymers from renewable resources. *Nature* **2016**, *540*, 354. [CrossRef]
21. Wang, H.; Qian, J.; Ding, F. Emerging chitosan-based films for food packaging applications. *J. Agric. Food Chem.* **2018**, *66*, 395–413. [CrossRef]
22. Ergene, C.; Yasuhara, K.; Palermo, E.F. Biomimetic antimicrobial polymers: Recent advances in molecular design. *Polym. Chem.* **2018**, *9*, 2407–2427. [CrossRef]
23. Kumar, M.N.V.R. A review of chitin and chitosan applications. *React. Funct. Polym.* **2000**, *46*, 1–27. [CrossRef]
24. Rinaudo, M. Chitin and chitosan: Properties and applications. *Prog. Polym. Sci.* **2006**, *31*, 603–632. [CrossRef]
25. Tang, H.; Zhang, P.; Kieft, T.L.; Ryan, S.J.; Baker, S.M.; Wiesmann, W.P.; Rogelj, S. Antibacterial action of a novel functionalized chitosan-arginine against gram-negative bacteria. *Acta Biomater.* **2010**, *6*, 2562–2571. [CrossRef] [PubMed]
26. Raafat, D.; von Bargen, K.; Haas, A.; Sahl, H.G. Insights into the mode of action of chitosan as an antibacterial compound. *Appl. Environ. Microbiol.* **2008**, *74*, 3764–3773. [CrossRef] [PubMed]
27. Muñoz-Bonilla, A.; Cerrada, M.L.; Fernández-García, M. Antimicrobial activity of chitosan in food, agriculture and biomedicine. In *Polymeric Materials with Antimicrobial Activity*; Muñoz-Bonilla, A., Cerrada, M.L., Fernández-García, M., Eds.; Royal Society of Chemistry: Cambridge, UK, 2013; pp. 22–54.
28. Rabea, E.I.; Badawy, M.E.-T.; Stevens, C.V.; Smagghe, G.; Steurbaut, W. Chitosan as antimicrobial agent: Applications and mode of action. *Biomacromolecules* **2003**, *4*, 1457–1465. [CrossRef] [PubMed]
29. Sahariah, P.; Masson, M. Antimicrobial chitosan and chitosan derivatives: A review of the structure-activity relationship. *Biomacromolecules* **2017**, *18*, 3846–3868. [CrossRef] [PubMed]
30. Verlee, A.; Mincke, S.; Stevens, C.V. Recent developments in antibacterial and antifungal chitosan and its derivatives. *Carbohydr. Polym.* **2017**, *164*, 268–283. [CrossRef] [PubMed]
31. Liu, W.; Qin, Y.; Liu, S.; Xing, R.; Yu, H.; Chen, X.; Li, K.; Li, P. Synthesis, characterization and antifungal efficacy of chitosan derivatives with triple quaternary ammonium groups. *Int. J. Biol. Macromol.* **2018**, *114*, 942–949. [CrossRef] [PubMed]
32. Cyervides-Munoz, E.; Pollet, E.; Ulrich, G.; de Jesus Sosa-Santillan, G.; Averous, L. Original method for synthesis of chitosan-based antimicrobial agent by quaternary ammonium grafting. *Carbohydr. Polym.* **2017**, *157*, 1922–1932. [CrossRef]
33. Martins, A.F.; Facchi, S.P.; Follmann, H.D.; Pereira, A.G.; Rubira, A.F.; Muniz, E.C. Antimicrobial activity of chitosan derivatives containing n-quaternized moieties in its backbone: A review. *Int. J. Mol. Sci.* **2014**, *15*, 20800–20832. [CrossRef]
34. Kulkarni, A.D.; Patel, H.M.; Surana, S.J.; Vanjari, Y.H.; Belgamwar, V.S.; Pardeshi, C.V. N,n,n-trimethyl chitosan: An advanced polymer with myriad of opportunities in nanomedicine. *Carbohydr. Polym.* **2017**, *157*, 875–902. [CrossRef]
35. Verheul, R.J.; Amidi, M.; van der Wal, S.; van Riet, E.; Jiskoot, W.; Hennink, W.E. Synthesis, characterization and in vitro biological properties of O-methyl free N,N,N-trimethylated chitosan. *Biomaterials* **2008**, *29*, 3642–3649. [CrossRef] [PubMed]
36. Sahariah, P.; Benediktssdottir, B.E.; Hjalmarsdottir, M.A.; Sigurjonsson, O.E.; Sorensen, K.K.; Thygesen, M.B.; Jensen, K.J.; Masson, M. Impact of chain length on antibacterial activity and hemocompatibility of quaternary n-alkyl and n,n-dialkyl chitosan derivatives. *Biomacromolecules* **2015**, *16*, 1449–1460. [CrossRef] [PubMed]
37. Sahariah, P.; Masson, M.; Meyer, R.L. Quaternary ammoniumyl chitosan derivatives for eradication of staphylococcus aureus biofilms. *Biomacromolecules* **2018**, *19*, 3649–3658. [CrossRef] [PubMed]
38. Wang, C.H.; Liu, W.S.; Sun, J.F.; Hou, G.G.; Chen, Q.; Cong, W.; Zhao, F. Non-toxic o-quaternized chitosan materials with better water solubility and antimicrobial function. *Int. J. Biol. Macromol.* **2016**, *84*, 418–427. [CrossRef]
39. Jung, J.; Wen, J.; Sun, Y. Amphiphilic quaternary ammonium chitosans self-assemble onto bacterial and fungal biofilms and kill adherent microorganisms. *Colloids Surf. B Biointerfaces* **2018**, *174*, 1–8. [CrossRef] [PubMed]
40. Shariatinia, Z. Carboxymethyl chitosan: Properties and biomedical applications. *Int. J. Biol. Macromol.* **2018**, *120*, 1406–1419. [CrossRef] [PubMed]
41. Liu, X.F.; Guan, Y.L.; Yang, D.Z.; Li, Z.; Yao, K.D. Antibacterial action of chitosan and carboxymethylated chitosan. *J. Appl. Polym. Sci.* **2001**, *79*, 1324–1335.

42. Kogan, G.; Skorik, Y.A.; Zitnanova, I.; Krizkova, L.; Durackova, Z.; Gomes, C.A.; Yatluk, Y.G.; Krajcovic, J. Antioxidant and antimutagenic activity of n-(2-carboxyethyl)chitosan. *Toxicol. Appl. Pharmacol.* **2004**, *201*, 303–310. [CrossRef]

43. Chen, X.-G.; Park, H.-J. Chemical characteristics of o-carboxymethyl chitosans related to the preparation conditions. *Carbohydr. Polym.* **2003**, *53*, 355–359. [CrossRef]

44. Anitha, A.; Divya Rani, V.V.; Krishna, R.; Sreeja, V.; Selvamurugan, N.; Nair, S.V.; Tamura, H.; Jayakumar, R. Synthesis, characterization, cytotoxicity and antibacterial studies of chitosan, o-carboxymethyl and n,o-carboxymethyl chitosan nanoparticles. *Carbohydr. Polym.* **2009**, *78*, 672–677. [CrossRef]

45. Mohamed, N.A.; Mohamed, R.R.; Seoudi, R.S. Synthesis and characterization of some novel antimicrobial thiosemicarbazone o-carboxymethyl chitosan derivatives. *Int. J. Biol. Macromol.* **2014**, *63*, 163–169. [PubMed]

46. Tan, Y.; Han, F.; Ma, S.; Yu, W. Carboxymethyl chitosan prevents formation of broad-spectrum biofilm. *Carbohydr. Polym.* **2011**, *84*, 1365–1370. [CrossRef]

47. Tan, Y.; Leonhard, M.; Moser, D.; Schneider-Stickler, B. Antibiofilm activity of carboxymethyl chitosan on the biofilms of non-candida albicans candida species. *Carbohydr. Polym.* **2016**, *149*, 77–82. [CrossRef] [PubMed]

48. Xiao, B.; Wan, Y.; Zhao, M.; Liu, Y.; Zhang, S. Preparation and characterization of antimicrobial chitosan-n-arginine with different degrees of substitution. *Carbohydr. Polym.* **2011**, *83*, 144–150. [CrossRef]

49. Sang, W.; Tang, Z.; He, M.Y.; Hua, Y.P.; Xu, Q. Synthesis and preservative application of quaternized carboxymethyl chitosan containing guanidine groups. *Int. J. Biol. Macromol.* **2015**, *75*, 489–494. [CrossRef] [PubMed]

50. Li, J.; Ye, Y.; Xiao, H.; He, B.; Qian, L. Microwave assisted preparation of antimicrobial chitosan with guanidine oligomers and its application in hygiene paper products. *Polymers* **2017**, *9*, 633. [CrossRef]

51. Sun, S.; An, Q.; Li, X.; Qian, L.; He, B.; Xiao, H. Synergistic effects of chitosan-guanidine complexes on enhancing antimicrobial activity and wet-strength of paper. *Bioresour. Technol.* **2010**, *101*, 5693–5700. [CrossRef]

52. Sahariah, P.; Oskarsson, B.M.; Hjalmarsdottir, M.A.; Masson, M. Synthesis of guanidinylated chitosan with the aid of multiple protecting groups and investigation of antibacterial activity. *Carbohydr. Polym.* **2015**, *127*, 407–417. [CrossRef]

53. Jenssen, H.; Hamill, P.; Hancock, R.E. Peptide antimicrobial agents. *Clin. Microbiol. Rev.* **2006**, *19*, 491–511. [CrossRef]

54. De la Fuente-Nunez, C.; Torres, M.D.; Mojica, F.J.; Lu, T.K. Next-generation precision antimicrobials: Towards personalized treatment of infectious diseases. *Curr. Opin. Microbiol.* **2017**, *37*, 95–102. [CrossRef]

55. Fjell, C.D.; Hiss, J.A.; Hancock, R.E.; Schneider, G. Designing antimicrobial peptides: Form follows function. *Nat. Rev. Drug Discov.* **2011**, *11*, 37–51. [CrossRef] [PubMed]

56. Li, P.; Zhou, C.; Rayatpisheh, S.; Ye, K.; Poon, Y.F.; Hammond, P.T.; Duan, H.; Chan-Park, M.B. Cationic peptidopolysaccharides show excellent broad-spectrum antimicrobial activities and high selectivity. *Adv. Mater.* **2012**, *24*, 4130–4137. [CrossRef] [PubMed]

57. Zhou, C.; Wang, M.; Zou, K.; Chen, J.; Zhu, Y.; Du, J. Antibacterial polypeptide-grafted chitosan-based nanocapsules as an "armed" carrier of anticancer and antiepileptic drugs. *ACS Macro Lett.* **2013**, *2*, 1021–1025. [CrossRef]

58. Barbosa, M.; Vale, N.; Costa, F.M.; Martins, M.C.; Gomes, P. Tethering antimicrobial peptides onto chitosan: Optimization of azide-alkyne "click" reaction conditions. *Carbohydr. Polym.* **2017**, *165*, 384–393. [CrossRef] [PubMed]

59. Sahariah, P.; Sorensen, K.K.; Hjalmarsdottir, M.A.; Sigurjonsson, O.E.; Jensen, K.J.; Masson, M.; Thygesen, M.B. Antimicrobial peptide shows enhanced activity and reduced toxicity upon grafting to chitosan polymers. *Chem. Commun.* **2015**, *51*, 11611–11614. [CrossRef] [PubMed]

60. Su, Y.; Tian, L.; Yu, M.; Gao, Q.; Wang, D.; Xi, Y.; Yang, P.; Lei, B.; Ma, P.X.; Li, P. Cationic peptidopolysaccharides synthesized by 'click' chemistry with enhanced broad-spectrum antimicrobial activities. *Polym. Chem.* **2017**, *8*, 3788–3800. [CrossRef]

61. Pranantyo, D.; Xu, L.Q.; Kang, E.T.; Chan-Park, M.B. Chitosan-based peptidopolysaccharides as cationic antimicrobial agents and antibacterial coatings. *Biomacromolecules* **2018**, *19*, 2156–2165. [CrossRef]

62. Bakkali, F.; Averbeck, S.; Averbeck, D.; Idaomar, M. Biological effects of essential oils—A review. *Food Chem. Toxicol.* **2008**, *46*, 446–475. [CrossRef] [PubMed]

63. Vergis, J.; Gokulakrishnan, P.; Agarwal, R.K.; Kumar, A. Essential oils as natural food antimicrobial agents: A review. *Crit. Rev. Food Sci. Nutr.* **2015**, *55*, 1320–1323. [CrossRef] [PubMed]

64. Calo, J.R.; Crandall, P.G.; O'Bryan, C.A.; Ricke, S.C. Essential oils as antimicrobials in food systems—A review. *Food Control* **2015**, *54*, 111–119. [CrossRef]

65. Yuan, G.; Chen, X.; Li, D. Chitosan films and coatings containing essential oils: The antioxidant and antimicrobial activity, and application in food systems. *Food Res. Int.* **2016**, *89*, 117–128. [CrossRef] [PubMed]

66. Priyadarshi, R.; Sauraj; Kumar, B.; Deeba, F.; Kulshreshtha, A.; Negi, Y.S. Chitosan films incorporated with apricot (*prunus armeniaca*) kernel essential oil as active food packaging material. *Food Hydrocoll.* **2018**, *85*, 158–166. [CrossRef]

67. Alves, V.L.C.D.; Rico, B.P.M.; Cruz, R.M.S.; Vicente, A.A.; Khmelinskii, I.; Vieira, M.C. Preparation and characterization of a chitosan film with grape seed extract-carvacrol microcapsules and its effect on the shelf-life of refrigerated salmon (*salmo salar*). *LWT* **2018**, *89*, 525–534. [CrossRef]

68. Nazzaro, F.; Orlando, P.; Fratianni, F.; Coppola, R. Microencapsulation in food science and biotechnology. *Curr. Opin. Biotechnol.* **2012**, *23*, 182–186. [CrossRef] [PubMed]

69. Pasanphan, W.; Chirachanchai, S. Conjugation of gallic acid onto chitosan: An approach for green and water-based antioxidant. *Carbohydr. Polym.* **2008**, *72*, 169–177. [CrossRef]

70. Li, K.; Guan, G.; Zhu, J.; Wu, H.; Sun, Q. Antibacterial activity and mechanism of a laccase-catalyzed chitosan–gallic acid derivative against escherichia coli and staphylococcus aureus. *Food Control* **2019**, *96*, 234–243. [CrossRef]

71. Souza, V.G.L.; Pires, J.R.A.; Vieira, É.T.; Coelhoso, I.M.; Duarte, M.P.; Fernando, A.L. Activity of chitosan-montmorillonite bionanocomposites incorporated with rosemary essential oil: From in vitro assays to application in fresh poultry meat. *Food Hydrocoll.* **2019**, *89*, 241–252. [CrossRef]

72. Lei, K.; Wang, X.; Li, X.; Wang, L. The innovative fabrication and applications of carvacrol nanoemulsions, carboxymethyl chitosan microgels and their composite films. *Colloids Surf. B Biointerfaces* **2018**, *175*, 688–696. [CrossRef]

73. Hasheminejad, N.; Khodaiyan, F.; Safari, M. Improving the antifungal activity of clove essential oil encapsulated by chitosan nanoparticles. *Food Chem.* **2019**, *275*, 113–122. [CrossRef]

74. Lee, K.H.; Lee, J.-S.; Kim, E.S.; Lee, H.G. Preparation, characterization, and food application of rosemary extract-loaded antimicrobial nanoparticle dispersions. *LWT* **2019**, *101*, 138–144. [CrossRef]

75. Kavaz, D.; Idris, M.; Onyebuchi, C. Physiochemical characterization, antioxidative, anticancer cells proliferation and food pathogens antibacterial activity of chitosan nanoparticles loaded with cyperus articulatus rhizome essential oils. *Int. J. Biol. Macromol.* **2019**, *123*, 837–845. [CrossRef] [PubMed]

76. Kumar-Krishnan, S.; Prokhorov, E.; Hernández-Iturriaga, M.; Mota-Morales, J.D.; Vázquez-Lepe, M.; Kovalenko, Y.; Sanchez, I.C.; Luna-Bárcenas, G. Chitosan/silver nanocomposites: Synergistic antibacterial action of silver nanoparticles and silver ions. *Eur. Polym. J.* **2015**, *67*, 242–251. [CrossRef]

77. Venkatesham, M.; Ayodhya, D.; Madhusudhan, A.; Veera Babu, N.; Veerabhadram, G. A novel green one-step synthesis of silver nanoparticles using chitosan: Catalytic activity and antimicrobial studies. *Appl. Nanosci.* **2012**, *4*, 113–119. [CrossRef]

78. Wongpreecha, J.; Polpanich, D.; Suteewong, T.; Kaewsaneha, C.; Tangboriboonrat, P. One-pot, large-scale green synthesis of silver nanoparticles-chitosan with enhanced antibacterial activity and low cytotoxicity. *Carbohydr. Polym.* **2018**, *199*, 641–648. [CrossRef] [PubMed]

79. Huang, X.; Bao, X.; Liu, Y.; Wang, Z.; Hu, Q. Catechol-functional chitosan/silver nanoparticle composite as a highly effective antibacterial agent with species-specific mechanisms. *Sci. Rep.* **2017**, *7*, 1860. [CrossRef] [PubMed]

80. De Lima, R.; Seabra, A.B.; Duran, N. Silver nanoparticles: A brief review of cytotoxicity and genotoxicity of chemically and biogenically synthesized nanoparticles. *J. Appl. Toxicol.* **2012**, *32*, 867–879. [CrossRef]

81. Mackevica, A.; Olsson, M.E.; Hansen, S.F. Silver nanoparticle release from commercially available plastic food containers into food simulants. *J. Nanopart. Res.* **2016**, *18*, 5. [CrossRef]

82. Wu, Z.; Huang, X.; Li, Y.-C.; Xiao, H.; Wang, X. Novel chitosan films with laponite immobilized Ag nanoparticles for active food packaging. *Carbohydr. Polym.* **2018**, *199*, 210–218. [CrossRef]

83. Tabesh, E.; Salimijazi, H.; Kharaziha, M.; Hejazi, M. Antibacterial chitosan-copper nanocomposite coatings for biomedical applications. *Mater. Today Proc.* **2018**, *5*, 15806–15812. [CrossRef]

84. Wahid, F.; Wang, H.S.; Lu, Y.S.; Zhong, C.; Chu, L.Q. Preparation, characterization and antibacterial applications of carboxymethyl chitosan/cuo nanocomposite hydrogels. *Int. J. Biol. Macromol.* **2017**, *101*, 690–695. [CrossRef]

85. Wahid, F.; Yin, J.J.; Xue, D.D.; Xue, H.; Lu, Y.S.; Zhong, C.; Chu, L.Q. Synthesis and characterization of antibacterial carboxymethyl chitosan/zno nanocomposite hydrogels. *Int. J. Biol. Macromol.* **2016**, *88*, 273–279. [CrossRef]

86. Rahman, P.M.; Mujeeb, V.M.; Muraleedharan, K. Flexible chitosan-nano zno antimicrobial pouches as a new material for extending the shelf life of raw meat. *Int. J. Biol. Macromol.* **2017**, *97*, 382–391. [CrossRef] [PubMed]

87. Montaser, A.S.; Wassel, A.R.; Al-Shaye'a, O.N. Synthesis, characterization and antimicrobial activity of schiff bases from chitosan and salicylaldehyde/tio2 nanocomposite membrane. *Int. J. Biol. Macromol.* **2019**, *124*, 802–809. [CrossRef] [PubMed]

88. Zhang, X.; Xiao, G.; Wang, Y.; Zhao, Y.; Su, H.; Tan, T. Preparation of chitosan-tio2 composite film with efficient antimicrobial activities under visible light for food packaging applications. *Carbohydr. Polym.* **2017**, *169*, 101–107. [CrossRef]

89. Kubacka, A.; Diez, M.S.; Rojo, D.; Bargiela, R.; Ciordia, S.; Zapico, I.; Albar, J.P.; Barbas, C.; Martins dos Santos, V.A.; Fernandez-Garcia, M.; et al. Understanding the antimicrobial mechanism of TiO$_2$-based nanocomposite films in a pathogenic bacterium. *Sci. Rep.* **2014**, *4*, 4134. [CrossRef] [PubMed]

90. Zhu, Z.; Cai, H.; Sun, D.-W. Titanium dioxide (TiO$_2$) photocatalysis technology for nonthermal inactivation of microorganisms in foods. *Trends Food Sci. Technol.* **2018**, *75*, 23–35. [CrossRef]

91. Munoz-Bonilla, A.; Cerrada, M.L.; Fernandez-Garcia, M.; Kubacka, A.; Ferrer, M.; Fernandez-Garcia, M. Biodegradable polycaprolactone-titania nanocomposites: Preparation, characterization and antimicrobial properties. *Int. J. Mol. Sci.* **2013**, *14*, 9249–9266. [CrossRef] [PubMed]

92. Almeida, A.P.C.; Canejo, J.P.; Fernandes, S.N.; Echeverria, C.; Almeida, P.L.; Godinho, M.H. Cellulose-based biomimetics and their applications. *Adv. Mater.* **2018**, *30*, 1703655. [CrossRef] [PubMed]

93. Echeverria, C.; Aguirre, L.E.; Merino, E.G.; Almeida, P.L.; Godinho, M.H. Carbon nanotubes as reinforcement of cellulose liquid crystalline responsive networks. *ACS Appl. Mater. Interfaces* **2015**, *7*, 21005–21009. [CrossRef] [PubMed]

94. Geng, Y.; Almeida, P.L.; Fernandes, S.N.; Cheng, C.; Palffy-Muhoray, P.; Godinho, M.H. A cellulose liquid crystal motor: A steam engine of the second kind. *Sci. Rep.* **2013**, *3*, 1028. [CrossRef]

95. Purwar, R.; Srivastava, C.M. Antimicrobial cellulose and cellulose derivative materials. In *Cellulose and Cellulose Derivatives: Synthesis, Modification and Applications*; Nova Science Publishers, Inc.: New York, NY, USA, 2015; pp. 455–472.

96. Demircan, D.; Zhang, B. Facile synthesis of novel soluble cellulose-grafted hyperbranched polymers as potential natural antimicrobial materials. *Carbohydr. Polym.* **2017**, *157*, 1913–1921. [CrossRef] [PubMed]

97. Wu, H.; Teng, C.; Liu, B.; Tian, H.; Wang, J. Characterization and long term antimicrobial activity of the nisin anchored cellulose films. *Int. J. Biol. Macromol.* **2018**, *113*, 487–493. [CrossRef] [PubMed]

98. Li, H.; Peng, L. Antimicrobial and antioxidant surface modification of cellulose fibers using layer-by-layer deposition of chitosan and lignosulfonates. *Carbohydr. Polym.* **2015**, *124*, 35–42. [CrossRef] [PubMed]

99. Li, F.; Biagioni, P.; Finazzi, M.; Tavazzi, S.; Piergiovanni, L. Tunable green oxygen barrier through layer-by-layer self-assembly of chitosan and cellulose nanocrystals. *Carbohydr. Polym.* **2013**, *92*, 2128–2134. [CrossRef]

100. Moghimi, R.; Aliahmadi, A.; Rafati, H. Antibacterial hydroxypropyl methyl cellulose edible films containing nanoemulsions of thymus daenensis essential oil for food packaging. *Carbohydr. Polym.* **2017**, *175*, 241–248. [CrossRef] [PubMed]

101. Heredia-Guerrero, J.A.; Ceseracciu, L.; Guzman-Puyol, S.; Paul, U.C.; Alfaro-Pulido, A.; Grande, C.; Vezzulli, L.; Bandiera, T.; Bertorelli, R.; Russo, D.; et al. Antimicrobial, antioxidant, and waterproof rtv silicone-ethyl cellulose composites containing clove essential oil. *Carbohydr. Polym.* **2018**, *192*, 150–158. [CrossRef] [PubMed]

102. Jamarani, R.; Erythropel, H.; Nicell, J.; Leask, R.; Marić, M. How green is your plasticizer? *Polymers* **2018**, *10*, 834. [CrossRef]

103. Liakos, I.; Rizzello, L.; Hajiali, H.; Brunetti, V.; Carzino, R.; Pompa, P.P.; Athanassiou, A.; Mele, E. Fibrous wound dressings encapsulating essential oils as natural antimicrobial agents. *J. Mater. Chem. B* **2015**, *3*, 1583–1589. [CrossRef]

104. Liakos, I.L.; Holban, A.M.; Carzino, R.; Lauciello, S.; Grumezescu, A.M. Electrospun fiber pads of cellulose acetate and essential oils with antimicrobial activity. *Nanomaterials* **2017**, *7*, 84. [CrossRef]

105. Fernandes, S.N.; Canejo, J.P.; Echeverria, C.; Godinho, M.H. Functional materials from liquid crystalline cellulose derivatives: Synthetic routes, characterization and applications. In *Liquid Crystalline Polymers: Volume 2—Processing and Applications*, Thakur, V.K., Kessler, M.R., Eds.; Springer International Publishing: Cham, Germany, 2015; pp. 339–358.

106. Tran, C.D.; Prosenc, F.; Franko, M.; Benzi, G. One-pot synthesis of biocompatible silver nanoparticle composites from cellulose and keratin: Characterization and antimicrobial activity. *ACS Appl. Mater. Interfaces* **2016**, *8*, 34791–34801. [CrossRef]

107. Tran, C.D.; Mututuvari, T.M. Cellulose, chitosan and keratin composite materials: Facile and recyclable synthesis, conformation and properties. *ACS Sustain. Chem. Eng.* **2016**, *4*, 1850–1861. [CrossRef] [PubMed]

108. Islam, M.S.; Akter, N.; Rahman, M.M.; Shi, C.; Islam, M.T.; Zeng, H.; Azam, M.S. Mussel-inspired immobilization of silver nanoparticles toward antimicrobial cellulose paper. *ACS Sustain. Chem. Eng.* **2018**, *6*, 9178–9188. [CrossRef]

109. Dairi, N.; Ferfera-Harrar, H.; Ramos, M.; Garrigós, M.C. Cellulose acetate/agnps-organoclay and/or thymol nano-biocomposite films with combined antimicrobial/antioxidant properties for active food packaging use. *Int. J. Biol. Macromol.* **2019**, *121*, 508–523. [CrossRef] [PubMed]

110. Moodley, J.S.; Krishna, S.B.N.; Pillay, K.; Govender, P. Green synthesis of silver nanoparticles from moringa oleifera leaf extracts and its antimicrobial potential. *Adv. Nat. Sci. Nanosci. Nanotechnol.* **2018**, *9*, 015011.

111. Krishnaraj, C.; Jagan, E.G.; Rajasekar, S.; Selvakumar, P.; Kalaichelvan, P.T.; Mohan, N. Synthesis of silver nanoparticles using acalypha indica leaf extracts and its antibacterial activity against water borne pathogens. *Colloids Surf. B. Biointerfaces* **2010**, *76*, 50–56. [CrossRef] [PubMed]

112. Vemu Anil, K.; Takashi, U.; Toru, M.; Yoshikata, N.; Yoshihiro, K.; Tatsuro, H.; Toru, M. Synthesis of nanoparticles composed of silver and silver chloride for a plasmonic photocatalyst using an extract from a weed solidago altissima (goldenrod). *Adv. Nat. Sci. Nanosci. Nanotechnol.* **2016**, *7*, 015002.

113. Tran, C.D.; Prosenc, F.; Franko, M. Facile synthesis, structure, biocompatibility and antimicrobial property of gold nanoparticle composites from cellulose and keratin. *J. Colloid Interface Sci.* **2018**, *510*, 237–245. [CrossRef]

114. Yang, X.; Yang, M.; Pang, B.; Vara, M.; Xia, Y. Gold nanomaterials at work in biomedicine. *Chem. Rev.* **2015**, *115*, 10410–10488. [CrossRef]

115. Echeverria, C.; Almeida, P.L.; Feio, G.; Figueirinhas, J.L.; Godinho, M.H. A cellulosic liquid crystal pool for cellulose nanocrystals: Structure and molecular dynamics at high shear rates. *Eur. Polym. J.* **2015**, *72*, 72–81. [CrossRef]

116. Echeverria, C.; Fernandes, S.N.; Almeida, P.L.; Godinho, M.H. Effect of cellulose nanocrystals in a cellulosic liquid crystal behaviour under low shear (regime i): Structure and molecular dynamics. *Eur. Polym. J.* **2016**, *84*, 675–684. [CrossRef]

117. Fernandes, S.N.; Almeida, P.L.; Monge, N.; Aguirre, L.E.; Reis, D.; de Oliveira, C.L.P.; Neto, A.M.F.; Pieranski, P.; Godinho, M.H. Mind the microgap in iridescent cellulose nanocrystal films. *Adv. Mater.* **2017**, *29*, 1603560. [CrossRef] [PubMed]

118. Xu, Q.; Jin, L.; Wang, Y.; Chen, H.; Qin, M. Synthesis of silver nanoparticles using dialdehyde cellulose nanocrystal as a multi-functional agent and application to antibacterial paper. *Cellulose* **2018**. [CrossRef]

119. De Castro, D.O.; Bras, J.; Gandini, A.; Belgacem, N. Surface grafting of cellulose nanocrystals with natural antimicrobial rosin mixture using a green process. *Carbohydr. Polym.* **2016**, *137*, 1–8. [CrossRef] [PubMed]

120. Zmejkoski, D.; Spasojević, D.; Orlovska, I.; Kozyrovska, N.; Soković, M.; Glamočlija, J.; Dmitrović, S.; Matović, B.; Tasić, N.; Maksimović, V.; et al. Bacterial cellulose-lignin composite hydrogel as a promising agent in chronic wound healing. *Int. J. Biol. Macromol.* **2018**, *118*, 494–503. [CrossRef] [PubMed]

121. Kucińska-Lipka, J.; Gubanska, I.; Janik, H. Bacterial cellulose in the field of wound healing and regenerative medicine of skin: Recent trends and future prospectives. *Polym. Bull.* **2015**, *72*, 2399–2419. [CrossRef]

122. Mohite, B.V.; Patil, S.V. In situ development of nanosilver-impregnated bacterial cellulose for sustainable released antimicrobial wound dressing. *J. Appl. Biomater. Funct. Mater.* **2016**, *14*, e53–e58. [CrossRef] [PubMed]

123. Pal, S.; Nisi, R.; Stoppa, M.; Licciulli, A. Silver-functionalized bacterial cellulose as antibacterial membrane for wound-healing applications. *ACS Omega* **2017**, *2*, 3632–3639. [CrossRef]

124. Wiegand, C.; Moritz, S.; Hessler, N.; Kralisch, D.; Wesarg, F.; Müller, F.A.; Fischer, D.; Hipler, U.C. Antimicrobial functionalization of bacterial nanocellulose by loading with polihexanide and povidone-iodine. *J. Mater. Sci. Mater. Med.* **2015**, *26*, 245. [CrossRef]

125. Li, J.; Cha, R.; Mou, K.; Zhao, X.; Long, K.; Luo, H.; Zhou, F.; Jiang, X. Nanocellulose-based antibacterial materials. *Adv. Health. Mater.* **2018**, *7*, 1800334. [CrossRef]

126. Aulin, C.; Netrval, J.; Wågberg, L.; Lindström, T. Aerogels from nanofibrillated cellulose with tunable oleophobicity. *Soft Matter* **2010**, *6*, 3298–3305. [CrossRef]

127. Littunen, K.; Snoei De Castro, J.; Samoylenko, A.; Xu, Q.; Quaggin, S.; Vainio, S.; Seppälä, J. Synthesis of cationized nanofibrillated cellulose and its antimicrobial properties. *Eur. Polym. J.* **2016**, *75*, 116–124. [CrossRef]

128. Saini, S.; Belgacem, M.N.; Bras, J. Effect of variable aminoalkyl chains on chemical grafting of cellulose nanofiber and their antimicrobial activity. *Mater. Sci. Eng. C* **2017**, *75*, 760–768. [CrossRef] [PubMed]

129. Tejero, R.; Gutiérrez, B.; López, D.; López-Fabal, F.; Gómez-Garcés, J.L.; Fernández-García, M. Copolymers of acrylonitrile with quaternizable thiazole and triazole side-chain methacrylates as potent antimicrobial and hemocompatible systems. *Acta Biomater.* **2015**, *25*, 86–96. [CrossRef] [PubMed]

130. Takahashi, H.; Caputo, G.A.; Vemparala, S.; Kuroda, K. Synthetic random copolymers as a molecular platform to mimic host-defense antimicrobial peptides. *Bioconjugate Chem.* **2017**, *28*, 1340–1350. [CrossRef] [PubMed]

131. Fernandes, S.C.M.; Sadocco, P.; Alonso-Varona, A.; Palomares, T.; Eceiza, A.; Silvestre, A.J.D.; Mondragon, I.; Freire, C.S.R. Bioinspired antimicrobial and biocompatible bacterial cellulose membranes obtained by surface functionalization with aminoalkyl groups. *ACS Appl. Mater. Interfaces* **2013**, *5*, 3290–3297. [CrossRef] [PubMed]

132. Sarwar, M.S.; Niazi, M.B.K.; Jahan, Z.; Ahmad, T.; Hussain, A. Preparation and characterization of pva/nanocellulose/ag nanocomposite films for antimicrobial food packaging. *Carbohydr. Polym.* **2018**, *184*, 453–464. [CrossRef]

133. Wu, Z.; Deng, W.; Luo, J.; Deng, D. Multifunctional nano-cellulose composite films with grape seed extracts and immobilized silver nanoparticles. *Carbohydr. Polym.* **2019**, *205*, 447–455. [CrossRef]

134. Spagnol, C.; Fragal, E.H.; Pereira, A.G.B.; Nakamura, C.V.; Muniz, E.C.; Follmann, H.D.M.; Silva, R.; Rubira, A.F. Cellulose nanowhiskers decorated with silver nanoparticles as an additive to antibacterial polymers membranes fabricated by electrospinning. *J. Colloid Interface Sci.* **2018**, *531*, 705–715. [CrossRef]

135. Lizundia, E.; Goikuria, U.; Vilas, J.L.; Cristofaro, F.; Bruni, G.; Fortunati, E.; Armentano, I.; Visai, L.; Torre, L. Metal nanoparticles embedded in cellulose nanocrystal based films: Material properties and post-use analysis. *Biomacromolecules* **2018**, *19*, 2618–2628. [CrossRef]

136. Gunalan, S.; Sivaraj, R.; Rajendran, V. Green synthesized zno nanoparticles against bacterial and fungal pathogens. *Prog. Nat. Sci. Mater. Int.* **2012**, *22*, 693–700. [CrossRef]

137. Cazón, P.; Velazquez, G.; Ramírez, J.A.; Vázquez, M. Polysaccharide-based films and coatings for food packaging: A review. *Food Hydrocoll.* **2017**, *68*, 136–148. [CrossRef]

138. Liu, Z.; Huang, M.; Li, A.; Yang, H. Flocculation and antimicrobial properties of a cationized starch. *Water Res.* **2017**, *119*, 57–66. [CrossRef] [PubMed]

139. Wu, H.; Liu, Z.; Yang, H.; Li, A. Evaluation of chain architectures and charge properties of various starch-based flocculants for flocculation of humic acid from water. *Water Res.* **2016**, *96*, 126–135. [CrossRef] [PubMed]

140. Huang, M.; Liu, Z.; Li, A.; Yang, H. Dual functionality of a graft starch flocculant: Flocculation and antibacterial performance. *J. Environ. Manag.* **2017**, *196*, 63–71. [CrossRef] [PubMed]

141. Huang, M.; Wang, Y.; Cai, J.; Bai, J.; Yang, H.; Li, A. Preparation of dual-function starch-based flocculants for the simultaneous removal of turbidity and inhibition of *Escherichia coli* in water. *Water Res.* **2016**, *98*, 128–137. [CrossRef] [PubMed]

142. Sen, F.; Uzunsoy, I.; Basturk, E.; Kahraman, M.V. Antimicrobial agent-free hybrid cationic starch/sodium alginate polyelectrolyte films for food packaging materials. *Carbohydr. Polym.* **2017**, *170*, 264–270. [CrossRef]

143. Tan, W.; Li, Q.; Wang, H.; Liu, Y.; Zhang, J.; Dong, F.; Guo, Z. Synthesis, characterization, and antibacterial property of novel starch derivatives with 1,2,3-triazole. *Carbohydr. Polym.* **2016**, *142*, 1–7. [CrossRef]

144. Tan, W.; Li, Q.; Wei, L.; Wang, P.; Cao, Z.; Chen, Y.; Dong, F.; Guo, Z. Synthesis, characterization, and antifungal property of starch derivatives modified with quaternary phosphonium salts. *Mater. Sci. Eng. C Mater. Biol. Appl.* **2017**, *76*, 1048–1056. [CrossRef]

145. Rivadeneira, J.; Di Virgilio, A.L.; Audisio, M.C.; Boccaccini, A.R.; Gorustovich, A.A. 45s5 bioglass® concentrations modulate the release of vancomycin hydrochloride from gelatin–starch films: Evaluation of antibacterial and cytotoxic effects. *J. Mater. Sci.* **2017**, *52*, 9091–9102. [CrossRef]

146. Chen, K.; Zhang, S.; Wang, H.; Wang, X.; Zhang, Y.; Yu, L.; Ke, L.; Gong, R. Fabrication of doxorubicin-loaded glycyrrhetinic acid-biotin-starch nanoparticles and drug delivery into hepg2 cells in vitro. *Starch Stärke* **2018**. [CrossRef]

147. Colle Resa, C.P.; Jagus, R.J.; Gerschenson, L.N. Effect of natamycin, nisin and glycerol on the physicochemical properties, roughness and hydrophobicity of tapioca starch edible films. *Mater. Sci. Eng. C Mater. Biol. Appl.* **2014**, *40*, 281–287. [CrossRef] [PubMed]

148. Lozano-Navarro, J.I.; Diaz-Zavala, N.P.; Velasco-Santos, C.; Melo-Banda, J.A.; Paramo-Garcia, U.; Paraguay-Delgado, F.; Garcia-Alamilla, R.; Martinez-Hernandez, A.L.; Zapien-Castillo, S. Chitosan-starch films with natural extracts: Physical, chemical, morphological and thermal properties. *Materials* **2018**, *11*, 120. [CrossRef] [PubMed]

149. Lozano-Navarro, J.I.; Diaz-Zavala, N.P.; Velasco-Santos, C.; Martinez-Hernandez, A.L.; Tijerina-Ramos, B.I.; Garcia-Hernandez, M.; Rivera-Armenta, J.L.; Paramo-Garcia, U.; Reyes-de la Torre, A.I. Antimicrobial, optical and mechanical properties of chitosan-starch films with natural extracts. *Int. J. Mol. Sci.* **2017**, *18*, 997. [CrossRef] [PubMed]

150. Pattanayaiying, R.; Sane, A.; Photjanataree, P.; Cutter, C.N. Thermoplastic starch/polybutylene adipate terephthalate film coated with gelatin containing nisin z and lauric arginate for control of foodborne pathogens associated with chilled and frozen seafood. *Int. J. Food Microbiol.* **2018**, *290*, 59–67. [CrossRef] [PubMed]

151. Pattanayaiying, R.; H-Kittikun, A.; Cutter, C.N. Incorporation of nisin z and lauric arginate into pullulan films to inhibit foodborne pathogens associated with fresh and ready-to-eat muscle foods. *Int. J. Food Microbiol.* **2015**, *207*, 77–82. [CrossRef] [PubMed]

152. Moreno, O.; Atarés, L.; Chiralt, A.; Cruz-Romero, M.C.; Kerry, J. Starch-gelatin antimicrobial packaging materials to extend the shelf life of chicken breast fillets. *LWT* **2018**, *97*, 483–490. [CrossRef]

153. Campos-Requena, V.H.; Rivas, B.L.; Pérez, M.A.; Garrido-Miranda, K.A.; Pereira, E.D. Release of essential oil constituent from thermoplastic starch/layered silicate bionanocomposite film as a potential active packaging material. *Eur. Polym. J.* **2018**, *109*, 64–71. [CrossRef]

154. Ramos, M.; Jiménez, A.; Peltzer, M.; Garrigós, M.C. Characterization and antimicrobial activity studies of polypropylene films with carvacrol and thymol for active packaging. *J. Food Eng.* **2012**, *109*, 513–519. [CrossRef]

155. Saberi, B.; Chockchaisawasdee, S.; Golding, J.B.; Scarlett, C.J.; Stathopoulos, C.E. Characterization of pea starch-guar gum biocomposite edible films enriched by natural antimicrobial agents for active food packaging. *Food Bioprod. Process.* **2017**, *105*, 51–63. [CrossRef]

156. Acosta, S.; Chiralt, A.; Santamarina, P.; Rosello, J.; González-Martínez, C.; Cháfer, M. Antifungal films based on starch-gelatin blend, containing essential oils. *Food Hydrocoll.* **2016**, *61*, 233–240. [CrossRef]

157. Sapper, M.; Chiralt, A. Starch-based coatings for preservation of fruits and vegetables. *Coatings* **2018**, *8*, 152. [CrossRef]

158. Valencia-Sullca, C.; Atarés, L.; Vargas, M.; Chiralt, A. Physical and antimicrobial properties of compression-molded cassava starch-chitosan films for meat preservation. *Food Bioprocess Technol.* **2018**, *11*, 1339–1349. [CrossRef]

159. Valencia-Sullca, C.; Vargas, M.; Atarés, L.; Chiralt, A. Thermoplastic cassava starch-chitosan bilayer films containing essential oils. *Food Hydrocoll.* **2018**, *75*, 107–115. [CrossRef]

160. Atarés, L.; Chiralt, A. Essential oils as additives in biodegradable films and coatings for active food packaging. *Trends Food Sci. Technol.* **2016**, *48*, 51–62. [CrossRef]

161. Moreno, O.; Atares, L.; Chiralt, A. Effect of the incorporation of antimicrobial/antioxidant proteins on the properties of potato starch films. *Carbohydr. Polym.* **2015**, *133*, 353–364. [CrossRef] [PubMed]

162. Buonocore, G.G.; Del Nobile, M.A.; Panizza, A.; Bove, S.; Battaglia, G.; Nicolais, L. Modeling the lysozyme release kinetics from antimicrobial films intended for food packaging applications. *J. Food Sci.* **2003**, *68*, 1365–1370. [CrossRef]

163. Conte, A.; Buonocore, G.G.; Sinigaglia, M.; Del Nobile, M.A. Development of immobilized lysozyme based active film. *J. Food Eng.* **2007**, *78*, 741–745. [CrossRef]

164. Matouskova, P.; Marova, I.; Bokrova, J.; Benesova, P. Effect of encapsulation on antimicrobial activity of herbal extracts with lysozyme. *Food Technol. Biotechnol.* **2016**, *54*, 304–316. [CrossRef]

165. Arshad, H.; Ali, T.M.; Abbas, T.; Hasnain, A. Effect of microencapsulation on antimicrobial and antioxidant activity of nutmeg oleoresin using mixtures of gum arabic, osa, and native sorghum starch. *Starch Stärke* **2018**, *70*, 1700320. [CrossRef]

166. Sarkar, P.; Bhunia, A.K.; Yao, Y. Impact of starch-based emulsions on the antibacterial efficacies of nisin and thymol in cantaloupe juice. *Food Chem.* **2017**, *217*, 155–162. [CrossRef]

167. Qin, Y.; Zhang, S.; Yu, J.; Yang, J.; Xiong, L.; Sun, Q. Effects of chitin nano-whiskers on the antibacterial and physicochemical properties of maize starch films. *Carbohydr. Polym.* **2016**, *147*, 372–378. [CrossRef] [PubMed]

168. Mohan, S.; Oluwafemi, O.S.; Songca, S.P.; Jayachandran, V.P.; Rouxel, D.; Joubert, O.; Kalarikkal, N.; Thomas, S. Synthesis, antibacterial, cytotoxicity and sensing properties of starch-capped silver nanoparticles. *J. Mol. Liq.* **2016**, *213*, 75–81. [CrossRef]

169. Jung, J.; Raghavendra, G.M.; Kim, D.; Seo, J. One-step synthesis of starch-silver nanoparticle solution and its application to antibacterial paper coating. *Int. J. Biol. Macromol.* **2018**, *107*, 2285–2290. [CrossRef] [PubMed]

170. Jung, J.; Kasi, G.; Seo, J. Development of functional antimicrobial papers using chitosan/starch-silver nanoparticles. *Int. J. Biol. Macromol.* **2018**, *112*, 530–536. [CrossRef] [PubMed]

171. Ni, S.; Zhang, H.; Dai, H.; Xiao, H. Starch-based flexible coating for food packaging paper with exceptional hydrophobicity and antimicrobial activity. *Polymers* **2018**, *10*, 1260. [CrossRef]

172. Guan, Y.; Qian, L.; Xiao, H.; Zheng, A. Preparation of novel antimicrobial-modified starch and its adsorption on cellulose fibers: Part I. Optimization of synthetic conditions and antimicrobial activities. *Cellulose* **2008**, *15*, 609–618. [CrossRef]

173. Pan, Y.; Xiao, H.; Cai, P.; Colpitts, M. Cellulose fibers modified with nano-sized antimicrobial polymer latex for pathogen deactivation. *Carbohydr. Polym.* **2016**, *135*, 94–100. [CrossRef]

174. Heydarifard, S.; Pan, Y.; Xiao, H.; Nazhad, M.M.; Shipin, O. Water-resistant cellulosic filter containing non-leaching antimicrobial starch for water purification and disinfection. *Carbohydr. Polym.* **2017**, *163*, 146–152. [CrossRef]

175. Villanueva, M.E.; Diez, A.M.; Gonzaález, J.A.; Peérez, C.J.; Orrego, M.; Piehl, L.; Teves, S.; Copello, G.J. Antimicrobial activity of starch hydrogel incorporated with copper nanoparticles. *ACS Appl. Mater. Interfaces* **2016**, *8*, 16280–16288. [CrossRef]

176. Gonzalez, K.; Garcia-Astrain, C.; Santamaria-Echart, A.; Ugarte, L.; Averous, L.; Eceiza, A.; Gabilondo, N. Starch/graphene hydrogels via click chemistry with relevant electrical and antibacterial properties. *Carbohydr. Polym.* **2018**, *202*, 372–381. [CrossRef]

177. Lee, K.Y.; Mooney, D.J. Alginate: Properties and biomedical applications. *Prog. Polym. Sci.* **2012**, *37*, 106–126. [CrossRef] [PubMed]

178. Senturk Parreidt, T.; Muller, K.; Schmid, M. Alginate-based edible films and coatings for food packaging applications. *Foods* **2018**, *7*, 170. [CrossRef] [PubMed]

179. Wang, G.; Zhu, J.; Chen, X.; Dong, H.; Li, Q.; Zeng, L.; Cao, X. Alginate based antimicrobial hydrogels formed by integrating diels–alder "click chemistry" and the thiol–ene reaction. *RSC Adv.* **2018**, *8*, 11036–11042. [CrossRef]

180. Han, Y.; Yu, M.; Wang, L. Physical and antimicrobial properties of sodium alginate/carboxymethyl cellulose films incorporated with cinnamon essential oil. *Food Pack. Shelf Life* **2018**, *15*, 35–42. [CrossRef]

181. Omonijo, F.A.; Kim, S.; Guo, T.; Wang, Q.; Gong, J.; Lahaye, L.; Bodin, J.C.; Nyachoti, M.; Liu, S.; Yang, C. Development of novel microparticles for effective delivery of thymol and lauric acid to pig intestinal tract. *J. Agric. Food Chem.* **2018**, *66*, 9608–9615. [CrossRef] [PubMed]

182. Zimet, P.; Mombrú, Á.W.; Faccio, R.; Brugnini, G.; Miraballes, I.; Rufo, C.; Pardo, H. Optimization and characterization of nisin-loaded alginate-chitosan nanoparticles with antimicrobial activity in lean beef. *LWT* **2018**, *91*, 107–116. [CrossRef]

183. Niaz, T.; Shabbir, S.; Noor, T.; Abbasi, R.; Raza, Z.A.; Imran, M. Polyelectrolyte multicomponent colloidosomes loaded with nisin z for enhanced antimicrobial activity against foodborne resistant pathogens. *Front. Microbiol.* **2017**, *8*, 2700. [CrossRef]

184. Motshekga, S.C.; Sinha Ray, S.; Maity, A. Synthesis and characterization of alginate beads encapsulated zinc oxide nanoparticles for bacteria disinfection in water. *J. Colloid Interface Sci.* **2018**, *512*, 686–692. [CrossRef]

185. Gomez Chabala, L.F.; Cuartas, C.E.E.; Lopez, M.E.L. Release behavior and antibacterial activity of chitosan/alginate blends with aloe vera and silver nanoparticles. *Mar. Drugs* **2017**, *15*, 328. [CrossRef]

186. Thaya, R.; Vaseeharan, B.; Sivakamavalli, J.; Iswarya, A.; Govindarajan, M.; Alharbi, N.S.; Kadaikunnan, S.; Al-Anbr, M.N.; Khaled, J.M.; Benelli, G. Synthesis of chitosan-alginate microspheres with high antimicrobial and antibiofilm activity against multi-drug resistant microbial pathogens. *Microb. Pathog.* **2018**, *114*, 17–24. [CrossRef]

187. Ben Amara, C.; Kim, L.; Oulahal, N.; Degraeve, P.; Gharsallaoui, A. Using complexation for the microencapsulation of nisin in biopolymer matrices by spray-drying. *Food Chem.* **2017**, *236*, 32–40. [CrossRef] [PubMed]

188. Safaei, M.; Taran, M. Optimized synthesis, characterization, and antibacterial activity of an alginate-cupric oxide bionanocomposite. *J. Appl. Polym. Sci.* **2018**, *135*, 45682. [CrossRef]

189. Alboofetileh, M.; Rezaei, M.; Hosseini, H.; Abdollahi, M. Antimicrobial activity of alginate/clay nanocomposite films enriched with essential oils against three common foodborne pathogens. *Food Control* **2014**, *36*, 1–7. [CrossRef]

190. Shankar, S.; Rhim, J.W. Antimicrobial wrapping paper coated with a ternary blend of carbohydrates (alginate, carboxymethyl cellulose, carrageenan) and grapefruit seed extract. *Carbohydr. Polym.* **2018**, *196*, 92–101. [CrossRef] [PubMed]

191. Nesic, A.; Onjia, A.; Davidovic, S.; Dimitrijevic, S.; Errico, M.E.; Santagata, G.; Malinconico, M. Design of pectin-sodium alginate based films for potential healthcare application: Study of chemico-physical interactions between the components of films and assessment of their antimicrobial activity. *Carbohydr. Polym.* **2017**, *157*, 981–990. [CrossRef]

192. Lopes, N.A.; Pinilla, C.M.B.; Brandelli, A. Pectin and polygalacturonic acid-coated liposomes as novel delivery system for nisin: Preparation, characterization and release behavior. *Food Hydrocoll.* **2017**, *70*, 1–7. [CrossRef]

193. Fan, C.; Guo, M.; Liang, Y.; Dong, H.; Ding, G.; Zhang, W.; Tang, G.; Yang, J.; Kong, D.; Cao, Y. Pectin-conjugated silica microcapsules as dual-responsive carriers for increasing the stability and antimicrobial efficacy of kasugamycin. *Carbohydr. Polym.* **2017**, *172*, 322–331. [CrossRef] [PubMed]

194. Otoni, C.G.; de Moura, M.R.; Aouada, F.A.; Camilloto, G.P.; Cruz, R.S.; Lorevice, M.V.; de FFSoares, N.; Mattoso, L.H. Antimicrobial and physical-mechanical properties of pectin/papaya puree/cinnamaldehyde nanoemulsion edible composite films. *Food Hydrocoll.* **2014**, *41*, 188–194. [CrossRef]

195. Ye, S.; Zhu, Z.; Wen, Y.; Su, C.; Jiang, L.; He, S.; Shao, W. Facile and green preparation of pectin/cellulose composite films with enhanced antibacterial and antioxidant behaviors. *Polymers* **2019**, *11*, 57. [CrossRef]

196. Radi, M.; Akhavan-Darabi, S.; Akhavan, H.-R.; Amiri, S. The use of orange peel essential oil microemulsion and nanoemulsion in pectin-based coating to extend the shelf life of fresh-cut orange. *J. Food Process. Preserv.* **2018**, *42*, e13441. [CrossRef]

197. Nisar, T.; Wang, Z.C.; Yang, X.; Tian, Y.; Iqbal, M.; Guo, Y. Characterization of citrus pectin films integrated with clove bud essential oil: Physical, thermal, barrier, antioxidant and antibacterial properties. *Int. J. Biol. Macromol.* **2018**, *106*, 670–680. [CrossRef] [PubMed]

198. Vishnuvarthanan, M.; Rajeswari, N. Food packaging: Pectin–laponite–Ag nanoparticle bionanocomposite coated on polypropylene shows low O_2 transmission, low Ag migration and high antimicrobial activity. *Environ. Chem. Lett.* **2018**. [CrossRef]

199. Necas, J.; Bartosikova, L. Carrageenan: A review. *Vet. Med.* **2013**, *58*, 187–205. [CrossRef]

200. Roy, S.; Shankar, S.; Rhim, J.-W. Melanin-mediated synthesis of silver nanoparticle and its use for the preparation of carrageenan-based antibacterial films. *Food Hydrocoll.* **2019**, *88*, 237–246. [CrossRef]

201. Zepon, K.M.; Marques, M.S.; da Silva Paula, M.M.; Morisso, F.D.P.; Kanis, L.A. Facile, green and scalable method to produce carrageenan-based hydrogel containing in situ synthesized agnps for application as wound dressing. *Int. J. Biol. Macromol.* **2018**, *113*, 51–58. [CrossRef] [PubMed]

202. Meindrawan, B.; Suyatma, N.E.; Wardana, A.A.; Pamela, V.Y. Nanocomposite coating based on carrageenan and zno nanoparticles to maintain the storage quality of mango. *Food Pack. Shelf Life* **2018**, *18*, 140–146. [CrossRef]

203. Oun, A.A.; Rhim, J.-W. Carrageenan-based hydrogels and films: Effect of zno and cuo nanoparticles on the physical, mechanical, and antimicrobial properties. *Food Hydrocoll.* **2017**, *67*, 45–53. [CrossRef]

204. Nouri, A.; Tavakkoli Yaraki, M.; Ghorbanpour, M.; Wang, S. Biodegradable kappa-carrageenan/nanoclay nanocomposite films containing rosmarinus officinalis l. Extract for improved strength and antibacterial performance. *Int. J. Biol. Macromol.* **2018**, *115*, 227–235. [CrossRef]

205. Martins, J.T.; Bourbon, A.I.; Pinheiro, A.C.; Souza, B.W.S.; Cerqueira, M.A.; Vicente, A.A. Biocomposite films based on κ-carrageenan/locust bean gum blends and clays: Physical and antimicrobial properties. *Food Bioprocess Technol.* **2013**, *6*, 2081–2092. [CrossRef]

206. Shankar, S.; Reddy, J.P.; Rhim, J.W.; Kim, H.Y. Preparation, characterization, and antimicrobial activity of chitin nanofibrils reinforced carrageenan nanocomposite films. *Carbohydr. Polym.* **2015**, *117*, 468–475. [CrossRef]

207. Pedram Rad, Z.; Mokhtari, J.; Abbasi, M. Fabrication and characterization of PCL/zein/gum arabic electrospun nanocomposite scaffold for skin tissue engineering. *Mater. Sci. Eng. C* **2018**, *93*, 356–366. [CrossRef] [PubMed]

208. Vogt, L.; Liverani, L.; Roether, J.A.; Boccaccini, A.R. Electrospun zein fibers incorporating poly(glycerol sebacate) for soft tissue engineering. *Nanomaterials* **2018**, *8*, 150. [CrossRef]

209. Tian, H.; Guo, G.; Fu, X.; Yao, Y.; Yuan, L.; Xiang, A. Fabrication, properties and applications of soy-protein-based materials: A review. *Int. J. Biol. Macromol.* **2018**, *120*, 475–490. [CrossRef] [PubMed]

210. Bibi, F.; Guillaume, C.; Gontard, N.; Sorli, B. Wheat gluten, a bio-polymer to monitor carbon dioxide in food packaging: Electric and dielectric characterization. *Sens. Actuators B Chem.* **2017**, *250*, 76–84. [CrossRef]

211. Gómez-Estaca, J.; Gavara, R.; Catalá, R.; Hernández-Muñoz, P. The potential of proteins for producing food packaging materials: A review. *Packag. Technol. Sci.* **2016**, *29*, 203–224. [CrossRef]

212. Rouse, J.G.; Van Dyke, M.E. A review of keratin-based biomaterials for biomedical applications. *Materials* **2010**, *3*, 999–1014. [CrossRef]

213. Pereda, M.; Aranguren, M.I.; Marcovich, N.E. Characterization of chitosan/caseinate films. *J. Appl. Polym. Sci.* **2008**, *107*, 1080–1090. [CrossRef]

214. Arrieta, M.P.; Peltzer, M.A.; del Carmen Garrigós, M.; Jiménez, A. Structure and mechanical properties of sodium and calcium caseinate edible active films with carvacrol. *J. Food Eng.* **2013**, *114*, 486–494. [CrossRef]

215. Audic, J.L.; Chaufer, B.; Daufin, G. Non-food applications of milk components and dairy co-products: A review. *Lait* **2003**, *83*, 417–438. [CrossRef]

216. Calderón-Aguirre, Á.-G.; Chavarría-Hernández, N.; Mendoza-Mendoza, B.; Vargas-Torres, A.; García-Hernández, E.; Rodríguez-Hernández, A.-I. Antilisterial activity and physical-mechanical properties of bioactive caseinate films. *CyTA J. Food* **2015**, *13*, 483–490. [CrossRef]

217. Arrieta, M.P.; Peltzer, M.A.; López, J.; del Carmen Garrigós, M.; Valente, A.J.M.; Jiménez, A. Functional properties of sodium and calcium caseinate antimicrobial active films containing carvacrol. *J. Food Eng.* **2014**, *121*, 94–101. [CrossRef]

218. Noori, S.; Zeynali, F.; Almasi, H. Antimicrobial and antioxidant efficiency of nanoemulsion-based edible coating containing ginger (*zingiber officinale*) essential oil and its effect on safety and quality attributes of chicken breast fillets. *Food Control* **2018**, *84*, 312–320. [CrossRef]

219. Imran, M.; Klouj, A.; Revol-Junelles, A.-M.; Desobry, S. Controlled release of nisin from HPMC, sodium caseinate, poly-lactic acid and chitosan for active packaging applications. *J. Food Eng.* **2014**, *143*, 178–185. [CrossRef]

220. Dickerson, M.B.; Sierra, A.A.; Bedford, N.M.; Lyon, W.J.; Gruner, W.E.; Mirau, P.A.; Naik, R.R. Keratin-based antimicrobial textiles, films, and nanofibers. *J. Mater. Chem. B* **2013**, *1*, 5505–5514. [CrossRef]

221. Fernández-d'Arlas, B. Improved aqueous solubility and stability of wool and feather proteins by reactive-extraction with H_2O_2 as bisulfide (SS) splitting agent. *Eur. Polym. J.* **2018**, *103*, 187–197. [CrossRef]

222. Hill, P.; Brantley, H.; Van Dyke, M. Some properties of keratin biomaterials: Kerateines. *Biomaterials* **2010**, *31*, 585–593. [CrossRef] [PubMed]

223. Heliopoulos, N.S.; Papageorgiou S.K.; Galeou, A.; Favvas, E.P.; Katsaros, F.K.; Stamatakis, K. Effect of copper and copper alginate treatment on wool fabric. Study of textile and antibacterial properties. *Surf. Coat. Technol.* **2013**, *235*, 24–31. [CrossRef]

224. Yu, D.; Cai, J.Y.; Liu, X.; Church, J.S.; Wang, L. Novel immobilization of a quaternary ammonium moiety on keratin fibers for medical applications. *Int. J. Biol. Macromol.* **2014**, *70*, 236–240. [CrossRef] [PubMed]

225. Nayak, K.K.; Gupta, P. Study of the keratin-based therapeutic dermal patches for the delivery of bioactive molecules for wound treatment. *Mater. Sci. Eng. C* **2017**, *77*, 1088–1097. [CrossRef]

226. Gaodbane, S.A.; Dunn, M.G. Physical and mechanical properties of cross-linked type i collagen scaffolds derived from bovine, porcine, and ovine tendons. *J. Biomed. Mater. Res. A* **2016**, *104*, 2685–2692. [CrossRef] [PubMed]

227. Balaure, P.C.; Holban, A.M.; Grumezescu, A.M.; Mogoşanu, G.D.; Bălşeanu, T.A.; Stan, M.S.; Dinischiotu, A.; Volceanov, A.; Mogoantă, L. In vitro and in vivo studies of novel fabricated bioactive dressings based on collagen and zinc oxide 3d scaffolds. *Int. J. Pharm.* **2019**, *557*, 199–207. [CrossRef]

228. Vijayakumar, S.; Vaseeharan, B. Antibiofilm, anti cancer and ecotoxicity properties of collagen based zno nanoparticles. *Adv. Powder Technol.* **2018**, *29*, 2331–2345. [CrossRef]

229. Zhang, H.; Peng, M.; Cheng, T.; Zhao, P.; Qiu, L.; Zhou, J.; Lu, G.; Chen, J. Silver nanoparticles-doped collagen–alginate antimicrobial biocomposite as potential wound dressing. *J. Mater. Sci.* **2018**, *53*, 14944–14952. [CrossRef]

230. You, C.; Li, Q.; Wang, X.; Wu, P.; Ho J.K.; Jin, R.; Zhang, L.; Shao, H.; Han, C. Silver nanoparticle loaded collagen/chitosan scaffolds promote wound healing via regulating fibroblast migration and macrophage activation. *Sci. Rep.* **2017**, *7*, 10489. [CrossRef] [PubMed]

231. Michalska-Sionkowska, M.; Walczak. M.; Sionkowska, A. Antimicrobial activity of collagen material with thymol addition for potential application as wound dressing. *Polym. Test.* **2017**, *63*, 360–366. [CrossRef]

232. Campos-Requena, V.H.; Rivas, B.L.; Pérez, M.A.; Figueroa, C.R.; Sanfuentes, E.A. The synergistic antimicrobial effect of carvacrol and thymol in clay/polymer nanocomposite films over strawberry gray mold. *LWT Food Sci. Technol.* **2015**, *64*, 390–396. [CrossRef]

233. Xu, J.; Zhou, F.; Ji, B.-P.; Pei, R.-S.; Xu, N. The antibacterial mechanism of carvacrol and thymol against escherichia coli. *Lett. Appl. Microbiol.* **2008**, *47*, 174–179. [CrossRef]

234. Gan, I.; Chow, W.S. Antimicrobial poly(lactic acid)/cellulose bionanocomposite for food packaging application: A review. *Food Pack. Shelf Life* **2018**, *17*, 150–161. [CrossRef]

235. Scaffaro, R.; Lopresti, F.; Marino, A.; Nostro, A. Antimicrobial additives for poly (lactic acid) materials and their applications: Current state and perspectives. *Appl. Microbiol. Biotechnol.* **2018**, *102*, 7739–7756. [CrossRef]

236. Auras, R.; Harte, B.; Selke, S. An overview of polylactides as packaging materials. *Macromol. Biosci.* **2004**, *4*, 835–864. [CrossRef]

237. Armentano, I.; Bitinis, N.; Fortunati, E.; Mattioli, S.; Rescignano, N.; Verdejo, R.; Lopez-Manchado, M.A.; Kenny, J.M. Multifunctional nanostructured pla materials for packaging and tissue engineering. *Prog. Polym. Sci.* **2013**, *38*, 1720–1747. [CrossRef]

238. Piergiovanni, L.; Limbo, S. Plastic packaging materials. In *Food Packaging Materials*; Springer International Publishing: Cham, Germany, 2016; pp. 33–49.

239. Drumright, R.E.; Gruber, P.R.; Henton, D.E. Polylactic acid technology. *Adv. Mater.* **2000**, *12*, 1841–1846. [CrossRef]

240. Arrieta, M.P.; Samper, M.D.; Aldas, M.; López, J. On the use of PLA-PHB blends for sustainable food packaging applications. *Materials* **2017**, *10*, 1008. [CrossRef] [PubMed]

241. Lim, L.T.; Auras, R.; Rubino, M. Processing technologies for poly(lactic acid). *Prog. Polym. Sci.* **2008**, *33*, 820–852. [CrossRef]

242. Arrieta, M.P.; Peltzer, M.A.; López, J.; Peponi, L. PLA-based nanocomposites reinforced with CNC for food packaging applications: From synthesis to biodegradation. In *Industrial Applications of Renewable Biomass Products: Past, Present and Future*; Springer: Cham, Switzerland, 2017; pp. 265–300.

243. Fathima, P.E.; Panda, S.K.; Ashraf. P.M. Varghese, T.O.; Bindu, J. Polylactic acid/chitosan films for packaging of indian white prawn (*fenneropenaeus indicus*). *Int. J. Biol. Macromol.* **2018**, *117*, 1002–1010. [CrossRef] [PubMed]

244. Ramos, M.; Arrieta, M.P.; Beltran, A.; Garrigós, M.C. Characterization of PLA, PCL and sodium caseinate active bio-films for food packaging applications. In *Food Packaging: Procedures, Management and Trends*; Nova Science Publishers Inc.: New York, NY, USA, 2012; pp. 63–78.

245. Armentano, I.; Fortunati, E.; Burgos, N.; Dominici, F.; Luzi, F.; Fiori, S.; Jiménez, A.; Yoon, K.; Ahn, J.; Kang, S.; et al. Bio-based PLA_PHB plasticized blend films: Processing and structural characterization. *LWT Food Sci. Technol.* **2015**, *64*, 980–988. [CrossRef]

246. Burgos, N.; Armentano, I.; Fortunati, E.; Dominici, F.; Luzi, F.; Fiori, S.; Cristofaro, F.; Visai, L.; Jiménez, A.; Kenny, J.M. Functional properties of plasticized bio-based poly(lactic acid)_poly(hydroxybutyrate) (PLA_PHB) films for active food packaging. *Food Bioprocess Technol.* **2017**, *10*, 770–780. [CrossRef]

247. Ahmed, J.; Arfat, Y.A.; Bher, A.; Mulla, M.; Jacob, H.; Auras, R. Active chicken meat packaging based on polylactide films and bimetallic Ag–Cu nanoparticles and essential oil. *J. Food Sci.* **2018**, *83*, 1299–1310. [CrossRef]

248. Torres, A.; Ilabaca, E.; Rojas, A.; Rodríguez, F.; Galotto, M.J.; Guarda, A.; Villegas, C.; Romero, J. Effect of processing conditions on the physical, chemical and transport properties of polylactic acid films containing thymol incorporated by supercritical impregnation. *Eur. Polym. J.* **2017**, *89*, 195–210. [CrossRef]

249. Villegas, C.; Torres, A.; Rios, M.; Rojas, A.; Romero, J.; de Dicastillo, C.L.; Valenzuela, X.; Galotto, M.J.; Guarda, A. Supercritical impregnation of cinnamaldehyde into polylactic acid as a route to develop antibacterial food packaging materials. *Food Res. Int.* **2017**, *99*, 650–659. [CrossRef]

250. Alvarado, N.; Romero, J.; Torres, A.; López de Dicastillo, C.; Rojas, A.; Galotto, M.J.; Guarda, A. Supercritical impregnation of thymol in poly(lactic acid) filled with electrospun poly(vinyl alcohol)-cellulose nanocrystals nanofibers: Development an active food packaging material. *J. Food Eng.* **2018**, *217*, 1–10. [CrossRef]

251. De Dicastillo, C.L.; Villegas, C.; Garrido, L.; Roa, K.; Torres, A.; Galotto, M.J.; Rojas, A.; Romero, J. Modifying an active compound's release kinetic using a supercritical impregnation process to incorporate an active agent into pla electrospun mats. *Polymers* **2018**, *10*, 479. [CrossRef]

252. Reneker, D.H.; Yarin, A.L.; Fong, H.; Koombhongse, S. Bending instability of electrically charged liquid jets of polymer solutions in electrospinning. *J. Appl. Phys.* **2000**, *87*, 4531–4547. [CrossRef]

253. Arrieta, M.P.; López de Dicastillo, C.; Garrido, L.; Roa, K.; Galotto, M.J. Electrospun PVA fibers loaded with antioxidant fillers extracted from durvillaea antarctica algae and their effect on plasticized PLA bionanocomposites. *Eur. Polym. J.* **2018**, *103*, 145–157. [CrossRef]

254. Valerini, D.; Tammaro, L.; Di Benedetto, F.; Vigliotta, G.; Capodieci, L.; Terzi, R.; Rizzo, A. Aluminum-doped zinc oxide coatings on polylactic acid films for antimicrobial food packaging. *Thin Solid Film.* **2018**, *645*, 187–192. [CrossRef]

255. Rokbani, H.; Daigle, F.; Ajji, A. Combined effect of ultrasound stimulations and autoclaving on the enhancement of antibacterial activity of ZnO and SiO₂/ZnO nanoparticles. *Nanomaterials* **2018**, *8*, 129. [CrossRef] [PubMed]

256. Jordá-Vilaplana, A.; Sánchez-Nácher, L.; Fombuena, V.; García-García, D.; Carbonell-Verdú, A. Improvement of mechanical properties of polylactic acid adhesion joints with bio-based adhesives by using air atmospheric plasma treatment. *J. Appl. Polym. Sci.* **2015**, *132*. [CrossRef]

257. Hu, S.; Li, P.; Wei, Z.; Wang, J.; Wang, H.; Wang, Z. Antimicrobial activity of nisin-coated polylactic acid film facilitated by cold plasma treatment. *J. Appl. Polym. Sci.* **2018**, *135*, 46844. [CrossRef]

258. Zhang, Q.; Tu, Q.; Hickey, M.E.; Xiao, J.; Gao, B.; Tian, C.; Heng, P.; Jiao, Y.; Peng, T.; Wang, J. Preparation and study of the antibacterial ability of graphene oxide-catechol hybrid polylactic acid nanofiber mats. *Colloids Surf. B Biointerfaces* **2018**, *172*, 496–505. [CrossRef]

259. Munteanu, B.; Sacarescu, L.; Vasiliu, A.-L.; Hitruc, G.; Pricope, G.; Sivertsvik, M.; Rosnes, J.; Vasile, C. Antioxidant/antibacterial electrospun nanocoatings applied onto PLA films. *Materials* **2018**, *11*, 1973. [CrossRef]

260. Bucci, D.Z.; Tavares, L.B.B.; Sell, I. PHB packaging for the storage of food products. *Polym. Test.* **2005**, *24*, 564–571. [CrossRef]

261. Ugur, A.; Sahin, N.; Beyatli, Y. Accumulation of poly-b-hydroxybutyrate in streptomyces species during growth with different nitrogen sources. *Turk. J. Biol.* **2002**, *26*, 171–174.

262. Lenz, R.W.; Marchessault, R.H. Bacterial polyesters: Biosynthesis, biodegradable plastics and biotechnology. *Biomacromolecules* **2005**, *6*, 1–8. [CrossRef] [PubMed]

263. Moire, L.; Rezzonico, E.; Poirier, Y. Synthesis of novel biomaterials in plants. *J. Plant Physiol.* **2003**, *160*, 831–839. [CrossRef] [PubMed]

264. Fabra, M.J.; Castro-Mayorga, J.L.; Randazzo, W.; Lagarón, J.M.; López-Rubio, A.; Aznar, R.; Sánchez, G. Efficacy of cinnamaldehyde against enteric viruses and its activity after incorporation into biodegradable multilayer systems of interest in food packaging. *Food Environ. Virol.* **2016**, *8*, 125–132. [CrossRef] [PubMed]

265. Arrieta, M.P.; Castro-López, M.D M.; Rayón, E.; Barral-Losada, L.F.; López-Vilariño, J.M.; López, J.; González-Rodríguez, M.V. Plasticized poly(lactic acid)-poly(hydroxybutyrate) (PLA-PHB) blends incorporated with catechin intended for active food-packaging applications. *J. Agric. Food Chem.* **2014**, *62*, 10170–10180. [CrossRef] [PubMed]

266. Garcia-Garcia, D.; Rayón, E.; Carbonell-Verdu, A.; Lopez-Martinez, J.; Balart, R. Improvement of the compatibility between poly(3-hydroxybutyrate) and poly(ε-caprolactone) by reactive extrusion with dicumyl peroxide. *Eur. Polym. J.* **2017**, *86*, 41–57. [CrossRef]

267. Torres-Giner, S.; Hilliou, L.; Melendez-Rodriguez, B.; Figueroa-Lopez, K.J.; Madalena, D.; Cabedo, L.; Covas, J.A.; Vicente, A.A.; Lagaron, J.M. Melt processability, characterization, and antibacterial activity of compression-molded green composite sheets made of poly(3-hydroxybutyrate-*co*-3-hydroxyvalerate) reinforced with coconut fibers impregnated with oregano essential oil. *Food Pack. Shelf Life* **2018**, *17*, 39–49. [CrossRef]

268. Arrieta, M.P.; López, J.; López, D.; Kenny, J.M.; Peponi, L. Effect of chitosan and catechin addition on the structural, thermal, mechanical and disintegration properties of plasticized electrospun PLA-PHB biocomposites. *Polym. Degrad. Stab.* **2016**, *132*, 145–156. [CrossRef]

269. Castro-Mayorga, J.L.; Freitas, F. Reis, M.A.M.; Prieto, M.A.; Lagaron, J.M. Biosynthesis of silver nanoparticles and polyhydroxybutyrate nanocomposites of interest in antimicrobial applications. *Int. J. Biol. Macromol.* **2018**, *108*, 426–435. [CrossRef]

270. Xing, Z.-C.; Chae, W.-P.; Baek, J.-Y.; Choi, M.-J.; Jung, Y.; Kang, I.-K. In vitro assessment of antibacterial activity and cytocompatibility of silver-containing phbv nanofibrous scaffolds for tissue engineering. *Biomacromolecules* **2010**, *11*, 1248–1253. [CrossRef]

271. Cherpinski, A.; Gozutok, M.; Sasmazel, H.T.; Torres-Giner, S.; Lagaron, J.M. Electrospun oxygen scavenging films of poly(3-hydroxybutyrate) containing palladium nanoparticles for active packaging applications. *Nanomaterials* **2018**, *8*, 469. [CrossRef] [PubMed]

272. Aflori, M. Embedding silver nanoparticles at PHB surfaces by means of combined plasma and chemical treatments. *Rev. Roum. Chim.* **2016**, *61*, 405–409.

273. Panaitescu, D.; Ionita, E.; Nicolae. C.-A.; Gabor, A.; Ionita, M.; Trusca, R.; Lixandru, B.-E.; Codita, I.; Dinescu, G. Poly (3-hydroxybutyrate) modified by nanocellulose and plasma treatment for packaging applications. *Polymers* **2018**, *10* 1249. [CrossRef]

274. Naveen, N.; Kumar, R.; Balaji, S.; Uma, T.; Natrajan, T.; Sehgal, P. Synthesis of nonwoven nanofibers by electrospinning–a promising biomaterial for tissue engineering and drug delivery. *Adv. Eng. Mater.* **2010**, *12*, B380–B387. [CrossRef]

275. Walczak, M.; Brzezinska, M.S.; Richert, A.; Kalwasińska, A. The effect of polyhexamethylene guanidine hydrochloride on biofilm formation on polylactide and polyhydroxybutyrate composites. *Int. Biodeterior. Biodegrad.* **2015**, *98*, 1–5. [CrossRef]

276. Zeikus, J.G.; Jain, M.K.; Elankovan, P. Biotechnology of succinic acid production and markets for derived industrial products. *Appl. Microbiol Biotechnol.* **1998**, *51*, 545–552. [CrossRef]

277. Burgard, A.; Burk, M.J.; Osterhout, R.; Van Dien, S.; Yim, H. Development of a commercial scale process for production of 1,4-butanediol from sugar. *Curr. Opin. Biotechnol.* **2016**, *42*, 118–125. [CrossRef]

278. Forte, A.; Zucaro, A.; Basosi, R.; Fierro, A. Lca of 1,4-butanediol produced via direct fermentation of sugars from wheat straw feedstock within a territorial biorefinery. *Materials* **2016**, *9*, 563. [CrossRef]

279. Phua, Y.J.; Chow, W.S.; Mohd Ishak, Z.A. Mechanical properties and structure development in poly(butylene succinate)/organo-montmorillonite nanocomposites under uniaxial cold rolling. *Express Polym. Lett.* **2011**, *5*, 93–103. [CrossRef]

280. Lee, P.C.; Lee, W.G.; Lee, S.Y.; Chang, H.N. Succinic acid production with reduced by-product formation in the fermentation of anaerobiospirillum succiniciproducens using glycerol as a carbon source. *Biotechnol. Bioeng.* **2001**, *72*, 41–48. [CrossRef]

281. Guettler, M.V.; Rumler, D.; Jainf, M.K. Actinobacillus succinogenes sp. Nov., a novel succinic-acid-producing strain frorn the bovine rurnen. *Int. J. Syst. Bacteriol.* **1999**, *49*, 207–216. [CrossRef] [PubMed]

282. Yim, H.; Haselbeck, R.; Niu, W.; Pujol-Baxley, C.; Burgard, A.; Boldt, J.; Khandurina, J.; Trawick, J.D.; Osterhout, R.E.; Stephen, R.; et al. Metabolic engineering of escherichia coli for direct production of 1,4-butanediol. *Nat. Chem. Biol.* **2011**, *7*, 445–452. [CrossRef] [PubMed]

283. Varadarajan, S.; Miller, D.J. Catalytic upgrading of fermentation-derived organic acids. *Biotechnol. Progr.* **1999**, *15*, 845–854. [CrossRef] [PubMed]

284. Cukalovic, A.; Stevens, C.V. Feasibility of production methods for succinic acid derivatives: A marriage of renewable resources and chemical technology. *Biofuelsbioprod. Bior.* **2008**, *2*, 505–529. [CrossRef]

285. Van Dien, S. From the first drop to the first truckload: Commercialization of microbial processes for renewable chemicals. *Curr. Opin. Biotechnol.* **2013**, *24*, 1061–1068. [CrossRef] [PubMed]

286. De Paula, F.C.; de Paula, C.B.; Contiero, J. Prospective biodegradable plastics from biomass conversion processes. In *Biofuels*; Biernat, K., Ed.; IntechOpen: Rijeka, Croatia, 2018.

287. Available online: http://www.pttmcc.com/new/content.php (accessed on 28 January 2019).

288. Bautista, M.; Martínez de Ilarduya, A.; Alla, A.; Vives, M.; Morató, J.; Muñoz-Guerra, S. Cationic poly(butylene succinate) copolyesters. *Eur. Polym. J.* **2016**, *75*, 329–342. [CrossRef]

289. Wang, H.; Ji, J.; Zhang, W.; Wang, W.; Zhang, Y.; Wu, Z.; Chu, P.K. Rat calvaria osteoblast behavior and antibacterial properties of o(2) and n(2) plasma-implanted biodegradable poly(butylene succinate). *Acta Biomater.* **2010**, *6*, 154–159. [CrossRef]

290. Petchwattana, N.; Naknaen, P. Utilization of thymol as an antimicrobial agent for biodegradable poly(butylene succinate). *Mater. Chem. Phys.* **2015**, *163*, 369–375. [CrossRef]

291. Rota, M.C.; Herrera, A.; Martínez, R.M.; Sotomayor, J.A.; Jordán, M.J. Antimicrobial activity and chemical composition of thymus vulgaris, thymus zygis and thymus hyemalis essential oils. *Food Control* **2008**, *19*, 681–687. [CrossRef]

292. Guarda, A.; Rubilar, J.F.; Miltz, J.; Galotto, M.J. The antimicrobial activity of microencapsulated thymol and carvacrol. *Int. J. Food Microbiol.* **2011**, *146*, 144–150. [CrossRef] [PubMed]

293. Wiburanawong, S.; Petchwattana, N.; Covavisaruch, S. Carvacrol as an antimicrobial agent for poly(butylene succinate): Tensile properties and antimicrobial activity observations. *Adv. Mater. Res.* **2014**, *931–932*, 111–115. [CrossRef]

294. Zhao, Q.; Chen, X.Y.; Martin, C. Scutellaria baicalensis, the golden herb from the garden of chinese medicinal plants. *Sci. Bull.* **2016**, *61*, 1391–1398. [CrossRef] [PubMed]

295. Song, J.; Ge, Z.; An, S.; Zhang, M. Study on the antibacterial plastic properties of natural pigments dyed biodegradable polyester pbs. In Proceedings of the 2011 International Symposium on Water Resource and Environmental Protection, Xi'an, China, 20–22 May 2011.

296. Petchwattana, N.; Covavisaruch, S.; Wibooranawong, S.; Naknaen, P. Antimicrobial food packaging prepared from poly(butylene succinate) and zinc oxide. *Measurement* **2016**, *93*, 442–448. [CrossRef]

297. Naknaen, P. Utilization possibilities of antimicrobial biodegradable packaging produced by poly(butylene succinate) modified with zinc oxide nanoparticles in fresh-cut apple slices. *Int. Food Res. J.* **2014**, *21*, 2413–2420.

298. Odriozola-Serrano, I.; Soliva-Fortuny, R.; Martín-Belloso, O. Antioxidant properties and shelf-life extension of fresh-cut tomatoes stored at different temperatures. *J. Sci. Food Agric.* **2008**, *88*, 2606–2614. [CrossRef]

299. Mart, O. Fresh-cut fruits. In *Handbook of Fruits and Fruit Processing*; Hui, H.Y., Ed.; Blackwell Publishing: Oxford, UK, 2006; pp. 129–144.

300. Tang, X.; Dai, J.; Sun, H.; Nabanita, S.; Petr, S.; Tang, L.; Cheng, Q.; Wang, D.; Wei, J. Copper-doped nano laponite coating on poly(butylene succinate) scaffold with antibacterial properties and cytocompatibility for biomedical application. *J. Nanomater.* **2018**, *2018*, 1–11. [CrossRef]

301. Tian, L.; Wang, P.; Zhao, Z.; Ji, J. Antimicrobial activity of electrospun poly(butylenes succinate) fiber mats containing pvp-capped silver nanoparticles. *Appl. Biochem. Biotechnol.* **2013**, *171*, 1890–1899. [CrossRef]

302. Kora, A.J.; Rastogi, L. Enhancement of antibacterial activity of capped silver nanoparticles in combination with antibiotics, on model gram-negative and gram-positive bacteria. *Bioinorg. Chem. Appl.* **2013**, *2013*, 871097. [CrossRef]

303. Zhang, H.; Chongqiang, Z. Transport of silver nanoparticles capped with different stabilizers in water saturated. *J. Mater. Environ. Sci.* **2014**, *5*, 231–236.

304. Tiwari, V.; Tiwari, M.; Solanki, V. Polyvinylpyrrolidone-capped silver nanoparticle inhibits infection of carbapenem-resistant strain of acinetobacter baumannii in the human pulmonary epithelial cell. *Front. Immunol.* **2017**, *8*, 973. [CrossRef] [PubMed]

305. Setyawati, M.I.; Yuan, X.; Xie, J.; Leong, D.T. The influence of lysosomal stability of silver nanomaterials on their toxicity to human cells. *Biomaterials* **2014**, *35*, 6707–6715. [CrossRef] [PubMed]

306. Gnanadhas, D.P.; Thomas, M.B.; Thomas, R.; Raichur, A.M.; Chakravortty, D. Interaction of silver nanoparticles with serum proteins affects their antimicrobial activity in vivo. *Antimicrob. Agents Chemother.* **2013**, *57*, 4945–4955. [CrossRef] [PubMed]

307. Russell, A.D. Whither triclosan? *J. Antimicrob. Chemother.* **2004**, *53*, 693–695. [CrossRef] [PubMed]

308. Hatcher, H.; Planalp, R.; Cho, J.; Torti, F.M.; Torti, S.V. Curcumin: From ancient medicine to current clinical trials. *Cell. Mol. Life Sci.* **2008**, *65*, 1631–1652. [CrossRef] [PubMed]

309. Torti, S.; Jiao, Y.; Kennedy, D.; Cho, J.; Planalp, R.; Bou-Abdallah, F.; Chasteen, N.; Wilkinson IV, J.; Wang, W.; Torti, F.M. Curcumin, a cancer chemopreventive agent, is a biologically relevant iron chelator. *Second Congr. Int. Bioiron Soc.* **2007**, *113*, 462–469.

310. Llorens, E.; Ibanez, H.; Del Valle, L.J.; Puiggali, J. Biocompatibility and drug release behavior of scaffolds prepared by coaxial electrospinning of poly(butylene succinate) and polyethylene glycol. *Mater. Sci. Eng. C Mater. Biol. Appl.* **2015**, *49*, 472–484. [CrossRef]

311. Meier, M.A.R.; Metzger, J.O.; Schubert, U.S. Plant oil renewable resources as green alternatives in polymer science. *Chem. Soc. Rev.* **2007**, *36*, 1788. [CrossRef]

312. Sonseca, A.; El Fray, M. Enzymatic synthesis of an electrospinnable poly(butylene succinate-*co*-dilinoleic succinate) thermoplastic elastomer. *Rsc Adv.* **2017**, *7*, 21258–21267. [CrossRef]

313. Wcisłek, A.; Sonseca Olalla, A.; McClain, A.; Piegat, A.; Sobolewski, P.; Puskas, J.; El Fray, M. Enzymatic degradation of poly(butylene succinate) copolyesters synthesized with the use of candida antarctica lipase b. *Polymers* **2018**, *10*, 688. [CrossRef]

314. Xia, Y.; Zhang, Z.; Kessler, M.R.; Brehm-Stecher, B.; Larock, R.C. Antibacterial soybean-oil-based cationic polyurethane coatings prepared from different amino polyols. *ChemSusChem* **2012**, *5*, 2221–2227. [CrossRef] [PubMed]

315. Garrison, T.F.; Zhang, Z.; Kim, H.-J.; Mitra, D.; Xia, Y.; Pfister, D.P.; Brehm-Stecher, B.F.; Larock, R.C.; Kessler, M.R. Thermo-mechanical and antibacterial properties of soybean oil-based cationic polyurethane coatings: Effects of amine ratio and degree of crosslinking. *Macromol. Mater. Eng.* **2014**, *299*, 1042–1051. [CrossRef]

316. Liang, H.; Liu, L.; Lu, J.; Chen, M.; Zhang, C. Castor oil-based cationic waterborne polyurethane dispersions: Storage stability, thermo-physical properties and antibacterial properties. *Ind. Crop. Prod.* **2018**, *117*, 169–178. [CrossRef]

317. Bakhshi, H.; Yeganeh, H.; Mehdipour-Ataei, S.; Shokrgozar, M.A.; Yari, A.; Saeedi-Eslami, S.N. Synthesis and characterization of antibacterial polyurethane coatings from quaternary ammonium salts functionalized soybean oil based polyols. *Mater. Sci. Eng. C Mater. Biol. Appl.* **2013**, *33*, 153–164. [CrossRef] [PubMed]

318. Lligadas, G. Renewable polyols for polyurethane synthesis via thiol-ene/yne couplings of plant oils. *Macromol. Chem. Phys.* **2013**, *214*, 415–422. [CrossRef]

319. González-Paz, R.J.; Lligadas, G.; Ronda, J.C.; Galià, M.; Cádiz, V. Thiol–yne reaction of alkyne-derivatized fatty acids: Biobased polyols and cytocompatibility of derived polyurethanes. *Polym. Chem.* **2012**, *3*, 2471. [CrossRef]

320. Lluch, C.; Esteve-Zarzoso, B.; Bordons, A.; Lligadas, G.; Ronda, J.C.; Galia, M.; Cadiz, V. Antimicrobial polyurethane thermosets based on undecylenic acid: Synthesis and evaluation. *Macromol. Biosci.* **2014**, *14*, 1170–1180. [CrossRef]

321. Patil, C.K.; Jirimali, H.D.; Paradeshi, J.S.; Chaudhari, B.L.; Alagi, P.K.; Hong, S.C.; Gite, V.V. Synthesis of biobased polyols using algae oil for multifunctional polyurethane coatings. *Green Mater.* **2018**, *6*, 165–177. [CrossRef]

322. Dastjerdi, R.; Montazer, M. A review on the application of inorganic nano-structured materials in the modification of textiles: Focus on anti-microbial properties. *Colloids Surf. B Biointerfaces* **2010**, *79*, 5–18. [CrossRef] [PubMed]

323. Deka, H.; Karak, N.; Kalita, R.D.; Buragohain, A.K. Bio-based thermostable, biodegradable and biocompatible hyperbranched polyurethane/ag nanocomposites with antimicrobial activity. *Polym. Degrad. Stab.* **2010**, *95*, 1509–1517. [CrossRef]

324. Das, B.; Mandal, M.; Upadhyay, A.; Chattopadhyay, P.; Karak, N. Bio-based hyperbranched polyurethane/fe3o4 nanocomposites: Smart antibacterial biomaterials for biomedical devices and implants. *Biomed. Mater.* **2013**, *8*, 035003. [CrossRef] [PubMed]

325. Das, B.; Chattopadhyay, P.; Upadhyay, A.; Gupta, K.; Mandal, M.; Karak, N. Biophysico-chemical interfacial attributes of Fe$_3$O$_4$ decorated mwcnt nanohybrid/bio-based hyperbranched polyurethane nanocomposite: An antibacterial wound healing material with controlled drug release potential. *New J. Chem.* **2014**, *38*, 4300–4311. [CrossRef]

326. Thakur, S.; Barua, S.; Karak, N. Self-healable castor oil based tough smart hyperbranched polyurethane nanocomposite with antimicrobial attributes. *RSC Adv.* **2015**, *5*, 2167–2176. [CrossRef]

327. Wu, S.; Li, J.; Zhang, G.; Yao, Y.; Li, G.; Sun, R.; Wong, C. Ultrafast self-healing nanocomposites via infrared laser and their application in flexible electronics. *ACS Appl. Mater. Interfaces* **2017**, *9*, 3040–3049. [CrossRef] [PubMed]

328. Duarah, R.; Singh, Y.P.; Gupta, P.; Mandal, B.B.; Karak, N. High performance bio-based hyperbranched polyurethane/carbon dot-silver nanocomposite: A rapid self-expandable stent. *Biofabrication* **2016**, *8*, 045013. [CrossRef]

329. Yücedag, F.; Atalay-Oral, C.; Erkal, S.; Sirkecioglu, A.; Karasartova, D.; Sahin, F.; Tantekin-Ersolmaz, S.B.; Güner, F.S. Antibacterial oil-based polyurethane films for wound dressing applications. *J. Appl. Polym. Sci.* **2010**, *115*, 1347–1357. [CrossRef]

330. Akram, D.; Sharmin, E.; Ahmad, S. Development and characterization of boron incorporated linseed oil polyurethanes. *J. Appl. Polym. Sci.* **2009**, *116*, 499–508. [CrossRef]

331. Sharmin, E.; Zafar, F.; Akram, D.; Ahmad, S. Plant oil polyol nanocomposite for antibacterial polyurethane coating. *Prog. Org. Coat.* **2013**, *76*, 541–547. [CrossRef]

materials

MDPI

Article

Interference of Biodegradable Plastics in the Polypropylene Recycling Process

María Dolores Samper [1,*], **David Bertomeu [1]**, **Marina Patricia Arrieta [2]**, **José Miguel Ferri [1]** and **Juan López-Martínez [1]**

[1] Instituto de Tecnología de Materiales, Univesitat Politècnica de València, 03801 Alcoy-Alicante, Spain; daberpe@alumni.upv.es (D.B.); joferaz@upvnet.upv.es (J.M.F.); jlopezm@mcm.upv.es (J.L.-M.)
[2] Instituto de Ciencia y Tecnología de Polímeros (ICTP-CSIC), 28006 Madrid, Spain; marina.arrieta@gmail.com
[*] Correspondence: masammad@upvnet.upv.es

Received: 24 August 2018; Accepted: 28 September 2018; Published: 2 October 2018

Abstract: Recycling polymers is common due to the need to reduce the environmental impact of these materials. Polypropylene (PP) is one of the polymers called 'commodities polymers' and it is commonly used in a wide variety of short-term applications such as food packaging and agricultural products. That is why a large amount of PP residues that can be recycled are generated every year. However, the current increasing introduction of biodegradable polymers in the food packaging industry can negatively affect the properties of recycled PP if those kinds of plastics are disposed with traditional plastics. For this reason, the influence that generates small amounts of biodegradable polymers such as polylactic acid (PLA), polyhydroxybutyrate (PHB) and thermoplastic starch (TPS) in the recycled PP were analyzed in this work. Thus, recycled PP was blended with biodegradables polymers by melt extrusion followed by injection moulding process to simulate the industrial conditions. Then, the obtained materials were evaluated by studding the changes on the thermal and mechanical performance. The results revealed that the vicat softening temperature is negatively affected by the presence of biodegradable polymers in recycled PP. Meanwhile, the melt flow index was negatively affected for PLA and PHB added blends. The mechanical properties were affected when more than 5 wt.% of biodegradable polymers were present. Moreover, structural changes were detected when biodegradable polymers were added to the recycled PP by means of FTIR, because of the characteristic bands of the carbonyl group (between the band 1700–1800 cm^{-1}) appeared due to the presence of PLA, PHB or TPS. Thus, low amounts (lower than 5 wt.%) of biodegradable polymers can be introduced in the recycled PP process without affecting the overall performance of the final material intended for several applications, such as food packaging, agricultural films for farming and crop protection.

Keywords: recycling; polypropylene; biodegradable polymers; degradation; inmiscibility

1. Introduction

The world plastic production has reached more than 330 million tons in the last few years. Among all plastics, polypropylene (PP) is the most demanded for plastic converter industries in Europe [1]. In fact, PP is one of the most used and consumed polymers in the world due to its good processing performance and versatility; it is used for a wide variety of applications: commodities, medical applications, automotive, etc. It is known as one of the "packaging plastics" and packaging products are mainly short-term applications which ultimately represent a big source of plastic waste. Thus, a large amount of PP waste is produced every year after its useful life. Fortunately, a huge part of plastic residues (more than 30%) are retrieved using industrial recycling, closing the loop of circular economy [1]. Particularly, recycled PP can be used in different ways like new packaging products, films or matrix of wood composites [2,3]. Moreover, recycled PP can be considered a safe material because the producers do not usually use hazardous materials in its process. However, we must take into account that some recycled materials can be hazardous such as granulated end-of-life tyres because they can contain polycyclic aromatic hydrocarbons (PAHs), some of which are identified as carcinogens. Also, recycled expended polystyrene (EPS) coming from building and construction sector can be considered a hazardous waste because EPS is highly combustible and flame retardant hexabromocyclododecane (HBCDD). It had been frequently added until it was included in the reach regulation list in 2015 because it is considered a persistent Organic Pollutant (POPs) [4]. The best option for disposal hazardous polymers waste is energy recovery, since it can meet partial energy demand and reduce disposal cost, including CO_2 emissions [5].

Another interesting approach to close the loop of circular economy is the use of biobased and biodegradable polymers, known as biopolymers, which have non-dependence on petrochemical resources and also do not represent an environmental potential hazard if they ultimately reach landfills. Therefore, in recent years a great interest on the use of biobased and biodegradable polymers has increased in order to replace the petrochemicals-based packaging materials and to reduce plastic waste in landfills in certain applications, mainly short-term packaging and agricultural films [6–8]. Thus, the use of biodegradable plastics is rising, mainly because there is an increasing concern about the reduction of the plastics' environmental impact. In fact, currently, industries and consumers demand these types of products on the market. According to Bastioli et al., Europe should take advantage of the great potential of these materials to add value to products by taking advantage of the new bio-economic feature of bioplastics, as well as to preserve and improve ecosystems and biodiversity [9,10].

Biopolymers [11], such as poly (lactic acid) (PLA) [12,13], polyhydroxybutyrate (PHB) [14] and thermoplastic starch (TPS) [15], are increasingly used in the food packaging and agricultural sector, in addition to other fields of application such as medical [16–18] or composite materials [18–20]. However, consumers have low information about where they have to throw away this kind of plastics after their useful life and they are commonly disposed of with traditional waste plastics [21]. Although the bioplastic products can also be recycled after their use by recycling in traditional ways [22], the current systems do not allow the recovery of high purity of plastic waste. Moreover, these new technologies increase the cost of the final product developed with recycled biopolymers. On the other hand, if biodegradable packaging residues are found in recycling channels, they could act as impurities for traditional plastics influencing the structural and thermal properties of recycled products. The separation and classification processes of these biodegradable products can be complex and expensive [23–25] and if the consumption of bio-based plastics continues to increase, as it has been predicted, current recycling systems will have to be considered a reorganization to avoid contamination of recycled plastics [26]. Currently, there have not been found in the literature works on the mixture of small % of biopolymers in a PP matrix that help to evaluate their inclusion in the recycling of this material. However, different studies of PP/bioplastics blends have been carried out using different compatibilizing agents. This is the case of PLA/PP blends, where studies have been conducted with different amounts of compatibilizing agents such as polypropylene-graft-maleic anhydride (PP-g-MAH) and styrene-ethylene-butylene-styrene-graft-maleic anhydride (SEBS-g-MAH), these compatibilizers improve some properties such as impact strength, especially using a 3 phr of PP-g-MAH [27–29]. Studies have also been carried out on PP/TPS blends with different percentages of TPS, observing that the increase in TPS causes a decrease in tensile strength, elongation at break and MFI [30,31]. In the case of PP/PHB mixtures, not too much information was found, Sadi et al. performed a work on the compatibility of PP/PHB blends with 20 wt.% of PHB using different compatibilizers, the PHB causes a significant decrease in the mechanical properties of PP that can be improved using poly (ethylene-co-methyl acrylate-co-glycidyl methacrylate) (P(E-MA-GMA)) [32].

It is expected that in the near future, the consumption of biodegradable polymers will grow up and the conventional recycled polymers may have a low amount of biodegradable polymers acting as impurities. The presence of small fractions of impurities can negatively influence the structural and mechanical properties of the conventional recycled materials, which might decrease their price and viability [33]. This lack of properties is due to the incompatibility of the polymeric components in the blend. In fact, blending approaches use a number of compatibilization strategies that in general are related to the addition of a third component that is miscible with both phases (i.e., co-solvent, nanoparticles), or one part of the third component that is miscible with one phase and another part with another phase (i.e., copolymers) [34,35]. Compatibilizers can be used to balance not only the loss of mechanical properties but also the morphological changes of the immiscibility of polymers, as suggested by MacAubas et al. and Fekete et al. [36,37]. However, considering the industrial plastic recycling process, it is expected that small fractions of impurities reach the PP recycling process without any kind of compatibilizers.

In this work, blends based on recycled PP and the most typically used biodegradable polymers in short-term applications were studied in order to simulate recycled PP contaminated with low amounts of PLA, TPS and PHB up to 15 wt.%. With this purpose, five different percentages of biodegradable plastics were blended with polypropylene and further processed by melt extrusion followed by an injection molding process to simulate the most typically used processing technologies at industrial level. Then, the effect of the biodegradable materials presence on the mechanical and thermal properties were evaluated. Therefore, the changes on the softening temperature (VICAT) and the melt flow index were studied. The mechanical properties were also analyzed to determine the influence of the biodegradable plastic presence on the extruded blends on the mechanical performance of the final materials. Furthermore, FTIR studies were carried out to easily determine the presence of biodegradable materials in recycled polypropylene, while scanning electron microscopy (SEM) was used to evaluate the polymer-polymer microstructural interaction. Additionally, since a huge amount of plastic still ends in landfill the blends were also exposed to composting conditions at a laboratory scale level in order to get information about the influence of biopolymers into recycled PP under environmental composting conditions. The results allowed to identify the maximum amount of biodegradable materials that can be blended with recycled PP as impurity without compromising the mechanical and thermal integrity of the PP based products.

2. Materials and Methods

2.1. Materials and Preparation of the Blends

Recycled PP, with reference PP1B, has been supplied by Acteco (Ibi, Spain), PLA 4032D by NatureWorks LLC (Minnetonka, MN, USA), TPS Mater Bi by Novamont (Novara, Italy) and PHB P226 by Biomer (Krailling, Germany).

The blends were made by mixing PP with different percentages of biodegradable polymers, that ranged from 0 to 15 wt.%, as can be seen in the Table 1, in a twin screw extruder (Dupra S.L., Castalla, Spain), processed at a temperature range of 200–220 °C at 50 rpm. The blend samples were then injected by an injection molding process using a Babyplast estandar 6.6 (Cronoplast S.L., Albrera, Spain) machine with a mold with normalized sample dimensions for tensile test according to ISO-527-2, specifically 5A samples.

Table 1. Samples acronym.

Sample	PP1B (wt.%)	PLA (wt.%)	PHB (wt.%)	TPS (wt.%)
PP	100.0	-	-	-
PP-2.5%PLA	97.5	2.5	-	-
PP-5%PLA	95.0	5.0	-	-
PP-7.5%PLA	92.5	7.5	-	-
PP-10%PLA	90.0	10.0	-	-
PP-15%PLA	85.0	15.0	-	-
PP-2.5%PHB	97.5	-	2.5	-
PP-5%PHB	95.0	-	5.0	-

Table 1. *Cont.*

Sample	PP1B (wt.%)	PLA (wt.%)	PHB (wt.%)	TPS (wt.%)
PP-7.5%PHB	92.5	-	7.5	-
PP-10%PHB	90.0	-	10.0	-
PP-15%PHB	85.0	-	15.0	-
PP-2.5%TPS	97.5	-	-	2.5
PP-5%TPS	95.0	-	-	5.0
PP-2.5%TPS	92.5	-	-	7.5
PP-10%TPS	90.0	-	-	10.0
PP-15%TPS	85.0	-	-	15.0

2.2. Scanning Electron Microscopy Analysis

The Scanning Electron Microscopy (SEM) images were took with a Phenon of FEI equipment (Eindhoven, The Netherlands) using 5 kV voltage, to observe the miscibility of the components in the blends subjected to a cryofracture process. Before the observation, the samples were coated with a gold-palladium alloy by a Sputter Mod Coater Emitech SC7620 (Quórum Technologies, East Sussex, UK).

2.3. Infrared Spectroscopy Analysis

The infrared spectroscopy analysis was conducted using a Perkin Elmer Spectrum BX spectrometer (Perkin-Elmer España S.L., Madrid, Spain). The test was made with 20 scans between 600 and 4000 cm^{-1} with a resolution of 32 cm^{-1} mode using an attenuated total reflectance (ATR) accessory, indicated for samples with poor transparency.

2.4. Thermal Characterization

2.4.1. Thermogravimetric Analysis (TGA)

Dynamic thermal degradation analysis was carried out using thermogravimetric analyzer TGA/SDTA 851 Mettler Toledo (Schwarzenbach, Switzerland). TGA measurements were run at 20 °C·min^{-1} constant heating rates. Temperature was raised from 30 to 600 °C under air conditions in order to study oxidative degradation process following the conditions used in a previous work [38]. The initial degradation temperature (T_0) was calculated at 5% mass loss, while temperatures at the maximum degradation rate (T_{max}) for each stage were determined from the first derivatives of the TGA curves (DTG).

2.4.2. Differential Scanning Calorimetry

The differential scanning calorimetry (DSC) was conducted with a Mettler Toledo 821 equipment (Mettler Toledo, Schwerzenbach, Switzerland) using samples of 4–6 mg. The heating and cooling programs were performed at a 20 °C·min^{-1} speed in a nitrogen atmosphere (60 mL·min^{-1}). The DSC program was carried out in three stages: the first heating took place from 30 to 200 °C, followed by a cooling process up to 30 °C to −20 °C·min^{-1} followed by a second heating up to 250 °C. The first heating was carried out to remove the thermal history of the materials. The melting temperature, T_m, and the melting enthalpy, ΔH_m, were obtained from the second heating.

2.5. Mechanical Properties

The tensile test properties were performed with a universal testing machine Ibertest ELIB 30 (SAE Ibertest, Madrid, Spain) at room temperature, according to ISO 527; the tests were performed with a load cell of 5 kN and at a speed of 10 mm·min^{-1}. From each sample type at least 5 specimens were tested and the mean of those tests was calculated.

The Shore D hardness was measured according to the UNE-EN ISO 868 standard using a hardness equipment Mod.673-D (Instruments J. Bot S.A., Barcelona, Spain). The results were the mean hardness of at least 5 measurements of samples with thickness of 4 mm.

2.6. Exposition to Composting Medium

The PP based blends were exposed to compost condition with the main objective to study the influence of PLA, PHB and TPS into PP based blends disintegration. The disintegration under composting conditions was performed at laboratory scale level according to the ISO 20200 standard [39]. Dogbone samples were buried at 4–6 cm depth in perforated plastic boxes containing a solid synthetic wet waste (10% of compost (Mantillo. Spain), 30% rabbit food, 10% starch, 5% sugar, 1% urea, 4% corn oil and 40% sawdust as well as approximately 50 wt.% of water content) and were incubated at aerobic conditions (58 ± 2 °C). PP based blends were recovered at 8, 21 and 30 days. A qualitative check of the physical disintegration in compost as a function of time was done by taken photographs, while the structural changes were followed by TGA measurements conducted from 30 to 600 °C at 20 °C·min^{-1} under oxidation conditions.

2.7. Other Techniques

The VICAT (VST) softening temperature was studied with the VICAT/HDT station DEFLEX 687-A2 (Metrotec S.A., San Sebastiár, Spain) according to the ISO 306, to 50 N with a heating rate of 50 °C·h^{-1}. The flow index measures of the different blends were performed according to ISO113, using 2.16 kg and 230 °C, with an extrusion plastometer (AtsFaarS.p.A., Vignate, Italy).

3. Results and Discussion

3.1. Miscibility

The miscibility between different polymers depends on the chemical structure of the polymers as well as on their crystalline nature and morphology of the starting polymers [23,40]. While miscibility is limited to a specific set of conditions, several polymers form immiscible blends. The incompatibility between two polymeric matrices causes the loss of the mechanical properties and even superficial lamination. This loss of properties also depends on the percentages of each component on the blend sample. In the present work, it seems that biodegradable polymers are acting as impurities, probably due to their different polarities. It is known that the relative affinity between two polymers can be estimated using the solubility parameter (δ) [41]. To consider that a mixture's components are compatible, their solubility parameter should be similar. In this sense, δ should be calculated taking into account the contribution that each group has in the overall structure of the molecule (Equation (1)).

$$\delta = \frac{\rho \Sigma_j F_j}{M_n} \qquad (1)$$

where δ ((cal·cm^{-3})$^{1/2}$) is the solubility parameter for each component, ρ (g·cm^{-3}) is the polymer density, Mn (g·mol^{-1}) is the molar mass of the repeated unit, Σ_j and F_j are the sum of the contributions of all groups (F, (cal·cm^{-3})$^{1/2}$·mol^{-1}).

The results of the calculated δ can be seen in Table 2, where δ was calculated according to the Small method, using Equation (1) and the values of F of Table 3, the solubility parameter results were very similar with available data in polymerdatabase.com for the polymers studied [42]. While the δ of PP is 16.4 MPa$^{1/2}$, that of PLA is 19.5 MPa$^{1/2}$ and it is the biodegradable polymer with the nearest δ. Although the solubility values of these two polymers are close, it is not enough to consider these two materials miscible. Regarding the biodegradable polymers TPS and PHB, whose solubility parameter are 8.4 MPa$^{1/2}$ and 21.4 MPa$^{1/2}$, respectively. Thus, they clearly indicate that there is an increased immiscibility with the PP matrix, since the solubility parameter of these polymers is more distant from PP. Therefore, according with the solubility parameter results it seems that the biodegradable materials studied here are not miscible with PP and, for that reason, they could generate a thermal and mechanical properties deterioration on the recycled PP. The microstructural analysis was performed by SEM. In Figure 1, we observed the SEM images of the different PP blends with 15 wt.% of different biodegradable polymers as an example, PLA (Figure 1b), PHB (Figure 1a) and TPS (Figure 1c). It can be clearly seen that the blends based on PP and biodegradable polymers studied here are immiscible, since a phase separation of the components in the different blends can be observed. In fact, all blends samples exhibit spherical droplets dispersed in the PP matrix. Some of the spherical droplets have been pulled out of the PP matrix during fracture, indicating very weak interfacial adhesion and immiscibility between both polymers, particularly in the case of PP-TPS blend which showed higher spherical droplets (Figure 1c). This could be due to the polyolefin structural differences in comparison with the biodegradable polymers, as predicted using the solubility parameter. When two polymers are immiscible and are blended together, a two-phase system is formed. Generally this material has low mechanical properties due to the stress concentration generated by the poor adhesion between the phases [43,44].

Table 2. Values of the solubility parameters calculated from the present constants.

Polymer	Structure	Calculated Small $\delta_{(cal)}$ (MPa$^{1/2}$)	Available Date $\delta_{(cal)}$ (MPa$^{1/2}$) [42]
PP		16.4	15.5–17.5
PLA		19.5	19.2–21.1
TPS		8.4	-
PHB		21.4	19.2

Table 3. Small's molar attraction constants for some functional groups [38].

Group	F ((cal·cm^{-3})$^{1/2}$ mol^{-1})
–CH$_3$	214
–CH$_2$–	133
–CH<	28
>C<	-93
–OH	83
–O–	70
–H (variable)	80–100
>C=O	275

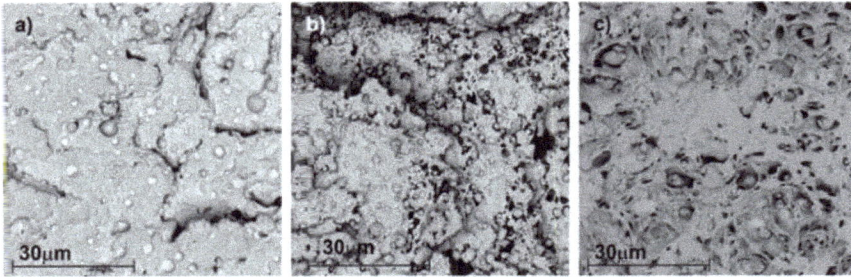

Figure 1. SEM Images at 3500× magnification of the samples. (**a**) PP-15PHB; (**b**) PP-15PLA; (**c**) PP-15TPS.

3.2. Detection of Biodegradable Materials in the Recycled PP Using the FTIR Technique

Through the FTIR technique, biodegradable materials can be easily detected in the recycled PP, since, as it was demonstrated in our previous work, some of the characteristic bands of the biodegradable polymers (PLA, PHB and TPS) do not overlap with the PP characteristic bands [38]. As shown in Figure 2 between 1700 and 1800 cm^{-1}, the PP/biodegradable polymers blends (with 15 wt.% of the different biodegradable polymers) exhibit a strong band that has no presented the neat recycled PP. This is due to PLA and PHB present the carbonyl group (–C=O) characteristic band at this wavelength. The asymmetric stretching of the carbonyl group in neat PLA is at higher wavelength (1754 cm^{-1}) and it is attributed to the amorphous carbonyl vibration. Meanwhile, the stretching vibration of crystalline carbonyl groups is centered at lower wavelengths (1726 cm^{-1}) in the spectrum of neat PP-15%PHB associated with the crystalline state of PHB [45]. TPS presents the same band due to the additives used for their manufacturing in the thermoplastic form [38]. Although FTIR technique does not allow to quantify the amount of biopolymer in the blends, it represents a simple and fast method to detect the presence of this kind of impurities in the recycled PP process which is easily scalable up to the plastic recycling industry.

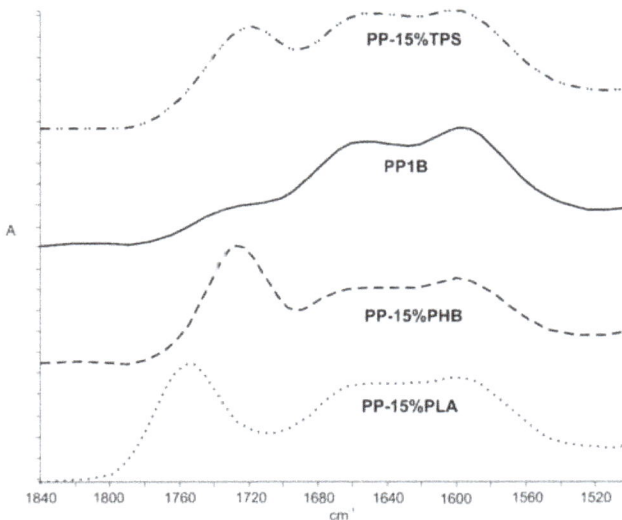

Figure 2. FTIR spectra: PP (PP1B) and PP with 15% biodegradable polymer.

3.3. Thermal Characterization

3.3.1. Thermogravimetric Analysis

Since the amount of different components in a polymeric blend sometimes can be estimated from TGA, the thermal decomposition of the blends was studied by means of TGA and DTG. In Figure 3, we show the TGA (Figure 3a) and DTG (Figure 3b) results of PP blends blended with 15 wt.% of biopolymers as example. Moreover, the thermal degradation is very important for the plastic processing industry since biopolyesters thermal degradation could lead to the formation of oligomers, such as oligomeric lactic acid (OLA) in the case of PLA and oligomers of 3-hydroxybutyrate (OHB) in the case of PHB, which can further act as plasticizers. TGA shows a complete weight loss of PP in a single degradation step. Meanwhile, PP blended with PLA and PHB were degraded in two steps, where the first one is assigned to the biopolyesters decomposition and the second one, at higher temperatures, was related to the PP thermal degradation. TGA revealed that all PP based blends showed minor thermal stability with respect to PP sample (PP T_0 = 357 °C). TPS was the biopolymer that less shifted the onset degradation temperature to lower values, around 10 °C for PP-15%TPS, T_0 = 346 °C. Higher reduction in onset thermal degradation was observed when biopolyesters were blended with PP, particularly in the case of PHB (PP-15%PLA T_0 = 315 °C and PP-15%PHB T_0 = 315 °C). Nevertheless, it should be highlighted that no degradation takes place in the temperature region from room temperature to 220 °C, which is the temperature range where the blend samples were processed. While FTIR allowed to identify the presence of biopolymers in the recycled PP, TGA allows to estimate the amount of biopolyesters in the blends. For instance, from Figure 3a, the loss of both biopolyesters could be estimated from TGA and, as it is expected, it is around 15%. This result confirms that there were not thermal degradation of biopolyesters during processing, and it is particularly important for PP-PHB blends since it is known that the foremost drawback for the industrial production of PHB based blends is its small processing window [21]. Different situation is observed for PP-TPS based blends, since the degradation take place in one step process like PP, avoiding the possibility to quantify the amount of TPS as impurities in the blend. Although the amount of TPS could not be quantified, the contamination of PP could be identified from DTG curve (Figure 3b) in which is possible to observe that the degradation starts prior to the degradation process of PP (see the shoulder in the insert Figure 3b), which has been attributed to the starch pyrolysis (between 300 and 360 °C) [46]. In addition, after the main degradation process of PP there is another degradation step between 360 and 500 °C that has been related to the oxidation of the partially decomposed starch in air atmosphere [47] and to the decomposition of the biodegradable co-polyester component in TPS [46]. Moreover, the maximum degradation temperature of PP was shifted from 423 °C in PP to 456 °C in PP-15%TPS suggesting somewhat positive interface interaction between PP and biodegradable TPS material. Biopolyesters also shifted the maximum degradation temperature of PP to higher values (T_{max}PP in PP-15%PLA 474 °C and in PP-15%PHB 430 °C).

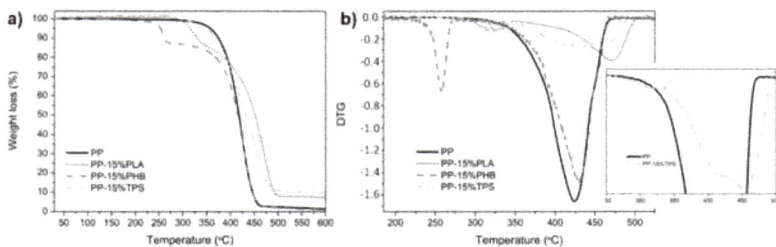

Figure 3. (**a**) TGA and (**b**) DTG thermograms of PP blends with 15 wt.% of biodegradable polymers.

3.3.2. Differential Scanning Calorimetry

In Table 4 and Figure 4, we show the effect of the presence of different biodegradable polymers on the thermal properties of the recycled PP measured by DSC. In Figure 5, we show the second heating of the samples with 15% of bio-based polymers, the recycled PP calorimetric curve had 2 melting peaks, the second correspond with melting peak of PP and the little first peak could be a contamination with another polymer like a HDPE or LDPE. This double melting peak can be also observed in the other PP with bio-based DSC curves. In Table 4 can be observed that the melting temperature, T_m, does not vary and it is between 163.4 and 165.1 °C. Compared to the T_c, obtained from the DSC cooling process, it can be observed that PP presents a crystallization temperature at 124.5 °C and in all samples containing biodegradable polymers, either PLA, PHB or TPS, the crystallization temperature decreased, being between 120.5 and 121.6 °C. In general, the presence of biodegradable polymers in the PP matrix caused a decrease of the crystallinity, as the enthalpy values of crystallization and melting decreased. The decrease of crystallinity may be due to the fact that biodegradable polymers in the blend make difficult the pack of PP chains, since the presence of biodegradable polymers acts as impurities, and thus, they would reduce the free volume of PP [48,49].

Table 4. DSC results of PP blends with biodegradable polymers.

Sample	T_c (°C)	ΔH_c (J·g^{-1})	T_m (°C)	ΔH_m (J·g^{-1})
PP	124.5	85.5	164.1	66.0
PP-5%TPS	121.1	89.0	164.0	66.4
PP-10%TPS	120.5	77.4	164.0	57.3
PP-15%TPS	120.8	67.5	163.7	49.3
PP-5%PHB	120.9	80.5	163.9	62.8
PP-10%PHB	120.8	71.0	164.3	65.1
PP-15%PHB	120.8	69.2	165.1	58.1
PP-5%PLA	121.6	80.2	163.4	56.9
PP-10%PLA	121.5	80.4	163.6	62.0
PP-15%PLA	120.9	70.3	165.1	53.5

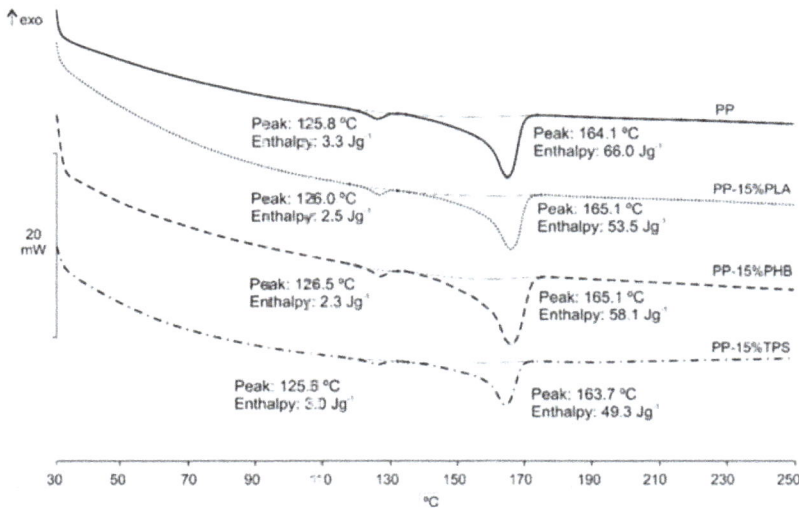

Figure 4. DSC curves of PP blends with 15 wt % of biodegradable polymers.

Therefore, the presence of biodegradable polymers not only affects the mechanical properties of the recycled PP as it will be discussed in the following sections, but also affects the thermal performance, especially the PP crystallinity considering that the melting temperature is only slightly modified.

3.4. Thermomechanical Characterization

Previously, we discussed the changes caused by the presence of biodegradable polymers on the thermal properties of the different blends studied but not only these properties are important in the polymeric materials recycling. Therefore, the changes on thermomechanical properties were also taken into account, as they are too important mainly for the polymer processing industry.

Figure 5 shows the graphical representation of the melt flow index (MFI). It can be seen that the TPS does not significantly modify the MFI of the PP, since it practically remains constant for all the percentages studied, despite being the polymer with the farthest solubility value compared to PP. In the case of blends made with PP with PLA or PHB, it is observed that the MFI increases as the percentage of biodegradable polymer in the blends increase. This increase is more pronounced for the samples made with PLA. Nevertheless, blends prepared with low amount (2.5 wt.%) of these biodegradable polymers, that is PP-2.5%PLA and PP-2.5%PHB, MFI is not practically affected. This could be related with the fact that increasing the polyester amount in the blend, the amount of ester groups, which are relatively easy to breakdown and have poor thermal stability [50], increases and more chain scission occurs leading to an increase in MFI.

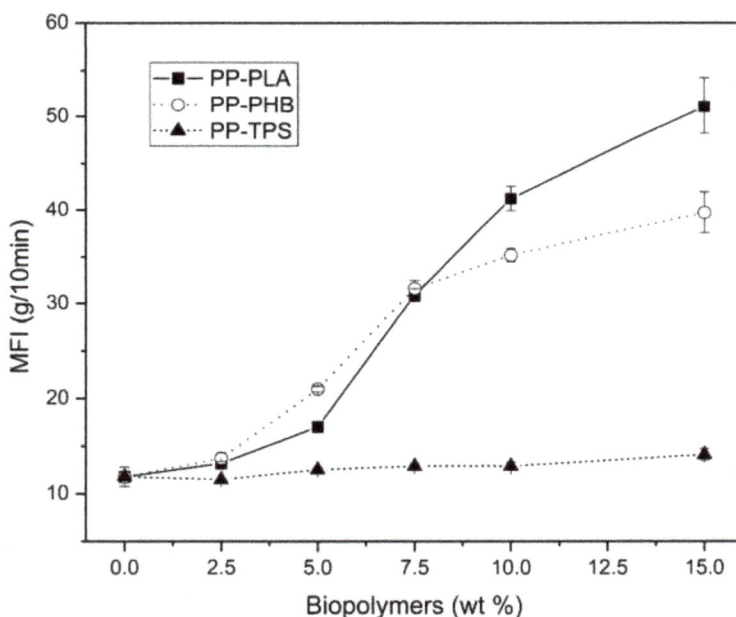

Figure 5. Plot of MFI vs. wt.% biodegradable polymer.

The blend thermal stability was studied by determining the softening temperature VICAT (VST). The results for all the studied systems, PP-PLA, PP-PHB and PP-TPS, show the same behavior, since as the amount of biodegradable polymer increases it causes a decreases of the VST (Figure 6). Depending on the system, the decrease of this property is more or less pronounced. While the PP-PLA blend system is the one with the highest VST (lower VST reduction), TPS is the biodegradable material that causes a greater decrease of this property. This behavior could be related to the fact that starchy materials are water sensitive and are able to show a rubber-like behavior depending on its moisture content [51].

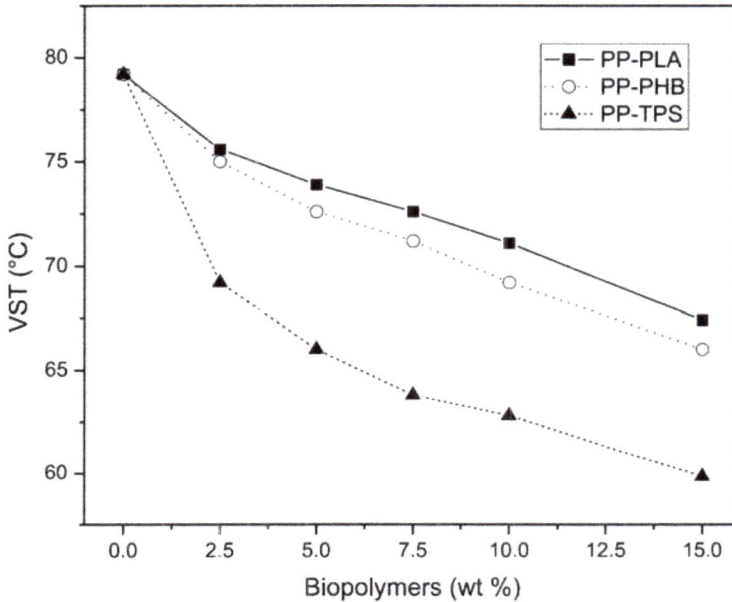

Figure 6. VST vs. wt.% biodegradable polymer.

3.5. Mechanical Characterization

The determination of the mechanical properties in blends is very important because of the incompatibility of different polymers negatively affects the material performance, causing a decrease on the mechanical properties [36]. A small alteration can be observed on tensile strength (Figure 7) and tensile modulus (Figure 8) due to the different biodegradable polymers presence, while these differences increased as the biodegradable polymers percentage increased up to 5 wt.%.

Figure 7 shows the variation in the tensile strength of the different blends made with PP and biodegradable polymers. It is observed that percentages lower than 5% of PLA and TPS do not significantly vary the tensile strength of the recycled PP, since it practically remains constant. However, larger quantities of these two polymers decrease the tensile strength. On the other hand, blends made with PP-PHB show a decrease in tensile strength in all percentages. This decrease in strength is due to the lack of interaction between the polymers blends at the interfaces. With regard to the elongation at break, the results showed (Table 5) that this property not change significantly with the addition of different bioplastics studied, maybe due to the recycled PP used in this work has a very low level of elongation. Although these results show scattered values, it seems that biopolymers are not acting as plasticizer for the PP matrix, in good agreement with thermal degradation results in which it was observed that there were not thermal degradation of biopolyesters during processing, which would lead to the formation of oligomers able to plasticize the polymeric matrix.

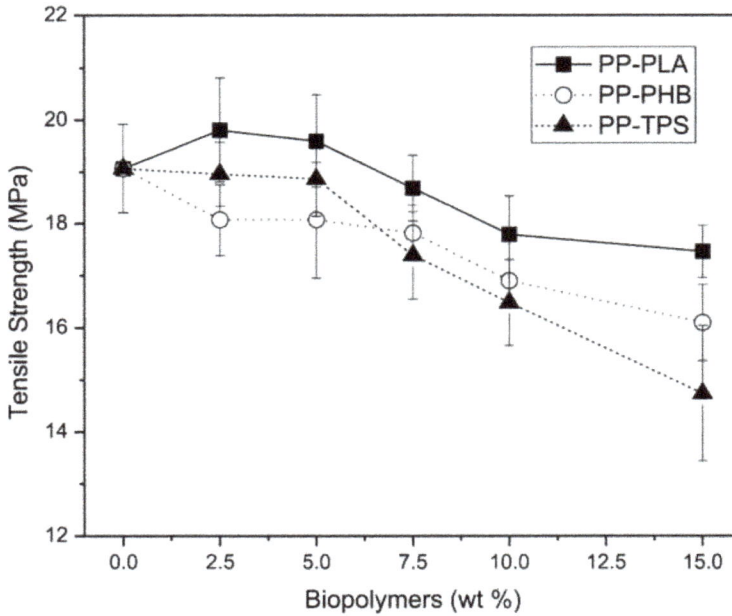

Figure 7. Variation of tensile strength vs. wt.% biodegradable polymer in polypropylene.

Table 5. Elongation results of PP blends with biodegradable polymers.

Sample	Elongation at Break (%)	Standard Desviation (%)
PP	4.87	0.61
PP-5%PLA	4.11	1.84
PP-15%PLA	3.29	1.26
PP-5%PHB	3.09	1.39
PP-15%PHB	3.54	1.78
PP-5%TPS	2.26	0.33
PP-15%TPS	5.68	1.5

The graphical representation of the tensile modulus (Figure 8) shows that the variation of this property depends on the biodegradable polymer used in the blend. In blends made with PP-TPS it is possible to observe that the elastic modulus remains constant when the amounts of TPS added are lower than 5 wt.%. For higher values it can be seen that the modulus decrease considerably up to 400 MPa. In the PP-PHB blends the modulus remains practically constant around 450 MPa for all the percentages studied, this may be because the PHB has an elastic modulus similar to that of the PP [52]. On the other side, in PP-PLA blends, it can be observed that the elastic modulus increases as the PLA content increases from around 450 for recycled PP to around 610 MPa for PP-15% PLA, this increase may be due to the fact that PLA possess a higher elastic modulus than PP [53].

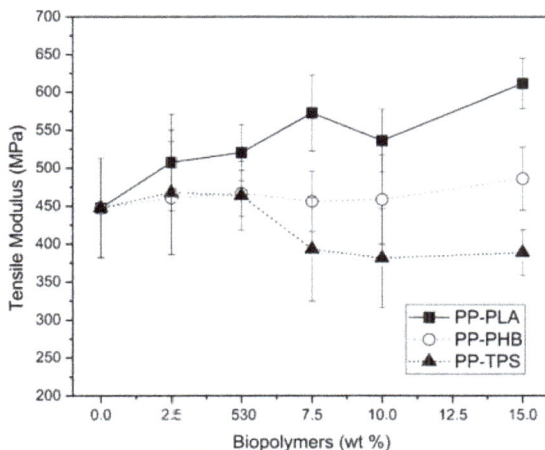

Figure 8. Variation of modulus of elasticity vs. wt.% biodegradable polymer in polypropylene.

The results obtained in the mechanical characterization were in good accordance with the calculated solubility parameters. PLA is a biodegradable polymer with the closest δ to that of the PP and the presence of low percentages of this polymer less or equal to 5 wt.% of PLA, did not cause a decrease on the mechanical properties, but in higher percentages the tensile strength and the tensile modulus decreased. Similar findings were found in the study carried out by Pivsa-Art et al. who analyze PLA-PP blends, the incorporation of 20 wt.% of PLA into PP matrix caused a slight increase in tensile strength and Young's modulus. However, to improve the mechanical properties of PLA-PP blends they used polypropylene grafted with maleic anhydride as compatibilizer [27]. Regarding the PHB and TPS presence in the recycled PP, they caused the mechanical properties deterioration. For instance, it happened to Sadi et al. who performed a study on PP blends with 20 wt.% of PHB. The mechanical properties of this blend were lower than that of PP, and thus, they studied the compatibilization with different copolymers, founding that the most effective approach was using poly (ethylene-co-methyl acrylate-co-glycidyl methacrylate) [32].

In a study conducted by Kaseem et al. on blends made with PP and TPS, the increase of TPS caused a decrease on the tensile strength since the immiscible TPS acted as a filler for PP matrix [30].

3.6. Desintegration under Composting Medium

Unfortunately, instead of reaching the recycling system, several PP-based products still go to landfill after their useful life, and thus in order to simulate this end of life option the materials were exposed to composting conditions at laboratory scale level. It is known that PLA, PHB or starch-based materials are totally disintegrated under composting medium exposed to thermophilic aerobic conditions [21,47], that is according to the ISO standard in less than three months [39]. In fact, blends containing PLA, PHB or TPS in their formulations requires between one and two months to be completely disintegrated under composting [54,55]. Meanwhile, since PP is not a biodegradable polymer it is not suitable to perform disintegration in a composting medium. Thus, the PP blends were subjected under controlled composting conditions during 1 month. The visual appearance of recovered samples at different time of exposition in composting conditions (8, 21 and 30 days, on the basis of previous work [38]) are shown in Figure 8. It was observed that the samples suffered somewhat physical changes at the surface after 21 days and mainly after 30 days, suggesting that the biodegradation of PLA, PHB and TPS is taking place. Therefore, TGA analysis were conducted to follow the loss of biodegradable materials during composting exposition as it was previously reported for polystyrene/biopolymers blends [38]. In Figure 9 are shown the TGA and DTG curves of PP-15%PLA (Figure 10a,b), PP-15%PHB (Figure 10c,d) and PP-15%TPS (Figure 10e,f) before and after 30 days exposed to composting conditions. As it was already commented in TGA results, in the case of PP-15%PLA and PP-15%PHB blends the thermal degradation takes place in two-step process in which the first step is related with the loss of the biodegradable material, PLA or PHB, and the second one corresponds to the degradation of the PP. After 30 days in composting the onset degradation temperature of PP-15%PLA blend was considerably reduced (Figure 10a), since the disintegration of PLA is taking place and thus there are shorter PLA chains, such as oligomers, which present lower thermal stability [13], which degrade faster than longer PLA polymeric chains. Similarly, the maximum degradation temperature corresponding to PLA (T_{max}PLA) at about 325 °C was shifted to 289 °C after 30 days (Figure 10b). The second maximum degradation temperature of PP-15%PLA was about 472 °C before composting, while after 30 days in composting it was shifted towards lower temperatures (462 °C) approaching to that of PP because there is less PLA impurities in PP matrix at this composting stage. In the case of PP-15%PHB (Figure 10c) the onset degradation temperature was shifted to higher temperatures, since PHB has lower thermal stability than PP. The maximum degradation temperature of PHB (T_{max}PHB = 257 °C) was shifted to higher values during composting reaching 270 °C after 30 days (Figure 10d), because of the blend behaves more similarly to PP while it loss the PHB. Similarly, the second maximum degradation was shifted from 430 °C to 442 °C during composting. Finally, for PP-15%TPS the thermal degradation take place in only one step and the onset degradation temperature was around 346 °C in PP-15%TPS blends (Figure 10e) and this value was maintained after 30 days in composting. Nevertheless, the shoulder observed just before the maximum degradation temperature corresponds to the cleavage of ether linkages in starch backbone of TPS [47] and it was slightly reduced (Figure 10f). Meanwhile, there was a second peak at higher temperatures, at about 450 °C in Figure 10f, which was shifted to 465 °C after 30 days in composting. The displacements of the maximum degradation temperatures were more pronounced in the case of PP-PLA and PP-TPS blends, since there were less amount of biodegradable material in both formulations after the exposition to the composting medium. Meanwhile, the more crystalline PHB was less disintegrated at this stage of disintegration. However, it should be mentioned that in all cases still remains biodegradable polymer in the formulations, showing that PP can limit the exposure of biodegradable polymers to the composting degradation suggesting that there is somewhat positive interface interaction between PP and biodegradable materials as it was observed in TGA result (Section 3.3.1) and/or the blend separation is forcing a discontinuous disintegration, and thus, biodegradable materials are less available for the hydrolysis and the further microorganisms attack in the composting medium.

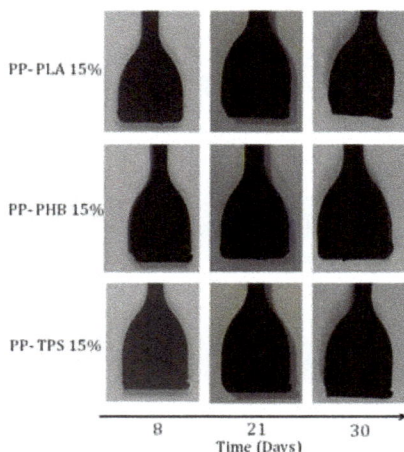

Figure 9. Visual appearance of recovered blend samples at different times of composting (8, 21 and 30 days).

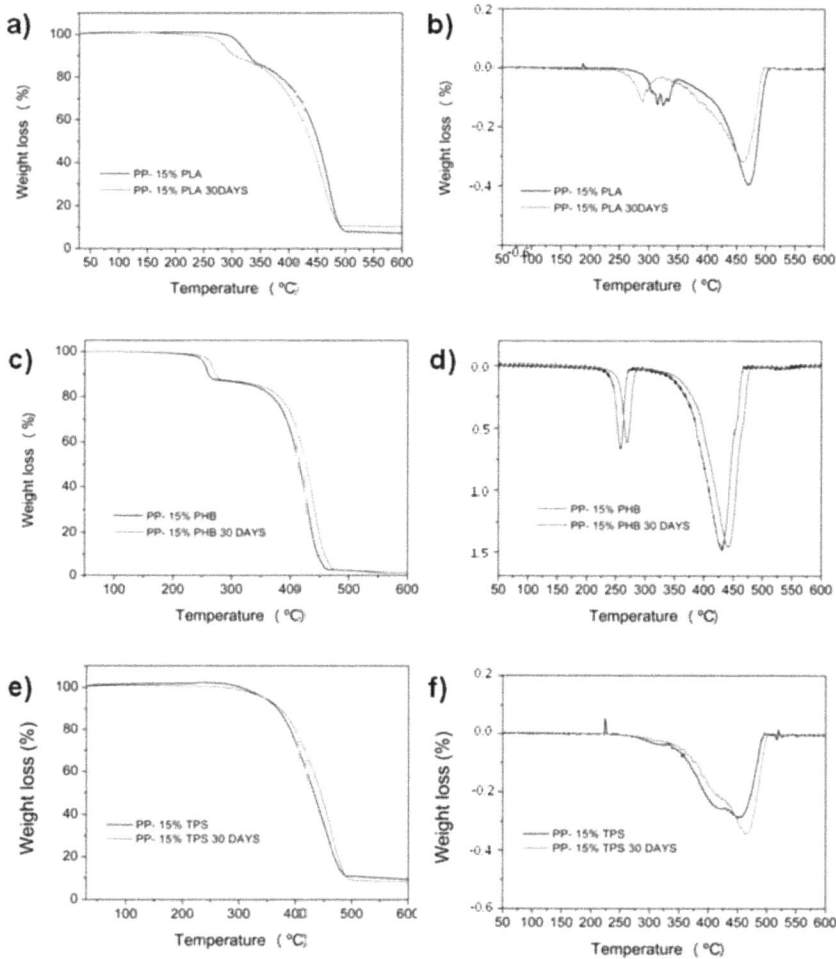

Figure 10. TGA (a,c,e) and DTG (b,d,f) thermograms of PP- biodegradable blends before and after 30 days exposed to a composting medium.

4. Conclusions

In this study, the microstructural, thermal and mechanical properties of blends based on recycled PP with different biodegradable polymers (PLA, PHB and TPS) as impurities were evaluated. The presence of biodegradable polymers in recycled PP caused a significant loss of mechanical, thermomechanical as well as thermal properties, especially when using percentages of biodegradable polymers higher than 5 wt.%. In addition, the effect of the presence of the biodegradable polymers resulted in evident features seen in SEM images, where the immiscibility of the blends was clearly observed by the presence of two separated phases. The exposition of PP-based blends to composting medium showed that although the PP-biodegradable polymer blends were mainly immiscible, they had somewhat positive interactions with PP matrix, since biodegradable polymers delay their disintegration process. As the thermal and mechanical properties of the recycled PP are affected by the presence of more than 5wt.% of PLA, PHB and TPS biodegradable polymers, it is very important to be able to detect biodegradable materials in PP recycling process. The FTIR technique allowed to easily detect the presence of biodegradable polymers in the recycled PP by the appearance of the –C=O characteristic band of PLA, PHB and TPS between 1700–1800 cm^{-1}. Meanwhile, TGA results and effective technique to quantify the presence of biopolyesters PLA and PHB in the recycled PP. Thus, the use of these techniques can help to detect and even quantify the contaminated part of the PP recycling chain with biopolymers, being very important for the PP-based materials final applications and/or to further eliminate the presence of impurities in recycled PP.

Author Contributions: J.L.-M. designed the experiments; D.B. collected mechanical results data; M.P.A. made the exposition to composting medium; M.D.S. calculated values of the solubility parameters; J.M.F. and J.L.-M. collected other experimental data; M.D.S wrote the paper with M.P.A.

Funding: This research was funded by Conselleria d'Educació, Investigació, Cultura y Esport de la Generalitat Valenciana, grant number APOSTD/2018/209.

Acknowledgments: This work has been supported by the Spanish Ministry of Economy and Competitiveness, PROMADEPCOL (MAT2017-84909-C2-2-R).

Conflicts of Interest: The authors declare no conflict of interest. The founding sponsors had no role in the design of the study; in the collection, analyses, or interpretation of data; in the writing of the manuscript, and in the decision to publish the results.

References

1. Plastics Europe, Plastics—The Facts 2017. Available online: https://www.plasticseurope.org/application/files/5715/1717/4180/Plastics_the_facts_2017_FINAL_for_website_one_page.pdf (accessed on 25 September 2018).
2. Ares, A.; Bouza, R.; Pardo, S.G.; Abad, M.J.; Barral, L. Rheological, Mechanical and Thermal Behaviour of Wood Polymer Composites Based on Recycled Polypropylene. *J. Polym. Environ.* **2010**, *18*, 318–325. [CrossRef]
3. Zadeh, K.M.; Ponnamma, D.; Al-Maadeed, M.A.A. Date palm fibre filled recycled ternary polymer blend composites with enhanced flame retardancy. *Polym. Test.* **2017**, *61*, 341–348. [CrossRef]
4. Bodar, C.; Spijker, J.; Lijzen, J.; Waaijers-van der Loop, S.; Luit, R.; Heugens, E.; Janssen, M.; Wassenaar, P.; Traas, T. Risk management of hazardous substances in a circular economy. *J. Environ. Manag.* **2018**, *212*, 108–114. [CrossRef] [PubMed]
5. Alam, O.; Wang, S.; Lu, W. Heavy metals dispersion during thermal treatment of plastic bags and its recovery. *J. Environ. Manag.* **2018**, *212*, 367–374. [CrossRef] [PubMed]
6. Bucci, D.Z.; Tavares, L.B.B.; Sell, I. PHB packaging for the storage of food products. *Polym. Test.* **2005**, *24*, 564–571. [CrossRef]
7. Siracusa, V.; Rocculi, P.; Romani, S.; Dalla Rosa, M. Biodegradable polymers for food packaging: A review. *Trends Food Sci. Technol.* **2008**, *19*, 634–643. [CrossRef]
8. Claro, P.; Neto, A.; Bibbo, A.; Mattoso, L.; Bastos, M.; Marconcini, J. Biodegradable blends with potential use in packaging: A comparison of PLA/chitosan and PLA/cellulose acetate films. *J. Polym. Environ.* **2016**, *24*, 363–371. [CrossRef]
9. Bastioli, C. Bio plastics for a new economy. *Chim. Oggi-Chem. Today* **2008**, *26*, 30–31.
10. Bastioli, C. Unlocking the potential of added value products in Europe: an Italian perspective. *Chim. Oggi-Chem. Today* **2013**, *31*, 64–67.
11. Averous, L. Biodegradable multiphase systems based on plasticized starch: A review. *J. Macromol. Sci. Polym. Rev.* **2004**, *C44*, 231–274. [CrossRef]
12. Armentano, I.; Fortunati, E.; Burgos, N.; Dominici, F.; Luzi, F.; Fiori, S.; Jimenez, A.; Yoon, K.; Ahn, J.; Kang, S.; Kenny, J.M. Processing and characterization of plasticized PLA/PHB blends for biodegradable multiphase systems. *Express Polym. Lett.* **2015**, *9*, 583–596. [CrossRef]
13. Arrieta, M.P.; Lopez, J.; Rayon, E.; Jimenez, A. Disintegrability under composting conditions of plasticized PLA-PHB blends. *Polym. Degrad. Stab.* **2014**, *108*, 307–318. [CrossRef]
14. Garcia-Garcia, D.; Ferri, J.M.; Montanes, N.; Lopez-Martinez, J.; Balart, R. Plasticization effects of epoxidized vegetable oils on mechanical properties of poly (3-hydroxybutyrate). *Polym. Int.* **2016**, *65*, 1157–1164. [CrossRef]

15. Russo, M.A.L.; O'Sullivan, C.; Rounsefell, B.; Halley, P.J.; Truss, R.; Clarke, W.P. The anaerobic degradability of thermoplastic starch: Polyvinyl alcohol blends: Potential biodegradable food packaging materials. *Bioresour. Technol.* **2009**, *100*, 1705–1710. [CrossRef] [PubMed]

16. Neumann, I.A.; Sydenstricker Flores-Sahagun, T.H.; Ribeiro, A.M. Biodegradable poly (L-lactic acid) (PLLA) and PLLA-3-arm blend membranes: The use of PLLA-3-arm as a plasticizer. *Polym. Test.* **2017**, *60*, 84–93. [CrossRef]

17. Peterson, G.I.; Dobrynin, A.V.; Becker, M.L. Biodegradable Shape Memory Polymers in Medicine. *Adv. Healthc. Mater.* **2017**, *6*, 1700694. [CrossRef] [PubMed]

18. Ray, S.; Kalia, V.C. Biomedical Applications of Polyhydroxyalkanoates. *Indian J. Microbiol.* **2017**, *57*, 261–269. [CrossRef] [PubMed]

19. Khalid, S.; Yu, L.; Meng, L.; Liu, H.; Ali, A.; Chen, L. Poly(lactic acid)/starch composites: Effect of microstructure and morphology of starch granules on performance. *J. Appl. Polym. Sci.* **2017**, *134*, 45504. [CrossRef]

20. Rubio-Lopez, A.; Artero-Guerrero, J.; Pernas-Sanchez, J.; Santiuste, C. Compression after impact of flax/PLA biodegradable composites. *Polym. Test.* **2017**, *59*, 127–135. [CrossRef]

21. Arrieta, M.P.; Samper, M.D.; Aldas, M.; López, J. On the use of PLA-PHB blends for sustainable food packaging applications. *Materials* **2017**, *10*, 1008. [CrossRef] [PubMed]

22. Cosate de Andrade, M.F.; Souza, P.M.S.; Cavalett, O.; Morales, A.R. Life Cycle Assessment of Poly(Lactic Acid) (PLA): Comparison Between Chemical Recycling, Mechanical Recycling and Composting. *J. Polym. Environ.* **2016**, *24*, 372–384. [CrossRef]

23. Navarro, R.; Ferrandiz, S.; Lopez, J.; Segui, V.J. The influence of polyethylene in the mechanical recycling of polyethylene terephtalate. *J. Mater. Proc. Technol.* **2008**, *195*, 110–116. [CrossRef]

24. Navarro, R.; Lopez, J.; Parres, F.; Ferrandiz, S. Process behavior of compatible polymer blends. *J. Appl. Polym. Sci.* **2012**, *124*, 2485–2493. [CrossRef]

25. Sanchez-Jimenez, P.E.; Perez-Maqueda, L.A.; Crespo-Amoros, J.E.; Lopez, J.; Perejon, A.; Criado, J.M. Quantitative Characterization of Multicomponent Polymers by Sample-Controlled Thermal Analysis. *Anal. Chem.* **2010**, *82*, 8875–8880. [CrossRef] [PubMed]

26. Alaerts, L.; Augustinus, M.; Van Acker, K. Impact of Bio-Based Plastics on Current Recycling of Plastics. *Sustainability* **2018**, *10*, 1487. [CrossRef]

27. Pivsa-Art, S.; Kord-Sa-Ard, J.; Pivsa-Art, W.; Wongpajan, R.; O-Charoen, N.; Pavasupree, S.; Hamada, H. Effect of Compatibilizer on PLA/PP Blend for Injection Molding. *Energy Procedia* **2016**, *89*, 353–360. [CrossRef]

28. Ploypetchara, N.; Suppakul, P.; Atong, D.; Pechyen, C. Blend of polypropylene/poly (lactic acid) for medical packaging application: physicochemical, thermal, mechanical, and barrier properties. *Energy Procedia* **2014**, *56*, 201–210. [CrossRef]

29. Yoo, T.W.; Yoon, H.G.; Choi, S.J.; Kim, M.S.; Kim, Y.H.; Kim, W.N. Effects of compatibilizers on the mechanical properties and interfacial tension of polypropylene and poly (lactic acid) blends. *Macromol. Res.* **2010**, *18*, 583–588. [CrossRef]

30. Kaseem, M.; Hamad, K.; Deri, F. Rheological and mechanical properties of polypropylene/thermoplastic starch blend. *Polym. Bull.* **2012**, *68*, 1079–1091. [CrossRef]

31. Rosa, D.; Guedes, C.; Carvalho, C. Processing and thermal, mechanical and morphological characterization of post-consumer polyolefins/thermoplastic starch blends. *J. Mater. Sci.* **2007**, *42*, 551–557. [CrossRef]

32. Sadi, R.K.; Kurusu, R.S.; Fechine, G.J.M.; Demarquette, N.R. Compatibilization of polypropylene/poly(3-hydroxybutyrate) blends. *J. Appl. Polym. Sci.* **2012**, *123*, 3511–3519. [CrossRef]

33. Parres, F.; Balart, R.; Lopez, J.; Garcia, D. Changes in the mechanical and thermal properties of high impact polystyrene (HIPS) in the presence of low polypropylene (PP) contents. *J. Mater. Sci.* **2008**, *43*, 3203–3209. [CrossRef]

34. Utracki, L.A.; Mukhopadhyay, P.; Gupta, R. Polymer blends: introduction. In *Polymer Blends Handbook*; Springer: Berlin, Germany, 2014; pp. 3–170.

35. Arrieta, M.P.; Fortunati, E.; Burgos, N.; Peltzer, M.A.; López, J.; Peponi, L. Chapter 7—Nanocellulose-Based Polymeric Blends for Food Packaging Applications. In *Multifunctional Polymeric Nanocomposites Based on Cellulosic Reinforcement*; Puglia, D., Fortunati, E., Kenny, J.M., Eds.; William Andrew Publishing: Norwich, NY, USA, 2016; pp. 205–252.

36. Fekete, E.; Foldes, E.; Pukanszky, M. Effect of molecular interactions on the miscibility and structure of polymer blends. *Eur. Polym. J.* **2005**, *41*, 727–736. [CrossRef]

37. MacAubas, P.H.P.; Demarquette, N.R. Time-temperature superposition principle applicability for blends formed of immiscible polymers. *Polym. Eng. Sci.* **2002**, *42*, 1509–1519. [CrossRef]

38. Samper, M.D.; Arrieta, M.P.; Ferrándiz, S.; López, J. Influence of biodegradable materials in the recycled polystyrene. *J. Appl. Polym. Sci.* **2014**, *131*, 41161.

39. *ISO-20200: U.U.-E. Determination of the Degree of Disintegration of Plastic Materials under Simulated Composting Conditions in a Laboratory-Scale Test*; International Organization for Standardization: Geneva, Switzerland, 2015.

40. Ferrandiz, S.; Arrieta, M.P.; Samper, M.D.; Lopez, J. Prediction of properties value in thermoplastic mixtures applying box equivalent model incompatibility in recycled polymer blends. *J. Optoelectron. Adv. Mater.* **2013**, *15*, 662–666.

41. Odelius, K.; Ohlson, M.; Hoeglund, A.; Albertsson, A.-C. Polyesters with small structural variations improve the mechanical properties of polylactide. *J. Appl. Polym. Sci.* **2013**, *127*, 27–33. [CrossRef]

42. Polymer Properties Database. Available online: https://polymerdatabase.com/polymer%20classes/Intro. html (accessed on 25 September 2018).

43. Goonoo, N.; Bhaw-Luximon, A.; Jhurry, D. Biodegradable polymer blends: miscibility, physicochemical properties and biological response of scaffolds. *Polym. Int.* **2015**, *64*, 1289–1302. [CrossRef]

44. Sundararaj, U.; Ghodgaonkar, P. Effect of compatibilization in immiscible polymer blends. *Abstr. Papers Am. Chem. Soc.* **1996**, *212*, 277.

45. Arrieta, M.P.; López, J.; López, D.; Kenny, J.; Peponi, L. Development of flexible materials based on plasticized electrospun PLA–PHB blends: Structural, thermal, mechanical and disintegration properties. *Eur. Polym. J.* **2015**, *73*, 433–446. [CrossRef]

46. Ferri, J.M.; Garcia-Garcia, D.; Carbonell-Verdu, A.; Fenollar, O.; Balart, R. Poly(lactic acid) formulations with improved toughness by physical blending with thermoplastic starch. *J. Appl. Polym. Sci.* **2018**, *135*, 45751. [CrossRef]

47. Sessini, V.; Arrieta, M.P.; Kenny, J.M.; Peponi, L. Processing of edible films based on nanoreinforced gelatinized starch. *Polym. Degrad. Stab.* **2016**, *132*, 157–168. [CrossRef]

48. Hiemenz, P.C.; Lodge, T.P. Linear Viscosity. In *Polymer Chemistry*, 2nd ed.; CRC Press: Boca Raton, FL, USA, 2007; pp. 419–464.

49. Wang, L.; Gramlich, W.M.; Gardner, D.J.; Han, Y.; Tajvidi, M. Spray-Dried Cellulose Nanofibril-Reinforced Polypropylene Composites for Extrusion-Based Additive Manufacturing: Nonisothermal Crystallization Kinetics and Thermal Expansion. *J. Compos. Sci.* **2018**, *2*, 7. [CrossRef]

50. Fan, Y.; Nishida, H.; Shirai, Y.; Tokiwa, Y.; Endo, T. Thermal degradation behaviour of poly(lactic acid) stereocomplex. *Polym. Degrad. Stab.* **2004**, *86*, 197–208. [CrossRef]

51. Sessini, V.; Raquez, J.-M.; Lourdin, D.; Maigret, J.E.; Kenny, J.M.; Dubois, P.; Peponi, L. Humidity-Activated Shape Memory Effects on Thermoplastic Starch/EVA Blends and Their Compatibilized Nanocomposites. *Macromol. Chem. Phys.* **2017**, *218*, 1700388. [CrossRef]

52. Gerard, T.; Budtova, T.; Podshivalov, A.; Bronnikov, S. Polylactide/poly(hydroxybutyrate-co-hydroxyvalerate) blends: Morphology and mechanical properties. *Express Polym. Lett.* **2014**, *8*, 609–617. [CrossRef]

53. Lanzotti, A.; Grasso, M.; Staiano, G.; Martorelli, M. The impact of process parameters on mechanical properties of parts fabricated in PLA with an open-source 3-D printer. *Rapid Prototyp. J.* **2015**, *21*, 604–617. [CrossRef]

54. Arrieta, M.P.; López, J.; Hernández, A.; Rayón, E. Ternary PLA–PHB–Limonene blends intended for biodegradable food packaging applications. *Eur. Polym. J.* **2014**, *50*, 255–270. [CrossRef]

55. Du, Y.-L.; Cao, Y.; Lu, F.; Li, F.; Cao, Y.; Wang, X.-L.; Wang, Y.-Z. Biodegradation behaviors of thermoplastic starch (TPS) and thermoplastic dialdehyde starch (TPDAS) under controlled composting conditions. *Polym. Test.* **2008**, *27*, 924–930. [CrossRef]

materials

MDPI

Review

Bio- and Fossil-Based Polymeric Blends and Nanocomposites for Packaging: Structure–Property Relationship

Francesca Luzi, Luigi Torre, José Maria Kenny and Debora Puglia *

Civil and Environmental Engineering Department, University of Perugia, UdR INSTM, Strada di Pentima 4, 05100 Terni, Italy; francesca.luzi@unipg.it (F.L.); luigi.torre@unipg.it (L.T.); jose.kenny@unipg.it (J.M.K.)
* Correspondence: debora.puglia@unipg.it; Tel.: +39-0744-492916

Received: 27 December 2018; Accepted: 29 January 2019; Published: 3 February 2019

Abstract: In the present review, the possibilities for blending of commodities and bio-based and/or biodegradable polymers for packaging purposes has been considered, limiting the analysis to this class of materials without considering blends where both components have a bio-based composition or origin. The production of blends with synthetic polymeric materials is among the strategies to modulate the main characteristics of biodegradable polymeric materials, altering disintegrability rates and decreasing the final cost of different products. Special emphasis has been given to blends functional behavior in the frame of packaging application (compostability, gas/water/light barrier properties, migration, antioxidant performance). In addition, to better analyze the presence of nanosized ingredients on the overall behavior of a nanocomposite system composed of synthetic polymers, combined with biodegradable and/or bio-based plastics, the nature and effect of the inclusion of bio-based nanofillers has been investigated.

Keywords: bio-based; fossil; hybrids; blends; packaging

---

1. Introduction

In the last decades, after the signing of the environmental treaty on the 11th December 1997, that became law on the 16th February 2005, the environmental issue related to greenhouse gas emissions and climate changes was definitively raised and made one of the most important worldwide concerns, particularly in the face of toxic waste, pollution, contamination, exhaustion of natural resources, and environmental deterioration. As a result, different studies have been followed at diverse stages to develop different and strategic alternatives [1]. A serious alarm is the field of packaging, which, every year, causes enormous quantities of petroleum-based wastes that are stored in particular areas around the planet (26% of the plastic manufacture volume has been applied in the packaging sector) [2], determining enormous negative effects and high recycling costs [3]. In the future years, especially in 2030 and 2050, it was estimated that the quantity of plastic wastes due to the packaging sector will be grown by two-fold and three-fold, respectively [4]. Notwithstanding the environmental impacts, plastic materials are extremely useful in the packaging sector, due to positive and synergic combination of main characteristics, such as transparency, strength ability, flexibility, thermal performance, permeability, and simple sterilization methods, all of which making them appropriate for the food packaging sector. Hitherto, petroleum-based polymers (i.e., ethylene vinyl alcohol (EVOH), polypropylene (PP), polyethylene (PE), polyurethane (PU), poly (ethylene terephthalate) (PET), polystyrene (PS), expanded polystyrene, polyamides (PA), and poly (vinyl chloride) (PVC)) have led in the packaging function, for excellent mechanical and physical characteristics. However, according to results from plastics recycling and recovery data obtained from European associations in 2014, only 39.5% of post-consumer plastic waste is going to be re-used, while 38.6% of post-consumer plastic waste is considered for energy recovery [5]. With the intention of minimizing the environmental impacts induced by post-consumer plastic waste, bio-based polymers should be selected to realize short-lifespan devices. Environmentally friendly systems are appropriate solutions to realize disposable systems [6,7]; on the other hand, green polymeric systems are only used for some specific applications, due to their limiting characteristics, such as high cost and scarce mechanical and thermomechanical properties with respect of traditional commodity polymers. Developing green and eco-friendly polymeric blends with acceptable characteristics can overcome these limitations, even if it has been recently demonstrated that biodegradable plastic blends need both accurate cautious postconsumer organization and additional design to consent fast biodegradation in numerous environment conditions (as their release into the environment can determine plastic pollution) [8].

Preparation of blends with synthetic polymers [9–11] is among the options to enhance some characteristics of biodegradable polymers, changing degradation rates and modulating the cost of the obtained materials; polymer blends, particularly olefins with biodegradable polymers, are gaining popularity as an approach for degradable packaging plastics, since the partial loss of form and bulk during disintegration may be sufficient to decrease the volume in landfill [12]. This blending approach began in the 1970s at the U.S.D.A. with Otey [13], who studied and investigated blends based on starch and ethylene/acrylic acid copolymers and still now starch, being cheap, continues to be an attractive substitute to realize systems for the packaging sector [14]. In addition, to expand the spectrum of sustainability incorporating resources and practices that move a step closer toward sustainability [15], growing the renewable amount or lessening the overall weight of petroleum-based plastics have been considered as suitable options. Today's sustainable plastics are not automatically biodegradable and even contain polyolefins made from renewable feedstocks [16,17].

In the present review, the blending of commodities and bio-based and/or biodegradable polymers will be taken into account (limiting the study to this class of materials and not considering blends where both components have a bio-based composition or origin), and special emphasis will be given to their functional behavior in terms of packaging application (compostability, gas/water/light barrier properties, migration, antioxidant performance). In addition, to better analyze the effect of green nanosized ingredients on the overall behavior of systems composed of synthetic polymers, combined with biodegradable and/or bio-based plastics, the effect of the inclusion of bio-based nanofillers has been investigated.

2. Bio-Based Nanofillers in the Packaging Sector

Recently, the growth of nanotechnology approaches and strategies has made their use become of interest in several sectors. Automotive, aerospace, biomedical, and packaging sectors have adopted and largely investigated the use of nanotechnology applications, as valid strategies to modulate and improve the characteristic main properties required in specific sectors [18]. Nanotechnology allows the realization of new systems to enhance material performances; of particular note is the recent development of nanocomposite systems that permitted the advancement of new polymeric-based formulations, with enhanced structural and functional properties (thermal, electrical, mechanical, and numerous other characteristics, in respect to the neat polymers [19–23]. Different nanocomposite-based systems have been realized by combining different polymers (petroleum-based and biodegradable/bio-based), and fillers at the nanoscale level. The nanofillers show strong reinforcing effects, several works have also analyzed their positive behavior in terms of barrier and mechanical properties, characteristics of essential importance in packaging and food packaging applications [24,25].

In this review, the current status of nanotechnology in packaging and also in food packaging systems are briefly reviewed and summarized. Nanofillers can be extracted from organic or inorganic sources; here the authors focused their attention to describe the main characteristics of nanofillers extracted from bio-based/natural sources (plant and animal origin and nanofillers from proteins) applied in the packaging sector. An explanation of different nanofillers, with an emphasis on the functionality, synthesis, characteristics, and structure is included.

2.1. Nanofillers from Polysaccharides—Plant Origin

In literature, different works have proposed the study of extraction and analysis of nanofillers from polysaccharides with a plant origin: cellulose nanofibers/nanocrystals, lignin, and starch nanoparticles. The lignocellulosic source is one of the most copious renewable materials existing in the world; these materials are natural, eco-friendly, sustainable, biodegradable, and considered as low-cost materials, with advantageous properties and with a significant value for packaging and industrial sectors. In comparison with petroleum-based natural sources, some interesting advantages are found: (i) low density and low cost, (ii) high variety, (iii) specific modulus and strength, (iv) reactive surfaces that can be changed and functionalized by a large variety of reactive chemical groups, (v) high applicability in nanocomposites, (vi) high recyclability in respect to inorganic fillers [25,26]. Lignocellulosic materials are generally composed by cellulose (40–50 wt %), hemicellulose (20–30 wt %), and lignin (about 10–25 wt %), and the quantities of the different components can be different according to the native lignocellulosic origin source [27].

2.1.1. Cellulose Nanofibers/Nanocrystals

Cellulose is the natural polymer largely diffused on Earth, with excellent biocompatibility, good chemical and thermal stability, and high hydrophilicity. These attractive characteristics have determined cellulose as an interesting material for different applications in packaging and in biomedical applications. The cellulosic nanofillers are categorized on the basis of preparation methods considered for their extraction from native cellulose; they can be found as bacterial cellulose (BC) synthesized through microorganisms, microfibrillated cellulose (MFC) or nanofibrillated cellulose (NFC), or cellulose nanocrystals/nanocrystalline cellulose, also named cellulose nanowhiskers, (CNC). MCF is pulled out by means of a mechanical retting/disintegration method, starting from a variety of cellulosic extracts, including wood and non-wood fibers [28], consequently obtaining cellulose microfibrils with a three-dimensional network, the obtained structures showed higher surface area than original cellulosic fibers or from cellulosic powder. This effect influences a number of extremely interesting characteristics, such as an exceptionally high-water holding capability and the capacity to realize a configuration with strong gels at low concentrations. Although microfibrillated cellulose is not soluble in water, it can show several characteristics of water-soluble cellulose derivatives. Simultaneously, it has some advantages, such as stability over the whole pH range, at elevated temperatures and at elevated salt concentrations [29]. MFC show lengths in micrometers and diameters in nanometers, characterizing them as long and thin reinforcements. This elevated aspect ratio characterizes the material high strength as functional in several applications, such as the reinforcement phase for composites and films, and as an agent to modulate the barrier performance. Chemically extracted CNC are characterized by acicular structure and rigid rod-like particles, monocrystalline domains of 100 to hundreds of nanometers in length, and 1–100 nm in diameter (Figure 1, Panel A a)) [30,31]; morphology and crystallinity degree depend, fundamentally, on the native source and the different parameters used for the extraction process [27]. The extraordinary mechanical characteristics (Young's modulus is higher than glass fibers and comparable to Kevlar (60–125 GPa)) give to cellulose crystal the role of a perfect filler material for the preparation of polymer composites. CNCs exhibit enormous applications in the biomedical sector and in bio-based material science [32,33].

BIOBASED NANOFILLERS

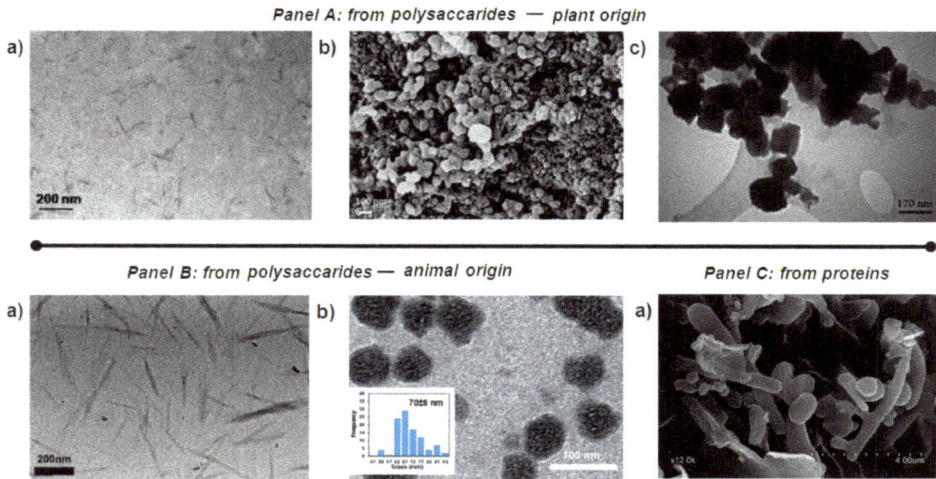

Panel A: from polysaccharides — plant origin

Panel B: from polysaccharides — animal origin   Panel C: from proteins

Figure 1. Morphological characterization of bio-based nanofillers. Panel A: Polysaccharides—plant origin: (**a**) Transmission Electron Microscopy (TEM) image of Cellulose Nanocrystals CNC [27]; (**b**) Field Emission Scanning Electron Microscopy (FESEM) image of lignin nanoparticles [34]; (**c**) TEM image of starch nanoparticles [35]. Panel B: Nanofillers from Polysaccharides—animal origin: (**a**) TEM image of Chitin nanocrystals [36], (**b**) TEM image of modified chitosan nanoparticles (CSNP) by poly (ethylene glycol) methyl ether methacrylate (PEGMA) (PEGMA-graft-CSNP) [37]. Panel C: From proteins: (**a**) Scanning Electron Microscopy (SEM) image of nanokeratin [38].

Cellulose nanocrystals were largely used and applied to realize nanocomposites with modulated properties in respect to the neat matrices. Several studies, reported in literature, analyzed their effect in biodegradable matrices and polymeric blends, even in the presence of natural active ingredients and antimicrobial nanoparticles [24,39–41]. Cellulose nanocrystals are also used as the reinforcement phase in conventional matrices [42–45], by also providing, at the same time, an enhancement in terms of barrier properties.

2.1.2. Lignin Nanoparticles

Lignin is the second most abundant aromatic polymeric material on earth. It is a cross-linked macromolecule in a three-dimensional shape, composed by three typologies of alternative phenols, yielding numerous functional groups and linkages, with a mutable chemistry as a consequence of its native source [34,46,47]. Lignin is an efficient phase to be included in polymers. Works on the applicability of micro-lignin in thermally processable plastic matrices, elastomers, and thermoset-based systems have been recently investigated [48–50]. Figure 1, Panel A, (b) shows the typical morphological aspect of lignin nanoparticles (LNP). LNP diameters are distributed in the range from 30 to 90 nm [47]. Recently, LNPs from a variety of native origins were synthesized/extracted by applying physicochemical procedures [34,51,52]. Furthermore, the lignin tendency to self-aggregate shows some disadvantages in terms of their dispersion in thermoplastic, thermoset, or elastomer-based systems. Therefore, many strategies have been considered and attempted in order to improve the dispersion of lignin particles into bio-based polymers [51].

2.1.3. Starch Nanoparticles

Starch is composed of amylose and amylopectin. The amylose is a linear and long molecule built up of 1,4-linked β-D-glucose, although in the amylopectin chains the glucose monomers are linked through α-1,6-linkages, determining an extremely branched arrangement. Therefore, the molecular structure of amylose is simpler than amylopectin, showing a linear structure with few α-1,6-branches [53]. Amylopectin generally is the main component of starch, composed by short chains and a high number of α-1,6-branches (5% of the molecule) [53], while amylose, in general, is randomly arranged among the amylopectin molecules in the amorphous regions. Amylose plays an important part on the structure of the amylopectin in the crystalline lamellae by cross-linking the two polysaccharides [54].

Starch consists in granules, with diameters ranging from 2 to 100 μm. In relation to their native extraction starch, they have different characteristics properties, different chemical composition, different shape and size [21], that can be small (3.1–3.7 nm) and large (15–19 nm), and disc-shaped and/or spheroidal [55].

The purification of starch granules with appropriate chemical treatments allow one to obtain nanoparticles (SNP). In general, many different approaches can be selected and applied to starch with the intention to obtain granules, while their conversion in NPs is usually carried out by applying acid hydrolysis. The structural differences in starch granule sizes affects the outcome of the process in terms of starch purification, characteristics, and nanocrystal yield [19,56]. The acidic hydrolysis process permits one to obtain crystals with a spherical shape and dimensions ranging 20–50 nm [19,56].

2.2. Nanofillers from Polysaccharides—Animal Origin

Chitin and chitosan nanofillers can be isolated from α-chitin powder extracted from lobster wastes. The extraction of chitin and chitosan represents the possibility to revalorize the oceanic biomass and the revalorization of food extracts from the fishing industry. Lobster wastes are eco-friendly, renewable, sustainable, low cost, and biodegradable, with advantageous properties and with a significant value for packaging and biomedical applications.

2.2.1. Chitin Nanoparticles/Nanofibers

Chitin is one of the most copious natural polymers obtained from shellfish waste (exoskeleton/shells), and it is used in combination with biopolymers to realize nanocomposites with modulated properties, in respect to the neat matrices, principally for food packaging and biomedical applications. Chitin is extracted from cuticles of insects and exoskeleton of arthropods at the micro/nanoscale (length: 200–300 nm; diameter: 10–20 nm) [57]. Different treatments can be used to extract crystalline chitin in nanosized fibrils, the different methods influenced also the dimension and the morphology of extracted materials [58–60]. The isolation of these nanosized structures can be performed by: (i) mechanical treatments/disintegration [57,61], resulting in chitin nanofibers (CHNF) and fibrils with high aspect ratio; (ii) acidic treatments [60–62], resulting in chitin nanocrystals (CHNC), with higher crystallinity degree in respect to CHNF and with a rod-like appearance (Figure 1, Panel B, (a)), [57].

Nanochitin has received a crucial position in nanocomposite materials as the filler phase, due to its intrinsic properties [63]. The physiochemical and biological properties (light weight, small size, natural and biodegradable character, chemical stability, and non-cytotoxicity) of chitin at nanoscale dimensions make this material a valid candidate for utilization in food packaging and biomedical sectors [63], especially thanks to its high antibacterial effect [58,64,65]. Chitin is characterized by antimicrobial activity, this character is related to its chemical organization consisting of (1,4)-b-N-acetyl-D-glucosamine-replicating units. Likewise to starch and cellulose nano-fillers, a broad variety of nanocomposites exploited the interesting mechanical characteristics of nanochitin in polymeric-based systems [57,66–69].

Salaberria and co-authors [57] analyzed the effect of chitin nanocrystals (CHNC) and nanofibers (CHNF) at 5 and 20 wt % in thermoplastic starch matrices. The authors observed that the improvement of the final characteristics of the nanocomposites (superior barrier, thermal, mechanical, and antifungal characteristics) depended essentially on the morphological characteristics of the nanofillers used as the reinforcement phase.

The chitin whiskers were found to enhance the water resistance and tensile strength of the neat polymer when assembled to soy protein [70]. Likewise, when chitosan whiskers were combined with chitosan films, it was noted that whiskers improved water resistance and tensile strength of the chitosan films [71]; however, when included in hydroxypropyl and carboxy methylcellulose, were capable of enhancing the mechanical and barrier functions of the films [72,73]. Research has also shown that nanosized chitin can improve barrier properties when embedded in a polymer matrix, such as starch or PVA [74,75].

2.2.2. Chitosan Nanoparticles

Nanochitosan is a green extract with exceptional physicochemical characteristics; it is characterized by bioactivity that does not damage humans [76]. It is largely utilized as a controlled release drug carrier, and, for gene transfer, Nanochitosan has been applied to realize systems with improved strength and wash ability of textiles, conferring antimicrobial effects [77]. Chitosan at the nanoscale can be realized considering precipitation or coagulation, ionic cross-linking, emulsion droplet coalescence, and covalent cross-linking procedure. Berthold and co-authors [78] obtained chitosan particles by considering sodium sulfate as a precipitation agent and incorporating a dispersant (Tween 80) to the chitosan acidic solution. Tian and Groves and co-authors [79] enhanced this procedure and obtained chitosan nanoparticles (CSNPs, 600–800 nm). Ohya and co-authors [80] proposed the use of glutaraldehyde to cross-link the free amino groups of chitosan; this was a water-in-oil (W=O) emulsifier, realizing 5-fluorouracil (5-FU) chitosan particles (size: 0.8–0.1 mm).

Kongkaoroptham and co-authors [37] proposed the variation of chitosan (CS) poly(ethylene glycol) methyl ether methacrylate (PEGMA) (Figure 1, Panel B, (b)), prepared by radiation-induced graft copolymerization, as a new compatible bio-based filler at the nanoscale level. The nanoparticle dimensions of PEGMA-graft-CSNPs were distributed from 30 to 100 nm. The strategy was studied to improve the compatibility, the mechanical, and the thermal characteristics of poly(lactic acid) (PLA). The mechanical properties of the PEGMA-graft-CSNP/PLA blends showed an improvement in terms of deformation at break, and a reduction of the tensile modulus brittle behavior to a more ductile behavior. In the case of poly(butylene adipate-co-terephthalate) (PBAT) films, antimicrobial packaging systems, combining different quantities of chitosan nanofibers (CS-NF), have been obtained by applying the solvent casting procedure; the biocomposites had high stiffness, strength, and glass transition temperatures, and low ductility, water vapor, and oxygen permeability. For all the nanocomposites, the migrated quantities in polar and non-polar food simulants were significantly under the overall limits recognized by the present legislation on food contact products. The produced films showed antimicrobial effect towards foodborne pathogens [81].

2.3. Nanofillers from Proteins

Proteins are another class of biomaterial largely investigated as a valid material to modulate functional properties in biomedical and packaging applications. Specifically, in this section, the authors focused their attention on the characteristics of keratin and gelatin.

2.3.1. Nanokeratin

Keratins are natural proteins [82] mostly diffused in poultry feather horns of animals, hair, and wool [83]. Keratin-based biomaterials have been essentially studied to realize hydrogels, films, scaffolds, and dressing, which were applied to get several biomedical disposals, as well as wound healing, cell culture, bone, and nerve regeneration, due to their essential biocompatibility and biodegradability [84,85]. Keratin obtained from human hair have been established in enhancing survivability in multiple animal models of bleeding and the efficacy in arresting hemorrhage [86,87]. Human hair keratin was utilized for the first time to realize hemostatic disposal, as documented in a

Chinese medical book named Ming Y. Bie Lu in the 5th century [83]. In the last years, keratin products extracted from human hair have been used and applied to realize nanoparticles, hydrogel, sponge, and fibers, to develop hemostatic agents [88].

Fabra and co-authors [38] investigated the combination of nanokeratin obtained from poultry feathers, applying chemical treatment with polyhydroxyalkanoate (PHA)-based materials, following different strategies. Nanobiocomposites with high-barrier properties, based on the mixture of PHAs with nanokeratin, showed different morphologies, such as spherical nanoparticles and fibrillar sizes (Figure 1 Panel C, (a)). They were productively designed and realized via both: (i) direct melting technique; and (ii) pre-incorporated into an electrospun masterbatch of PHA, which was then melt compounded with PHA. Improved barrier characteristics for nanocomposites were experientially investigated and were seen to be related to PHA grade. Secondly, nanokeratin films, extracted using the solvent casting technique, were hydrophobized by coating them with electrospun PHA fibers. The multilayered selection was characterized by good adhesion and led to the improvement of the ultimate barrier properties.

2.3.2. Nanogelatin

Gelatin is a biodegradable protein extracted from natural sources by applying acid- or base-catalyzed hydrolysis of collagen. Gelatin is largely used in biomedical application. It is a polyampholyte macro-molecule characterized by the presence of anionic, cationic, and hydrophobic groups [89]. Gelatin molecules have repeating strings of alanine, amino acid triplets, proline, and glycine that influence the triple helical structure of gelatin. The high reliability of gelatin is due to its exceptional triplehelix organization, due to three polypeptide chains [90].

The properties of gelatin nanoparticles (GNPs) can be maximized by applying a particular extraction procedure. GNPs can be extracted by applying: (i) two-step desolvation, the method is characterized by the presence of a desolvating agent in an aqueous gelatin solution, resulting in conformational modification from triple helical coiled macromolecular arrangement to nanoparticles. Kumari and co-authors organized GNPs in a size range of 110–257 nm by this procedure [91]. (ii) Simple coacervation: Mohanty and co-authors [92] successfully obtained stable spherical nanoparticles (45 ± 5 nm) by measured presence of ethanol to aqueous gelatin solution. (iii) Solvent evaporation: this methodology is characterized by a single emulsion, oil-in-water (w/o), or double-emulsion, (water-in-oil)-in-water (w/o)/w, procedure. Water solutions that contain drug and gelatin are incorporated with ultrasonic treatment or high-speed homogenization with the oil phase. The water-in-water mixture method is utilized to produce insulin-loaded GNPs (250 nm) in mild conditions for the activity of insulin [93]. (iv) Microemulsion: It is a new and successful procedure used to prepare gelatin particles. In this process, gelatin in aqueous solution is combined to the solution of surfactant [sodium bis (2-ethylhexyl) sulfosuccinate (AOT)] in n-hexane and then glutaraldehyde (GA) to cross-link the nanostructures, followed by evaporation of n-hexane for recovery of GNPs [94]. The nanostructures had dimensions ranging 5–50 nm. (v) Nanoprecipitation: Implies the precipitation of pre-formed GNPs from an organic solution and the dispersion of the organic solvent in the water solution with prevalence of a surfactant [95–97].

The diverse aspects of gelatin nanostructures influence the particle properties, like polydispersity index, zeta potential, drug release, and entrapment efficacy characteristics [98]. In the field of packaging [99], combination with chitosan has been found: Kumar et al. [100] efficiently made hybrid nanocomposite films realized with chitosan, polyethylene glycol, gelatin, and silver nanoparticles (AgNPs), applying a solvent casting procedure, and the researches on packaging of red grapes underlined that the shelf life of the fruit was prolonged for a supplementary two weeks in case of the hybrid film. Other than blending, gelatin as nanoparticles can find application in the food sector as nanocarrier systems for the controlled delivery of a variety of food supplements and additives.

3. Conventional Matrices for Packaging

The conventional matrices are used in different sectors: packaging, textiles, construction, electronic, transportation, etc. Figure 2 shows the global plastic production worldwide (Panel A) and in Europe (Panel B) in relation to the different application sectors. Packaging is one of the main important sectors that supports the use and application of polymeric materials derived from fossil sources, with an annual global demand at around 36% [1], followed by the building and construction sector, with an annual request of 16%, while the total consumption of plastic in Europe is at around 39.9% [101]. The packaging sector is influenced by the demands requested by various stakeholders (consumers, producers, and retailers), who have specific necessities and do not, each time, recognize the packaging as an additional significance to the product [102]. The conventional characteristics of packaging are centered on the protection of food products from degradation processes (mainly induced by several factors such as temperature, light, moisture, and oxygen conditions of the environment), to enclose the food, and to supply consumers with ingredient and nutritional information and the description of several items of information regarding the conservation of food products [103,104]. In addition, the packaging system needs to improve the shelf life of packaged food, preventing the deterioration and the organoleptic/external (color and esthetic characteristics) qualities of products [19,23]. These concepts have constantly been related with an inert substrate, acting as a barrier substrate between the food and the outside atmosphere, reducing and eliminating the passage of dangerous substances from the packaging to the food [20,24].

Conventional matrices have long been crucial materials in packaging and in food packaging, due to numerous reasons, with their easy processability, low cost and essentially for their mechanical performances. In the first half of the 20th century, thermally processable polymeric matrices were developed, studied, and applied to packaging to restore glass, paper, and metals (foils and laminates, aluminum, tin-free steel, and tinplate) [104].

The petroleum-based matrices were developed and studied from the first half of the XX century. As an example, polyethylene (PE) was studied/synthesized in the 1930s and polypropylene (PP) in the 1950s, while polyethylene terephthalate (PET) and linear low-density polyethylene (LLDPE) were studied in the 1970s [105]. The common and most used food-packaging polymers are polyethylene terephthalate (PET) (applied in food, beverage, and other liquid containers), polyethylene (PE) (cooking oil, milk, and water containers), polystyrene (PS) (mushroom and eggs), polypropylene (PP), polyvinylchloride (PVC) (spice ice tea, yogurt and margarine), and polyamide (PA) (stretchy packaging of fresh food, such as cheese and meat) [102].

Table 1 summarizes the main important physical characteristics—glass transition temperature (T_g); melting temperature (T_m); tensile strength (T); deformation at break (ε_b, (%)); optical properties (OP), as well as transmission of visible light, haze, gloss, and permeability (water vapor (H_2O), Oxygen gas (O_2), and carbon dioxide (CO_2))—of petroleum based polymers applied in packaging and the food packaging sector, highlighting at the same time the main applications.

Polyethylene terephthalate (PET) is fundamentally applied in sectors requiring low permeability to gases, high mechanical performances (deformation at break and strength), wide temperature resistance (resistance at freezing temperatures, high softening point), and adequate transparency. High density polyethylene (HDPE) is an interesting polymeric matrix that represents more than over half of the food packaging available on the market. HDPE is mainly used in contact with a variety of foods, in addition it is easily shaped; low density polyethylene (LDPE) is extensively used in film to cover foods. High resistance to tear, low heat seal temperature, and low permeability to water (its deformation at break exceeds that of other commercial and largely used plastics) are the main required properties for the specific application.

Polyvinyl chloride (PVC) shows good gas barrier performances and high chemical hardiness, this polymer is often considered as a valid option to cover meat and fatty foods. PP is largely utilized for containers and walled cups. Satisfied barrier properties, optical transparency and esthetic quality, strength, and high temperature resistance allow this polymeric matrix to be suitable for the realization of microwave/freeze containers, and sterilized or hot-filled containers [106,107]. Polypropylene (PP) is gradually substituting polystyrene (PS) in rigid thermally-processed cups (e.g., yoghurt) [105]. Nevertheless, the low barrier properties of PP makes it acceptable for packing limited shelf-life food (e.g., cheese); the high thermal properties suggested the possibility to realize PP in thermal insulation applications by using foams (e.g., disposable packages) and reduced weight, but still not flexible and rigid dishes.

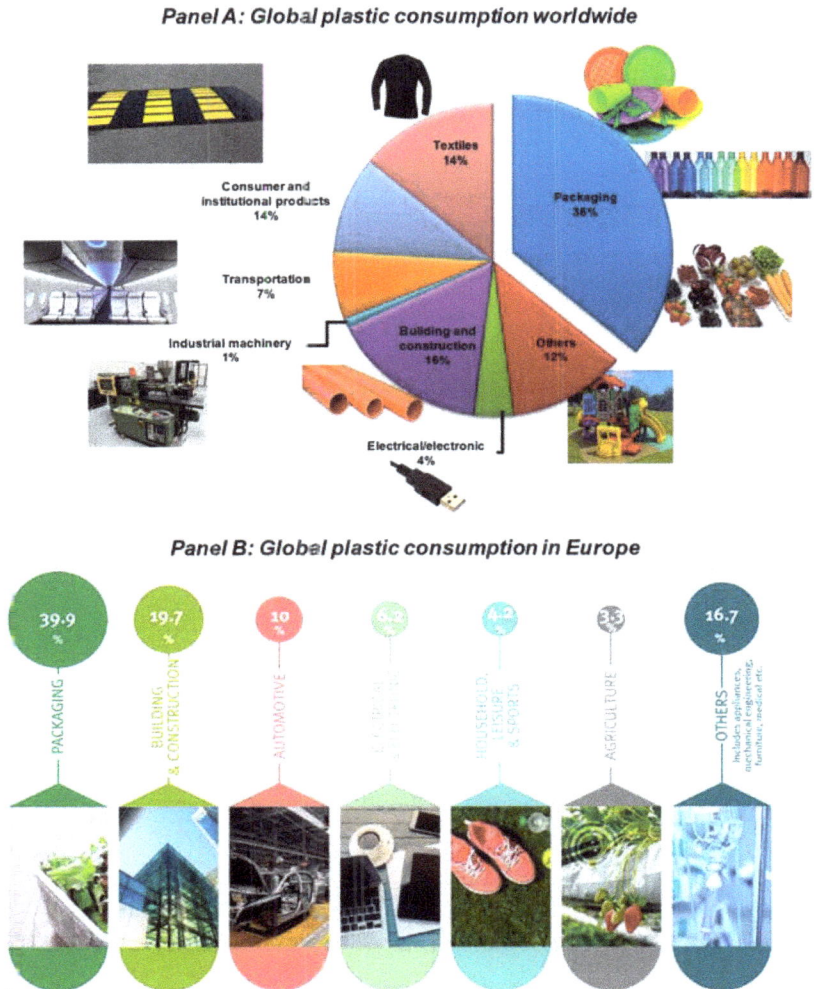

Figure 2. Panel **A**: Global worldwide plastic production and related application sectors; Panel **B**: Global plastic consumption in Europe [101].

Polyamides (PA) are engineered semicrystalline thermoplastics extensively utilized in the food packaging segment, thanks to good barrier characteristics to gas permeation, interesting mechanical and chemical resistance, good printability, and interesting optical properties in terms of transparency. The performances of the final plastic objects are, certainly, strongly related to the chemical structure of the PA and also related to the selected process utilized to realize the products [108]. The aid of a comonomer can actually influence both intramolecular and intermolecular interactions, incorporating relevant modifications in the crystallization phenomenon of a PA.

Poly(vinyl alcohol-co-ethylene) (EVOH) is a semicrystalline random copolymer with elevated transparency, optical characteristics, and outstanding gas barrier performances to hydrocarbons [109], particularly when low content of ethylene (below 38 mol % ethylene) is considered [110], and excellent chemical resistance [111]. EVOH-based formulations have been progressively applied in food packaging, being the sector characterized by severe value/standard in terms of gas and chemical resistance, water hydrocarbon permeation, and aroma migration [112]. The physical (barrier and mechanical) performances of EVOH under dry environments are attributed to the elevated intra- and inter-molecular cohesive energy and semi-crystalline microstructure [113]. The disadvantage of EVOH copolymers is their moisture sensitivity, which influences negatively their characteristics (barrier, thermal, and mechanical properties) at high relative humidity [114].

To limit this disadvantage, EVOH in the food packaging sector is often utilized in multilayer polymeric systems, and combined with hydrophobic polymeric matrices, to improve the final characteristics of packaging systems [110].

PVA is one of the most popular synthetic polymers for packaging sectors, thanks to its good compatibility, processability, and acceptable thermal properties. It also possesses good chemical resistance and high mechanical properties, although its disadvantages include limited barrier and thermal properties and relatively high cost [115]. The properties of PVA generally depend on its molecular weight and degree of hydrolysis: many hydroxyl groups on the PVA surface makes it one of the most hydrophilic polymers with high moisture sensitivity, and hence its resulting blends and composite materials have become popular for packaging applications [116,117]. In general, full-hydrolysis PVA is not considered to be a thermoplastic polymer, mainly due to its melting temperature being very close to the degradation temperature in the absence of plasticizers. Therefore, it is essential to use plasticizers for PVA in order to control the relevant melting temperature, fluidity, and thermal stability, especially for screw extrusion and injection molding processes widely used for the packaging sector.

The triumph of polymeric-based systems in the packaging sector is related to the incessant progress of better performing neat polymers and polymeric based blends, with relevant progresses in polymeric matrix processes. In fact, a clever process control of the thickness of different systems permits the production of lighter packages, reducing the material cost [105]. The lamination and the coextrusion processes represent a valid opportunity to design and realize multilayer-based systems, selecting individual layers to ensure specific properties (mechanical resistance, barrier, and aesthetical); however, biaxial orientation increases, considerably, the mechanical and barrier characteristics of a film or bottle, positively influencing some attractive behaviors.

Table 1. Physical properties of polymers applied in packaging and food packaging.

Polymer	Packaging Types	Thermal		Mechanical			Permeability		
		T_g (°C)	T_m (°C)	T (MPa)	ε_B (%)	OP	H_2O: WVTR (g/m²/day)	O_2 (g/m²/day)	CO_2 (ml μm m^{-2} day^{-1} atm^{-1})
PET	Bottles, microwaveable and ovenable trays, boil in-the-bag products	70–87	243–268	48–72	20–300	++	15–20	100–150	300–600
HDPE	Jars and other rigid containers, pallets, films, or layers for dry food	−125 to −90	135	22–31	100 ≥ 1000	+++	7–10	1600–2000	12,000–14,000
PVC	Wrapping films, bottles, trays, containers	60–100	n.d.	40–51	40–75	++	0.5–1.0	2–4	400–10,000
LDPE	Films (wrapping, carrier bags, pouches), bottles	−125 to −100	112–135	8–31	200–900	++	10–20	6500–8500	20,000–40,000
PP	Cups and containers for frozen and microwaveable food, lids, thin-walled containers (yoghurt)	−10	167–177	31–41	100–600	−	10–12	3500–4900	10,000–11,000
PS	Disposable cups, plates and trays, boxes (egg cartons), rigid containers (yoghurt)	100	n.d	35–51	1–4	++		4500–6000	14,000–30,000
PA	flexible packaging of perishable food, such as cheese and meat	50–60	220	40–52	5–10	++	300–400	50–75	n.d
PVA	Films for moisture barrier, confectionery products	70–75	215–220	25–30	220–250	++	n.d.	n.d.	n.d
EVOH	Thin films for dry/fatty food, multilayer	60–65	180–150	45–110	180–250	+++	1000	0.5	n.d

T_g—glass transition temperature; T_m—melting temperature; T—tensile strength; ε_B—elongation at break (%); OP—overall optical properties including haze, gloss, and transmission of visible light; permeability (H_2O, water vapor; O_2, oxygen gas; CO_2, carbon dioxide gas) for polymeric films. Oxygen permeability and water vapor transmission rate were evaluated (WVTR) in (g/m²/24 h) in tropical conditions (90% Relative Humidity (RH) at 38 °C); n.d.: not defined. +: low. ++: medium. +++: high. Polyethylene terephthalate (PET); high density polyethylene (HDPE); polyvinyl chloride (PVC); low density polyethylene (LDPE); polypropylene (PP); polystyrene (PS); polyamide (PA); poly (vinyl alcohol) (PVA); poly (vinyl alcohol-co-ethylene) (EVOH). Partially reprinted from: [42,105,108,109,112,118–122].

The process technology adopted using polymers are usually more energy-friendly in respect to the use of other materials (the working process temperature profile is above 300 °C, while the temperature profile used to process the plastic is typically below 300 °C). The packaging materials incorporate, sometimes, plasticizers, stabilizers, and fillers. Plasticizers are used to modulate the mechanical performance (ductility, strengthen, flexibility, and toughness of polymers; though induced a reduction of stiffness and hardness) [43,123], while stabilizers are included into the polymers to moderate the reduction of mechanical characteristics induced by UV light and oxygenation, and fillers tend to improve or preserve the mechanical and barrier performances in respect to the neat/control [24,124]. In addition, technological developments have permitted the reduction of the weight of packages (at around 28% in the last few years [105]), which in turn induces relevant savings not only in terms of transportation costs, but also in terms of environmental issues. Regarding this issue, in the last decades, the growing environmental contamination brought by the high impact of plastic wastes based on petroleum extracts has attracted the attention of industrial and academic researchers to design and develop some innovative polymeric systems. In this context, bio-based polymers have attracted some interest in respect to conventional ones. Green polymers symbolize a strategic option to design and realize new eco-friendly and sustainable systems that are able to reduce/minimize the plastic wastes stored every day in landfills, and the emission of greenhouse gases (GHG), which strongly depend on fossil extraction, production, use, and end-life [24,125]. There is an increasing interest worldwide to substitute traditional plastics with bio-based ones, mainly in packaging sector. The utilization of bio-based materials and resources is seen as one of the numerous strategies able to reduce the environmental impacts induced by the use of petroleum-based extracts.

Future prospective in the polymeric package sector will be influenced by: (i) How these resources will guarantee the increasingly additional rigorous necessities for packages; (ii) the processability of novel bio-based/biodegradable polymers; and (iii) legal, market, and environmental issues.

4. Bio-Based Matrices for Packaging Applications

Presently, packaging films are typically constituted by petroleum-based synthetic polymeric matrices that monitor the market, due to their reduced price and simple accessibility. As already analyzed, these polymeric matrices comprise polyolefins (polyethylene), ethylene vinyl alcohol (EVOH), that are responsible for a relevant barrier towards water and oxygen. Nevertheless, they are hindered by margins in petroleum resources and the lack of bio-disintegration, which magnify the ecological and cost-effective concerns.

Bio-based polymers are considered a valid replacement of petroleum synthetic polymeric matrices; they are obtained by the processing of renewable resources (vegetable and animal wastes) and offer several positive aspects, such as environmental advantages, disintegrability and degradability, improved possibility to recycle the polymeric wastes, no presence of toxic components, and high biocompatibility, in respect to the conventional petrochemical polymeric matrices. The interesting barrier properties of different bio-based polymeric films for package designs has been acknowledged in several research review articles [126,127]. Due to the fact that numerous biopolymers show water affinity, their barrier and mechanical characteristics are subjected to the humidity and ambient atmosphere, which may decrease their overall performance and the quality of packages when compared with petroleum-based polymers (Figure 3).

Furthermore, molecular weight, physical properties (crystallization phenomenon and crystallization degree), visco-elasticity, and rheological characteristics may induce disadvantages, thus several modifications or adjustments during the processing steps are necessary to modulate the final performances. Consequently, biopolymers should be modified, studying new polymeric blends combining two or more different polymeric matrices or fillers, at the micro/nanoscale level, to expand their characteristics/properties, especially dealing with nature and processability behavior.

Bioplastics can be classified into three important classes, in function of their source: non-biodegradable bio-based bioplastics (e.g., polyethylene terephthalate (PET), PA); biodegradable bio-based bioplastics (e.g., PLA, polyhydroxyalkanoates (PHA) or starch, other polysaccharides or proteins); or fossil-based biodegradable plastics (e.g., polycaprolactone (PCL)) [128] (Table 2).

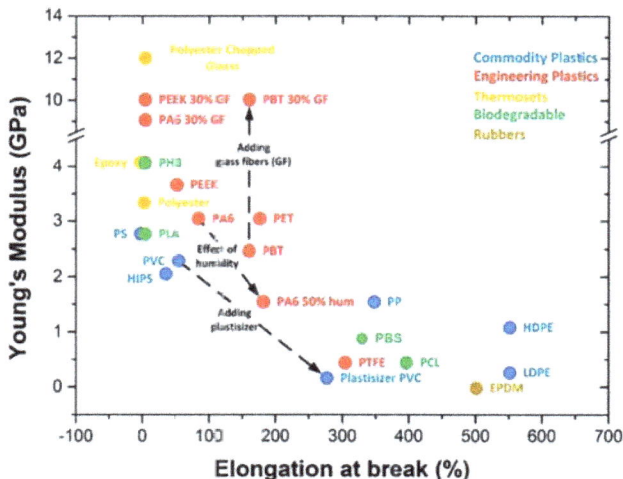

Figure 3. A comparison of the intrinsic properties (mechanical characteristics) of usually utilized plastic materials, engineered polymers, rubbers, bio-based polymers, thermosets, and plastic materials [127].

Table 2. Bioplastics as food contact materials [129].

Bioplastic	Main Food Applications
Starch-based polymers	Substitute for polystyrene (PS). Used in food packaging, disposable tableware and cutlery, coffee machine capsules, bottles.
Cellulose-based polymers	Low water vapor barrier, poor mechanical properties, bad processability, brittleness (pure cellulosic polymer), Regulated under 2007/42/EC. Coated, compostable cellulose films. Used in the packaging of bread, fruits, meat, dried products, etc.
Polylactide (PLA)	Possible alternative of low- and high-density polyethylene (LDPE and HDPE), polystyrene (PS), and poly terephthalate (PET). Transparent, rigid containers, bags, jars, films.
Polyhydroxyalkanoates (PHA)	Family of many, chemically different polymers Brittleness, stiffness, thermal instability.
Bio-based polypropylene (PP) and polyethylene (PE)	Mainly based on sugar cane. Identical physicochemical properties.
Partially bio-based (PET)	Alternative to conventional PET. Up to 30% bio-based raw materials. Used in bottles.
Bio-based polyethylene furanoate (PEF)	Better barrier function than PET. Up to 100% bio-based raw materials. May be used in the future in bottles, fibers, films.
Aliphatic (co)polyesters	Includes polybutylene succinate (PBS), polyethylene succinate (PES), and polyethylene adipate (PEA). Used in disposable cutlery.
Aliphatic-aromatic (co)polyesters	Includes polybutylene adipate terephthalate (PBAT), polybutylene, and succinate terephthalate (PBST). Used as fast food disposable packaging, PBAT for plastic films.
Polycaprolactone (PCL)	Biodegradable polyester. Low melting temperature, easily biodegradable. Used in medical applications
Polyvinyl alcohol (PVOH)	Used for coatings, adhesives, and as additive in paper and board production.

Nevertheless, the general principle to be considered to classify biodegradable materials is the raw material origin and their production step. In accordance to this, biodegradable polymers are categorized into three groups: "1st class" or biomass derived polymers, including cellulose acetate, cellulose, chitin, and starch; "2nd class" or biopolymers synthesized utilizing microorganisms and plants, as poly(hydroxy alkanoates (PHAs); and "3rd class" or synthetic polymeric matrices, as polylactide (PLA), poly(butylene succinate) (PBS), bio-polyolefins, bio-poly(ethylene terephtalic acid) (bio-PET), and synthetic polymeric matrices chemically produced from renewable sources [130] (Figure 4a,b).

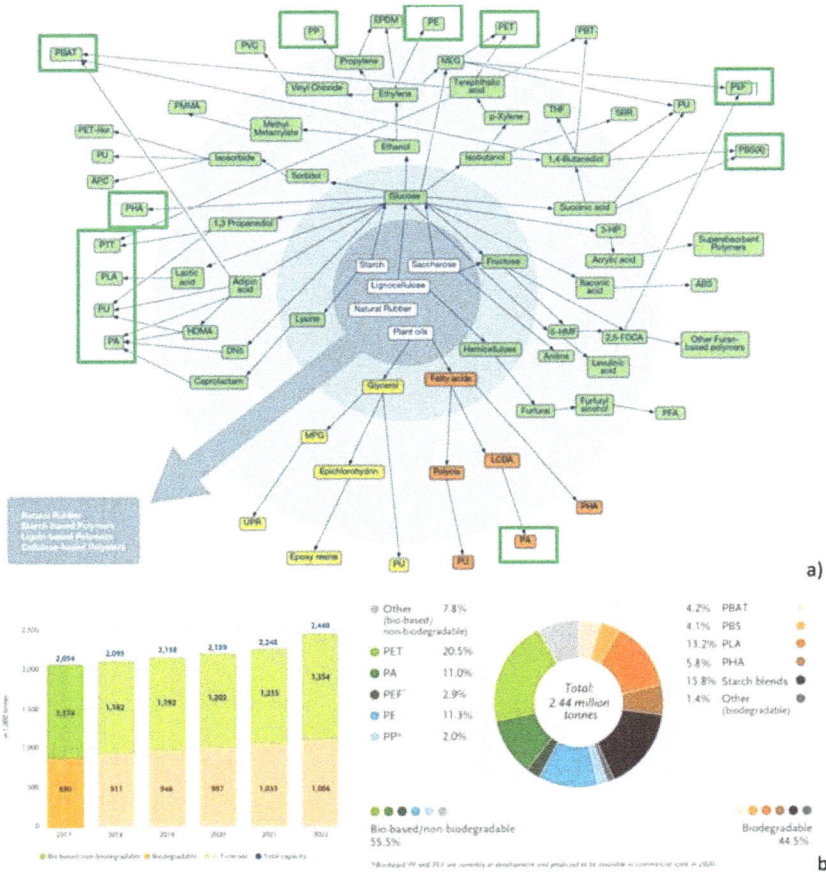

Figure 4. Commercially realized pathways from biomass via different building blocks and monomers to bio-based polymers (**a**); estimated global production capacities of bioplastics (biodegradable, bio-based/non-biodegradable) for 2022 by material type (**b**) [131].

Typically, 1st class is utilized avoiding the refining step, whereas 2nd class polymeric matrices are obtained from natural extracts, and they play a central position in conditions that necessitate biodegradability. The utilization of 1st and 2nd class polymeric matrices permits a more proficient manufacture, with materials having preferred and modified physical and functional characteristics, but limited flexibility in the chemical structure. Monomers utilized in 3rd class polymers are obtained by modifying natural molecules or by the chemical modification of natural macromolecules combining and applying chemical and biochemical technologies. Several of these 3rd class polymeric matrices, such as bio-PET and bio-polyolefines, will not go into natural cycles after utilization, so their concurrence to the reduction of environmental impact is mainly related to the decrease of the carbon footprint. Studies and researches regarding progress, innovation, and applications of the three classes of polymers are described in the following sections, their main important characteristics and chemistries are also described.

A selection of polysaccharides and their modified products have been utilized to realize biodegradable films to be used in packaging and edible coatings. These carbohydrates composed by glycosidic bonds are one of the most important structural components of vegetables (e.g., cellulose) and animal exoskeletons (e.g., chitin), and they can have, in addition, a crucial function in the "green" energy storage (e.g., starch) [132]. With growing consideration and research on this area, it has been practicable to open the way to overtake their intrinsic limitations and identify solutions [133]. In the following paragraph, some of the principle utilized polysaccharides in food packages are summarized.

4.1. Starch

It is composed of amylose and amylopectin polymers (α-d-glucose monomers), containing hydroxyl groups (–OH). These groups show strong intermolecular interactions, which give, on the basis of tighter arrangement, increased crystallinity and melting temperatures. As a consequence, starch molecules are thermally affected, which strongly limits their application as a packaging material (Table 3) [134,135]. Alternatively, the presence of hydroxyl groups supports the break of hydrogen bonds that induce the disintegration into small fragments. Starch products are food sources for microorganisms in determined environment conditions, as elevated humidity (~55–60% RH), in the presence of satisfactory oxygen feedstock, and suitable temperatures to make the biodegradation easy [136,137]. In addition, starch hydrophilicity creates materials with a poor water barrier [138]. At a modest or elevated relative humidity, starch-based substances tend to soak up moisture from the environment. Firstly, this behavior will induce swelling of the matrix, with disruption of hydrogen bonds, after that the increase of free volume spaces will also enhance the chain mobility. Therefore, moisture and gas barriers are negatively influenced and, consequently, are drastically compromised. According to this, starch may not be adequate to realize packages for dry and oxygen-sensitive foods. Other than low barrier properties, starch materials have the inclination to show reduced mechanical characteristics, reducing, consequently, again their utilization [139]. Furthermore, even if starch price is relatively low, adaptation of its performances may improve the final cost of different materials. Numerous parameters should be considered for the design of starch food packages.

Table 3. Characteristics and food use of polysaccharides extracted from animals and vegetables [140].

Polysaccharide	Properties	Main Food Applications
Starch	Biodegradable Transparent Odorless and tasteless Retrogradation high elongation and tensile strength	Flexible packaging: • Extruded bags • Nets for fresh fruit and vegetables • Rigid packaging • Thermoformed trays and containers for packaging fresh food.
Cellulose	Biodegradable Good mechanical properties Transparent Highly sensitive to water Resistance to fats and oils Need to perform modification, use of plasticizer, or polymer blend	Cellophane membranes.

Table 3. *Cont.*

Polysaccharide	Properties	Main Food Applications
Chitin	Biodegradable Antibacterial and fungistatic properties Biocompatible and non-toxic Highly transparent	Coffee capsules Food bags Packaging films
Chitosan	Biodegradable Biocompatible and non-toxic Antifungal and antibacterial properties Good mechanical properties Barrier to gases High water vapor permeability Brittle—need to use plasticizer	Edible membranes and coatings (strawberries, cherries, mango, guava, among others) Packaging membranes for vegetables and fruit

Regardless of several published research articles that developed and studied TPS (thermoplastic starch) products, the weaknesses, such as reduced barrier and mechanical performances, are still unsolved. As a consequence, additional strategies, such as the compatibilization, chemical and physical modifications, and development of polymeric blends, have been suggested to solve the disadvantages of using TPS. Polymeric blends combining two or more polymers are a valid possibility to modulate and improve TPS performances. This approach is considered cost-effective, since no modification is required and a wide frame of polymeric material processability can be opened.

4.2. Cellulose

Cellulose (Table 3) shows regular arrangement and structure of hydroxyl groups, it is characterized by the tendency to organize crystalline microfibrils with strong hydrogen bonds. It has specific interesting characteristics, such as high mechanical strength, low density, high durability, low price, no presence of toxic elements or substances, biodegradability, interesting and easy chemical modification, and stability [141]. In the textile sector, the application of cellulose is largely utilized in packaging and fibers, and can be classified into two groups: modified and regenerated cellulose [142]. Different chemical modifications, such as etherification and esterification, are frequently and typically considered to improve the thermal processability cellulosic materials. Numerous modified celluloses are commercially available, the main ones are cellulose esters (for melt compounding), cellulose acetate, and regenerated cellulose for fibers [143]. The application of plasticizers and polymeric blends are also evaluated considering that the chemical and mechanical characteristics are largely influenced by the blend composition and technical processability followed and applied for their realization [144].

4.3. Chitin/Chitosan

Chitin (Table 3) is a natural polymeric material that composes the exoskeleton of arthropods, it is also present in the cell walls of yeasts and fungi. In addition, it is an acetylated polysaccharide made up of N-acetyl-d-glucosamine. Chitin is present also on the market after a chemical treatment (extraction from crabs and prawns wastes). Chitosan is indeed extracted after the treatment of chitin, applying a deacetylation process in which several parameters that should affect the extraction and also modify the main characteristics (e.g., temperature, native origin of chitin, and alkali concentration) need to be controlled. Chitosan does not show affinity to water, in fact it is insoluble in water, but may be without problems dissolved in acidic aqueous solutions. The interesting film-forming characteristics permit the realization of coatings and membranes capable to be utilized for food conservation [145,146].

The membrane of chitosan is characterized by modest water permeation and low oxygen permeability, crucial in the conservation and preservation of some food that are sensitive to the presence of oxygen in the packaging [147,148].

The development of polymeric blends composed by chitosan and other polymeric materials, such as starch [149] and proteins [150], represents a valid approach to guarantee an enhancement in terms of mechanical performances, improved characteristics in terms of lower water solubility, and water vapor permeability. Several other approaches, for instance coating, dipping, casting, Layer-by-Layer (LbL) assembly, and extrusion, have been considered to realize chitosan systems with several characteristics. The promising food sector of chitosan systems as antibacterial active compounds and sensing and barrier systems have had enormous progresses [151].

4.4. Biopolyesters from Microorganisms—Polyhydroxyalcanoates (PHAs)

PHAs are bio-based polyesters produced by microorganisms as a carbon source under nutrient stress conditions. PHAs are realized after the fermentation process of different materials (sugars, organic, agricultural, and municipal solid waste, etc.) [152]. This class of biopolyesters is biocompatible, biodegradable, and thermally processable. PHAs are utilized in food packaging application to realize coatings, films, boxes, foam materials, and fibers [153]. The final characteristics of the PHAs are influenced by monomer composition, carbon nature, and source and kind of microorganism utilized during the fermentation. Within PHAs, polyhydroxybutyrate (PHB) homopolymer has a high degree of crystallinity, resulting in a stiff and brittle nature. PHAs are resistant to hydrolytic degradation, PHAs show interesting and exceptional film-forming ability and offer a low permeation to gases (oxygen and water vapor), good UV resistance, but low chemical resistance towards acids and bases [154]. Two main restrictions for their use and application in large-scale are related to the high cost of polymers and characteristics (they suffer problems during processing because of the limited temperature processing range and a moderately low-impact performance related to high crystallinity values). Some disadvantages of PHB for industrial applications may be overtaken by copolymerization with hydroxyvalerate or hydroxyhexanoate. Monomers can be used in 150 different combinations to produce copolymers with diverse properties. PHB polymeric matrices showed some different characteristics, modulating the copolymers ratio. PHB can be less crystalline and more elastic in the presence of long alkyl side chains, such as hydroxyvalerate (HV) and hydroxybutyrate (HB), that provide, respectively, P(3HB-co-3HV) and P(3HB-co-4HB) [125]. The polymeric blends based on PHB, combined with other polymeric matrices, can influence and encourage its use (Figure 5a) [153]. For example, PHB has been combined with poly(vinyl butyral), poly(vinyl acetate), poly(ethylene oxide), cellulose acetate butyrate, poly(vinyl phenol), chitosan, and chitin. Distinctive characteristics of this set of polymeric matrices are outstanding rigidity, heat, and chemical performances. Isotactic polypropylene (PP) showed characteristics comparable to the cited PHB copolymers. Isotactic PP showed remarkable water vapor resistance, in respect to the bio-based polymeric matrices present in the market. For all these characteristics, bio-based polymeric matrices may be used in several applications, including packaging.

4.5. Biopolyesters from Biotechnology and Conventional Synthesis from Synthetic Monomers

4.5.1. PLA

(Poly lactic) polymers are obtained by the fermentation process applied to agricultural wastes and byproducts, such as starch-rich materials (wheat, corn starch, and maize). The procedure implied the transformation of corn natural resources into dextrose and following fermentation into lactic acid. PLA is a biodegradable aliphatic polyester and thermally-processable matrix, having interesting properties for the packaging sector [155]. The monomers based on lactic acid are also polycondensed or synthetized by ring-opening polymerization of lactide [156], and the properties are strongly related to the content of optical isomers of the lactic acid. High crystalline polymers are composed of 100% L-PLA monomers, while the different concentration of 90/10 D/L copolymers is considered to facilitate the processing above the glass transition. PLA is the first biodegradable polymer present on the market and largely commercialized. (Poly lactic) can be utilized to realize injection-molded coatings, films, and objects. In this context, PLA has substituted polyethylene terephthalate (PET), low-density polyethylene (LDPE), high-density polyethylene (HDPE), and polystyrene (PS) in the packaging sector. PLA is several times utilized to realize single use material: blister packages, cold drink cups, lids, containers, thermoformed trays, bottles, as well as flexible films [157]. Even though PLA is, today, advantageous from an economic point of view, and it also has many favorable characteristics for packaging uses (i.e., simple processability, high transparency), it also shows some disadvantages, such as mechanical characteristics and poor barrier properties, which inhibit its industrial utilization [158]. Significant academic and industrial energies have been dedicated to the improvement of PLA characteristics to enlarge the application of PLA in different commercial applications of crucial importance in the industrial sector, in order to develop ecofriendly disposable materials. In addition, modulation of PLA properties by combining PLA with different polymeric matrices, is realized by melt blending, being this procedure is quite simple. Furthermore, the technologies required for the process are available at industrial level with limited costs. The technique permits the realization of simple packages with modulated characteristics by changing the content of different polymeric matrices, which are utilized to develop the blends.

4.5.2. PBS

Conventionally, poly(butylene succinate) (PBS) is obtained from succinic acid and 1,4-butanediol through catalytic hydrogenation of maleic anhydride. Succinic acid can also be synthesized by microbial fermentation. PBS can be degraded via ester linkages, and retains exceptional mechanical characteristics that permit it to be prepared by using conventional melt processes. Its uses consist of bags, hygiene commodities, and mulching. Their mechanical performance is inferior to polymeric matrices extracted from petroleum. Consequently, aliphatic polyesters could be utilized as transparent thin systems for packaging bags, filaments, blown bottles, agriculture, and thermo-processed or injection-molded systems. In order to enhance their properties, new polymeric blends or chemical modified copolymeric systems (aliphatic-aromatic copolyesters: poly(butylene terephthalate-co-succinate) (PBTS) and polybutylene adipate terephthalate (PBAT) can be realized. Copolymers based on polyesters are frequently composed by terephthalic acid to gain better processability and mechanical performance. Aliphatic-aromatic blends and copolyesters merge the fine useful characteristics of aromatic polyesters and the interesting biodegradability of aliphatic polyesters. Furthermore, the combination of characteristics permits the modulation of the performance of the final product, and address them to the utilization in the industry dedicated to the production of packages. The aliphatic or aromatic polyesters are based on petrochemical extracts and are usually prepared through conventional polycondensation methods.

4.5.3. PBAT

Polybutylene adipate terephthalate (PBAT) is composed of a linear polymeric chain made by two comonomers (forming a copolyester): a rigid unit consisting of terephthalic acid and 1,4-butanediol monomers, and the elastic unit composed by adipic acid and 1,4-butanediol monomers. PBAT shows interesting thermal and mechanical characteristics when the amount of terephthalic acid is more than 35 mol %; nevertheless, the increasing amount is followed by an important reduction, considering the rate and the percentage of biodegraded quantities [159]. This aspect is related to the existence of aromatic polymeric chains, which build these materials up to be more resistant against microbial attacks [160]. PBAT is also extremely flexible and soft, to make it applicable in films (mulch), blown bottles, and injection-molded and thermoformed systems and filaments. It is extensively utilized for disposals characterized by a short life, such as food films and compostable systems. This copolyester is commercially utilized for the realization of single-use not reusable packages, organic waste bin liners, compostable bags, and protective plastic systems [161].

4.6. Bio-Polyamides

Castor oil is the raw material used as a renewable feedstock for the production of commercially available bio-based polyamides. Castor beans are characterized by an unusually high amount of ricinoleic acid (40–60%). The presence of hydroxy groups and double bonds of the acid determine numerous opportunities for its chemical modification. Chemical modifications permit the synthesys of different blocks, such as aminoundecane (decamethylenediamine) and sebacic acid [162]. As in the case of standard polyamides, bio-PAs are frequently produced by polycondensation of dicarboxylic acids with diamines, by ring-opening polymerization of lactams, or polycondensation of amino acids [163]. Monomers are mostly obtained from fossil oil, but can even be from biomass (Figure 5b). There is an available patented method that considers PA6 production by ring-opening polymerization of ε-caprolactam, obtained by glucose fermentation, from totally renewable feedstocks, such as sugar [163]. The same bio-based methodology can be utilized to obtain PA 66. In this case, adipic acid, extracted during the fermentation process of glucose or plant-oil and hexamethylenediamine (HMD), which can be quickly extracted from biomass, can be used.

4.7. Bio-Polyolefins

Bio-based polypropylene can be produced by using bio-ethanol (Figure 5c); however, this process is relatively more complex and involves several ways of obtaining the propylene monomer C_3H_6 from various renewable resources [164,165]. Major applications of bio-based polyolefins include packaging, building and construction, automotive and transportation, and others including blow molded bottles, stretch and shrink films, and detergents. Packaging emerged as the leading application segment on account of shifting demand from synthetic polymers to bio-based polyolefins in 2013. Automotive and transportation is the second largest market for bio-based polyolefins, owing to its increased application in manufacturing automotive parts. Moreover, building and construction is expected to witness strong growth in the bio-based polyolefins market.

4.8. Bio-Poly (Ethylene Terephtalic Acid) (Bio-PET or PEF)

The furanic–aliphatic polymeric class has been largely investigated in recent years. These polymers based on ester groups are characterized by the presence, in their backbone, of aliphatic and furan units (i.e., obtained from an aliphatic monomer and FDCA (2,5-furandicarboxylic acid)) (Figure 5d). They have been custom-made by means of several aliphatic monomers, counting those with a linear carbon sequence or those with an extra rigid cyclic arrangement [166]. Poly(ethylene 2,5-furandicarboxylate) (PEF) is currently considered as an attractive sustainable substitute of poly(ethylene terephthalate) (PET), due to the fact that it has better barrier and interesting thermal characteristics (e.g., lower melting phenomenon and higher glass transition temperature) than PET. The reduced permeability of PEF to CO_2, O_2, and H_2O is a great benefit for packages. Ii is expected to enter the market in 2020 to replace PET [167].

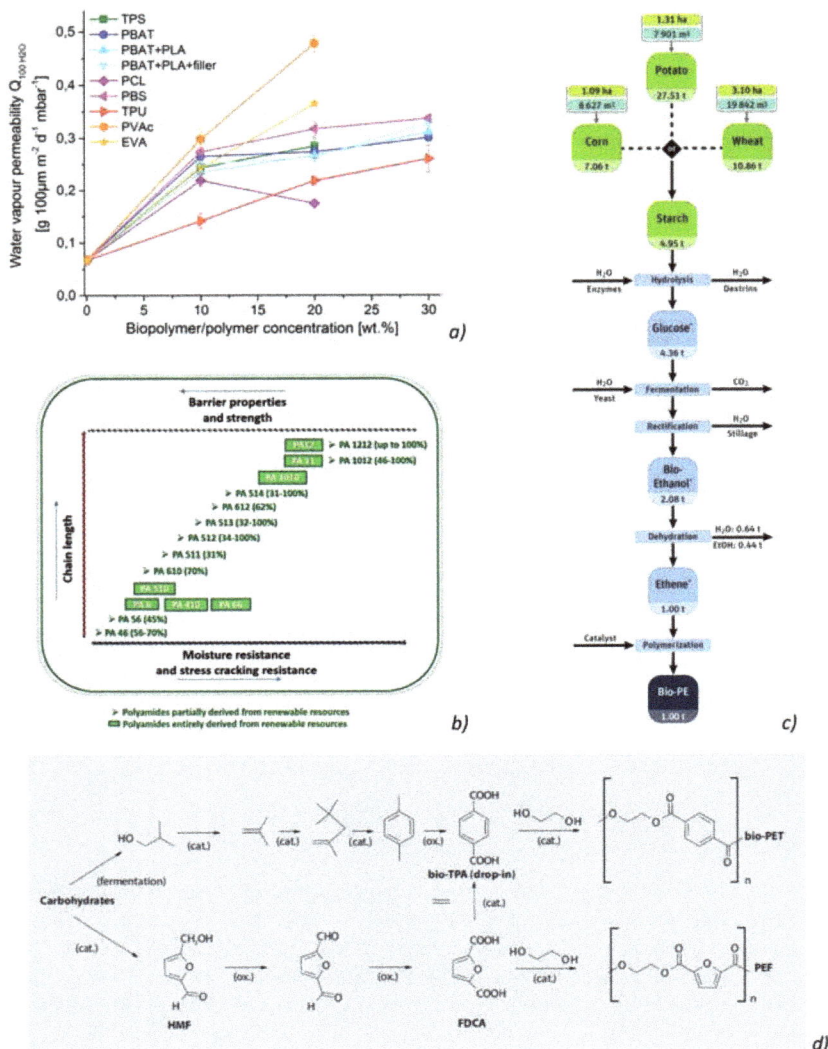

Figure 5. Water vapor permeability for Poly(3-hydroxybutyric acid-co-3-hydroxyvaleric acid) (PHBV) blended with nine different biopolymers and polymers at different concentrations (**a**) [168]; polyamides completely and partially obtained from renewable sources (**b**) [163]; processing route for bio-polyethylene (bio-PE) (**c**) [169]; (**d**) transformation of sugar to Polyethylene 2,5-furandicarboxylate (PEF) and bio- polyethylene terephthalate (bio-PET), a new bio-based plastic similar to PET [170].

5. Hybrid Blends Based on Bio-Based and Fossil Fuel-Derived Polymers

Natural-based polymeric matrices are largely susceptible to moisture and, according to this, do not afford a good barrier to gas diffusion. Hybrid blends containing renewable polymers in combination with man-made polymers and additives have shown great potential in solving some of these limitations. In fact, most of the commercial food packaging is based on hybrid materials, giving the required properties and functionality to a variety of foods [171].

Bio-based plastics are characterized by costs much higher than traditional thermally-processable polymers (e.g., PP and HDPE), consequently, it is not advantageous to use them without combining these two products. In general, different parameters influence the final microstructure characteristics and mechanical performances of not totally miscible polymeric blends, thus it is evident that the overall performance of polymeric blends is strongly connected to blend composition and its phase morphology [172].

One of the most important aspects of such kind of blending is the recyclability, due to the fact that bio-based plastics could, on one hand, interfere with the current recycling of plastics, and hence hinder the closure of plastic cycles [173] (end-of-life compostable bioplastics may pollute recycled plastic streams if not properly separated and managed, which is undesirable given the current attention on a transition towards a circular economy), while on the other could enhance the recalcitrance of fossil-based mixtures to disintegration and compostability [174–176]. Nevertheless, even if the mechanical recycling of bio-based non-biodegradable plastics, such as bio-PP, bio-PE, and bio-PET, is chemically matched to their fossil counterparts, thus totally compatible with the current recycling methodology, a limiting issue is that these materials (as well as blends of bio-based and fossil plastics) are not compatible with sorting in single polymer streams, even in larger quantities. Therefore, if not mechanically recycled, they can only be incinerated with energy recovery (or anaerobic digested with biogas production).

Materials obtained by mixing synthetic polymers with natural polymers can potentially compromise environmental, economic, and social aspects by reducing the recyclability when compared to traditional polymers [177,178]. Therefore, a main concern associated with the strategy of introducing natural/biodegradable polymers into conventional polymers is related to the chance that these blends have lower levels of recyclability compared to the original polymers, which would then have an adverse impact on both environmental and industrial issues.

It should be also recognized that use of bio-based or biodegradable plastics is often justified by asserting that they biodegrade faster than their conventional petrochemical counterparts; on the other hand, it has to be considered that new demand for biomass inputs could negatively expand uses of land, fossil fuels, chemical inputs, and water. Additionaly, bioconversion may require the use of potentially toxic petroleum-based solvents and results from life cycle assessments could give high energy consumption and emissions for bio-based polymers. When deciding to fully or partially replace conventional petroleum-based plastics with bio-based plastics, it is important to understand the flow of these materials and their adverse impacts in all parts of their life cycles in order to select a material that is more sustainable [179].

Given, as recognized, that the characteristics of biopolymers must be improved substantially if we want a penetration in the market, the change of these materials toke the attention of researches. In a dissimilar manner to the design of new polymers and polymerization routes, polymer blending is a comparatively low-priced and fast process to modify the characteristics of polymeric materials. As a consequence, this procedure may take part in a critical manner in improving the attractiveness of bio-based polymeric matrices [180]. Further studies are requested to optimize the miscibility of these systems to exploit their potentials. It is widely reported in the literature that different chemical modification routes, such as grafting, transesterification, and copolymerization, have been developed to obtain polymeric materials and blends with valuable characteristics.

5.1. Starch

In the case of starch-based blends [181], numerous studies have reported, even recently [182], the blending of thermoplastic starch with other man-made plastics (e.g., polycaprolactone, polyamide, polyolefins) [183]. While the blending approach reduces some limits of the TPS systems, the successful approach in improving adhesion can be found by considering non-reactive mixing (blending TPS with graft block or random copolymers) or reactive compatibilization (by polymerization, grafting, or branching).

As an example, a comparison between the effects of carboxylic acids (stearic, palmitic, and myristic acids) and maleic anhydryde (MA)-grafted poly(propylene), used as a compatibilizer agent for PP/TPS polymeric blends, was conducted: Carboxylic acids revealed comparable compatibilization performance with respect to MA-grafted poly(propylene), regarding improved adhesion and better mechanical properties (Figure 6a) [184]. It was studied that carboxylic acids induced enhanced crystallinity in PP/TPCS blends. Polymeric blends based on the combination of traditional polymeric matrices and starch can also speed up the disintegration of traditional based systems used to realize packages [185,186]: The presence of different amounts of starch into traditional polymeric matrices can decrease the price and enhance the degradation of the blends, due to the fact that, if polymeric blends are buried in soil, the starch phase is degraded by microorganism attacks. This enhanced degradability due to the presence of porosity and voids induces the loss of integrity of polymeric systems (Figure 6b). Nguyen et al. [187] analyzed the behavior of LLDPE/TPS systems, obtaining a degree of disintegration of 63%. The systems disintegrated into methane, H_2O, CO_2, and biomass after five months in a composting environment. The results confirmed that LLDPE/TPS blends were disintegrated more rapidly than pure LLDPE, an enhancement in terms of porosity structure with a loss of integrity of the polymeric matrices was evidenced, finally achieving broken plastic small pieces. Consequently, if the PE part of the material is not modified to facilitate its disintegration and the subsequent biodegradation, only the starch part of the blend will biodegrade with fragmentation of the material as a result. However, it is highly questionable and hard to accept, as polyolefins present in the blends do not undergo biodegradation and are still present in the form of a polymer material when the biodegradable part of the blend degrades to water, CO_2, and biomass. In the environment, the packaging product made of this kind of blend is only fragmented to small fractions of plastic that can be definitely dangerous for living organisms [188]. The recycling opportunity still represent a possible solution to this problem, as it has already been demonstrated that the replacement of products based on LDPE with LDPE/TPS blends may not result in substantial changes in the recyclability issues associated with the reprocessing of the involved systems [189].

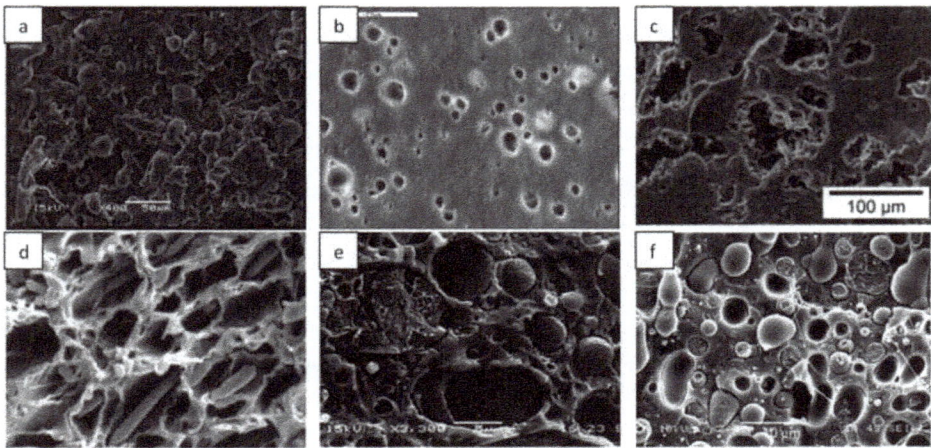

Figure 6. SEM micrographs of (**a**) PP/TPS/MA blend [184]; (**b**) TPS/PE compatibilized with PP-g-MA 20%—scale 20 μm [186]; (**c**) 30TPS/70PA12 blend [190]; (**d**) LDPE/PHB 80:20—scale 40 μm [191]; (**e**) PS45/PHB45-(block copolymer) C10 blend [192]; (**f**) PLA/PS (0.3 volume fraction) blend [193].

In the case of PE, further investigation is mandatory to utilize LDPE-starch blends as bio-based packages, because of the decrease of the mechanical characteristics of PE when the starch amount increases [194]. Generally, the mechanical performance is reduced with the presence of starch because of the limited compatibility between PE and starch. The reduced compatibility is strongly related to the fact that starch granules are extremely hydrophilic, due to the presence of hydroxyl groups at their surfaces, while LDPE generally does not show affinity to water. In several researches, if the content of starch is increased at around 20 wt % the deformability of the PE matrix is changed and converted into a fragile material (in PE/starch based formulations) [195]; even gas diffusion and the water vapor transmission rate (WVTR) is changed in relation with the starch ratio, according to Arvanitoyanis et al. [196], that found how WVTR is decreased proportionally when the amylose content increases, due to amylopectin degradation during the thermal processing.

Pedroso and coworkers [197] and Rosa et al. [183] observed immiscibility between TPS and LDPE, Euaphantasate et al. [198] produced LLDPE/TPS systems by extrusion, presenting different amounts of starch, from 10 to 40 wt %; the microstructure shows the presence of two polymeric phases and micro-voids.

Rare papers can be found for the realization of TPS blended with high melting synthetic polymers. In the case of polyamides, Landreau et al. [199] and Teyssandier et al. [190] investigated TPS-based systems combined with polyamide 11 (PA11) and polyamide 12 (PA12) (Figure 6c). In the first research, microstructures and characteristics of TPS/PA11 polymeric systems were investigated. The proposed formulations were modified, applying a compatibilization procedure by considering carboxymethyl cellulose (CMC) in an amount of 1 g CMC/100 g dry polymeric matrix. The presence of sodium neutralized anionic groups allowed the interaction between CMC and PA11, perhaps by hydrogen bonding of the amide groups and metal complexation. The polymeric blends showed interesting mechanical characteristics (high tensile strength and high modulus), even if TPS was the major element. In the second work, the authors analyzed the compatibilization of TPS/PA12 blends by the presence of poly(ethylene-co-butyl acrylate-co-maleic anhydride) terpolymer (Lotader 3410) and bisphenol A diglycidyl ether (DGEBA). The crystallization phenomenon of PA12 in TPS/PA12 polymeric blends was determined to be deeply dependent on the DGEBA amount; however, no effect was observed for the presence of Lotader 3410 t.

Moreover, Tureèková et al. [200] analyzed the preparation of polymeric-based systems composed by an aromatic-aliphatic copolyester, having wheat starch and 35 mol % of aromatic ester units, plasticized by 15, 20, or 30 wt % of glycerol. Blends prepared from copolyester and native starch, at ratios of 55:45, 65:35, or 75:25 by weight, were analyzed; tensile strength values varied between 9 and 30 MPa. Deformation at break of the polymeric blends was very low, while wettability characterization of the analyzed systems would not reduce possible utilization as package products.

In the case of polystyrene, Tomy et al. [201] investigated the characteristics of oxidized and native corn starch/polystyrene-based systems under reactive extrusion, using zinc octanoate as a catalyst. The systems were realized by reactive extrusion, and then processed by compression molding. The data undoubtedly showed that the used catalyst induced cross-linking between PS and starch, and that the induced oxidation of the starch magnified its reactivity. Systems did not show signs of swelling or antimicrobial growth. Yongjun et al [202] realized copolymer by reactive grafting of starch with polystyrene (starch-g-PS) by using ionic liquid 1-ethyl-3 methylimidazolium acetate ([EMIM]Ac) as the solvent and potassium persulfate as the initiator. The obtained analyses showed that ionic liquid suspension of starch, before polystyrene grafting, is a useful procedure for the synthesis of amphiphilic, polysaccharide-based graft copolymers with high grafting amount.

Only one paper is available that considered the tuned degradation of PVC when blended with TPS [203]; thanks to the extensive variety of uses for PVC, a complete comprehension of disintegration steps is of fundamental significance from an ecological viewpoint.

Lastly, an extensive number of results are accessible on the potential use of thermoplastic starch in a blend with EVOH [204–207]. Generally, starch systems show relatively interesting barrier characteristics at low moisture levels and plasticizer amount, with respect of traditional membranes, such as EVOH (copolymer of vinyl alcohol and ethylene) or polyamide, which are frequently mentioned as control materials for oxygen barriera. The difficulty with starch membranes is to verify the appropriate stability between interesting barrier and mechanical characteristics. Commonly, great amount of plasticizer amount is chosen to gain a soft material, but in this situation the barrier characteristics are going to be obviously reduced. Even if starch shows higher diffusion to carbon dioxide and oxygen than ethylene-vinyl alcohol, it is indeed considerably economically advantageous to use it; thus, comparable or higher barrier materials, in comparison with EVOH-based multilayer formulations, can be obtained by considering greater thicknesses.

It has been proven that the blending of starch and PVA enhanced the disintegration of starch-filled renewable polymers [115,208]. In general, PVA/starch blend films show certain limitations, such as high affinity to water and weak mechanical characteristics. In particular, barrier and tensile performance decrease with higher starch amount, resulting from their partial compatibility, especially in the absence of plasticizers. Consequently, the main approaches to overcome these limitations comprise the use of chemicals (such as cross-linkers and surfactants) to modify the compatibility during blending, the use of modified PVA and starch instead of native PVA and starch, respectively, as well as the incorporation of nanofillers to improve their properties. Zhao et al. [209] modified starch to tackle this problem by using methylated corn-starch (MCS) and then blended it with PVA. The water absorption capacity of PVA/MCS decreased by a factor of two when compared with that of PVA/native starch. On the other hand, Jayasekara et al. [210] modified surface compatibility of PVA/starch films with the aid of chitosan since the surface roughness of PVA was lower than that of starch. However, PVA/starch blend surfaces had an intermediate roughness between those of individual PVA and starch, which remained unchanged with the addition of chitosan.

5.2. Chitosan

Because of the large applications of chitosan in various fields, blends with synthetic polymers, having a wide range of physicochemical properties, have been prepared in various occasions, with solution blending investigated by many workers [211,212]. Polyvinyl alcohol and polyethylene are among the synthetic polymers that have been frequently blended with chitosan.

The combination of good mechanical properties and hydrophilicity of PVA with the biological activity of chitosan offers a good opportunity to produce beneficial blend films with high antimicrobial effects, high formability, good strength, and high barrier proprieties, despite the fact that the elongation at break may be a limiting factor for packaging applications [213].

The progress of scientific investigations on polyethylene/chitosan composites has gained significant consideration, especially those prepared by melt processing, because of their superior control of the final material's properties with respect of solvent evaporation methods. The result of the work by Lima et al. [214] specified that chitosan tunes the viscosity, loss modulus, storage and torque modulus (i.e., melt viscosity), and the mixture chitosan/PE-g-MA compatibilizer has a comparable, even if negligible, consequence. In the case of higher fillers amounts, (more than 15 wt %) the PE-g-MA influenced the rheological behavior of the mixtures, maybe improving matrix–filler interactions and working as an active compatibilizer [215].

The selection of the polymers to be blended with the chitosan depends on the property to be conferred or boosted. For example, the affinity to water characteristic of chitosan is changed by blending with polymeric matrices, such as PEG and PVA. Chitosan was similarly blended with a great number of polymeric matrices, such as polyamides, pol (acrylic acid), gelatin, silk fibroin, and cellulose to improve mechanical characteristics.

Chitosan/nylon 11 blends, at different ratios, were produced [216] and results revealed that the physical properties of nylon 11 were greatly affected by the addition of chitosan in the blended films and that good biodegradability of the resulting blends was observed. Smitha et al. [217] have considered the characterization of crosslinked blends of chitosan/nylon 66 at different weight compositions. The obtained results showed good indication for dehydration of dioxane and moderate water sorption (50–90%) of the blends, with no significant effects on the mechanical stability of the blends.

High temperature melting PET was also tested in combination with chitosan [218]. The PET/Chitosan blends showed a noteworthy antibacterial characteristic towards Gram-positive and Gram-negative bacteria, which improved at higher content of chitosan films. In addition, tensile strength and deformation at break of PET/chitosan films reduced with increasing chitosan amount. Consequently, the investigations of PET/chitosan systems underlined the possibility to utilize the proposed systems in the food industry as antimicrobial packages.

Mascarenhas et al. [219] blended chitosan with polystyrene to enhance mechanical and physical characteristics of chitosan, and to improve its functionality towards some specific applications: The versatility of the blends, such as film-forming ability, hydrophilicity, biodegradability, and biocompatibility are comparable with the existing blends.

Carrasco-Guigón et al. [220] realized PP/chitosan-based composites, applying the extrusion procedure and using chitosan at 9% (w/w) and PP-g-MA at 5% (w/w) as a compatibilizer. The presence of chitosan increased the wettability of the films, thanks to its high affinity to the water, while crystallinity and mechanical characteristics of PP reduced with the presence of chitosan (this behaviour could be predictable, due to the limited mechanical performance of chitosan bio-based polymeric matrix). In addition, due to the typical antimicrobial activity of chitosan, this material can be used to develop food containers to increase the conservation of food for a longer time in respect to traditional polymeric matrices.

Fernandez-Saiz et al. [221] reported, for the first time, about water barrier blends of chitosan with EVOH copolymers by solution casting: Optimal properties in terms of microstructure, optical characteristics, biocide, and water barrier activity could theoretically support the development of new bio-coatings based on chitosan salts and EVOH. The chitosan/copolymer blend can preserve the transparency and the dimensional stability, even in the presence of humidity, of the neat EVOH, but showing improved water barrier and exceptional biocide characteristics when compared to chitosan.

Polyvinyl chloride (PVC)/chitosan packaging has been also investigated [146]: Ouattara et al. [222] realized chitosan-based antibacterial packages. The diffusion of acetic or propionic acid from thin films (44–54 mm) was analyzed after water immersion at various pH (5.7, 6.4, or 7.0) and temperatures (4, 10, or 24 °C). Due to the antimicrobial characteristics of chitosan, these polymeric blends can be used as antibacterial packages [223]. Nevertheless, chitosan-based films may absorb water and swell on extended time contact. Therefore, the checking of geometric features, such as porosity, is recommended.

5.3. Cellulose

Cellulose, which is crystalline and an insoluble material in water, becomes appropriate for the realization of thin systems if it is transformed in cellophane [126]. Other than cellophane, ethyl, hydroxyl-ethyl, cellulose acetate, and hydroxyl-ethyl cellulose are commercially-available treated celluloses with good toughness, transparency, flexibility, and resistance to fats and oils [224]. Cellulose acetate can be used in combination with other bio-based polymers. Suvorova et al. [225,226] reported about the use of cellulose diacetate (CDA) mixed with potato or corn starch: Enhancements in barrier and mechanical properties were found when methyl cellulose was compounded with starch-whey protein, other polysaccharides, or lipids [227,228]. A recent study aimed to identify polycaprolactone as a candidate for blending with methyl cellulose, which can be synergistically employed in a layered system. PCL significantly lowers the water vapor permeability and increases the puncture resistance when compared to methyl cellulose (7.3×10^{-11} gm/m^{2}s Pa) [229].

Examples of CA in combination with petroleum based polymers can be found in blends or laminated PET, even if usage today is much reduced with the availability of the lower cost biaxially oriented polypropylene (BOPP) [230], while the use of PVC combined with CA is strictly related to ultrafiltration purposes [231], and few studies can be found where authors considered the blending as a way of aiming to reduce the amount of (PVC) waste products in the environment and to increase their biodegradability [232,233]. Its combination with polyamide is even restricted to membrane applications [234].

Because CA is expandable in a similar way to polystyrene, there are available studies in which surface roughness and foam morphology of cellulose acetate sheets have been compared with PS [235], and a few papers are present on their blending [236,237].

5.4. PHB

Neat PHB shows several drawbacks, such as a high degree of crystallinity and thermal instable behavior. Consequently, its blending with petrochemical-based polymeric matrices showing low degradation values, but superior mechanical characteristics can be an answer to gain environmental and eco-sustainable advantages. The blending approach of PHB and PET might guarantee variations in the performance of the produced disposals, such as the improved mechanical characteristics of the PET or PHB biodegradability (Figure 6d) [191]. Dias et al. [238] noted that comparable temperatures during the melting phenomenon were detected for PHB/PET blends without any important interactions between them. Numerous works regarding the investigation of the main characteristics of LDPE/PHB blends have been carried out and are reported in the literature [239,240]. There is confirmation that LDPE/PHB blends are not miscible matrices with fine distribution of phase boundaries between dispersed phase and matrix. In accordance to Pankova et al. [241], the morphology of these blends showed that the minor constituent (PHB) forms band-like fibrils entrenched in the LDPE polymer. If the content of PHB is higher than 16 wt %, the blend system undertakes microstructure modification from oriented PHB structure to isotropic one, where the PHB fibrils convert into a network. The dissimilarity between the two microstructures reveals the dissimilar values of water permeability of the produced systems. Consequently, the amount of PHB in the blends adjusts the microstructure and, as a result, determines the water barrier characteristics. The same effect was found in the case of PP/PHB blends [242], characterized by a reduction of stiffness and crystallinity in the presence of a PHB polymer, and at the same time the blend showed an increase of flexibility in relation to the content of PP in the blend. Olkhov et al. [243] investigated the replacement of ether functional groups belonging to PHB, having reduced affinity to water with more hydrophilic groups (amide) belonging to polyamide, while Abdelwahab et al. studied how to advance the compatibility of PHB and PS, which have analogous processing temperatures but are basically incompatible, by using commercial compatibilizing ingredients based on PS random copolymers, containing methyl methacrylate or maleic anhydride comonomers [192] (Figure 6e). PHB/ Polyvinyl acetate (PVAc) or PVA blends were widely analyzed: PVAc and its modification, as the PVA, are miscible with PHB in the melting region. In the case of PHB/PVAc (74/26) blend, a pronounced enhancement in the deformation at break, related to the reduction of crystallinity values of the blend, was noted, determined by the presence of PVAc. Other than this, PVAc addition inhibited the second crystallization phenomenon of PHB at room temperature and; therefore, the blends demonstrated unchanging physical characteristics during storage at room temperature [244].

5.5. PLA

Data provided by the literature include evaluation of properties of melt-blended PET containing small amounts of PLA, or blends containing greater amounts of PLA obtained by solution casting: These polymeric blends are characterized by reduced mechanical properties in relationship to the high content of PLA in PET, making it unappealing for industrial use [245]. The main reasons for these reductions can be found in: (a) PET processing temperature ($\sim$260–300 °C), well over the melting temperature of PLA ($\sim$160 °C), causing the PLA degradation and chain scission during blend compounding; and (b) no miscibility of the two polymeric matrices, which has been observed even for blends containing only small contents of PLA (5 wt %) [246]. The different polarity is also in charge of the limited compatibility of PLA with polyolefins; therefore, it is common to add a compatibilizer, such as polyethylene-grafted maleic anhydride (PE-g-MAH), in order to increase the characteristics of blends [247]. Alternatively, taking into account difficulties that need to be overcome in processing, modification by selecting a less expensive polymer, like PVC, is quite limited, so PLA, and a few numbers of articles concerning PVC/PLA-based blends have been published. It was proved that PLA stabilized the thermal degradation of PVC; however, a better compatibility was found only when MAH was considered in the blend (phase separation disappeared in the presence of MAH and the formation of MAH-g-PVC was confirmed by the increase of glass transition temperature of PVC) [248].

In the case of PS-containing blends, it is well recognized that polystyrene possessed limited degradation and narrow affinity to the water, which made the disposal of PS difficult after use. In order to modify the degradation process of PS, PS was blended with TPS, realized using several plasticizers, and the results indicated that blending of PLA with PS might be one of the best ways to find an equilibrium between cost effective PLA and also find reduced properties of PS [249,250]. Imre et al. [193] found a correlation between structure and interactions (Figure 6f), by determination of dispersed particles size, calculation of the Flory–Huggins interaction solubility parameters, and by the quantitative estimation of the composition dependence of tensile strength.

In addition to the identification of a possible process for refining of polylactide (PLA) toughness, polyamide (PA) with elevated toughness and strength was utilized to realize PLA/PA blends [251]. There is a limited number of works proposed in the literature that considered a PLA blend with polyamides, and the system result is still not totally understood [252]. Stoclet et al. [253] analyzed the microstructure and mechanical characteristics of this system and assumed that PLA/PA11 is a low interfacial tension blend with quite good compatibility, while Dong et al. [254] analyzed the result of introducing ethylene glycidyl methacrylate-graft-styrene-co-acrylonitrile (EGMA-g-AS) rubber particles in PLA or PA11 phases on the mechanical characteristics of PLA/PA11 blends. They detected a 78- and 5.2-fold enlargement in deformation at break and impact strength, respectively, in ternary blends, in which EGMA-g-AS is mainly dispersed in the PLA phase. They also concluded that the presence of EGMA-g-AS did not modify the dispersion of polymeric phase microstructure of PLA/PA11 blends. In the proposed works, several aspects of the microstructure, interfacial and coalescence characteristics of PLA/PA11 continue to be uncertain or contradictory. The possibility of having a bio-based PA has been also evaluated [255,256], where it was confirmed that low molecular weight epoxy resin could have a substantial role as a reactive compatibilizer in PLA/PA 610 blends.

Notwithstanding its good thermal and mechanical properties, the highly hydrophobic nature of PLA leads to low hydrolytic degradation rates. In general, hydrophobicity means the poor ability for holding-up water. As water uptake is an essential step to degradation process, PLA is often blended with synthetic biopolymers, such as PVA, in order to enhance its biodegradability. Li et al. [257] recommended that the blending of PVA with PLA could lead to promising ecofriendly materials for packaging applications with high performance, such as good mechanical properties and thermoplasticity. Furthermore, Shuai et al. [258] stated that poly(L-lactide) was immiscible with PVA with completely different T_g peaks being observed from differential scanning calorimetry (DSC) results, which is ascribed to the absence of hydrogen bonding in amorphous regions.

5.6. PBS

PBS is a costly polymer with limited mechanical properties, consequently it is not good for final use. Because polyethylene terephthalate is a material that is hardly degradable by microorganisms, the possible blending of PET with PBS could enhance the disintegrability of a non-degradable matrix in the blend. Threepopnatkul et al. [259] found that PET/PBS blends were totally immiscible. At the same time, an overall decrease of tensile properties, such as strength, modulus, and elongation at break, in the presence of PBS, in comparison with neat PET thin film, was measured. In the case of blending with polyolefins, due to the fact that HDPE has comparable mechanical properties with PBS, the blending of these two matrices not only modified the cost but also improved the overall mechanical performance of the final blend. Aontee et al. [172] found that a HDPE and PBS blend was immiscible, showing many different morphologies depending on HDPE content (from spherical domain to worm-like and elongated structure, up to coalescence; moreover, it was observed that yielding and breaking strength gradually decreased with increasing HDPE content. The use of PVC was recently considered by Chuayjuljit et al. [260], who designed how to recover the stiffness and induce degradability of poly(vinyl chloride) by inserting poly(butylene succinate) (PBS) in the blends: The obtained blends (from 10 to 50 wt % of PBS phase) showed an increase in the impact strength, elongation at break, and inclination to biodegrade when compared to the neat PVC, only with increasing content of the biodegradable PBS phase.

5.7. PBAT

Polybutylene adipate-co-terephthalate (PBAT) has great deformation at break comparable to a thermally processable elastomer, a characteristic that should be considered in case of blending with PET, having interesting thermal and mechanical characteristics, low gas diffusion, chemical resistance, and good transparency, but limited degradation in presence of water and microbials. As shown by Thongsonget al. [261], with increasing content of PBAT, thermal resistance of PBAT/PET film was decreased in comparison with neat PET film. In parallel, the increase of PBAT would result in lower values for elastic modulus and tensile strength, while the deformation at break would considerably improve, particularly when the PBAT amount is above 40 wt %. Nevertheless, the limited presence of 10 wt % of PBAT could even enhance deformation at break and tensile strength, to give values more elevated than neat PET system.

5.8. Bio-PE-Based Blends

A limited number of papers discussed the blending of bio-based polyolefins, and only one example was found with bio-based PLA. As reported in Brito et al. [262], the authors investigated the effect of ethylene-glycidyl methacrylate (E-GMA) and ethylene-methyl acrylate-glycidyl methacrylate (EMA-GMA) copolymers as compatibilizer agents in PLA/Bio-PE blend. The data underlined that the use of the E-GMA and EMA-GMA copolymers significantly enhanced the impact strength of the PLA/Bio-PE blend, thanks to the reaction between hydroxyl or carboxyl groups in PLA and the epoxy groups in the copolymer matrices.

6. Nanocomposites of Hybrid Blends and Bio-Based Nanofillers

Generally, the functional characteristics of bio-based polymers in relation to their physical and functional characteristics are required to be modified to accommodate foodstuff necessities, applying diverse strategies and modifications, as chemical or physical changes (crosslinking) or blending with different polymers and or fillers/components (compatibilizers or plasticizers). Consequently, several works have been made to design polymeric-based materials, extracted from renewable sources, with similar properties to those of traditional ones, with the final scope of reducing the environmental problems induced by the use of plastics [263,264]. The properties to be considered are numerous and may incorporate water vapor and gas diffusivity into polymeric chains, mechanical characteristics, thermal processability, sealing capability, resistance (chemical attacks (acid), grease, water, UV light, etc.), machinability (on the packaging line), antifogging ability, optical characteristics (transparency), printability, and economic aspects. It is significant to understand that no single natural material will assure all possible business areas and utilization. Consequently, an increasing attention is seen in the development of packages involving multilayer systems, in analogy to the conservative approach to create multilayer high-barrier films (e.g., a bio-based laminate implying plasticized chitosan or a starch based system combined with PHA or PLA), the final substance offers the barrier characteristics to gases and mechanical strength similar to a laminate with an external layer of polyamide (PA) or ethylene-vinyl alcohol (EVOH) assembled with LDPE [265]. Therefore, the characteristics of any biopolymer are analyzed focusing the attention on the main fundamental characteristics:

(a) Gas barrier properties: It was studied that biodegradable polymeric matrices show fine resistance to the oxygen transmission and, presently, several strategies are achieved to adjust their barrier characteristics. Nevertheless, the oxygen transmission rate (OTR) is superior in bio packaging with respect to conventionally-utilized polymeric matrices, with the consequence that the shelf-life of foods decreased. Similarly, the CO_2 barrier performance of polymeric matrices is also largely significant in packages for fresh foodstuffs. Conventional high barrier systems contain the presence of numerous layers to create a system with the necessary features; similarly, two or more bio-based polymeric matrices can be arranged to design a polymeric system taking the essential requirements [266].

(b) Water vapor barrier characteristics: The main restrictions of bio-based packages are their hydrophilic performance. Even if water vapor barrier performances of bio-polymeric matrices are related with conventional matrices, a composite biopolymer can be designed and innovated with transmittance rates rather similar to the conventional plastics. The biodegradable package is a high-quality alternative for products that necessitate high water vapor diffusion. Deep study has been performed to modulate the water diffusion characteristics of bio-systems also realizing numerous layers or developing nanocomposite approach [267].

(c) Light barrier characteristics: The sunlight energy accelerates the corrosion and deterioration processes that unfavorably influence photosensitive foods. Sunlight, in addition, operates as a catalyst to enhance the rancidity of lipids. Additionally, light provokes the oxidative modification in the polymeric, resulting in the deterioration of polymeric materials, counting both bio-based and conventional packages. To prevent the photo degradation of polymeric matrices and foodstuffs, UV stabilizer and absorber components can be included in the package systems [268].

(d) Compostability: It is one of the main significant conditions of bio polymeric matrices, as it changes their attitude to the disintegration and modifies the degradation behavior of wastes into advantageous soil products. The disintegration of polymeric materials in composting soil depends essentially on characteristics of the polymeric sources. The first stage of disintegration in composting soil is typically based on the hydrolytic process that damages the polymeric materials, and hydrolytic level is related to the water vapor diffusion of the substance [269–271].

It has been recommended that intrinsic weaknesses of bio-based packages, such as reduced barrier characteristics, hydrophilicity, low heat deflection temperatures, restricted processing gap, and low mechanical performance, may be overcome by a nanocomposite approach [140,272]. The potential of layered clay mineral nanoparticles has been established, thanks to their commercial accessibility, relatively high barrier and mechanical performance enhancements, low price, and moderately simple processability. Nevertheless, it may be underlined that although nanoclays are largely available natural sources, they are not biodegradable or renewable. Several other nanoparticles, comprising nano-metal oxide or metal have been utilized to enhance the matrix thermal performance and add some functions, such as strong antibacterial action.

In this scenario, bio-based nanofillers, obtained from no totally crystalline bio-polymeric matrices, such as cellulose, chitin, or starch, or from amorphous lignin, have recently obtained significant attention owing to their elevated accessibility and low density. Nevertheless, while literature on fully biodegradable nanocomposite blends containing nanofillers from plant and animal origin is quite extensive, research on hybrid systems is relatively unexplored. According to this, a deep revision of current results and overall performance of hybrid (bio-based plus synthetic matrices) polymeric nanocomposites containing different nanofillers (mainly bio-based ones) will follow.

6.1. Starch Based Hybrid Nanocomposites

Naderizadeha et al. [273] realized starch-based films by blending with PVA, produced by solvent casting. Two nanofillers (sodium montmorillonite (Na-MMT) and SNCs (starch nanocrystals)) were used and their achievements on overall performance of nanocomposites were investigated. This study unfolded the synergistic influence of SNC and Na-MMT to modulate mechanical characteristics, mostly in 3 wt % of the total amount of nanofillers in 50:50 (w/w).

To enhance the mechanical and moisture diffusion characteristics of starch-based systems, attention has been directed to produce nanocomposites by adding nanoscale particles [274]. For example, previous studies have revealed that whatever the clay type, the presence of nanoclays into PVA/starch, PCL/starch [275,276], PE/starch [277], and PP/starch blends [278] determined an enhancement in the material stiffness, mechanical characteristics, water resistance, and thermal stability. Tang and co-workers [279] developed several PVA–starch/nano-silicon dioxide (SiO$_2$) biodegradable blend films and discovered that the mechanism responsible for the improved tensile strength and water resistance of the composites was the formation of a strong chemical bond between nano-SiO$_2$ and PVA–starch blend. They also analyzed the biodegradability of nano-SiO$_2$ reinforced starch/PVOH nanocomposites and the results underlined that nanoparticles had no important control on biodegradability of films. Spiridon et al. [280], on the other hand, analyzed degradation of clay/starch/PVOH nanocomposites and determined that films biodegradation is related to both typology and number of nanoparticles, and the nanoparticles delayed the biodegradation degree.

The PVA/starch/clay nanocomposite as a food packaging material has been also developed [281] and, in active packaging, some additives have been also included in the formulation to promote an antimicrobial or bacterial adhesion inhibitory effect [282]. Quite recently, Tang and Alavi [283] reviewed the use of starch in combination with PVOH and established that thermally-processable starch/PVOH/montmorillonite micro and nanocomposites could demonstrate intercalated and exfoliated structures throughout the extrusion process technique.

6.2. Cellulose-Based Hybrid Nanocomposites

Carboxymethyl cellulose (CMC) added to polyvinyl alcohol (PVA) biopolymers have been extensively utilized for advanced biodegradable films in the packaging sector: CMC are compatible and miscible bio-based matrices, due to the existence of multifunctional groups on polymeric chains; as a result the blend strategy of these matrices can facilitate the production of bio-based matrices with characteristics that consent their use in the promising field of bio-packages. PVA/CMC blend can be utilized as a novel bio-blend matrix to develop and realize bio-based nanocomposite systems with improved characteristics. El Achaby et al. [284] established that the presence of 5 wt % CNC (cellulose nanocrystals) in PVA/CMC improved the tensile modulus and strength by 141% and 83%, respectively, and the water vapor diffusivity was weakened by 87%. Moreover, the developed systems reinforced with filler at thenanoscale level preserved similar transparency of the PVA/CMC system (transparency level in the visible region was estimated at around ~90%), highlighting that the CNC were well distributed at the nanoscale. The proposed nanocomposites showed interesting adhesion characteristics and the great number of functional groups present in the CNC's surface and in the macromolecular chains of the PVA/CMC system are useful to enhance the interfacial relationship between the polymeric blend and CNC. As a result, these eco-friendly organized bio-nanocomposite systems with higher characteristics are estimated to be of practical use in the food packaging sector. By using MMT, Taghizadeh et al. [285] analyzed that MMT content drastically influenced the rate of starch solubilization and explained the reduction of the degradation rate in MMT/PVA-/CMC. Additionally, at 5% (*w/w*) of MMT, the systems are characterized by the lowest water absorption capability (WAC) % levels.

6.3. Chitosan-Based Hybrid Nanocomposites

Cano and co-authors [45] designed and studied the combination of cellulose nanocrystals (CNC) in polymeric blend of poly(vinyl alcohol) (PVA) and pea starch (2:1 content) blend films, to enhance the physical and functional characteristics and the stability of produced formulations throughout storage, reducing some negative aspects of starch-based films. Specific amounts (1, 3, and 5 wt %) of CNC were selected to modulate the barrier properties of PVA/pea starch based film used as control. No changes in water vapor permeability (WVP) were registered adding the different content of CNC; this behavior can be related to the enhancement in the hydrophilic nature of the films, as also observed by the overall migration levels in polar and non-polar food simulants. The nanocomposites are characterized by lightly rigid and more stretchable behavior, crystallization phenomenon of PVA was moderately reduced by CNC addition [45].

Multifunctional poly(vinyl alcohol) (PVA) and (10 wt %) chitosan (CH) films reinforced with (3 wt %) CNC obtained from kiwi *Actinidia deliciosa* lignocellulosic wastes, obtained after the pruning were produced, and combined with (5 wt %) carvacrol, used as the active ingredient, for prospective industrial application [286]. Luzi and co-author developed and studied novel PVA and PVA/10CH systems with modulate characteristics of main significance in thefood packaging sector [286]. The microstructure analysis highlighted that no changes in the PVA and PVA/CH cross section areas were found due to the addition of nanocellulose (CNC) and/or carvacrol (Carv), underlining the positive synergism among the ingredients. The authors studied the optical and colorimetric characteristics, the data highlighted that no changes were detected on transmittance level and colorimetric appearance of PVA and PVA/CH blend, which were kept also in the case of four components-based formulations (Figure 7, Panel A). The different produced PVA-based formulations combined with carvacrol showed an important antioxidant effect, whilst the presence of carvacrol and chitosan induced an antimicrobial effect [286]. Yang and co-authors analyzed the effect of introducing lignin nanofillers (LNP) at two different weight ratios (1 and 3 wt %) on chitosan/polyvinyl alcohol (PVA) hybrid nanocomposite films, realized by using a solvent casting technique [48]. Antibacterial studies (Figure 7, Panel B) revealed the capability to decrease the microbial growth of Gram-negative bacteria, suggesting the possibility of using these films against the growth of microbial plant/fruit pathogens in the food package sector. Additionally, the results from the presence and combined effect of CH and LNP in the antioxidation property (Figure 7, Panel C) predicted their use not only in the food packaging sector, but also in the biomedical sector (drug delivery, tissue engineering, wound healing), where new antiseptic innovations are frequently necessary [48].

Panel A: Visual Image

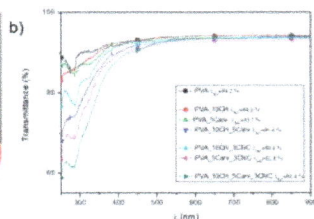

Panel B: Antimicrobial properties   **Panel C: Antioxidant properties**

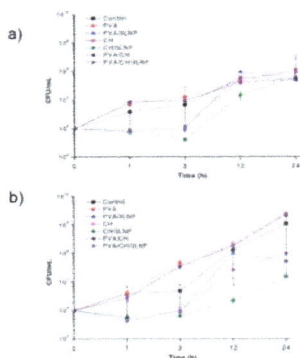

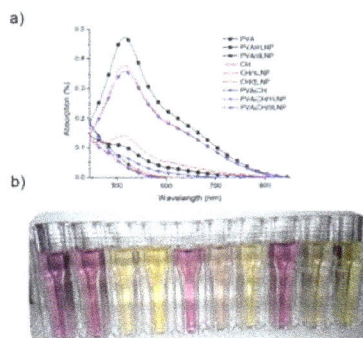

Figure 7. Panel **A**: Illustration of PVA/10CH/5Carv/3CNC formulation (**a**); and UV-Vis curves of PVA formulations (**b**) [286]. Panel **B**: Antibacterial study of PVA, PVA/CH, and PVA/CH/LNP nanocomposites, on the increase of plant pathogenic bacteria *Pectobacterium carotovorum* subsp. *odoriferum* (Pco) (CFBP 1115) 1×10^6 CFU/mL (**a**); and *Xanthomonas arboricola* pv. *pruni* (Xap) (CFBP 3894) 1×10^6 CFU/mL (**b**) [48]. Panel **C**: Antioxidant properties migrating substances of PVA/CH/LNP nanocomposites immersed in methanol solution for 24 h, measuring the absorbance level at 517 nm (**a**); and colorimetric deviation of the DPPH methanolic solution (**b**) [48].

Another application of chitosan hybrid nanocomposites was analyzed by Sadeghi and co-authors [287]. In that work, ethylene vinyl alcohol copolymer (EVOH) and chitosan were combined with nano zinc oxide (nano-ZnO) and glycerol used as plasticizer, these systems were deeply analyzed and developed for food packaging systems. The functional characteristics of EVOH nanocomposites underlined improved barrier, mechanical, and transparency characteristics. The antiseptic characteristics were enhanced by adding chitosan. The addition of plasticizer determined a reduction of barrier properties and, at the same time, an increase of deformation at break (ε_b) was detected. The addition of nano-ZnO improved barrier, mechanical, and antibacterial characteristics [287].

6.4. Polyester-Based Hybrid Nanocomposites

Nuñez et al. [288] focused the investigation on the possible efficiency of sepiolite clay on the final performance (tensile toughening) of matrices in ternary blending systems, having PLA as the polymeric phase and low-density PE as the distributed phase. Results highlighted that the blends realized without clay are characterized by high thermal stability and tensile durability, compared to those realized with sepiolite. The nanocomposite blends showed comparable thermal profile, lower tensile strength and Young's modulus values, and enhanced deformation at break and tensile toughness with respect of PLA nanocomposites. These data evidenced how clay dispersion, type of microstructure of the various blends, localization of the sepiolite in several phases, thermomechanical degradation of the PLA through melt blending, and grafting level of the utilized compatibilizer ingredients were critical for the successful performance of the materials. As'habi et al. [289] considered the blending of poly lactic acid (PLA)/linear low-density polyethylene (LLDPE) polymeric nanocomposites, realized using two commercial-grade nanoclays: due to the selective localization of the nanoclays in the PLA phase, the systems that were realized applying a two stages mixing procedure showed interesting biodegradability, microstructure, and improved melt resistance in comparison to the one step mixing process.

7. Production, Market, and Future Perspective of Hybrid Bio-Based Polymers

Every year, 125 million tons of plastic materials are utilized worldwide, of which 25% is utilized for packaging aims. It is obvious that the possible business for packaging systems is massive if they can be realized with good functionality, processability, and at an interesting and advantageous price. This is the high request for the producers of biopolymers oriented to the packaging sector, that have to deal with conventional plastics, which are available at low price, are easy to be realized, and with modulated and improved packaging characteristics [290]. Table 4 summarizes a list of foodstuffs, their barrier necessities, the classic packaging systems and green selections and, at the same time, with their technology readiness levels; while in Figure 8, examples of commercial bio-based plastics and polymers for packaging and disposals are provided.

The large amount of green food packaging systems are quite expensive in respect to fossil-based systems [291]. The price for commodity plastics is largely centered in the range \$1.32–\$3.3/kg. Unluckily, no exact evaluation of price for traditional and bio-based systems is accessible. It was estimated that bio-based materials are three to five times more expensive in comparison to traditional packaging systems [292]. The higher drivers of the cost to realize green materials include the cost for mobilizing biomass wastes, cost for technical and scientific innovations, and the lack of economies-of-scale (introduction of a eco-friendly bio-based creation chain, from natural wastes and biomass wastes to final bio-based material is a difficult and expensive procedure, moreover novelties are required to adjust present bioconversion procedures to new typologies of feedstock, or improve new procedures).

Table 4. Barrier properties of selected foodstuffs with classic and bio-based packaging systems [293].

Packed Product	Barrier Requirements	Classic Packaging Solution	Bio-Based Packaging Solution	Technology Readiness Level
Meat/fish	High barrier against oxygen and gas (aroma);	Trays (PS, PP, PVC with EVOH + LDPE or PVC as coating) + foil (PVC) or lid, bags for short term storage; waxed paper (wrapping), paperboard external packaging; transparent films (PP, PE)	Multilayer packaging materials, functional bio-based coating (modified starches) + antimicrobial and anti-fogging systems	On the market (as pilot packaging on selected markets); still more expensive than conventional solutions
Fresh cheese	High barrier properties; grease, water, O_2, CO_2 and N_2, aroma and light. MAP (80% N_2, 20% CO_2)	Transparent films/foils; bags (e.g., LDPE/ EVA /PVdC /EVA), trays, wrapping films (PE, laminated), plastic cups (HDPE, PP, PS) + high barrier lid (PA/LDPE)	Eco-paper for short term storage (wrapping); PHA/modified PLA films	On the market, still more expensive than conventional plastics
Dairy products/ Liquids	High barrier properties; water vapor (scavenging moisture), O_2, light high/moderate for grease and aroma	Waxed paper, LDPE, PVC, or aluminum-coated/laminated paper or paperboard, plastic films (BOPP), metal cans	Paper/paperboard coated with bio-based materials	Close to market
Salad (flexible packaging)	High oxygen barrier, water resistant	Transparent laminated PP films	PLA films (perforated) Coated paper with bio-based films + transparent window	On the market, still more expensive than conventional plastics
Fruits/ vegetables	Medium barrier properties (water vapor)	Perforated PP, OPP, LDPE; PVC films/bags, trays, pouches, overwraps; PS/PP trays	Molding pulp—trays PLA films (perforated) Edible coatings (polysaccharides: xanthan gum, starch, cellulose, HPC, MC, CMC, proteins: chitosan, corn zein, wheat gluten) + low barrier packaging films	On the market (molded pulp trays); on the market (PLA as pilot packaging in selected markets, e.g., for tomatoes); still more expensive than conventional solutions
Take-away food	Grease, thermal insulation	Polystyrene foam trays	Paperboard with grease barrier coating on the inside	On the market

BOPP—biaxially oriented polypropylene, CMC/carboxymethyl cellulose, EVA = ethylene vinyl acetate, EVOH = ethylene vinyl alcohol, HDPE = high-density polyethylene, HIPS = high-impact polystyrene, HPC = hydroxypropyl cellulose, LDPE = low-density polyethylene, MC = methyl cellulose, OPP = oriented polypropylene, PA = polyamide, PE = polyethylene, PET = polyethylene terephthalate, PHA = polyhydroxyalkanoate, PLA = poly lactic acid, PP = polypropylene, PS = polystyrene, PVC = polyvinyl chloride, PVdC = polyvinylidene chloride.

The "Bio-Based Polymers Producer Database," which is incessantly updated by the Nova-Institute, exhibits that Europe's running situation in creating bio-based polymeric matrices is restricted to a few number of polymeric matrices. The European community has so far determined a solid role, principally in the area of starch blending materials (polymeric blending based on the combination of polymers with starch or thermoplastic starch) and it is anticipated to continue to be solid in this specific sector for the following few years. Commercial examples of starch-based blends with conventional polymers (Linear low-density polyethylene (LLDPE), polypropylene (PP), High Impact Polystyrene (HIPS)) can be found, such as Cereplast and Teknor Apex products (Starch/LLDPE, Starch/PP, Starch/HIPS from 30/70 up to 50/50 wt %, Starch/LLDPE and Starch/PP up to 50/50 wt %), TPS/synthetic copolyesters/additives from BIOP and TPS/polyolefins hybrid from Biograde and Cardia Bioplastics. Similarly, for polybutylene terephthalate (PBT), new progresses in the realization of bio-based 1,4 butanediol (BDO) have established that the ecofriendly course to the polymer is commercially operable and its realization is deliberated to be introduced by 2020. The market size is quite differentiated, taking into account that some sectors taken by conventional plastics (PS, PVC, PET) have been now targeted by eco-friendly compostable plastic materials, such as poly lactic acid (PLA), polybutylene succinate (PBS), and polyhydroxyalkanoates (PHA's). Exact objective markets contain the catering-service manufacturing in single uses, such as plates, disposable eating utensils, foamed cups, containers, and bowls (Figure 9).

PACKAGING

DISPOSABLES

FILMS

CONSUMER GOODS / DURABLES

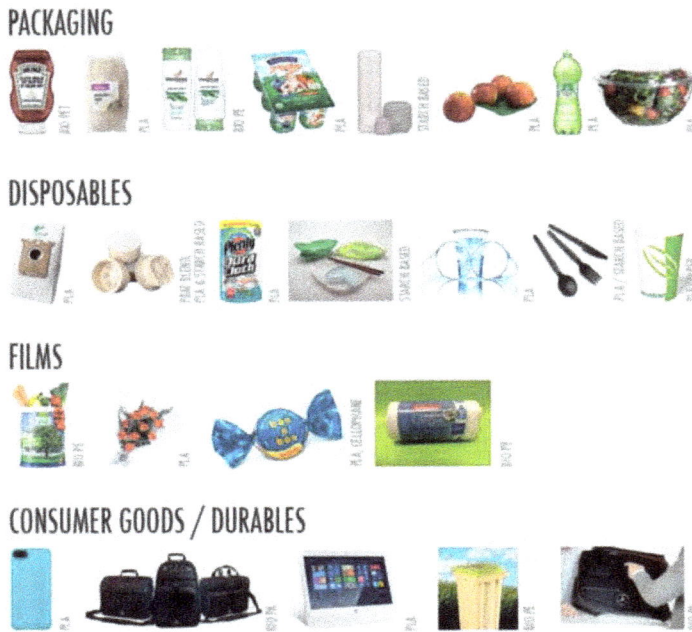

Figure 8. Bio-based plastic products.

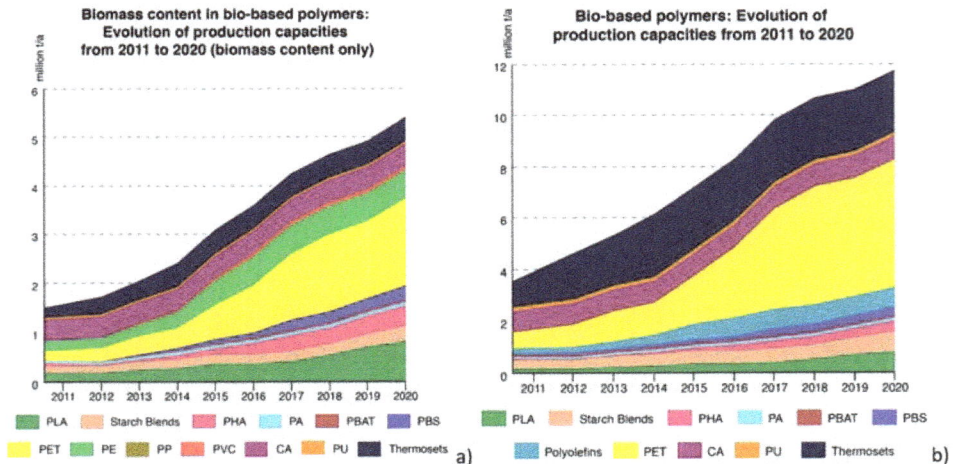

Figure 9. Bio-based polymers: Biomass content utilized in bio-based polymeric matrices (**a**); and development of production capacities from 2011 to 2020 (**b**) [282].

Other markets are represented by green or moderately bio-based non-biodegradable thermally-processable materials, such as eco-friendly PVC, PP, PE, or PET: In 2010, the inclusion of a 100% bio-material comparable to HDPE by Braskem changed the bioplastics sector, requiring at the same time a renewable resource component and a compostability behavior. Obtained from sugar cane ethanol, this material is chemically comparable to petrochemical extracts, thus can be utilized in the same sector. Similarly, bio-PVC can be realized totally from natural sources. Since 2007, Solvay Indupa, the Brazilian arm of Belgium-based chemical colossal Solvay, established strategies to utilize Brazilian sugarcane ethanol as a PVC raw material. At the moment, Bio-PET contains bio-based glycol extracted from sugar cane ethanol and the amount of renewable carbon was estimated at around 20%.

The main markets for this creation are ketchup bottles (Heinz), soda bottles (Coca Cola), and Pantene shampoo packages (Proctor and Gamble). Nevertheless, the price of bio ethylene glycol is yet significantly superior in respect to its traditional equivalent based one, and so market diffusion is limited. Even if nylon 6 and nylon 6,6, based on renewable components, are not currently presented on the market area, Rennovia has established its process to realize both hexamethylene diamine and adipic acid monomers to facilitate bio nylon 6,6 and nylon 6 [294].

On the other hand, even though it is obvious that important development has been completed in reducing numerous characteristic issues with aliphatic polyesters, there is still a requirement for further advancements. Almost all the processes that are practicable nowadays utilize, also, traditional petrochemical and non-compostable ingredients, or do not present the development economics needed (the content of these ingredients that can be utilized are also restricted if compostability is yet a condition). In detail, in some blanket positive assessment of bio-based and biodegradable plastics, it is frequently forgotten that energy from petroleum fuels is also utilized in their fabrication—be it in the sowing of crops, harvesting, fermentation, transport, etc. It is, consequently, necessary to evaluate a product's complete life cycle, because only after this evaluation is it feasible to carry out scientifically sound life cycle assessment comparisons and reach a logical conclusion about a product's sustainability.

Finally, sustainability of the polymeric matrix production will center on the successful presence of novel polymeric matrices that are extracted/obtained from yearly renewable sources. These resources must have reasonable end of life possibilities (e.g., they should be able to be reused by chemical or physical means, biodegraded to inoffensive mixtures, or minimally destroyed to recuperate the energy amount). Alternatively, "unsustainable polymers" may yet contribute as key roles in advancing sustainability, as lately underlined in a research work in Chemical and Engineering News [283], centered on food packaging systems with whole multilayered polymeric systems, where it has been conclusively remarked how the cost of plastic packaging is certainly high, but the cost of not using it may be higher.

8. Conclusions

Biopolymers, considered as green polymeric matrices or plastics realized from natural feedstock by synthetic routes, frequently have poorer characteristics and performances in respect to traditional polymeric matrices. One route to be monitored for achieving characteristic combinations necessary for the packaging sector is their blending, in the presence or not of nanosized fillers. The present review underlined how hybrid blends containing renewable polymers, in concurrence with synthetic polymers and additives, have great potential in enhancing the moisture and gas barrier properties of bio-based materials, and how they are not economically practicable to be utilized without polymeric blends with low-priced plastics of comparable necessary characteristics. On the other hand, the overall performance of polymeric blends is undoubtedly correlated to blend compositions and phase morphologies that need to be optimized by using compatibilization methods or a nanocomposite approach. In order to consider their potentials and enter new markets, other than the packaging sector where a feeble interest has already risen, scientific research should strongly focus its efforts on extending their use and improving general performance in other parallel or different sectors.

Author Contributions: F.L. and D.P. wrote the paper, L.T. and J.M.K. supervisioned the content.

Conflicts of Interest: The authors declare no conflicts of interest.

References

1. Geyer, R.; Jambeck, J.R.; Law, K.L. Production, use, and fate of all plastics ever made. *Sci. Adv.* **2017**, *3*, e1700782. [CrossRef] [PubMed]
2. Ellen MacArthur Foundation and McKinsey and Company. *The New Plastics Economy-Rethinking the Future of Plastics*; Ellen MacArthur Foundation: Cowes, UK, 2016.
3. Pongrácz, E. The environmental impacts of packaging. *Environ. Conscious Mater. Chem. Process.* **2007**, *2*, 237.
4. Europe, F.D. *Environmental Sustainability Vision towards 2030*; Achievements, Challenges and Opportunities: Brussels, Belgium, 2012.
5. Hestin, M.; Faninger, T.; Milios, L. *Increased EU Plastics Recycling Targets: Environmental, Economic and Social Impact Assessment*; Deloitte: New York, NY, USA, 2015.
6. Robertson, G. State-of-the-art biobased food packaging materials. In *Environmentally Compatible Food Packaging*; Woodhead Publishing: Sawston, CA, USA, 2008.
7. Babu, R.; O'Connor, K.; Seeram, R. Current progress on bio-based polymers and their future trends. *Prog. Biomater.* **2013**, *2*, 8. [CrossRef] [PubMed]
8. Narancic, T.; Verstichel, S.; Reddy Chaganti, S.; Morales-Gamez, L.; Kenny, S.T.; De Wilde, B.; Babu Padamati, R.; O'Connor, K.E. Biodegradable plastic blends create new possibilities for end-of-life management of plastics but they are not a panacea for plastic pollution. *Environ. Sci. Technol.* **2018**, *52*, 10441–10452. [CrossRef] [PubMed]
9. Wang, S.; Yu, J.; Yu, J. Influence of maleic anhydride on the compatibility of thermal plasticized starch and linear low-density polyethylene. *J. Appl. Polym. Sci.* **2004**, *93*, 686–695. [CrossRef]
10. Bikiaris, D.; Prinos, J.; Koutsopoulos, K.; Vouroutzis, N.; Pavlidou, E.; Frangis, N.; Panayiotou, C. LDPE/plasticized starch blends containing PE-g-MA copolymer as compatibilizer. *Polym. Degrad. Stab.* **1998**, *59*, 287–291. [CrossRef]
11. Biresaw, G.; Carriere, C.J. Compatibility and mechanical properties of blends of polystyrene with biodegradable polyesters. *Compos. Part A Appl. Sci. Manuf.* **2004**, *35*, 313–320. [CrossRef]
12. Swift, G. Degradability of commodity plastics and specialty polymers. An overview. *Degrad. Commod. Plast. Spe. Polym.* **1990**, *433*, 2–12.
13. Otey, F.H.; Westhoff, R.P.; Russell, C.R. Biodegradable films from starch and ethylene-acrylic acid copolymer. *Ind. Eng. Chem. Prod. Res. Dev.* **1977**, *16*, 305–308. [CrossRef]
14. Swanson, C.L.; Shogren, R.L.; Fanta, G.F.; Imam, S.H. Starch-plastic materials—Preparation, physical properties, and biodegradability (a review of recent USDA research). *J. Environ. Polym. Degrad.* **1993**, *1*, 155–166. [CrossRef]
15. Schneiderman, D.K.; Hillmyer, M.A. 50th anniversary perspective: There is a great future in sustainable polymers. *Macromolecules* **2017**, *50*, 3733–3749. [CrossRef]
16. Dartee, M. It's time to get to know your way around bioplastics. *Plast. Technol.* **2010**, *56*, 18–22.
17. Sam, S.T.; Nuradibah, M.A.; Ismail, H.; Noriman, N.Z.; Ragunathan, S. Recent advances in polyolefins/natural polymer blends used for packaging application. *Polym. Plast. Technol. Eng.* **2014**, *53*, 631–644. [CrossRef]
18. Rallini, M.; Kenny, J.M. Nanofillers in Polymers. In *Modification of Polymer Properties*; Elsevier: New York, NY, USA, 2017; pp. 47–86.
19. Luzi, F.; Fortunati, E.; Di Michele, A.; Pannucci, E.; Botticella, E.; Santi, L.; Kenny, J.M.; Torre, L.; Bernini, R. Nanostructured starch combined with hydroxytyrosol in poly(vinyl alcohol) based ternary films as active packaging system. *Carbohydr. Polym.* **2018**, *193*, 239–248. [CrossRef] [PubMed]
20. Seoane, I.T.; Luzi, F.; Puglia, D.; Cyras, V.P.; Manfredi, L.B. Enhancement of paperboard performance as packaging material by layering with plasticized polyhydroxybutyrate/nanocellulose coatings. *J. Appl. Polym. Sci.* **2018**, *135*, 46872. [CrossRef]
21. Le Corre, D.; Bras, J.; Dufresne, A. Starch nanoparticles: A review. *Biomacromolecules* **2010**, *11*, 1139–1153. [CrossRef] [PubMed]
22. Peponi, L.; Puglia, D.; Torre, L.; Valentini, L.; Kenny, J.M. Processing of nanostructured polymers and advanced polymeric based nanocomposites. *Mater. Sci. Eng.* **2014**, *85*, 1–46. [CrossRef]
23. Khosravi-Darani, K.; Bucci, D.Z. Application of Poly (hydroxyalkanoate) In Food Packaging: Improvements by Nanotechnology. *Chem. Biochem. Eng. Q.* **2015**, *29*, 275–285. [CrossRef]

24. Luzi, F.; Fortunati, E.; Jiménez, A.; Puglia, D.; Pezzolla, D.; Gigliotti, G.; Kenny, J.M.; Chiralt, A.; Torre, L. Production and characterization of PLA_PBS biodegradable blends reinforced with cellulose nanocrystals extracted from hemp fibres. *Ind. Crops Prod.* **2016**, *93*, 276–289. [CrossRef]

25. Yang, W.; Fortunati, E.; Luzi, F.; Kenny, J.M.; Torre, L.; Puglia, D. Lignocellulosic Based Bionanocomposites for Different Industrial Applications. *Curr. Org. Chem.* **2018**, *22*, 1205–1221. [CrossRef]

26. Mohanty, A.K.; Misra, M.; Drzal, L.T. Sustainable bio-composites from renewable resources: Opportunities and challenges in the green materials world. *J. Polym. Environ.* **2002**, *10*, 19–26. [CrossRef]

27. Luzi, F.; Fortunati, E.; Puglia, D.; Lavorgna, M.; Santulli, C.; Kenny, J.M.; Torre, L. Optimized extraction of cellulose nanocrystals from pristine and carded hemp fibres. *Ind. Crops Prod.* **2014**, *56*, 175–186. [CrossRef]

28. Alila, S.; Besbes, I.; Vilar, M.R.; Mutjé, P.; Boufi, S. Non-woody plants as raw materials for production of microfibrillated cellulose (MFC): A comparative study. *Ind. Crops Prod.* **2013**, *41*, 250–259. [CrossRef]

29. Lavoine, N.; Desloges, I.; Dufresne, A.; Bras, J. Microfibrillated cellulose–Its barrier properties and applications in cellulosic materials: A review. *Carbohydr. Polym.* **2012**, *90*, 735–764. [CrossRef] [PubMed]

30. Brinchi, L.; Cotana, F.; Fortunati, E.; Kenny, J.M. Production of nanocrystalline cellulose from lignocellulosic biomass: Technology and applications. *Carbohydr. Polym.* **2013**, *94*, 154–169. [CrossRef] [PubMed]

31. Habibi, Y.; Lucia, L.A.; Rojas, O.J. Cellulose nanocrystals: Chemistry, self-assembly, and applications. *Chem. Rev.* **2010**, *110*, 3479–3500. [CrossRef] [PubMed]

32. Dufresne, A. Nanocellulose: A new ageless bionanomaterial. *Mater. Today* **2013**, *16*, 220–227. [CrossRef]

33. Rescignano, N.; Fortunati, E.; Montesano, S.; Emiliani, C.; Kenny, J.M.; Martino, S.; Armentano, I. PVA bio-nanocomposites: A new take-off using cellulose nanocrystals and PLGA nanoparticles. *Carbohydr. Polym.* **2014**, *99*, 47–58. [CrossRef]

34. He, X.; Luzi, F.; Yang, W.; Xiao, Z.; Torre, L.; Xie, Y.; Puglia, D. Citric acid as green modifier for tuned hydrophilicity of surface modified cellulose and lignin nanoparticles. *ACS Sustain. Chem. Eng.* **2018**, *6*, 9966–9978. [CrossRef]

35. Nasseri, R.; Mohammadi, N. Starch-based nanocomposites: A comparative performance study of cellulose whiskers and starch nanoparticles. *Carbohydr. Polym.* **2014**, *106*, 432–439. [CrossRef]

36. Zhu, L.; Liang, K.; Ji, Y. Prominent reinforcing effect of chitin nanocrystals on electrospun polydioxanone nanocomposite fiber mats. *J. Mech. Behave. Biomed. Mater.* **2015**, *44*, 35–42. [CrossRef] [PubMed]

37. Kongkaoroptham, P.; Piroonpan, T.; Hemvichian, K.; Suwanmala, P.; Rattanasakulthong, W.; Pasanphan, W. Poly (ethylene glycol) methyl ether methacrylate-graft-chitosan nanoparticles as a biobased nanofiller for a poly (lactic acid) blend: Radiation-induced grafting and performance studies. *J. Appl. Polym. Sci.* **2015**, *132*. [CrossRef]

38. Fabra, M.J.; Pardo, P.; Martínez-Sanz, M.; Lopez-Rubio, A.; Lagarón, J.M. Combining polyhydroxyalkanoates with nanokeratin to develop novel biopackaging structures. *J. Appl. Polym. Sci.* **2016**, *133*. [CrossRef]

39. Fortunati, E.; Luzi, F.; Puglia, D.; Dominici, F.; Santulli, C.; Kenny, J.M.; Torre, L. Investigation of thermo-mechanical, chemical and degradative properties of PLA-limonene films reinforced with cellulose nanocrystals extracted from Phormium tenax leaves. *Eur. Polym. J.* **2014**, *56*, 77–91. [CrossRef]

40. Fortunati, E.; Luzi, F.; Puglia, D.; Petrucci, R.; Kenny, J.M.; Torre, L. Processing of PLA nanocomposites with cellulose nanocrystals extracted from Posidonia oceanica waste: Innovative reuse of coastal plant. *Ind. Crops Prod.* **2015**, *67*, 439–447. [CrossRef]

41. Fortunati, E.; Armentano, I.; Zhou, Q.; Iannoni, A.; Saino, E.; Visai, L.; Berglund, L.A.; Kenny, J.M. Multifunctional bionanocomposite films of poly(lactic acid), cellulose nanocrystals and silver nanoparticles. *Carbohydr. Polym.* **2012**, *87*, 1596–1605. [CrossRef]

42. Fortunati, E.; Puglia, D.; Luzi, F.; Santulli, C.; Kenny, J.M.; Torre, L. Binary PVA bio-nanocomposites containing cellulose nanocrystals extracted from different natural sources: Part I. *Carbohydr. Polym.* **2013**, *97*, 825–836. [CrossRef]

43. Fortunati, E.; Luzi, F.; Puglia, D.; Terenzi, A.; Vercellino, M.; Visai, L.; Santulli, C.; Torre, L.; Kenny, J.M. Ternary PVA nanocomposites containing cellulose nanocrystals from different sources and silver particles: Part II. *Carbohydr. Polym.* **2013**, *97*, 837–848. [CrossRef] [PubMed]

44. Fortunati, E.; Benincasa, P.; Balestra, G.M.; Luzi, F.; Mazzaglia, A.; Del Buono, D.; Puglia, D.; Torre, L. Revalorization of barley straw and husk as precursors for cellulose nanocrystals extraction and their effect on PVA_CH nanocomposites. *Ind. Crops Prod.* **2016**, *92*, 201–217. [CrossRef]

45. Cano, A.; Fortunati, E.; Cháfer, M.; González-Martínez, C.; Chiralt, A.; Kenny, J.M. Effect of cellulose nanocrystals on the properties of pea starch–poly (vinyl alcohol) blend films. *J. Mater. Sci.* **2015**, *50*, 6979–6992. [CrossRef]

46. Habibi, Y. Key advances in the chemical modification of nanocelluloses. *Chem. Soc. Rev.* **2014**, *43*, 1519–1542. [CrossRef] [PubMed]

47. Yang, W.; Kenny, J.M.; Puglia, D. Structure and properties of biodegradable wheat gluten bionanocomposites containing lignin nanoparticles. *Ind. Crops Prod.* **2015**, *74*, 348–356. [CrossRef]

48. Yang, W.; Owczarek, J.S.; Fortunati, E.; Kozanecki, M.; Mazzaglia, A.; Balestra, G.M.; Kenny, J.M.; Torre, L.; Puglia, D. Antioxidant and antibacterial lignin nanoparticles in polyvinyl alcohol/chitosan films for active packaging. *Ind. Crops Prod.* **2016**, *94*, 800–811. [CrossRef]

49. Yang, W.; Fortunati, E.; Bertoglio, F.; Owczarek, J.S.; Bruni, G.; Kozanecki, M.; Kenny, J.M.; Torre, L.; Visai, L.; Puglia, D. Polyvinyl alcohol/chitosan hydrogels with enhanced antioxidant and antibacterial properties induced by lignin nanoparticles. *Carbohydr. Polym.* **2018**, *181*, 275–284. [CrossRef] [PubMed]

50. Thakur, V.K.; Thakur, M.K.; Raghavan, P.; Kessler, M.R. Progress in green polymer composites from lignin for multifunctional applications: A review. *ACS Sustain. Chem. Eng.* **2014**, *2*, 1072–1092. [CrossRef]

51. Beisl, S.; Miltner, A.; Friedl, A. Lignin from micro-to nanosize: Production methods. *Int. J. Mol. Sci.* **2017**, *18*, 1244. [CrossRef] [PubMed]

52. Tian, D.; Hu, J.; Bao, J.; Chandra, R.P.; Saddler, J.N.; Lu, C. Lignin valorization: Lignin nanoparticles as high-value bio-additive for multifunctional nanocomposites. *Biotechnol. Biofuels* **2017**, *10*, 192.

53. Bertoft, E. Understanding starch structure: Recent progress. *Agronomy* **2017**, *7*, 56. [CrossRef]

54. Chaudhary, D.; Adhikari, B. Effect of temperature and plasticizer molecular size on moisture diffusion of plasticized-starch biopolymer. *Starch Stärke* **2010**, *62*, 364–372. [CrossRef]

55. Mäkelä, M.J.; Korpela, T.; Laakso, S. Studies of starch size and distribution in 33 barley varieties with a celloscope. *Starch Stärke* **1982**, *34*, 329–334. [CrossRef]

56. De la Concha, B.B.S.; Agama-Acevedo, E.; Nuñez-Santiago, M.C.; Bello-Perez, L.A.; Garcia, H.S.; Alvarez-Ramirez, J. Acid hydrolysis of waxy starches with different granule size for nanocrystal production. *J. Cereal Sci.* **2018**, *79*, 193–200. [CrossRef]

57. Salaberria, A.M.; Diaz, R.H.; Labidi, J.; Fernandes, S.C.M. Role of chitin nanocrystals and nanofibers on physical, mechanical and functional properties in thermoplastic starch films. *Food Hydrocoll.* **2015**, *46*, 93–102. [CrossRef]

58. Salaberria, A.M.; Fernandes, S.C.M.; Diaz, R.H.; Labidi, J. Processing of α-chitin nanofibers by dynamic high pressure homogenization: Characterization and antifungal activity against *A. niger*. *Carbohydr. Polym.* **2015**, *116*, 286–291. [CrossRef]

59. Wang, C.; Xiong, Y.; Fan, B.; Yao, Q.; Wang, H.; Jin, C.; Sun, Q. Cellulose as an adhesion agent for the synthesis of lignin aerogel with strong mechanical performance, Sound-absorption and thermal Insulation. *Sci. Rep.* **2016**, *6*, 32383. [CrossRef]

60. Goodrich, J.D.; Winter, W.T. α-Chitin nanocrystals prepared from shrimp shells and their specific surface area measurement. *Biomacromolecules* **2007**, *8*, 252–257. [CrossRef] [PubMed]

61. Fan, Y.; Fukuzumi, H.; Saito, T.; Isogai, A. Comparative characterization of aqueous dispersions and cast films of different chitin nanowhiskers/nanofibers. *Int. J. Biol. Macromol.* **2012**, *50*, 69–76. [CrossRef] [PubMed]

62. Gopalan Nair, K.; Dufresne, A. Crab shell chitin whisker reinforced natural rubber nanocomposites. 1. Processing and swelling behavior. *Biomacromolecules* **2003**, *4*, 657–665. [CrossRef]

63. Ifuku, S.; Saimoto, H. Chitin nanofibers: Preparations, modifications, and applications. *Nanoscale* **2012**, *4*, 3308–3318. [CrossRef] [PubMed]

64. Butchosa, N.; Brown, C.; Larsson, P.T.; Berglund, L.A.; Bulone, V.; Zhou, Q. Nanocomposites of bacterial cellulose nanofibers and chitin nanocrystals: Fabrication, characterization and bactericidal activity. *Green Chem.* **2013**, *15*, 3404–3413. [CrossRef]

65. Zhang, H.; Li, R.; Liu, W. Effects of chitin and its derivative chitosan on postharvest decay of fruits: A review. *Int. J. Mol. Sci.* **2011**, *12*, 917–934. [CrossRef] [PubMed]

66. Averous, L. Biodegradable multiphase systems based on plasticized starch: A review. *J. Macromol. Sci. Polym. Rev.* **2004**, *C44*, 231–274. [CrossRef]

67. Ezekiel Mushi, N.; Butchosa, N.; Zhou, Q.; Berglund, L.A. Nanopaper membranes from chitin–protein composite nanofibers—Structure and mechanical properties. *J. Appl. Polym. Sci.* **2014**, *131*, 40121. [CrossRef]

68. Gopalan Nair, K.; Dufresne, A.; Gandini, A.; Belgacem, M.N. Crab shell chitin whiskers reinforced natural rubber nanocomposites. 3. Effect of chemical modification of chitin whiskers. *Biomacromolecules* **2003**, *4*, 1835–1842. [CrossRef] [PubMed]

69. Morin, A.; Dufresne, A. Nanocomposites of chitin whiskers from Riftia tubes and poly (caprolactone). *Macromolecules* **2002**, *35*, 2190–2199. [CrossRef]

70. Lu. Y.; Weng, L.; Zhang, L. Morphology and properties of soy protein isolate thermoplastics reinforced with chitin whiskers. *Biomacromolecules* **2004**, *5*, 1046–1051. [CrossRef] [PubMed]

71. Sriupayo, J.; Supaphol, P.; Blackwell, J.; Rujiravanit, R. Preparation and characterization of α-chitin whisker-reinforced chitosan nanocomposite films with or without heat treatment. *Carbohydr. Polym.* **2005**, *62*, 130–136. [CrossRef]

72. De Moura, M.R.; Avena-Bustillos, R.J.; McHugh, T.H.; Krochta, J.M.; Mattoso, L.H.C. Properties of novel hydroxypropyl methylcellulose films containing chitosan nanoparticles. *J. Food Sci.* **2008**, *73*, N31–N37. [CrossRef]

73. Li, M.-C.; Wu, Q.; Song, K.; Cheng, H.N.; Suzuki, S.; Lei, T. Chitin nanofibers as reinforcing and antimicrobial agents in carboxymethyl cellulose films: Influence of partial deacetylation. *ACS Sustain. Chem. Eng.* **2016**, *4*, 4385–4395. [CrossRef]

74. Mincea, M.; Negrulescu, A.; Ostafe, V. Preparation, modification, and applications of chitin nanowhiskers: A review. *Rev. Adv. Mater. Sci.* **2012**, *30*, 225–242.

75. Wu, J.; Zhang, K.; Girouard, N.; Meredith, J.C. Facile route to produce chitin nanofibers as precursors for flexible and transparent gas barrier materials. *Biomacromolecules* **2014**, *15*, 4614–4620. [CrossRef]

76. Huang, K.-S.; Sheu, Y.-R.; Chao, I.-C. Preparation and properties of nanochitosan. *Polym. Plast. Technol. Eng.* **2009**, *48*, 1239–1243. [CrossRef]

77. Ding, D.-R.; Shen, Y. Antibacterial finishing with chitosan derivatives and their nano particles. *Dye. Finish.* **2005**, *14*, 002.

78. Berthold, A.; Cremer, K.; Kreuter, J. Preparation and characterization of chitosan microspheres as drug carrier for prednisolone sodium phosphate as model for anti-inflammatory drugs. *J. Control. Release* **1996**, *39*, 17–25. [CrossRef]

79. Tian, X.X.; Groves, M.J. Formulation and biological activity of antineoplastic proteoglycans derived from Mycobacterium vaccae in chitosan nanoparticles. *J. Pharm. Pharmacol.* **1999**, *51*, 151–157. [CrossRef] [PubMed]

80. Ohya, Y.; Shiratani, M.; Kobayashi, H.; Ouchi, T. Release behavior of 5-fluorouracil from chitosan-gel nanospheres immobilizing 5-fluorouracil coated with polysaccharides and their cell specific cytotoxicity. *J. Macromol. Sci.—Pure Appl. Chem.* **1994**, *31*, 629–642. [CrossRef]

81. Díez-Pascual, A.M.; Díez-Vicente, A.L. Antimicrobial and sustainable food packaging based on poly (butylene adipate-co-terephthalate) and electrospun chitosan nanofibers. *RSC Adv.* **2015**, *5*, 93095–93107. [CrossRef]

82. Lee, H.; Noh, K.; Lee, S.C.; Kwon, I.-K.; Han, D.-W.; Lee, I.-S.; Hwang, Y.-S. Human hair keratin and its-based biomaterials for biomedical applications. *Tissue Eng. Reg. Med.* **2014**, *11*, 255–265. [CrossRef]

83. Wang, J.; Hao, S.; Luo, T.; Yang, Q.; Wang, B. Development of feather keratin nanoparticles and investigation of their hemostatic efficacy. *Mater. Sci. Eng. C* **2016**, *68*, 768–773. [CrossRef]

84. Li, L.; Wang, N.; Jin, X.; Deng, R.; Nie, S.; Sun, L.; Wu, Q.; Wei, Y.; Gong, C. Biodegradable and injectable in situ cross-linking chitosan-hyaluronic acid based hydrogels for postoperative adhesion prevention. *Biomaterials* **2014**, *35*, 3903–3917. [CrossRef]

85. Pace, L.A.; Plate, J.F.; Smith, T.L.; Van Dyke, M.E. The effect of human hair keratin hydrogel on early cellular response to sciatic nerve injury in a rat model. *Biomaterials* **2013**, *34*, 5907–5914. [CrossRef]

86. Poranki, D.; Whitener, W.; Howse, S.; Mesen, T.; Howse, E.; Burnell, J.; Greengauz-Roberts, O.; Molnar, J.; Van Dyke, M. Evaluation of skin regeneration after burns in vivo and rescue of cells after thermal stress in vitro following treatment with a keratin biomaterial. *J. Biomater. Appl.* **2014**, *29*, 26–35. [CrossRef] [PubMed]

87. Wang, S.; Wang, Z.; Foo, S.E.M.; Tan, N.S.; Yuan, Y.; Lin, W.; Zhang, Z.; Ng, K.W. Culturing fibroblasts in 3D human hair keratin hydrogels. *ACS Appl. Mater. Interfaces* **2015**, *7*, 5187–5198. [CrossRef] [PubMed]

88. Burnett, L.R.; Rahmany, M.B.; Richter, J.R.; Aboushwareb, T.A.; Eberli, D.; Ward, C.L.; Orlando, G.; Hantgan, R.R.; Van Dyke, M.E. Hemostatic properties and the role of cell receptor recognition in human hair keratin protein hydrogels. *Biomaterials* **2013**, *34*, 2632–2640. [CrossRef] [PubMed]

89. Elzoghby, A.O. Gelatin-based nanoparticles as drug and gene delivery systems: Reviewing three decades of research. *J. Control. Release* **2013**, *172*, 1075–1091. [CrossRef] [PubMed]

90. Flory, P.J.; Weaver, E.S. Helix [unk] coil transitions in dilute aqueous collagen solutions1. *J. Am. Chem. Soc.* **1960**, *82*, 4518–4525. [CrossRef]

91. Kumari, A.; Yadav, S.K.; Yadav, S.C. Biodegradable polymeric nanoparticles based drug delivery systems. *Colloids Surf. B Biointerfaces* **2010**, *75*, 1–18. [CrossRef] [PubMed]

92. Mohanty, B.; Aswal, V.K.; Kohlbrecher, J.; Bohidar, H.B. Synthesis of gelatin nanoparticles via simple coacervation. *J. Surf. Sci. Technol.* **2005**, *21*, 149.

93. Zhao, Y.-Z.; Li, X.; Lu, C.-T.; Xu, Y.-Y.; Lv, H.-F.; Dai, D.-D.; Zhang, L.; Sun, C.-Z.; Yang, W.; Li, X.-K. Experiment on the feasibility of using modified gelatin nanoparticles as insulin pulmonary administration system for diabetes therapy. *Acta Diabetol.* **2012**, *49*, 315–325. [CrossRef] [PubMed]

94. Gupta, A.K.; Gupta, M.; Yarwood, S.J.; Curtis, A.S.G. Effect of cellular uptake of gelatin nanoparticles on adhesion, morphology and cytoskeleton organisation of human fibroblasts. *J. Control. Release* **2004**, *95*, 197–207. [CrossRef]

95. Fessi, H.; Puisieux, F.; Devissaguet, J.P.; Ammoury, N.; Benita, S. Nanocapsule formation by interfacial polymer deposition following solvent displacement. *Int. J. Pharm.* **1989**, *55*, R1–R4. [CrossRef]

96. Galindo-Rodriguez, S.; Allemann, E.; Fessi, H.; Doelker, E. Physicochemical parameters associated with nanoparticle formation in the salting-out, emulsification-diffusion, and nanoprecipitation methods. *Pharm. Res.* **2004**, *21*, 1428–1439. [CrossRef] [PubMed]

97. Ganachaud, F.; Katz, J.L. Nanoparticles and nanocapsules created using the Ouzo effect: Spontaneous emulsification as an alternative to ultrasonic and high-shear devices. *ChemPhysChem* **2005**, *6*, 209–216. [CrossRef]

98. Sahoo, N.; Sahoo, R.K.; Biswas, N.; Guha, A.; Kuotsu, K. Recent advancement of gelatin nanoparticles in drug and vaccine delivery. *Int. J. Biol. Macromol.* **2015**, *81*, 317–331. [CrossRef] [PubMed]

99. Arfat, Y.A.; Ahmed, J.; Hiremath, N.; Auras, R.; Joseph, A. Thermo-mechanical, rheological, structural and antimicrobial properties of bionanocomposite films based on fish skin gelatin and silver-copper nanoparticles. *Food Hydrocoll.* **2017**, *62*, 191–202. [CrossRef]

100. Kumar, S.; Shukla, A.; Baul, P.P.; Mitra, A.; Halder, D. Biodegradable hybrid nanocomposites of chitosan/gelatin and silver nanoparticles for active food packaging applications. *Food Package. Shelf Life* **2018**, *16*, 178–184. [CrossRef]

101. Plastics—The Facts 2017. Available online: www.plasticseurope.org/application/files/5715/1717/4180/Plastics_the_facts_2017_FINAL_for_website_one_page.pdf (accessed on 10 December 2018).

102. Barlow, C.Y.; Morgan, D.C. Polymer film packaging for food: An environmental assessment. *Resour. Conserv. Recycl.* **2013**, *78*, 74–80. [CrossRef]

103. Marsh, K.; Bugusu, B. Food packaging—Roles, materials, and environmental issues. *J. Food Sci.* **2007**, *72*, R39–R55. [CrossRef]

104. Ramos, M.; Valdés, A.; Mellinas, A.; Garrigós, M. New trends in beverage packaging systems: A review. *Beverages* **2015**, *1*, 248–272. [CrossRef]

105. Hilliou, L.; Covas, J.A. Production and Processing of Polymer-Based Nanocomposites. In *Nanomaterials for Food Packaging*; Elsevier: New York, NY, USA, 2018; pp. 111–146.

106. Brandsch, J.; Piringer, O. Characteristics of plastic materials. In *Plastic Packaging Materials for Food: Barrier Function, Mass Transport, Quality Assurance, and Legislation*; John Wiley & Sons: Hoboken, NJ, USA, 2000; pp. 9–45.

107. Selke, S.E.M.; Culter, J.D. *Plastics Packaging: Properties, Processing, Applications, and Regulations*; Carl Hanser Verlag GmbH Co KG: Munich, Germany, 2016.

108. Scarfato, P.; Di Maio, L.; Incarnato, L.; Acierno, D.; Mariano, A. Influence of co-monomer structure on properties of co-polyamide packaging films. *Packag. Technol. Sci.* **2002**, *15*, 9–16. [CrossRef]

109. Cabedo, L.; Lagarón, J.M.; Cava, D.; Saura, J.J.; Giménez, E. The effect of ethylene content on the interaction between ethylene-vinyl alcohol copolymers and water—II: Influence of water sorption on the mechanical properties of EVOH copolymers. *Polym. Test.* **2006**, *25*, 860–867. [CrossRef]

110. López-Rubio, A.; Lagaron, J.M.; Giménez, E.; Cava, D.; Hernandez-Muñoz, P.; Yamamoto, T.; Gavara, R. Morphological Alterations Induced by Temperature and Humidity in Ethylene−Vinyl Alcohol Copolymers. *Macromolecules* **2003**, *36*, 9467–9476. [CrossRef]

111. Lagaron, J.M.; Powell, A.K.; Bonner, G. Permeation of water, methanol, fuel and alcohol-containing fuels in high-barrier ethylene–vinyl alcohol copolymer. *Polym. Test.* **2001**, *20*, 569–577. [CrossRef]

112. Luzi, F.; Puglia, D.; Dominici, F.; Fortunati, E.; Giovanale, G.; Balestra, G.M.; Torre, L. Effect of gallic acid and umbelliferone on thermal, mechanical, antioxidant and antimicrobial properties of poly (vinyl alcohol-co-ethylene) films. *Polym. Degrad. Stab.* **2018**, *152*, 162–176. [CrossRef]

113. Blanchard, A.; Gouanvé, F.; Espuche, E. Effect of humidity on mechanical, thermal and barrier properties of EVOH films. *J. Membr. Sci.* **2017**, *540*, 1–9. [CrossRef]

114. Lagarón, J.M.; Giménez, E.; Gavara, R.; Saura, J.J. Study of the influence of water sorption in pure components and binary blends of high barrier ethylene–vinyl alcohol copolymer and amorphous polyamide and nylon-containing ionomer. *Polymer* **2001**, *42*, 9531–9540. [CrossRef]

115. Abdullah, Z.W.; Dong, Y.; Davies, I.J.; Barbhuiya, S. PVA, PVA blends, and their nanocomposites for biodegradable packaging application. *Polym. Plast. Technol. Eng.* **2017**, *56*, 1307–1344. [CrossRef]

116. Zhou, J.; Ma, Y.; Ren, L.; Tong, J.; Liu, Z.; Xie, L. Preparation and characterization of surface crosslinked TPS/PVA blend films. *Carbohydr. Polym.* **2009**, *76*, 632–638. [CrossRef]

117. Rahman, W.; Sin, L.T.; Rahmat, A.R.; Samad, A.A. Thermal behaviour and interactions of cassava starch filled with glycerol plasticized polyvinyl alcohol blends. *Carbohydr. Polym.* **2010**, *81*, 805–810. [CrossRef]

118. López-Rubio, A.; Lagarón, J.M.; Hernandez-Munoz, P.; Almenar, E.; Catalá, R.; Gavara, R.; Pascall, M.A. Effect of high pressure treatments on the properties of EVOH-based food packaging materials. *Innov. Food Sci. Emerg. Technol.* **2005**, *6*, 51–58. [CrossRef]

119. Liu, X.; Wu, Q.; Berglund, L.A.; Fan, J.; Qi, Z. Polyamide 6-clay nanocomposites/polypropylene-grafted-maleic anhydride alloys. *Polymer* **2001**, *42*, 8235–8239. [CrossRef]

120. Birley, A.W. Plastics used in food packaging and the rôle of additives. *Food Chem.* **1982**, *8*, 81–84. [CrossRef]

121. Kirwan, M.J.; Plant, S.; Strawbridge, J.W. Plastics in Food Packaging. In *Food and Beverage Packaging Technology*; Coles, R., Kirwan, M., Eds.; Blackwell Publishing Ltd.: Hoboken, NJ, USA, 2011; pp. 157–211.

122. Finnigan, B. Barrier polymers. In *The Wiley Encyclopedia of Packaging Technology*; Yam, K.L., Ed.; John Wiley and Sons, Inc.: New York, NY, USA, 2009; pp. 103–109.

123. Armentano, I.; Fortunati, E.; Burgos, N.; Dominici, F.; Luzi, F.; Fiori, S.; Jiménez, A.; Yoon, K.; Ahn, J.; Kang, S. Processing and characterization of plasticized PLA/PHB blends for biodegradable multiphase systems. *Express Polym. Lett.* **2015**, *9*, 583–596. [CrossRef]

124. Fortunati, E.; Yang, W.; Luzi, F.; Kenny, J.; Torre, L.; Puglia, D. Lignocellulosic nanostructures as reinforcement in extruded and solvent casted polymeric nanocomposites: An overview. *Eur. Polym. J.* **2016**, *80*, 295–316. [CrossRef]

125. Peelman, N.; Ragaert, P.; De Meulenaer, B.; Adons, D.; Peeters, R.; Cardon, L.; Van Impe, F.; Devlieghere, F. Application of bioplastics for food packaging. *Trends Food Sci. Technol.* **2013**, *32*, 128–141. [CrossRef]

126. Tang, X.Z.; Kumar, P.; Alavi, S.; Sandeep, K.P. Recent advances in biopolymers and biopolymer-based nanocomposites for food packaging materials. *Crit. Rev. Food sci. Nutr.* **2012**, *52*, 426–442. [CrossRef] [PubMed]

127. Rastogi, V.; Samyn, P. Bio-based coatings for paper applications. *Coatings* **2015**, *5*, 887–930. [CrossRef]

128. Rubie van Crevel, Bio-Based Food Packaging in Sustainable Development. 2016, Forest Products Team, Forestry Policy and Resources Division. Available online: http://www.fao.org/forestry/45849-023667e93ce5f79f4df3c74688c2067cc.pdf (accessed on 10 December 2018).

129. Abbrescia, M.; Colaleo, A.; Iaselli, G.; Loddo, F.; Maggi, M.; Marangelli, B.; Natali, S.; Nuzzo, S.; Pugliese, G.; Ranieri, A. New developments on front-end electronics for the CMS Resistive Plate Chambers. *Nucl. Instrum. Methods Phys. Res. Sect. A* **2000**, *456*, 143–149. [CrossRef]

130. Nakajima, H.; Dijkstra, P.; Loos, K. The recent developments in biobased polymers toward general and engineering applications: Polymers that are upgraded from biodegradable polymers, analogous to petroleum-derived polymers, and newly developed. *Polymers* **2017**, *9*, 523. [CrossRef]

131. Assessment Conducted by Nova-Institut Concludes That First-Generation Fermentable Sugar Is Appropriate for a Sustainable Raw Material Strategy of the European Chemical Industry. Available online: https://www.european-bioplastics.org/tag/nova-institut/ (accessed on 10 December 2018).

132. Thakur, V.K.; Thakur, M.K. *Handbook of Sustainable Polymers: Processing and Applications*; CRC Press: Boca Raton, FL, USA, 2016.

133. Satam, C.C.; Irvin, C.W.; Lang, A.W.; Jallorina, J.C.R.; Shofner, M.L.; Reynolds, J.R.; Meredith, J.C. Spray-Coated Multilayer Cellulose Nanocrystal—Chitin Nanofiber Films for Barrier Applications. *ACS Sustain. Chem. Eng.* **2018**, *6*, 10637–10644. [CrossRef]

134. Narayan, R. Biobased & biodegradable polymer materials: Rationale, drivers and technology examples. *ACS Polym. Prepr.* **2005**, *46*, 319–320.

135. Vilar, M. *Starch-Based Materials in Food Packaging: Processing, Characterization and Applications*; Academic Press: Cambridge, MA, USA, 2017.

136. Castro-Aguirre, E.; Iñiguez-Franco, F.; Samsudin, H.; Fang, X.; Auras, R. Poly (lactic acid)—Mass production, processing, industrial applications, and end of life. *Adv. Drug Deliv. Rev.* **2016**, *107*, 333–366. [CrossRef] [PubMed]

137. Gadhave, R.V.; Das, A.; Mahanwar, P.A.; Gadekar, P.T. Starch Based Bio-Plastics: The Future of Sustainable Packaging. *Open J. Polym. Chem.* **2018**, *8*, 21. [CrossRef]

138. Glenn, G.M.; Orts, W.; Imam, S.; Chiou, B.-S.; Wood, D.F. Starch plastic packaging and agriculture applications. In *Starch Polymers*; Elsevier: New York, NY, USA, 2014; pp. 421–452.

139. Ferreira, A.; Alves, V.; Coelhoso, I. Polysaccharide-based membranes in food packaging applications. *Membranes* **2016**, *6*, 22. [CrossRef] [PubMed]

140. Swain, S.K.; Mohanty, F. Polysaccharides-Based Bionanocomposites for Food Packaging Applications. In *Bionanocomposites for Packaging Applications*; Springer: Berlin, Germany, 2018; pp. 191–208.

141. Credou, J.; Berthelot, T. Cellulose: From biocompatible to bioactive material. *J. Mater. Chem. B* **2014**, *2*, 4767–4788. [CrossRef]

142. Tajeddin, B. Cellulose-Based Polymers for Packaging Applications. In *Lignocellulosic Polymer Composites: Processing, Characterization, and Properties*; John Wiley & Sons: Hoboken, NJ, USA, 2014; pp. 477–498.

143. Shaghaleh, H.; Xu, X.; Wang, S. Current progress in production of biopolymeric materials based on cellulose, cellulose nanofibers, and cellulose derivatives. *RSC Adv.* **2018**, *8*, 825–842. [CrossRef]

144. Hu, B. Biopolymer-based lightweight materials for packaging applications. *Lightw. Mater. Biopolym. Biofibers* **2014**, *1175*, 239–255.

145. Arvanitoyannis, I.S. The use of chitin and chitosan for food packaging applications. In *Environmentally Compatible Food Packaging*; Woodhead Publishing: Sawston, CA, USA, 2008; Volume 137.

146. Srinivasa, P.C.; Tharanathan, R.N. Chitin/Chitosan—Safe, Ecofriendly Packaging Materials with Multiple Potential Uses. *Food Rev. Int.* **2007**, *23*, 53–72. [CrossRef]

147. Van den Broek, L.A.M.; Knoop, R.J.I.; Kappen, F.H.J.; Boeriu, C.G. Chitosan films and blends for packaging material. *Carbohydr. Polym.* **2015**, *116*, 237–242. [CrossRef]

148. Rinaudo, M. Chitin and chitosan: Properties and applications. *Prog. Polym. Sci.* **2006**, *31*, 603–632. [CrossRef]

149. Xu, Y.X.; Kim, K.M.; Hanna, M.A.; Nag, D. Chitosan–starch composite film: Preparation and characterization. *Ind. Crops Prod.* **2005**, *21*, 185–192. [CrossRef]

150. Kurek, M.; Galus, S.; Debeaufort, F. Surface, mechanical and barrier properties of bio-based composite films based on chitosan and whey protein. *Food Package. Shelf Life* **2014**, *1*, 56–67. [CrossRef]

151. Wang, H.; Qian, J.; Ding, F. Emerging Chitosan-Based Films for Food Packaging Applications. *J. Agric. Food Chem.* **2018**, *66*, 395–413. [CrossRef] [PubMed]

152. Vijayendra, S.V.N.; Shamala, T.R. Film forming microbial biopolymers for commercial applications—A review. *Crit. Rev. Biotechnol.* **2014**, *34*, 338–357. [CrossRef] [PubMed]

153. Arrieta, P.M.; Samper, D.M.; Aldas, M.; López, J. On the Use of PLA-PHB Blends for Sustainable Food Packaging Applications. *Materials* **2017**, *10*. [CrossRef]

154. Plackett, D.; Siró, I. 18—Polyhydroxyalkanoates (PHAs) for food packaging. In *Multifunctional and Nanoreinforced Polymers for Food Packaging*; Lagarón, J.-M., Ed.; Woodhead Publishing: Sawston, CA, USA, 2011; pp. 498–526.

155. Rhim, J.-W.; Hong, S.-I.; Ha, C.-S. Tensile, water vapor barrier and antimicrobial properties of PLA/nanoclay composite films. *LWT Food Sci. Technol.* **2009**, *42*, 612–617. [CrossRef]

156. Rasal, R.M.; Janorkar, A.V.; Hirt, D.E. Poly (lactic acid) modifications. *Prog. Polym. Sci.* **2010**, *35*, 338–356. [CrossRef]

157. Auras, R.; Harte, B.; Selke, S. An Overview of Polylactides as Packaging Materials. *Macromol. Biosci.* **2004**, *4*, 835–864. [CrossRef] [PubMed]

158. Murariu, M.; Dubois, P. PLA composites: From production to properties. *Adv. Drug Deliv. Rev.* **2016**, *107*, 17–46. [CrossRef] [PubMed]

159. Vroman, I.; Tighzert, L. Biodegradable Polymers. *Materials* **2009**, *2*. [CrossRef]

160. Kijchavengkul, T.; Auras, R.; Rubino, M.; Selke, S.; Ngouajio, M.; Fernandez, R.T. Biodegradation and hydrolysis rate of aliphatic aromatic polyester. *Polym. Degrad. Stab.* **2010**, *95*, 2641–2647. [CrossRef]

161. Rychter, P.; Kawalec, M.; Sobota, M.; Kurcok, P.; Kowalczuk, M. Study of Aliphatic-Aromatic Copolyester Degradation in Sandy Soil and Its Ecotoxicological Impact. *Biomacromolecules* **2010**, *11*, 839–847. [CrossRef] [PubMed]

162. Biron, M. *Industrial Applications of Renewable Plastics: Environmental, Technological, and Economic Advances*; William Andrew: Norwich, NY, USA, 2016.

163. Kyulavska, M.; Toncheva-Moncheva, N.; Rydz, J. Biobased Polyamide Ecomaterials and Their Susceptibility to Biodegradation. In *Handbook of Ecomaterials*; Martínez, L.M.T., Kharissova, O.V., Kharisov, B.I., Eds.; Springer International Publishing: Cham, Switzerland, 2017; pp. 1–34.

164. Shi, B.; Wideman, G.; Wang, J.; Shlepr, M. Bio-based polyolefin composites and functional films for reducing total carbon footprint. *J. Compos. Mater.* **2014**, *49*, 2349–2355. [CrossRef]

165. Kamigaito, M.; Satoh, K. Bio-based Hydrocarbon Polymers. In *Encyclopedia of Polymeric Nanomaterials*; Kobayashi, S., Müllen, K., Eds.; Springer Berlin Heidelberg: Berlin/Heidelberg, Germany, 2015; pp. 109–118.

166. Sousa, A.F.; Vilela, C.; Fonseca, A.C.; Matos, M.; Freire, C.S.R.; Gruter, G.-J.M.; Coelho, J.F.J.; Silvestre, A.J.D. Biobased polyesters and other polymers from 2,5-furandicarboxylic acid: A tribute to furan excellency. *Polym. Chem.* **2015**, *6*, 5961–5983. [CrossRef]

167. Gross, R.A.; Cheng, H.N.; Smith, P.B. *Green Polymer Chemistry: Biobased Materials and Biocatalysis*; American Chemical Society: Washington, DC, USA, 2015.

168. Jost, V.; Miesbauer, O. Effect of different biopolymers and polymers on the mechanical and permeation properties of extruded PHBV cast films. *J. Appl. Polym. Sci.* **2018**, *135*, 46153. [CrossRef]

169. Institute for Bioplastics and Biocomposites. *Biopolymers—Facts and Statistics, Hannover 2016*; IFBB—Institute for Bioplastics and Biocomposites: London, UK, 2016.

170. Storz, H.; Vorlop, K.-D. Bio-based plastics: Status, challenges and trends. *Landbauforschung* **2013**, *63*, 321–332.

171. Imam, S.; Glenn, G.; Chiou, B.S.; Shey, J.; Narayan, R.; Orts, W. Types, production and assessment of biobased food packaging materials. In *Environmentally Compatible Food Packaging*; Woodhead Publishing: Sawston, CA, USA, 2008; Volume 28.

172. Aontee, A.; Sutapun, W. *Effect of Blend Ratio on Phase Morphology and Mechanical Properties of High Density Polyethylene and Poly (Butylene Succinate) Blend*; Trans Tech Publications: London, UK, 2013; pp. 555–559.

173. Lackner, M. *Bioplastics—Biobased Plastics as Renewable and/or Biodegradable Alternatives to Petroplastics, Kirk-Othmer Encyclopedia of Chemical Technology*; John Wiley & Sons: Hoboken, NJ, USA, 2015.

174. Alaerts, L.; Augustinus, M.; Van Acker, K. Impact of Bio-Based Plastics on Current Recycling of Plastics. *Sustainability* **2018**, *10*, 1487. [CrossRef]

175. Lambert, S.; Wagner, M. Environmental performance of bio-based and biodegradable plastics: The road ahead. *Chem. Soc. Rev.* **2017**, *46*, 6855–6871. [CrossRef]

176. Gironi, F.; Piemonte, V. Bioplastics and petroleum-based plastics: Strengths and weaknesses. *Energy Sources Part A* **2011**, *33*, 1949–1959. [CrossRef]

177. Peres, A.M.; Pires, R.R.; Oréfice, R.L. Evaluation of the effect of reprocessing on the structure and properties of low density polyethylene/thermoplastic starch blends. *Carbohydr. Polym.* **2016**, *136*, 210–215. [CrossRef]

178. Okan, M.; Aydin, H.M.; Barsbay, M. Current approaches to waste polymer utilization and minimization: A review. *J. Chem. Technol. Biotechnol.* **2019**, *94*, 8–21. [CrossRef]

179. Álvarez-Chávez, C.R.; Edwards, S.; Moure-Eraso, R.; Geiser, K. Sustainability of bio-based plastics: General comparative analysis and recommendations for improvement. *J. Clean. Prod.* **2012**, *23*, 47–56. [CrossRef]

180. Imre, B.; Pukánszky, B. Compatibilization in bio-based and biodegradable polymer blends. *Eur. Polym. J.* **2013**, *49*, 1215–1233. [CrossRef]

181. Khan, B.; Bilal Khan Niazi, M.; Samin, G.; Jahan, Z. Thermoplastic Starch: A Possible Biodegradable Food Packaging Material—A Review. *J. Food Process Eng.* **2016**, *40*, e12447. [CrossRef]

182. Tabasum, S.; Younas, M.; Zaeem, M.A.; Majeed, I.; Majeed, M.; Noreen, A.; Iqbal, M.N.; Zia, K.M. A review on blending of corn starch with natural and synthetic polymers, and inorganic nanoparticles with mathematical modeling. *Int. J. Biol. Macromol.* **2019**, *122*, 969–996. [CrossRef] [PubMed]

183. Rosa, D.S.; Guedes, C.G.F.; Carvalho, C.L. Processing and thermal, mechanical and morphological characterization of post-consumer polyolefins/thermoplastic starch blends. *J. Mater. Sci.* **2007**, *42*, 551–557. [CrossRef]

184. Martins, A.B.; Santana, R.M.C. Effect of carboxylic acids as compatibilizer agent on mechanical properties of thermoplastic starch and polypropylene blends. *Carbohydr. Polym.* **2016**, *135*, 79–85. [CrossRef] [PubMed]

185. Debiagi, F.; Mello, L.R.P.F.; Mali, S. Chapter 6—Thermoplastic Starch-Based Blends: Processing, Structural, and Final Properties. In *Starch-Based Materials in Food Packaging*; Villar, M.A., Barbosa, S.E., García, M.A., Castillo, L.A., López, O.V., Eds.; Academic Press: Cambridge, MA, USA, 2017; pp. 153–186.

186. Cerclé, C.; Sarazin, P.; Favis, B.D. High performance polyethylene/thermoplastic starch blends through controlled emulsification phenomena. *Carbohydr. Polym.* **2013**, *92*, 138–148. [CrossRef]

187. Nguyen, D.M.; Vu, T.T.; Grillet, A.-C.; Ha Thuc, H.; Ha Thuc, C.N. Effect of organoclay on morphology and properties of linear low density polyethylene and Vietnamese cassava starch biobased blend. *Carbohydr. Polym.* **2016**, *136*, 163–170. [CrossRef]

188. Kuciel, S.; Kuźniar, P.; Nykiel, M. Biodegradable polymers in the general waste stream—The issue of recycling with polyethylene packaging materials. *Polimery* **2018**, *63*, 1–8. [CrossRef]

189. Soroudi, A.; Jakubowicz, I. Recycling of bioplastics, their blends and biocomposites: A review. *Eur. Polym. J.* **2013**, *49*, 2839–2858. [CrossRef]

190. Teyssandier, F.; Cassagnau, P.; Gérard, J.F.; Mignard, N. Reactive compatibilization of PA12/plasticized starch blends: Towards improved mechanical properties. *Eur. Polym. J.* **2011**, *47*, 2361–2371. [CrossRef]

191. Burlein, G.A.D.; Rocha, M.C.G. Mechanical and morphological properties of LDPE/PHB blends filled with castor oil pressed cake. *Mater. Res.* **2014**, *17*, 97–105. [CrossRef]

192. Abdelwahab, M.A.; Martinelli, E.; Alderighi, M.; Grillo Fernandes, E.; Imam, S.; Morelli, A.; Chiellini, E. Poly[(R)-3-hydroxybutyrate)]/poly(styrene) blends compatibilized with the relevant block copolymer. *J. Polym. Sci. Part A Polym. Chem.* **2012**, *50*, 5151–5160. [CrossRef]

193. Imre, B.; Renner, K.; Pukánszky, B. Interactions, structure and properties in poly (lactic acid)/thermoplastic polymer blends. *Express Polym. Lett.* **2014**, *8*, 2–14. [CrossRef]

194. Sabetzadeh, M.; Bagheri, R.; Masoomi, M. Study on ternary low density polyethylene/linear low density polyethylene/thermoplastic starch blend films. *Carbohydr. Polym.* **2015**, *119*, 126–133. [CrossRef] [PubMed]

195. Shujun, W.; Jiugao, Y.; Jinglin, Y. Preparation and characterization of compatible thermoplastic starch/polyethylene blends. *Polym. Degrad. Stab.* **2005**, *87*, 395–401. [CrossRef]

196. Arvanitoyannis, I.; Biliaderis, C.G.; Ogawa, H.; Kawasaki, N. Biodegradable films made from low-density polyethylene (LDPE), rice starch and potato starch for food packaging applications: Part 1. *Carbohydr. Polym.* **1998**, *36*, 89–104. [CrossRef]

197. Pedroso, A.G.; Rosa, D.S. Mechanical, thermal and morphological characterization of recycled LDPE/corn starch blends. *Carbohydr. Polym.* **2005**, *59*, 1–9. [CrossRef]

198. Euaphantasate, N.; Prachayawasin, P.; Uasopon, S.; Methacanon, P. Moisture sorption characteristic and their relative properties of thermoplastic starch/linear low density polyethylene films for food packaging. *J. Met. Mater. Min.* **2008**, *18*, 103–109.

199. Landreau, E.; Tighzert, L.; Bliard, C.; Berzin, F.; Lacoste, C. Morphologies and properties of plasticized starch/polyamide compatibilized blends. *Eur. Polym. J.* **2009**, *45*, 2609–2618. [CrossRef]

200. TureČkovÁ, J.; Prokopova, I.; Niklova, P.; ŠImek, J.A.N.; ŠMejkalovÁ, P.; KeclÍK, F. Biodegradable copolyester/starch blends-preparation, mechanical properties, wettability, biodegradation course (in English). *Polimery* **2008**, *53*, 639–643. [CrossRef]

201. Gutiérrez, T.J.; Alvarez, V.A. Properties of native and oxidized corn starch/polystyrene blends under conditions of reactive extrusion using zinc octanoate as a catalyst. *React. Funct. Polym.* **2017**, *112*, 33–44. [CrossRef]

202. Men, Y.; Du, X.; Shen, J.; Wang, L.; Liu, Z. Preparation of corn starch-g-polystyrene copolymer in ionic liquid: 1-Ethyl-3-methylimidazolium acetate. *Carbohydr. Polym.* **2015**, *121*, 348–354. [CrossRef] [PubMed]

203. Ali, M.I.; Ahmed, S.; Javed, I.; Ali, N.; Atiq, N.; Hameed, A.; Robson, G. Biodegradation of starch blended polyvinyl chloride films by isolated Phanerochaete chrysosporium PV1. *Int. J. Environ. Sci. Technol.* **2014**, *11*, 339–348. [CrossRef]

204. Jiang, W.; Qiao, X.; Sun, K. Mechanical and thermal properties of thermoplastic acetylated starch/poly (ethylene-co-vinyl alcohol) blends. *Carbohydr. Polym.* **2006**, *65*, 139–143. [CrossRef]

205. George, E.R.; Sullivan, T.M.; Park, E.H. Thermoplastic starch blends with a poly (ethylene-co-vinyl alcohol): Processability and physical properties. *Polym. Eng. Sci.* **1994**, *34*, 17–23. [CrossRef]

206. Orts, W.J.; Nobes, G.A.R.; Glenn, G.M.; Gray, G.M.; Imam, S.; Chiou, B.S. Blends of starch with ethylene vinyl alcohol copolymers: Effect of water, glycerol, and amino acids as plasticizers. *Polym. Adv. Technol.* **2007**, *18*, 629–635. [CrossRef]

207. Simmons, S.; Thomas, E.L. Structural characteristics of biodegradable thermoplastic starch/poly (ethylene–vinyl alcohol) blends. *J. Appl. Polym. Sci.* **1995**, *58*, 2259–2285. [CrossRef]

208. Wu, Z.; Wu, J.; Peng, T.; Li, Y.; Lin, D.; Xing, B.; Li, C.; Yang, Y.; Yang, L.; Zhang, L. Preparation and application of starch/polyvinyl alcohol/citric acid ternary blend antimicrobial functional food packaging films. *Polymers* **2017**, *9*, 102. [CrossRef]

209. Guohua, Z.; Ya, L.; Cuilan, F.; Min, Z.; Caiqiong, Z.; Zongdao, C. Water resistance, mechanical properties and biodegradability of methylated-cornstarch/poly (vinyl alcohol) blend film. *Polym. Degrad. Stab.* **2006**, *91*, 703–711. [CrossRef]

210. Jayasekara, R.; Harding, I.; Bowater, I.; Christie, G.B.Y.; Lonergan, G.T. Preparation, surface modification and characterisation of solution cast starch PVA blended films. *Polym. Test.* **2004**, *23*, 17–27. [CrossRef]

211. El-Hefian, E.A.; Nasef, M.M.; Yahaya, A. H. Chitosan-Based Polymer Blends: Current Status and Applications. *J. Chem. Soc. Pak.* **2014**, *36*, 11–27.

212. Kausar, A. Scientific potential of chitosan blending with different polymeric materials: A review. *J. Plast. Film Sheet.* **2017**, *33*, 384–412. [CrossRef]

213. Li, H.-Z.; Chen, S.-C.; Wang, Y.-Z. Preparation and characterization of nanocomposites of polyvinyl alcohol/cellulose nanowhiskers/chitosan. *Compos. Sci. Technol.* **2015**, *115*, 60–65. [CrossRef]

214. Lima, P.S.; Brito, R.S.F.; Santos, B.F.F.; Tavares, A.A.; Agrawal, P.; Andrade, D.L.; Wellen, R.M.R.; Canedo, E.L.; Silva, S.M.L. Rheological properties of HDPE/chitosan composites modified with PE-g-MA. *J. Mater. Res.* **2017**, *32*, 775–787. [CrossRef]

215. Quiroz-Castillo, J.M.; Rodríguez-Félix, D.E.; Grijalva-Monteverde, H.; del Castillo-Castro, T.; Plascencia-Jatomea, M.; Rodríguez-Félix, F.; Herrera-Franco, P.J. Preparation of extruded polyethylene/chitosan blends compatibilized with polyethylene-graft-maleic anhydride. *Carbohydr. Polym.* **2014**, *101*, 1094–1100. [CrossRef] [PubMed]

216. Kuo, P.-C.; Sahu, D.; Yu, H.H. Properties and biodegradability of chitosan/nylon 11 blending films. *Polym. Degrad. Stab.* **2006**, *91*, 3097–3102. [CrossRef]

217. Smitha, B.; Dhanuja, G.; Sridhar, S. Dehydration of 1, 4-dioxane by pervaporation using modified blend membranes of chitosan and nylon 66. *Carbohydr. Polym.* **2006**, *66*, 463–472. [CrossRef]

218. Masoomi, M.; Tavangar, M.; Razavi, S.M.R. Preparation and investigation of mechanical and antibacterial properties of poly (ethylene terephthalate)/chitosan blend. *RSC Adv.* **2015**, *5*, 79200–79206. [CrossRef]

219. Mascarenhas, N.P.; Gonsalves, R.A.; Goveas, J.J.; Shetty, T.C.S.; Crasta, V. Preparation and characterization of chitosan-polystyrene polymer blends. *AIP Conf.* **2016**, *1731*, 140039.

220. Carrasco-Guigón, F.J.; Rodríguez-Félix, D.E.; Castillo-Ortega, M.M.; Santacruz-Ortega, H.C.; Burruel-Ibarra, S.E.; Encinas-Encinas, J.C.; Plascencia-Jatomea, M.; Herrera-Franco, P.J.; Madera-Santana, T.J. Preparation and Characterization of Extruded Composites Based on Polypropylene and Chitosan Compatibilized with Polypropylene-Graft-Maleic Anhydride. *Materials* **2017**, *10*, 105. [CrossRef] [PubMed]

221. Fernandez-Saiz, P.; Ocio, M.J.; Lagaron, J.M. Antibacterial chitosan-based blends with ethylene–vinyl alcohol copolymer. *Carbohydr. Polym.* **2010**, *80*, 874–884. [CrossRef]

222. Ouattara, B.; Simard, R.E.; Piette, G.; Begin, A.; Holley, R.A. Diffusion of acetic and propionic acids from chitosan-based antimicrobial packaging films. *J. Food Sci.* **2000**, *65*, 768–773. [CrossRef]

223. Dutta, P.K.; Tripathi, S.; Mehrotra, G.K.; Dutta, J. Perspectives for chitosan based antimicrobial films in food applications. *Food Chem.* **2009**, *114*, 1173–1182. [CrossRef]

224. Campos, C.A.; Gerschenson, L.N.; Flores, S.K. Development of edible films and coatings with antimicrobial activity. *Food Bioprocess Technol.* **2011**, *4*, 849–875. [CrossRef]

225. Douglass, E.F.; Avci, H.; Boy, R.; Rojas, O.J.; Kotek, R. A review of cellulose and cellulose blends for preparation of bio-derived and conventional membranes, nanostructured thin films, and composites. *Polym. Rev.* **2018**, *58*, 102–163. [CrossRef]

226. Suvorova, A.I.; Tyukova, I.S.; Trufanova, E.I. Biodegradable starch-based polymeric materials. *Russ. Chem. Rev.* **2000**, *69*, 451. [CrossRef]

227. Yoo, S.; Krochta, J.M. Starch–methylcellulose–whey protein film properties. *Int. J. Food Sci. Technol.* **2012**, *47*, 255–261. [CrossRef]

228. Paunonen, S. Strength and barrier enhancements of cellophane and cellulose derivative films: A review. *BioResources* **2013**, *8*, 3098–3121. [CrossRef]

229. Khan, R.A.; Salmieri, S.; Dussault, D.; Sharmin, N.; Lacroix, M. Mechanical, barrier, and interfacial properties of biodegradable composite films made of methylcellulose and poly (caprolactone). *J. Appl. Polym. Sci.* **2012**, *123*, 1690–1697. [CrossRef]

230. Nabar, Y.U.; Gupta, A.; Narayan, R. Isothermal crystallization kinetics of poly (ethylene terephthalate)–cellulose acetate blends. *Polym. Bull.* **2005**, *53*, 117–125. [CrossRef]

231. El-Gendi, A.; Abdallah, H.; Amin, A.; Amin, S.K. Investigation of polyvinylchloride and cellulose acetate blend membranes for desalination. *J. Mol. Struct.* **2017**, *1146*, 14–22. [CrossRef]

232. Miyashita, Y.; Suzuki, T.; Nishio, Y. Miscibility of cellulose acetate with vinyl polymers. *Cellulose* **2002**, *9*, 215–223. [CrossRef]

233. Abdel-Naby, A.S.; Al-Ghamdi, A.A. Poly (vinyl chloride) blend with biodegradable cellulose acetate in presence of N-(phenyl amino) maleimides. *Int. J. Biol. Macromol.* **2014**, *70*, 124–130. [CrossRef] [PubMed]

234. Khaparde, D. Preparation and prediction of physical properties of cellulose acetate and polyamide polymer blend. *Carbohydr. Polym.* **2017**, *173*, 338–343. [CrossRef] [PubMed]

235. Hopmann, C.; Hendriks, S.; Spicker, C.; Zepnik, S.; van Lück, F. Surface roughness and foam morphology of cellulose acetate sheets foamed with 1, 3, 3, 3-tetrafluoropropene. *Polym. Eng. Sci.* **2017**, *57*, 441–449. [CrossRef]

236. Meenakshi, P.; Noorjahan, S.E.; Rajini, R.; Venkateswarlu, U.; Rose, C.; Sastry, T.P. Mechanical and microstructure studies on the modification of CA film by blending with PS. *Bull. Mater. Sci.* **2002**, *25*, 25–29. [CrossRef]

237. Meireles, C.d.S.; Filho, G.R.; de Assunção, R.M.N.; Zeni, M.; Mello, K. Blend compatibility of waste materials—Cellulose acetate (from sugarcane bagasse) with polystyrene (from plastic cups): Diffusion of water, FTIR, DSC, TGA, and SEM study. *J. Appl. Polym. Sci.* **2007**, *104*, 909–914. [CrossRef]

238. Dias, D.S.; Crespi, M.S.; Kobelnik, M.; Ribeiro, C.A. Calorimetric and SEM studies of PHB–PET polymeric blends. *J. Therm. Anal. Calorim.* **2009**, *97*, 581–584. [CrossRef]

239. Ol'khov, A.A.; Iordanskii, A.L.; Zaikov, G.E.; Shibryaeva, L.S.; Litwinow, I.A.; Vlasov, S.V. Morphological Features of Poly (3-Hydroxybutyrate)/Low Density Polyethylene Blends. *Int. J. Polym. Mater. Polym. Biomater.* **2000**, *47*, 457–468. [CrossRef]

240. Ol'khov, A.A.; Iordanskii, A.L.; Zaikov, G.E.; Shibryaeva, L.S.; Litvinov†, I.A.; Vlasov, S.V. Morphologically Special Features of Poly(3-Hydroxybutyrate)/Low-Density Polyethylene Blends. *Polym. Plast. Technol. Eng.* **2000**, *39*, 783–792. [CrossRef]

241. Fabbri, P.; Bassoli, E.; Bon, S.B.; Valentini, L. Preparation and characterization of poly (butylene terephthalate)/graphene composites by in-situ polymerization of cyclic butylene terephthalate. *Polymer* **2012**, *53*, 897–902. [CrossRef]

242. Pachekoski, W.M.; Marcondes Agnelli, J.A.; Belem Thermal, L.P. Mechanical and morphological properties of poly (hydroxybutyrate) and polypropylene blends after processing. *Mat. Res.* **2009**, *12*, 159–164. [CrossRef]

243. Olkhov, A.A.; Pankova, Y.N.; Goldshtrakh, M.A.; Kosenko, R.Y.; Markin, V.S.; Ischenko, A.A.; Iordanskiy, A.L. Structure and properties of films based on blends of polyamide–polyhydroxybutyrate. *Inorg. Mater.* **2016**, *7*, 471–477. [CrossRef]

244. El-Hadi, A.; Schnabel, R.; Straube, E.; Müller, G.; Riemschneider, M. Effect of Melt Processing on Crystallization Behavior and Rheology of Poly(3-hydroxybutyrate) (PHB) and its Blends. *Macromol. Mater. Eng.* **2002**, *287*, 363–372. [CrossRef]

245. You, X.; Snowdon, M.R.; Misra, M.; Mohanty, A.K. Biobased Poly(ethylene terephthalate)/Poly(lactic acid) Blends Tailored with Epoxide Compatibilizers. *ACS Omega* **2018**, *3*, 11759–11769. [CrossRef]

246. McLauchlin, A.R.; Ghita, O.R. Studies on the thermal and mechanical behavior of PLA-PET blends. *J. Appl. Polym. Sci.* **2016**, *133*. [CrossRef]

247. Sangermano, M.; Marchi, S.; Valentini, L.; Bon, S.B.; Fabbri, P. Transparent and Conductive Graphene Oxide/Poly(ethylene glycol) diacrylate Coatings Obtained by Photopolymerization. *Macromol. Mater. Eng.* **2010**, *296*, 401–407. [CrossRef]

248. Hachemi, R.; Belhaneche-Bensemra, N.; Massardier, V. Elaboration and characterization of bioblends based on PVC/PLA. *J. Appl. Polym. Sci.* **2013**, *131*. [CrossRef]

249. Kaseem, M.; Ko, Y.G. Melt Flow Behavior and Processability of Polylactic Acid/Polystyrene (PLA/PS) Polymer Blends. *J. Polym. Environ.* **2017**, *25*, 994–998. [CrossRef]

250. Vital, A.; Vayer, M.; Tillocher, T.; Dussart, R.; Boufnichel, M.; Sinturel, C. Morphology control in thin films of PS:PLA homopolymer blends by dip-coating deposition. *Appl. Surf. Sci.* **2017**, *393*, 127–133. [CrossRef]

251. Feng, F.; Ye, L. Structure and Property of Polylactide/Polyamide Blends. *J. Macromol. Sci. Part B* **2010**, *49*, 1117–1127. [CrossRef]

252. Patel, R.; Ruehle, D.A.; Dorgan, J.R.; Halley, P.; Martin, D. Biorenewable blends of polyamide-11 and polylactide. *Polym. Eng. Sci.* **2013**, *54*, 1523–1532. [CrossRef]

253. Stoclet, G.; Seguela, R.; Lefebvre, J.M. Morphology, thermal behavior and mechanical properties of binary blends of compatible biosourced polymers: Polylactide/polyamide11. *Polymer* **2011**, *52*, 1417–1425. [CrossRef]

254. Dong, W.; Cao, X.; Li, Y. High-performance biosourced poly(lactic acid)/polyamide 11 blends with controlled salami structure. *Polym. Int.* **2013**, *63*, 1094–1100. [CrossRef]

255. Walha, F.; Lamnawar, K.; Maazouz, A.; Jaziri, M. Biosourced blends based on poly (lactic acid) and polyamide 11: Structure-properties relationships and enhancement of film blowing processability. *Adv. Polym. Technol.* **2017**, *37*, 2061–2074. [CrossRef]

256. Pai, F.-C.; Lai, S.-M.; Chu, H.-H. Characterization and Properties of Reactive Poly(lactic acid)/Polyamide 610 Biomass Blends. *J. Appl. Polym. Sci.* **2013**, *130*, 2563–2571. [CrossRef]

257. Li, H.-Z.; Chen, S.-C.; Wang, Y.-Z. Thermoplastic PVA/PLA Blends with Improved Processability and Hydrophobicity. *Ind. Eng. Chem. Res.* **2014**, *53*, 17355–17361. [CrossRef]

258. Shuai, X.; He, Y.; Asakawa, N.; Inoue, Y. Miscibility and phase structure of binary blends of poly(L-lactide) and poly(vinyl alcohol). *J. Appl. Polym. Sci.* **2001**, *81*, 762–772. [CrossRef]

259. Threepopnatkul, P.; Wongnarat, C.; Intolo, W.; Suato, S.; Kulsetthanchalee, C. Effect of TiO2 and ZnO on Thin Film Properties of PET/PBS Blend for Food Packaging Applications. *Energy Procedia* **2014**, *56*, 102–111. [CrossRef]

260. Chuayjuljit, S.; Kongthan, J.; Chaiwutthinan, P.; Boonmahitthisud, A. Poly(vinyl chloride)/Poly(butylene succinate)/wood flour composites: Physical properties and biodegradability. *Polym. Compos.* **2016**, *39*, 1543–1552. [CrossRef]

261. Thongsong, W.; Kulsetthanchalee, C.; Threepopnatkul, P. Effect of polybutylene adipate-co-terephthalate on properties of polyethylene terephthalate thin films. *Mater. Today* **2017**, *4*, 6597–6604. [CrossRef]

262. Brito, G.F.; Agrawal, P.; Mélo, T.J.A. Mechanical and Morphological Properties of PLA/BioPE Blend Compatibilized with E-GMA and EMA-GMA Copolymers. *Macromol. Symp.* **2016**, *367*, 176–182. [CrossRef]

263. Muller, J.; González-Martínez, C.; Chiralt, A. Combination of Poly(lactic) Acid and Starch for Biodegradable Food Packaging. *Materials* **2017**, *10*. [CrossRef]

264. Rhim, J.-W.; Kim, Y.-T. Chapter 17—Biopolymer-Based Composite Packaging Materials with Nanoparticles. In *Innovations in Food Packaging*, 2nd ed.; Han, J.H., Ed.; Academic Press: San Diego, CA, USA, 2014; pp. 413–442.

265. Ahmed, S. (Ed.) *Bio-Based Materials for Food Packaging*; Springer: Berlin, Germany, 2018.

266. Swain, S.K. Gas Barrier Properties of Biopolymer-based Nanocomposites: Application in Food Packaging. *Adv. Mater. Agric. Food Environ. Saf.* **2014**, 369–384. [CrossRef]

267. Lagaron, J.M.; Núñez, E. Nanocomposites of moisture-sensitive polymers and biopolymers with enhanced performance for flexible packaging applications. *J. Plast. Film Sheet.* **2011**, *28*, 79–89. [CrossRef]

268. Wolf, C.; Angellier-Coussy, H.; Gontard, N.; Doghieri, F.; Guillard, V. How the shape of fillers affects the barrier properties of polymer/non-porous particles nanocomposites: A review. *J. Membr. Sci.* **2018**, *556*, 393–418. [CrossRef]

269. Majid, I.; Thakur, M.; Nanda, V. Biodegradable Packaging Materials. *Ref. Mod. Mater. Sci. Mater. Eng.* **2018**. [CrossRef]

270. Sadeghi, G.M.M.; Mahsa, S. *Compostable Polymers and Nanocomposites—A Big Chance for Planet Earth, Recycling Materials Based on Environmentally Friendly Techniques*; Achilias, D.S., Ed.; IntechOpen: Rijeka, Croatia, 2015.

271. Gutiérrez, T.J. Biodegradability and Compostability of Food Nanopackaging Materials. In *Composites Materials for Food Packaging*; Cirillo, G., Kozlowski, M.A., Spizzirri, U.G., Eds.; Springer: Berlin, Germany, 2018; pp. 269–296.

272. Zinoviadou, K.G.; Gougouli, M.; Biliaderis, C.G. Chapter 9—Innovative Biobased Materials for Packaging Sustainability. In *Innovation Strategies in the Food Industry*; Galanakis, C.M., Ed.; Academic Press: Cambridge, MA, USA, 2016; pp. 167–189.

273. Naderizadeh, S.; Shakeri, A.; Mahdavi, H.; Nikfarjam, N.; Taheri Qazvini, N. Hybrid Nanocomposite Films of Starch, Poly(vinyl alcohol) (PVA), Starch Nanocrystals (SNCs), and Montmorillonite (Na-MMT): Structure–Properties Relationship. *Starch Stärke* **2018**, 1800027. [CrossRef]

274. Ashori, A. Effects of graphene on the behavior of chitosan and starch nanocomposite films. *Polym. Eng. Sci.* **2013**, *54*, 2258–2263. [CrossRef]

275. Vertuccio, L.; Gorrasi, G.; Sorrentino, A.; Vittoria, V. Nano clay reinforced PCL/starch blends obtained by high energy ball milling. *Carbohydr. Polym.* **2009**, *75*, 172–179. [CrossRef]

276. Ikeo, Y.; Aoki, K.; Kishi, H.; Matsuda, S.; Murakami, A. Nano clay reinforced biodegradable plastics of PCL starch blends. *Polym. Adv. Technol.* **2006**, *17*, 940–944. [CrossRef]

277. Liao, H.-T.; Wu, C.-S. Synthesis and characterization of polyethylene-octene elastomer/clay/biodegradable starch nanocomposites. *J. Appl. Polym. Sci.* **2005**, *97*, 397–404. [CrossRef]

278. DeLeo, C.; Pinotti, C.A.; do Carmo Gonçalves, M.; Velankar, S. Preparation and Characterization of Clay Nanocomposites of Plasticized Starch and Polypropylene Polymer Blends. *J. Polym. Environ.* **2011**, *19*, 689. [CrossRef]

279. Tang, S.; Zou, P.; Xiong, H.; Tang, H. Effect of nano-SiO2 on the performance of starch/polyvinyl alcohol blend films. *Carbohydr. Polym.* **2008**, *72*, 521–526. [CrossRef]

280. Spiridon, I.; Popescu, M.C.; Bodârlău, R.; Vasile, C. Enzymatic degradation of some nanocomposites of poly(vinyl alcohol) with starch. *Polym. Degrad. Stab.* **2008**, *93*, 1884–1890. [CrossRef]

281. Tang, X.; Alavi, S. Structure and Physical Properties of Starch/Poly Vinyl Alcohol/Laponite RD Nanocomposite Films. *J. Agric. Food Chem.* **2012**, *60*, 1954–1962. [CrossRef]

282. Alix, S.; Mahieu, A.; Terrie, C.; Soulestin, J.; Gerault, E.; Feuilloley, M.G.J.; Gattin, R.; Edon, V.; Ait-Younes, T.; Leblanc, N. Active pseudo-multilayered films from polycaprolactone and starch based matrix for food-packaging applications. *Eur. Polym. J.* **2013**, *49*, 1234–1242. [CrossRef]

283. Tang, X.; Alavi, S. Recent advances in starch, polyvinyl alcohol based polymer blends, nanocomposites and their biodegradability. *Carbohydr. Polym.* **2011**, *85*, 7–16. [CrossRef]

284. El Achaby, M.; El Miri, N.; Aboulkas, A.; Zahouily, M.; Bilal, E.; Barakat, A.; Solhy, A. Processing and properties of eco-friendly bio-nanocomposite films filled with cellulose nanocrystals from sugarcane bagasse. *Int. J. Biol. Macromol.* **2017**, *96*, 340–352. [CrossRef]

285. Taghizadeh, M.T.; Sabouri, N.; Ghanbarzadeh, B. Polyvinyl alcohol:starch:carboxymethyl cellulose containing sodium montmorillonite clay blends; mechanical properties and biodegradation behavior. *SpringerPlus* **2013**, *2*, 376. [CrossRef] [PubMed]

286. Luzi, F.; Fortunati, E.; Giovanale, G.; Mazzaglia, A.; Torre, L.; Balestra, G.M. Cellulose nanocrystals from Actinidia deliciosa pruning residues combined with carvacrol in PVA_CH films with antioxidant/antimicrobial properties for packaging applications. *Int. J. Biol. Macromol.* **2017**, *104*, 43–55. [CrossRef] [PubMed]

287. Sadeghi, K.; Shahedi, M. Physical, mechanical, and antimicrobial properties of ethylene vinyl alcohol copolymer/chitosan/nano-ZnO (ECNZn) nanocomposite films incorporating glycerol plasticizer. *J. Food Meas. Charact.* **2016**, *10*, 137–147. [CrossRef]

288. Nuñez, K.; Rosales, C.; Perera, R.; Villarreal, N.; Pastor, J.M. Poly(lactic acid)/low-density polyethylene blends and its nanocomposites based on sepiolite. *Polym. Eng. Sci.* **2011**, *52*, 988–1004. [CrossRef]

289. As'habi, L.; Jafari, S.H.; Khonakdar, H.A.; Boldt, R.; Wagenknecht, U.; Heinrich, G. Tuning the processability, morphology and biodegradability of clay incorporated PLA/LLDPE blends via selective localization of nanoclay induced by melt mixing sequence. *Express Polym. Lett.* **2013**, *7*, 21–39. [CrossRef]

290. Weber, C.J.; Haugaard, V.; Festersen, R.; Bertelsen, G. Production and applications of biobased packaging materials for the food industry. *Food Addit. Contam.* **2002**, *19*, 172–177. [CrossRef] [PubMed]

291. Wg1 Brochure: Moving From Oil-Based To Biobased Packaging—Key Material Aspects For Consideration Cellulose. *Chem. Technol.* **2015**, *49*, 709–713.

292. Enabling Biobased Nylons, Polyurethanes, Plasticizers, and Other Sustainable Materials. Available online: http://www.rennovia.com/markets/ (accessed on 10 December 2018).

293. Bio-Based Polymers—Production Capacity Will Triple from 3.5 Million Tonnes in 2011 to nearly 12 Million Tonnes in 2020. Available online: http://www.bio-based.eu/market_study/media/13-03-06FRMSBiopolymerslongnova.pcf (accessed on 10 December 2018).

294. Tullo, A.H. The cost of plastic packaging. *Chem. Eng. News* **2016**, *94*, 32–37.

materials

MDPI

Article

Calcium Chloride Modified Alginate Microparticles Formulated by the Spray Drying Process: A Strategy to Prolong the Release of Freely Soluble Drugs

Marta Szekalska * , Katarzyna Sosnowska , Anna Czajkowska-Kośnik and Katarzyna Winnicka *

Department of Pharmaceutical Technology, Medical University of Białystok, Mickiewicza 2c,
15222 Białystok, Poland; katarzyna.sosnowska@umb.edu.pl (K.S.); anna.czajkowska@umb.edu.pl (A.C.-K.)
* Correspondence: marta.szekalska@umb.edu.pl (M.S.); kwin@umb.edu.pl (K.W.); Tel.: +48-85-748-5616 (M.S.)

Received: 18 July 2018; Accepted: 21 August 2018; Published: 24 August 2018

Abstract: Alginate (ALG) cross-linking by $CaCl_2$ is a promising strategy to obtain modified-release drug delivery systems with mucoadhesive properties. However, current technologies to produce $CaCl_2$ cross-linked alginate microparticles possess major disadvantages, such as a poor encapsulation efficiency of water-soluble drugs and a difficulty in controlling the process. Hence, this study presents a novel method that streamlines microparticle production by spray drying; a rapid, continuous, reproducible, and scalable technique enabling obtainment of a product with low moisture content, high drug loading, and a high production yield. To model a freely water-soluble drug, metformin hydrochloride (MF) was selected. It was observed that MF was successfully encapsulated in alginate microparticles cross-linked by $CaCl_2$ using a one-step drying process. Modification of ALG provided drug release prolongation—particles obtained from 2% ALG cross-linked by 0.1% $CaCl_2$ with a prolonged MF rate of dissolution of up to 12 h. Cross-linking of the ALG microparticles structure by $CaCl_2$ decreased the swelling ratio and improved the mucoadhesive properties which were evaluated using porcine stomach mucosa.

Keywords: polymer cross-linking; alginate modification; calcium chloride; microparticles; spray drying; prolonged drug release

1. Introduction

Polysaccharides are polymeric carbohydrate molecules commonly exploited in the design of pharmaceutical formulations with prolonged drug release. These polymers possess many advantages such as non-toxicity, wide availability, simplicity to receive, and gelling ability using different cross-linking agents [1]. Sustained release formulations enable a prolonged release profile, which reduces the frequency of drug applications, minimizes side effects, and improves patient's compliance [2,3]. Dosage forms with mucoadhesive properties provide continuous contact with the mucosal membrane and as a consequence, a prolonged drug residence time and improved drug absorption and bioavailability can be achieved [4,5]. Mucoadhesive microparticles are an example of multi-unit carriers, where the active substance is incorporated in a natural or synthetic polymer matrix. Additionally, microparticles are characterized by a high surface area of drug release and short diffusion pathway, which enables the improvement of the therapeutic efficacy and a reduction of the drug toxicity [6,7].

Sodium alginate (ALG) is a non-toxic polymer which naturally occurs in seaweeds. It may also be synthesized by *Azotobacter* and *Pseudomonas* bacteria [8,9]. ALG is a polysaccharide consisting of β-D-mannuronic and α-L-guluronic acid residues linked by (1–4) glycosidic bonds. ALG is a biocompatible, nonirritant polymer with favorable swelling, gelling, and mucoadhesive properties, thus, it has a wide range of applications in drug delivery technology. ALG hydrogels can be obtained by various cross-linking methods, such as ionic modification ("egg-box" model) based on the binding of cations by the guluronate. In ionic cross-linking, $CaCl_2$ is commonly exploited [10,11]. The cross-linking of water-soluble ALG enables an improvement in the polymer stability. Hence, $CaCl_2$ cross-linked ALG has been widely examined as a material for pharmaceutical formulations. Additionally, an important feature of $CaCl_2$ cross-linking is that the process can be conducted in an aqueous environment at a low temperature, and that both macromolecular and low molecular weight therapeutic agents can be encapsulated [12–15]. Cross-linked ALG is also utilized in biomedicine—in cell encapsulation or tissue engineering [16–18]. The commonly used and well-described method to obtain ALG microparticles cross-linked by $CaCl_2$ is the internal gelation by emulsification method [19–22]. Major disadvantages of microparticles obtained by this technique include poor encapsulation efficiency of water-soluble drugs, low production yield, and organic solvent residues [8,23–25].

The spray-drying technique is a useful and valuable method to formulate spherical particles with low moisture content, high drug loading, and high production yield [25–28]. Therefore, in the present study, the opportunity of using a one-step drying process to formulate modified ALG microparticles was evaluated. As the slow release of active substances with high water solubility is a great challenge in pharmaceutical formulation designing, in the next step the influence of cross-linking on the metformin hydrochloride (MF) release, swelling, and mucoadhesive properties of designed microparticles was determined [2,3]. MF—a biguanide methyl derivative widely used as a first-line drug in non-insulin dependent diabetes mellitus was used as a model of a freely soluble drug [29].

2. Materials and Methods

2.1. Materials

Metformin hydrochloride (MF) was a product of Debao Fine Chemical CO (Henan, China). Sodium alginate (ALG) (with viscosity 132.6 mPa·s of 2% solution), mucin, and gelatin were purchased from Sigma Aldrich (Steinheim, Germany). Potassium dihydrogen phosphate, sodium hydroxide, hydrochloric acid, methanol, propan-1,2-diol, acetonitrile, and calcium chloride were from Chempur (Piekary Śląskie, Poland). Water was purified by osmosis system, Milli-Q Reagent Water System (Billerica, MA, USA). Porcine stomach mucosa was derived from the local veterinary service.

2.2. Formulation of $CaCl_2$ Modified ALG Microparticles

After preliminary studies, to prepare the microparticles: 2% (*w/w*) ALG, and drug:polymer ratio 2:1 was selected, see Table 1 [30], in addition to 0.5%, 0.1% and 0.05% (*w/w*) $CaCl_2$ were applied. In the first step, ALG solutions were prepared and poured into the $CaCl_2$ solution, after which their viscosity was determined using a rotational viscometer (Viscotester E Plus—Thermo Haake, Karlsruhe, Germany) [31]. The parameters of the spray-drying process (Mini Spray Dryer B-290, Büchi, Flawil, Switzerland) were set as follows: Flow rate of 4.5 mL/min, spray rate of 37 m^3/h, and spray flow of 600 L/h. The inlet and outlet temperatures were 115 °C and 46 °C, respectively. As a control, non-modified ALG microparticles (formulation C) were applied.

Table 1. Components of designed microparticles formulations.

Formulation	ALG [1] (%)	MF [2]:ALG [1] Ratio	$CaCl_2$ (%)
C	2	2:1	–
PCA1	2	–	0.1
PCA2	2	–	0.05
CA1	2	2:1	0.1
CA2	2	2:1	0.05

[1] Sodium alginate, [2] Metformin hydrochloride.

2.3. Evaluation of Microparticles

2.3.1. Shape and Size

To characterize shape and morphology of the microparticles, a scanning electron microscope (SEM) (Hitachi S4200, Tokyo, Japan) was utilized. The size distribution of microparticles suspended in propane-1,2-diol was examined by a Zetasizer NanoZS90 (Malvern Instruments, Malvern, UK) using at least three repetitions for each sample.

2.3.2. High Performance Liquid Chromatography (HPLC) Assay

MF concentration was studied by the HPLC method using an Agilent Technologies 1200 system (Agilent, Waldbronn, Germany) and a Waters Spherisorb® 5.0 μM ODS 4.6 × 250 mm, 5 μm column (Waters Corporation, Milford, MA, USA). As the mobile phase, an acetonitrile: methanol: phosphate buffer pH 3.0 (20:20:60, *v/v*) with a flow rate of 1.0 mL/min was exploited [30].

2.3.3. Drug Encapsulation

To assess MF loading, 20 mg of microparticles was dissolved in 10 mL of phosphate buffer pH 6.8. After 24 h of agitation in a water bath, solutions were filtrated, analyzed by the HPLC method [30], and drug loading (L) was calculated from the expression:

$$L = Q_m / W_m \times 100 \tag{1}$$

where Q_m—drug encapsulated in the microparticles, and W_m—microparticles weight.
Drug encapsulation efficiency (EE) was computed from the formula:

$$EE = Q_a / Q_t \times 100 \tag{2}$$

where Q_a—actual drug content, Q_t—theoretical drug content.
Yield of production (Y) was determined based on the equation:

$$Y = W_m / W_t \times 100 \tag{3}$$

where W_m—microparticles weight, W_t—theoretical calculated drug and polymer weight.

2.3.4. Zeta Potential

Directly after microparticles were suspended in propane-1,2-diol, Zeta potential values were determined by Zetasizer NanoZS90 (Malvern Instruments, Malvern, UK) using Zetasizer Software 6.20.

2.3.5. Swelling Characteristics

The swelling ratio (SR) was tested at 37 ± 1 °C in 0.1 M HCl (pH = 1.2) based on the expression [32]:

$$SR = W_S - W_0 / W_0 \times 100 \tag{4}$$

where W_0—microparticles weight, W_S—swollen microparticles weight.

2.3.6. Mucoadhesiveness

Mucoadhesive properties were evaluated at 37 ± 1 °C by a TA. XT. Plus Texture Analyzer (Stable Micro Systems, Godalming, UK). Gelatin, mucin, and porcine stomach mucosa were used as mucoadhesive layers [33]. Process parameters, chosen during preliminary tests, were as follows: Pretest speed 0.5 mm/s, test speed 0.1 m/s, contact time 180 s, post-test speed 0.1 mm/s, applied force 1 N. Mucoadhesiveness was expressed as the detachment force (F_{max}) and the work of mucoadhesion (W_{ad}).

2.4. MF Dissolution

MF dissolution from microparticle formulations (in the amount equivalent to 500 mg of MF) was performed at 37 ± 1 °C in a basket apparatus (Erweka Dissolution Tester Type DT 600HH, Heusenstamm, Germany) in 500 mL of 0.1 M HCl (pH = 1.2) [34]. MF concentration in the release medium was studied by the HPLC technique (as described in the point 2.3.2. HPLC analysis).

2.5. Mathematical Modeling of the MF Release Profile

To explain the drug release mechanism, data obtained from MF release tests were studied under different mathematical models [35,36]. Zero order kinetic:

$$F = k \times t \tag{5}$$

first order kinetic:

$$\ln F = k \times t \tag{6}$$

Higuchi model:

$$F = kt^{1/2} \tag{7}$$

Korsmeyer-Peppas model:

$$F = kt^n \tag{8}$$

Hixson-Crowell model:

$$1 - (1 - F)^{1/3} = kt \tag{9}$$

where F—the fraction of released drug, k—the constant connected with release, and t—the time.

2.6. Index of Similarity and Dissimilarity

To compare release profiles of designed microparticles, a model utilizing a difference factor f_1 and similarity factor f_2 was applied. The difference index f_1 was calculated by the formula:

$$f_1 = \{(\sum = 1n \mid R_t - T_t \mid) \, (\sum = 1nR_t)\} \times 100 \tag{10}$$

where n—number of samples, R_t and T_t—data of drug dissolution of control and test sample at the same time (t). The difference index f_1 expresses the percent variation between release from the studied and control sample. The similarity index f_2 indicates the potential release similarity and it is calculated by the following equation:

$$f_2 = 50 \times \text{Log} \, \{[1 + (1/n)\sum = 1n(R_t - T_t) \times 2] - 0.5 \times 100\} \tag{11}$$

f_2 between 50–100 indicates release profiles similarity between samples [37–39].

2.7. Differential Scanning Calorimetry (DSC)

Measurements were performed using an automatic thermal analyzer system (DSC TEQ2000, TA Instruments, New Castle, DE, USA). Samples (in the amount of 5 mg) were placed in aluminum pans and heated in the range 25–280 °C under a 20 mL/min nitrogen flow [40].

2.8. Statistics

Data were assessed by Statistica 10.0 (StatSoft, Tulsa, OK, USA) using one-way analysis of variance (ANOVA) or a Kruskal-Wallis test. Obtained results were presented as the mean and standard deviation.

3. Results and Discussion

ALG is a polymeric material commonly utilized in the design of sustained drug delivery systems via its in situ gelation when in contact with the acidic stomach environment. As a result of hydrogen bonding, ALG alternates into insoluble alginic acid, which controls water penetration and prevents matrix disintegration. Additionally, to improve the mechanical strength of ALG and to reduce the solubility in water, a cross-linking process by ionic interactions with divalent cations such as Ca^{2+} can be performed [41]. The gelation process is a result of the linkage by Ca^{2+} of guluronate regions in one ALG backbone with a similar region in another ALG molecule, which creates a cross-linked structure. Application of a cross-linking agent leads to a reduction in the solubility of the polymer matrix under acidic conditions, which could be an effective approach to sustain the release of freely water-soluble drugs [42–44]. A Schematic structure of the non-modified and $CaCl_2$ cross-linked formulation with MF is presented in Figure 1.

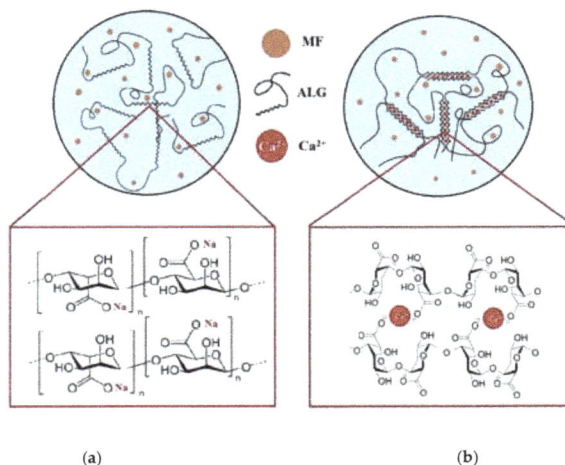

(a) (b)

Figure 1. Structure of non-modified (a) and $CaCl_2$ cross-linked (b) ALG microparticles containing MF.

3.1. Microparticles Characteristics

To receive $CaCl_2$ modified ALG microparticles, 2% ALG and 0.05% or 0.1% $CaCl_2$ solutions were chosen for the one-step spray-drying process. A 2% ALG solution modified with 0.5% $CaCl_2$ possessed a high viscosity and its spray drying was limited, see Table 2.

Table 2. Viscosity of $CaCl_2$ cross-linked ALG solutions.

Solution	Viscosity (mPa·s) [1]
2% ALG	132.6 ± 2.7
0.5% CaCl$_2$ + 2% ALG	4700.4 ± 14.5
0.1% CaCl$_2$ + 2% ALG	1600.6 ± 12.4
0.05% CaCl$_2$ + 2% ALG	791.7 ± 5.3

[1] n = 3.

The quality evaluation of obtained microparticles included analysis of particle size, MF percent loading, drug encapsulation efficiency, production yield, and Zeta potential, see Table 3. The morphology of microparticles formulation CA1 is presented in Figure 2.

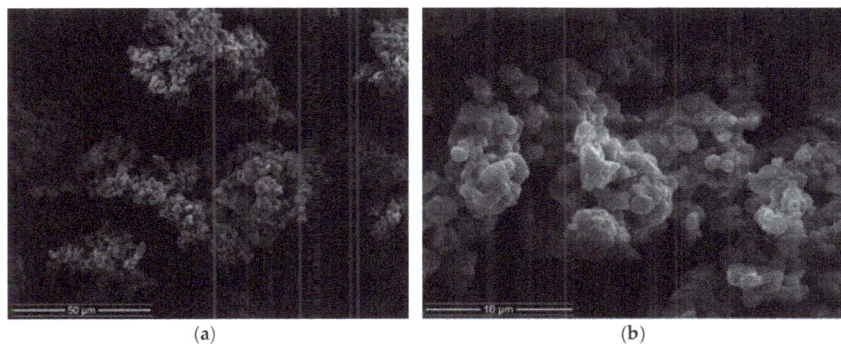

Figure 2. Images of microparticles CA1 (**a**) ×2000, (**b**) ×10000.

Obtained data revealed that the cross-linking agent induced only a slight increase in the mean diameter of the microparticles (from 3.0 ± 1.6 μm in formulation C to 3.5 ± 1.2 μm in CA1) and ALG modification by CaCl$_2$ cross-linking resulted in a small decline in production yield. Interestingly, ALG modification did not influence MF loading percentage in the microparticles, and all formulations were characterized by a drug loading of above 70%, as shown in Table 3. Additionally, it was observed that the cross-linking process resulted in a slight reduction in the encapsulation efficiency (from 113.4 ± 2.3 in formulation C to 91.5 ± 2.1% in CA1).

ALG is an anionic polymer with a negative charge, but positively charged MF and Ca^{2+} ions changed the Zeta potential [45–47] and designed microparticles possessed a positive charge, as shown in Table 3.

Table 3. Quality evaluation of non-modified microparticles C and cross-linked microparticles CA1 and CA2 containing MF.

Microparticles	Zeta Potential (mV)	Production Yield (%)	Encapsulation Efficiency (%)	Loading Percentage (%)	Particle Size (μm)
C	−1.3 ± 0.7	61.7 ± 2.1	113.4 ± 2.3	75.6 ± 1.5	3.0 ± 1.6
CA1	2.6 ± 0.4	57.1 ± 1.6	91.5 ± 2.1	77.1 ± 1.7	3.5 ± 1.2
CA2	2.6 ± 0.9	56.1 ± 2.4	93.9 ± 2.7	73.9 ± 3.4	3.4 ± 1.4

3.2. Swelling and Mucoadhesive Properties

The ALG swelling ability is a peculiar parameter affecting mucoadhesiveness and drug release. In contact with moisture, ALG begins to hydrate and swell, and as a consequence, forms a hydrogel which regulates the influx of the aqueous medium and drug dissolution [44]. As a result of gelling, the water influx is decreased; thus, the drug release is prolonged. Dissolution of freely water-soluble drugs from hydrophilic carriers is generally regulated by the drug's diffusive properties via the hydrogel matrix. Therefore, the swelling properties might significantly affect the drug release profiles [47].

The swelling behavior of designed microparticles was examined in an acidic environment and presented as the swelling ratio (SR), see Figure 3. At an acidic pH, ALG carboxylate groups on the surface of the microparticles are protonated, and water-insoluble alginic acid is formed, which impedes the penetration of fluid into the deeper layers of the particle matrix [48,49]. After contact with this medium, non-modified ALG microparticles take up the fluid, which penetrates into the matrix and leads to an increase of SR to 180 min. Formation of insoluble alginic acid leads to a limitation of swelling, and even shrinkage of particles, which was observed as a decreased SR of formulation C at 240 min. CaCl$_2$ cross-linked microparticles exhibited lower SR values than non-modified formulation C, and a linear increase up to 300 min was observed. The lower amount of swelling in cross-linked microparticles is the result of insoluble calcium alginate formation upon contact with the acidic medium, which leads to a lower water inflow into the matrix [48–50]. It was also found that microparticles CA1 and CA2 possessed higher values of SR, compared to the placebo formulations (PCA1 and PCA2), which is due to the occurrence of freely soluble molecules of MF and a faster influx of the aqueous medium into the microparticle structure.

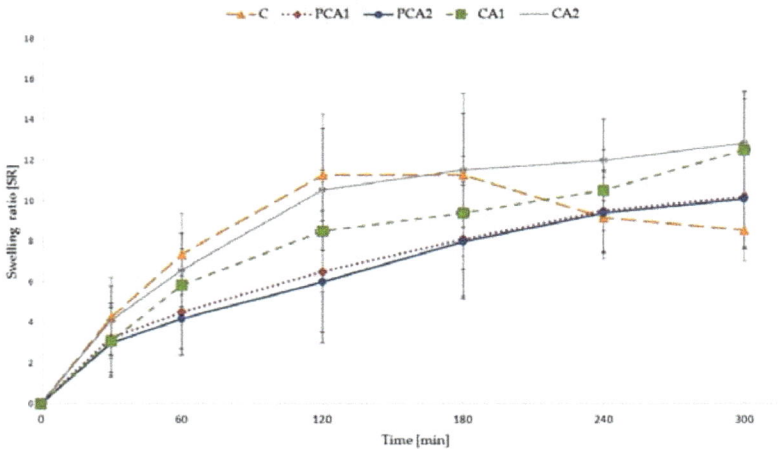

Figure 3. Swelling characteristics of non-modified C and cross-linked formulations PCA1, PCA2, CA1, and CA2.

The swelling process extends the contact time of the drug carrier with the mucosa and improves drug bioavailability. After contact with the hydrated mucus, ALG absorbs moisture and polymer chains with groups containing hydrogen bonds are relaxed [51]. This process, initiating deep contact of the polymer with the mucus layer enables the linkage of microparticles with mucosa and leads to the phenomenon of mucoadhesion [51,52]. Mucoadhesive properties of designed microparticles are presented in Table 4. All formulations adhered to tested materials, and mucoadhesiveness was considerably ($p < 0.05$) influenced by the type of adhesive layer, and the presence and concentration of $CaCl_2$. When a gelatin disc and mucosa gel were used, similar values of F_{max} and W_{ad} were noted, as shown in Table 4. In the case of porcine stomach mucosa, the highest F_{max} values—median from 0.6 N (C) to 1.1 N (CA1) and W_{ad} values from 454.8 µJ (C) to 583.2 µJ (CA1)—were observed, see Table 4 and Figure 4. Porcine stomach mucosa is a valuable model of the adhesive layer due to its similarity to human mucosa in terms of histology, ultrastructure, and composition, and it can be used to reflect the behavior of dosage forms in vivo [53]. It was demonstrated that in an acidic environment $CaCl_2$ cross-linking reduces ALG interaction with the mucous membrane as a result of poor swelling ability [54–56]. However, in this study improvement of the mucoadhesieve properties of $CaCl_2$ cross-linked ALG microparticles was observed. This fact can be explained by the presence of Ca^{2+} ions, which interact with the negatively charged mucin [57,58]. An increase in the positive charge of the polymer leads to better interactions with sialic acid and other anionic groups present in mucin and the formation of additional bonds with the negatively charged membrane [59,60]. With a higher concentration of cross-linking agent, increased work of adhesion was observed. Interestingly, freely water-soluble MF did not influence the mucoadhesive properties of the formulations.

Table 4. Mucoadhesive properties of designed microparticles.

Formulation	Kind of Adhesive Material					
	Gelatine Disc		Mucin Gel		Porcine Stomach Mucosa	
	F_{max} (N) [1]	W_{ad} (µJ) [2]	F_{max} (N) [1]	W_{ad} (µJ) [2]	F_{max} (N) [1]	W_{ad} (µJ) [2]
Control [3]	0.02 ± 0.01	15.2 ± 0.7	0.03 ± 0.01	18.1 ± 3.5	0.07 ± 0.01	29.4 ± 3.3
C	0.5 ± 0.2	283.3 ± 51.2	0.6 ± 0.3	342.3 ± 29.3	0.6 ± 0.2	467.5 ± 17.4
PCA1	0.7 ± 0.1	291.3 ± 14.4	0.5 ± 0.2	362.9 ± 18.8	1.3 ± 0.3	519.7 ± 16.9
PCA2	0.6 ± 0.2	272.3 ± 13.1	0.5 ± 0.1	353.4 ± 21.4	1.2 ± 0.1	504.6 ± 21.3
CA1	0.7 ± 0.2	269.4 ± 14.1	0.6 ± 0.2	359.2 ± 18.5	1.1 ± 0.7	583.4 ± 15.7
CA2	0.6 ± 0.1	254.6 ± 16.7	0.7 ± 0.3	347.1 ± 32.1	1.3 ± 0.2	500.5 ± 13.5

[1] Maximum detachment force, [2] Work of adhesion, [3] Cellulose paper.

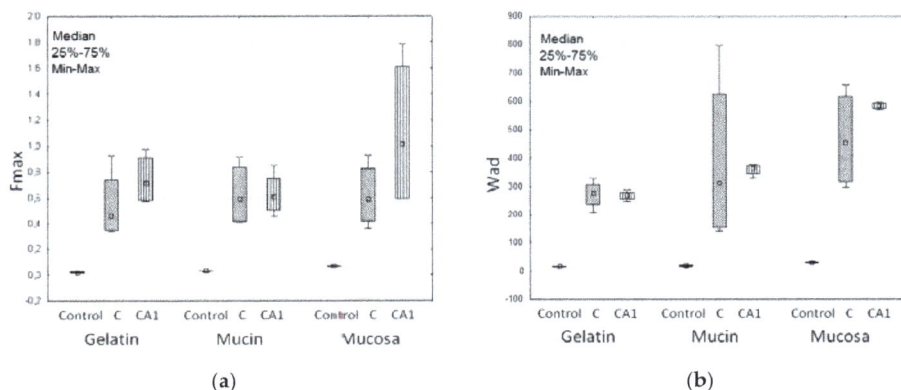

Figure 4. Mucoadhesive features: (**a**) Maximum detachment force (F_{max}) and (**b**) work of adhesion (W_{ad}) of non-modified formulation C, cross-linked formulation CA1 and cellulose paper (Control) (median; n = 6).

3.3. MF Dissolution

Dissolution of the active substance from microparticles depends on the drug's water solubility, influx of the medium into the structure of the dosage form, and polymer swelling [49,50]. In drug dissolution from all microparticles, a burst effect, caused by rapid dissolution of freely soluble MF bound at the microparticle surface, was noted and can be seen in Figure 5. Additionally, it was observed that MF was released faster from non-modified microparticles C ($82.5 \pm 3.6\%$ MF was released after 2 h). This fact is related to the higher SR properties of the unmodified polymer in an acidic environment, which expedites the influx of the aqueous medium into the matrix. Modification of the ALG structure by $CaCl_2$ affected MF dissolution—the release behavior from formulation CA2 (obtained with 0.05% of $CaCl_2$) was comparable to non-modified microparticles C, but formulation CA1 (with 0.1% of $CaCl_2$) significantly prolonged MF release. It was shown that in formulation CA1 $60.1 \pm 3.8\%$, MF was released in the first 2 h and sustained to 12 h, whereby it reached $97.5 \pm 2.7\%$. Ca^{2+} cross-linking leads to a more stable and more intact structure, improves the mechanical resistance of the polymeric network, and reduces its swelling ability [61].

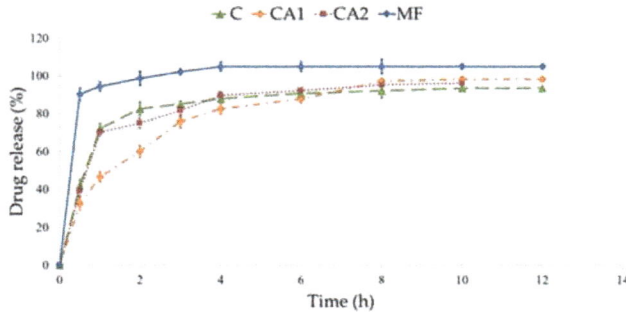

Figure 5. MF dissolution from non-modified microparticles C, cross-linked CA1, CA2, and commercial product used as a control (MF).

MF dissolution was also analyzed by different mathematical equations, see Table 5. It was shown that from designed microparticles, MF was released according to first-order kinetics. In the model of Highuchi, where the best-fit curve with a high R^2 was observed, it was proved that MF release was diffusion controlled. In the model of Korsmeyer-Peppas, values of index n were from 0.08 to 0.12 and confirmed diffusion-dependent MF dissolution. In comparison to formulation C, the Hixson-Crowell model exhibited a better linear relationship with a regression index from 0.94 to 0.96, indicating that the dominant mechanism of MF release from modified microparticle formulations is diffusion coupled with erosion determined by the presence of a cross-linking agent [62].

Table 5. Mathematical characteristics of MF dissolution.

Formulation	Zero Order		First Order		Highuchi		Korsmeyer-Peppas			Hixson-Crowell	
	R^2	K	R^2	K	R^2	K	R^2	K	n	R^2	K
C	0.52	4.62	0.73	0.21	0.66	18.76	0.59	0.28	0.12	0.65	0.19
CA1	0.86	1.68	0.96	0.55	0.95	29.37	0.88	0.34	0.08	0.94	0.58
CA2	0.92	1.67	0.98	0.17	0.97	22.98	0.94	0.31	0.08	0.96	0.18

R^2: correlation coefficient, K: dissolution constant, n: the dissolution index

During the design of oral formulations, it is crucial to explain the drug release mechanism, but also to evaluate if the development of a new technological process affects the modification of the drug release profile compared to the conventional dosage form. Therefore, the similarity and dissimilarity factors were determined to measure the similarity of MF dissolution profiles and to express the potential product similarity [63]. The difference and similarity indices f_1 and f_2, following the international (FDA, EMA) guidelines for the dissolution profile comparison were used [64,65]. According to these guidelines, release profiles are comparable if they possess a value of f_1 in the range 0–15 and f_2 in the range 50–100 [66].

MF release profiles of formulations CA1 and CA2 were compared with non-modified formulation C and a commercial product with non-modified MF release was used as a reference sample. When a non-modified tablet containing MF was used as a control, it was observed that all designed formulations have a similarity factor of <50 and difference factor of >15. Therefore, it was concluded that both non-modified microparticles and formulations modified by $CaCl_2$ were characterized by different drug release profiles. To assess the impact of a cross-linking agent on the MF dissolution process, an independent model approach was applied. It was found that in formulation CA2, values of difference and similarity factors were 4.05 and 72.27, respectively, which indicates a similar MF dissolution profile to formulation C. On the other hand, the formulation CA1 was characterized by the difference factor 17.55 and similarity factor 42.88, which indicates that $CaCl_2$ concentration affected MF dissolution.

3.4. Differential Scanning Calorimetry (DSC)

To evaluate the thermal characteristics of the materials and excipients used, DSC technology was utilized [67]. The MF thermogram exhibits a sharp endothermic peak at 233.02 °C, which corresponds to the melting point of pure MF, see Figure 6. The DSC curves of ALG and PCA1 were similar, and broad endothermal peaks between 100 °C and 150 °C were observed, which indicates the loss of water content in the polymer. Additionally, a sharp exothermic peak related to ALG decomposition at 248 °C was observed. The thermogram of PCA1 shows an exothermic peak registered at 252.6 °C, which might suggest an interaction between $CaCl_2$ and ALG. On the other hand, the MF peak demonstrated a slight decrease in the melting temperature in the CA1 formulation (218.13 °C) in comparison to MF (226.57 °C), which is probably as a result of interactions between MF and the polymer, and formation of prim polymer matrix [68].

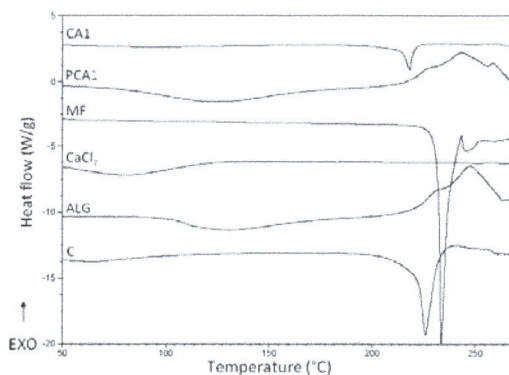

Figure 6. Differential scanning calorimetry (DSC) curves of pure sodium alginate (ALG), metformin (MF), $CaCl_2$, non-modified microparticles C, microparticles PCA1, and CA1.

4. Conclusions

Drug solubility exerts an evident impact on the mechanism and release kinetics. The rapid dissolution of active compounds characterized by being freely soluble in water is one of the main drawbacks when designing pharmaceutical formulations. To prolong the release of MF used as a model freely water-soluble drug, physical modification of ALG microparticles by $CaCl_2$ cross-linking using a novel one-step drying process was applied. The developed method involves considerably fewer unit operations than traditional emulsification techniques. The in vitro drug release data showed a significant difference among cross-linked and non-cross-linked formulations. This study demonstrates that $CaCl_2$ cross-linked ALG microparticles can be successfully used to prolong MF release. Additionally, it was observed that ALG cross-linking using $CaCl_2$ decreased the swelling ratio and improved the mucoadhesive properties of microparticles evaluated using porcine stomach mucosa.

Author Contributions: Conceptualization, M.S. and K.W.; Data curation, K.S. and A.C.-K.; Investigation, M.S.; Methodology, M.S., K.S., A.C.-K. and K.W.; Project administration, M.S.; Software, K.S. and A.C.-K.; Supervision, K.W.; Writing—original draft, M.S.; Writing—review & editing, K.W. M.S. conducted the research and collected the data. K.S. accomplished the statistical analysis. A.C.-K. performed the dissolution tests. K.W. and M.S. developed the research, resolved the results and wrote the article.

Funding: This research and the APC were funded by Medical University of Białystok grant number N/ST/ZB/17/007/2215.

Acknowledgments: Equipment used in this study was from OP DEP 2007–2013, Priority Axis I.3, contract No. POPW.01.03.00-20-008/09. Research was financed by Medical University of Białystok grant N/ST/ZB/17/007/2215.

Conflicts of Interest: No conflict of interest is declared.

References

1. Huang, G.; Mei, X.; Xiao, F.; Chen, X.; Tang, Q.; Peng, D. Applications of important polysaccharides in drug delivery. *Curr. Pharm. Des.* **2015**, *25*, 3692–3696. [CrossRef]
2. Asada, T.; Yoshihara, N.; Ochiai, Y.; Kimura, S.I.; Iwao, Y.; Itai, S. Formulation of a poorly water-soluble drug in sustained-release hollow granules with a high viscosity water-soluble polymer using a fluidized bed rotor granulator. *Int. J. Pharm.* **2018**, *541*, 246–252. [CrossRef] [PubMed]
3. Chakraborty, S.; Khandai, M.; Sharma, A.; Patra, C.N.; Patro, V.J.; Sen, K.K. Effects of drug solubility on the release kinetics of water soluble and insoluble drugs from HPMC based matrix formulations. *Acta Pharm.* **2009**, *59*, 313–323. [CrossRef] [PubMed]
4. Boddupalli, B.M.; Mohammed, Z.N.K.; Nath, R.A.; Banji, D. Mucoadhesive drug delivery system: An overview. *J. Adv. Pharm. Technol. Res.* **2010**, *1*, 381–387. [CrossRef] [PubMed]
5. Mythri, G.; Kavitha, K.; Kumar, M.R.; Jagadeesh Singh, S.D. Novel mucoadhesive polymers—A review. *J. App. Pharm. Sci.* **2011**, *1*, 37–42.
6. Patwekar, S.; Baramade, M.K. Controlled release approach to novel multiparticulate drug delivery system. *Int. J. Pharm. Pharm. Sci.* **2012**, *4*, 757–763.
7. Dey, N.S.; Majumdar, S.; Rao, M.E.B. Multiparticulate drug delivery systems for controlled release. *Trop. J. Pharm. Res.* **2008**, *7*, 1067–1075. [CrossRef]
8. Sachan, K.N.; Pushkar, S.; Jha, A.; Bhattcharya, A. Sodium alginate: The wonder polymer for controlled drug delivery. *J. Pharm. Res.* **2009**, *2*, 1191–1199.
9. Laurienzo, P. Marine polysaccharides in pharmaceutical applications: An overview. *Mar. Drugs* **2010**, *8*, 2435–2465. [CrossRef] [PubMed]
10. Wong, T.W. Alginate graft copolymers and alginate–co-excipient physical mixture in oral drug delivery. *J. Pharm. Pharmacol.* **2011**, *63*, 1497–1512. [CrossRef] [PubMed]
11. Murata, Y.; Jinno, D.; Liu, D.; Isobe, T.; Kofuji, K.; Kawashima, S. The drug release profile from calcium-induced alginate gel beads coated with an alginate hydrolysate. *Molecules* **2007**, *12*, 2559–2566. [CrossRef] [PubMed]
12. Kim, E.S.; Lee, J.S.; Lee, H.G. Calcium-alginate microparticles for sustained release of catechin prepared via an emulsion gelation technique. *Food Sci. Biotechnol.* **2016**, *25*, 1337–1343. [CrossRef]
13. Lira, A.A.; Rossetti, F.C.; Nanclares, D.M.; Neto, A.F.; Bentley, M.V.; Marchetti, J.M. Preparation and characterization of chitosan-treated alginate microparticles incorporating all-trans retinoic acid. *J. Microencapsul.* **2009**, *26*, 243–250. [CrossRef] [PubMed]
14. Szekalska, M.; Puciłowska, A.; Szymańska, E.; Ciosek, P.; Winnicka, K. Alginate: Current use and future perspectives in pharmaceutical and biomedical applications. *Int. J. Polym. Sci.* **2016**, *2016*. [CrossRef]
15. Agüero, A.; Zaldivar-Silva, D.; Peña, L.; Dias, M.L. Alginate microparticles as oral colon drug delivery device: A review. *Carbohydr. Polym.* **2017**, *168*, 32–43. [CrossRef] [PubMed]
16. Sarei, F.; Dounighi, N.M.; Zolfagharian, H.; Khaki, P.; Bidhendi, S.M. Alginate nanoparticles as a promising adjuvant and vaccine delivery system. *Indian J. Pharm. Sci.* **2013**, *75*, 442–449. [CrossRef] [PubMed]
17. Venkatesan, J.; Bhatnagar, I.; Manivasagan, P.; Kang, K.H.; Kim, S.K. Alginate composites for bone tissue engineering: A review. *Int. J. Biol. Macromol.* **2015**, *72*, 269–281. [CrossRef] [PubMed]
18. Sun, J.; Tan, H. Alginate-based biomaterials for regenerative medicine applications. *Materials* **2013**, *6*, 1285–1309. [CrossRef] [PubMed]
19. Khanna, O.; Larson, J.C.; Moya, M.L.; Opara, E.C.; Brey, E.M. Generation of alginate microspheres for biomedical applications. *J. Vis. Exp.* **2012**, *66*, 3388. [CrossRef] [PubMed]
20. Nagpal, M.; Maheshwari, D.; Rakha, P.; Dureja, H.; Goyal, S.; Dhingra, G. Formulation development and evaluation of alginate microspheres of ibuprofen. *J. Young Pharm.* **2012**, *4*, 13–16. [CrossRef] [PubMed]

21. Ahmed, M.M.; El-Rasoul, S.A.; Auda, S.H.; Ibrahim, M.A. Emulsification/internal gelation as a method for preparation of diclofenac sodium–sodium alginate microparticles. *Saudi Pharm. J.* **2013**, *21*, 61–69. [CrossRef] [PubMed]

22. Shukla, S.; Jain, D.; Verma, K.; Verma, S. Formulation and in vitro characterization of alginate microspheres loaded with diloxanide furoate for colon-specific drug delivery. *Asian J. Pharm.* **2010**, *6*, 199–204.

23. Suganya, V.; Anuradha, V. Microencapsulation and nanoencapsulation: A review. *Int. J. Pharm. Clin. Res.* **2017**, *9*, 233–239. [CrossRef]

24. Giri, T.K.; Choudhary, C.; Alexander, A.; Badwaik, H.; Tripathi, D.K. Prospects of pharmaceuticals and biopharmaceuticals loaded microparticles prepared by double emulsion technique for controlled delivery. *Saudi Pharm. J.* **2013**, *21*, 125–141. [CrossRef] [PubMed]

25. Sosnik, A.; Seremeta, K.P. Advantages and challenges of the spray-drying technology for the production of pure drug particles and drug-loaded polymeric carriers. *Adv. Colloid Interfacce Sci.* **2015**, *223*, 40–54. [CrossRef] [PubMed]

26. Patel, B.B.; Patel, J.K.; Chakraborty, S. Review of patents and application of spray drying in pharmaceutical, food and flavor industry. *Recent Pat. Drug Deliv. Formul.* **2014**, *8*, 63–78. [CrossRef] [PubMed]

27. Bagheri, L.; Madadlou, A.; Yarmand, M.; Mousavi, M.E. Spray-dried alginate microparticles carrying caffeine-loaded and potentially bioactive nanoparticles. *Food Res. Int.* **2014**, *62*, 1113–1119. [CrossRef]

28. Sarta-Maria, M.; Scher, H.; Jeoh, T. Microencapsulation of bioactives in cross-linked alginate matrices by spray drying. *J. Microencapsul.* **2012**, *29* 286–295. [CrossRef] [PubMed]

29. Foretz, M.; Guigas, B.; Bertrand, L.; Pollak, M.; Viollet, B. Metformin: From mechanisms of action to therapies. *Cell Metab.* **2014**, *20*, 953–966. [CrossRef] [PubMed]

30. Szekalska, M.; Wróblewska, M.; Sosnowska, K.; Winnicka, K. Influence of sodium alginate on hypoglycemic activity of metformin hydrochloride in the microspheres obtained by the spray drying. *Int. J. Polym. Sci.* **2016**, *2016*. [CrossRef]

31. Kulig, D.; Zimoch-Korzycka, A.; Jarmoluk, A.; Marycz, K. Study on alginate–chitosan complex formed with different polymers ratio. *Polymers* **2016**, *8*, 167. [CrossRef]

32. León, O.; Muñoz-Bonilla, A.; Soto, D.; Pérez, D.; Rangel, M.; Colina, M.; Fernández-García, M. Removal of anionic and cationic dyes with bioedsorbent oxidized chitosans. *Carbohydr. Polym.* **2018**, *194*, 375–383. [CrossRef] [PubMed]

33. Szymańska, E.; Winnicka, K.; Amelian, A.; Cwalina, U. Vaginal chitosan tablets with clotrimazole-design and evaluation of mucoadhesive properties using porcine vaginal mucosa, mucin and gelatin. *Chem. Pharm. Bull.* **2014**, *62*, 160–167. [CrossRef] [PubMed]

34. Council of Europe. *The European Pharmacopeia*, 9th ed.; Council of Europe: Strasburg, France, 2016; Volume 1, p. 302.

35. Costa, P.; Sousa Lobo, J.M. Modeling and comparison of dissolution profiles. *Eur. J. Pharm. Sci.* **2001**, *13*, 123–133. [CrossRef]

36. Siepmann, J.; Peppas, N.A. Modeling of drug release from delivery systems based on hydroxypropyl methylcellulose (HPMC). *Adv. Drug Deliv. Rev.* **2001**, *48*, 139–157. [CrossRef]

37. Soni, T.; Nagda, C.; Gandhi, T.; Chotai, N.P. Development of discriminating method for dissolution of aceclofenac marketed formulations. *Dissolut. Technol.* **2008**, *15*, 31–35. [CrossRef]

38. Diaz, D.D.; Colgan, S.T.; Langer, C.S.; Bandi, N.T.; Likar, M.D.; Alstine, L.V. Dissolution similarity requirements: How similar or dissimilar are the global regulatory expectations? *AAPS J.* **2016**, *18*, 15–22. [CrossRef] [PubMed]

39. Gray, V.; Kelly, G.; Xia, M.; Butler, C.; Thomas, S.; Mayock, S. The Science of USP 1 and 2 dissolution: Present challenges and future relevance. *Pharm. Res.* **2009**, *26*, 1289–1302. [CrossRef] [PubMed]

40. Mazurek-Wądołkowska, E.; Winnicka, K.; Czajkowska-Kośnik, A.; Czyżewska, U.; Miltyk, W. Application of differential scanning calorimetry in evaluation of solid state interactions in tablets containing acetaminophen. *Acta Pol. Pharm.* **2013**, *70*, 787–793. [PubMed]

41. Santana, A.A.; Kieckbusch, T.G. Physical evaluation of biodegradable films of calcium alginate plasticized with polyols. *Braz. J. Chem. Eng.* **2013**, *30*, 835–884. [CrossRef]

42. Mandal, S.; Basu, S.K.; Sa, B. Sustained release of a water-soluble drug from alginate matrix tablets prepared by wet granulation method. *AAPS PharmSciTech* **2009**, *10*, 1348–1356. [CrossRef] [PubMed]

43. Jain, D.; Bar-Shalom, D. Alginate drug delivery systems: Application in context of pharmaceutical and biomedical research. *Drug Dev. Ind. Pharm.* **2014**, *40*, 1576–1584. [CrossRef] [PubMed]

44. Colombo, P.; Bettini, R.; Santi, P.; Peppas, N.A. Swellable matrices for controlled drug delivery: Gel-layer behaviour, mechanisms and optimal performance. *Pharm. Sci. Technol. Today* **2000**, *3*, 198–204. [CrossRef]

45. Clogston, J.D.; Patri, A.K. Zeta potential measurement. *Methods Mol. Biol.* **2011**, *697*, 63–70. [CrossRef] [PubMed]

46. Sriamornsak, P.; Konthong, S.; Nunthanid, J. Fabrication of calcium pectinate microparticles from pomelo pectin by ionotropic gelation. *JAASP. J.* **2012**, *1*, 203–209.

47. Li, H.; Hardy, R.J.; Gu, X. Effect of drug solubility on polymer hydration and drug dissolution from polyethylene oxide (PEO) matrix tablets. *AAPS PharmSciTech* **2008**, *9*, 437–443. [CrossRef] [PubMed]

48. Qin, Y.; Hu, H.; Luo, A. The conversion of calcium alginate fibers into alginic acid fibers and sodium alginate fibers. *J. Appl. Polym. Sci.* **2006**, *101*, 4216–4221. [CrossRef]

49. Rojewska, M.; Olejniczak-Rabinek, M.; Bartkowiak, A.; Snela, A.; Prochaska, K.; Lulek, J. The wettability and swelling of selected mucoadhesive polymers in simulated saliva and vaginal fluids. *Colloids Surf. B Biointerfaces* **2017**, *156*, 366–374. [CrossRef] [PubMed]

50. Daemi, H.; Barikani, M. Synthesis and characterization of calcium alginate nanoparticles, sodium homopolymannuronate salt and its calcium nanoparticles. *Sci. Iran.* **2012**, *19*, 2023–2028. [CrossRef]

51. Shaikh, R.; Raj Singh, T.R.; Garland, M.J.; Woolfson, A.D.; Donnelly, R.F. Mucoadhesive drug delivery systems. *J. Pharm. Bioallied Sci.* **2011**, *3*, 89–100. [CrossRef] [PubMed]

52. Agarwal, S.; Aggarwal, S. Mucoadhesive polymeric platform for drug delivery; a comprehensive review. *Curr. Drug Deliv.* **2015**, *12*, 139–156. [CrossRef] [PubMed]

53. Jackson, S.J.; Perkins, A.C. In vitro assessment of the mucoadhesion of cholestyramine to porcine and human gastric mucosa. *Eur. J. Pharm. Biopharm.* **2001**, *52*, 121–127. [CrossRef]

54. Abdelbary, A.; El-Gazayerly, O.N.; El-Gendy, N.A.; Ali, A.A. Floating tablet of trimetazidine dihydrochloride: An approach for extended release with zero-order kinetics. *AAPS PharmSciTech* **2010**, *11*, 1058–1067. [CrossRef] [PubMed]

55. Segale, L.; Giovannelli, L.; Mannina, P.; Pattarino, F. Calcium alginate and calcium alginate-chitosan beads containing celecoxib solubilized in a self-emulsifying phase. *Scientifica* **2016**, *2016*. [CrossRef] [PubMed]

56. Davidovich-Pinhas, M.; Bianco-Peled, H. A quantitative analysis of alginate swelling. *Carbohydr. Polym.* **2010**, *79*, 1020–1027. [CrossRef]

57. Shtenberg, Y.; Goldfeder, M.; Prinz, H.; Shainsky, J.; Ghantous, Y.; Abu El-Naaj, I.; Schroeder, A.; Bianco-Peled, H. Mucoadhesive alginate pastes with embedded liposomes for local oral drug delivery. *Int. J. Biol. Macromol.* **2018**, *111*, 62–69. [CrossRef] [PubMed]

58. Zhang, Z.H.; Sun, Y.S.; Pang, H.; Munyendo, W.L.; Lv, H.X.; Zhu, S.L. Preparation and evaluation of berberine alginate beads for stomach-specific delivery. *Molecules* **2011**, *14*, 10347–10356. [CrossRef] [PubMed]

59. Pawar, V.K.; Kansal, S.; Garg, G.; Awasthi, R.; Singodia, D.; Kulkarni, G.T. Gastroretentive dosage forms: A review with special emphasis on floating drug delivery systems. *Drug Deliv.* **2011**, *18*, 97–110. [CrossRef] [PubMed]

60. Lehr, C.-M.; Bouwstra, J.A.; Schacht, E.H.; Junginger, H.E. In vitro evaluation of mucoadhesive properties of chitosan and some other natural polymers. *Int. J. Pharm.* **1992**, *78*, 43–48. [CrossRef]

61. Mi, F.L.; Sung, H.W.; Shyu, S.S. Drug release from chitosan–alginate complex microcapsules reinforced by a naturally occurring cross-linking agent. *Carbohydr. Polym.* **2002**, *48*, 61–72. [CrossRef]

62. Stevenson, C.L.; Bennett, D.B.; Lechuga-Ballesteros, D. Pharmaceutical liquid crystals: The relevance of partially ordered systems. *J. Pharm. Sci.* **2005**, *94*, 1861–1880. [CrossRef] [PubMed]

63. Shirkhorshidi, A.S.; Aghabozorgi, S.; Wah, T.Y. A comparison study on similarity and dissimilarity measures in clustering continuous data. *PLoS ONE* **2015**, *10*, e0144059. [CrossRef] [PubMed]

64. Department of Health and Human Services, Food and Drug Administration, Center for Drug Evaluation and Research (CDER), U.S. Government Printing Office. *Extended Release Oral Dosage Forms: Development, Evaluation, and Application of In Vitro/In Vivo Correlations*; Guidance for Industry; Department of Health and Human Services, Food and Drug Administration, Center for Drug Evaluation and Research (CDER), U.S. Government Printing Office: Washington, DC, USA, 1997. Available online: https://www.fda.gov/downloads/drugs/guidances/ucm070239.pdf.U.S (accessed on 10 June 2018).

65. European Medicines Agency. *Investigation of Bioequivalence*; European Medicines Agency: London, UK, 29 January 2010.
66. Gohel, M.C.; Sarvaiya, K.G.; Mehta, N.R.; Soni, C.D.; Vyas, V.U.; Dave, R.K. Assessment of similarity factor using different weighting approaches. *Dissolut. Technol.* **2005**, *12*, 22–37. [CrossRef]
67. Soares, J.P.; Santos, J.E.; Chierice, G.O.; Cavalheiro, E.T.G. Thermal behavior of alginic acid and its sodium salt. *Eclética Química* **2004**, *2*, 53–56. [CrossRef]
68. Mucha, M.; Pawlak, A. Thermal analysis of chitosan and its blends. *Thermochim. Acta* **2005**, *427*, 69–76. [CrossRef]

materials

Article

Composite Film Based on Pulping Industry Waste and Chitosan for Food Packaging

Ji-Dong Xu, Ya-Shuai Niu, Pan-Pan Yue, Ya-Jie Hu, Jing Bian, Ming-Fei Li, Feng Peng * and Run-Cang Sun

Beijing Key Laboratory of Lignocellulosic Chemistry, Beijing Forestry University, Beijing 100083, China; xujidong@bjfu.edu.cn (J.-D.X.); 18813076766@163.com (Y.-S.N.); ypp1109@bjfu.edu.cn (P.-P.Y.); huyajie0311@163.com (Y.-J.H.); bianjing31@bjfu.edu.cn (J.B.); limingfei@bjfu.edu.cn (M.-F.L.); rcsun3@bjfu.edu.cn (R.-C.S.)
* Correspondence: fengpeng@bjfu.edu.cn; Tel.: +86-10-62337250

Received: 23 October 2018; Accepted: 9 November 2018; Published: 13 November 2018

Abstract: Wood auto-hydrolysates (WAH) are obtained in the pulping process by the hydrothermal extraction, which contains lots of hemicelluloses and slight lignin. WAH and chitosan (CS) were introduced into this study to construct WAH-based films by the casting method. The FT-IR results revealed the crosslinking interaction between WAH and CS due to the Millard reaction. The morphology, transmittance, thermal properties and mechanical properties of composite WAH/CS films were investigated. As the results showed, the tensile strength, light transmittances and thermal stability of the WAH-based composite films increased with the increment of WAH/CS content ratio. In addition, the results of oxygen transfer rate (OTR) and water vapor permeability (WVP) suggested that the OTR and WVP values of the films decreased due to the addition of CS. The maximum value of tensile strengths of the composite films achieved 71.2 MPa and the OTR of the films was low as $0.16 \text{ cm}^3 \cdot \mu\text{m} \cdot \text{m}^{-2} \cdot 24 \text{ h}^{-1} \cdot \text{kPa}^{-1}$, these properties are better than those of other hemicelluloses composite films. These results suggested that the barrier composite films based on WAH and CS will become attractive in the food packaging application for great mechanical properties, good transmittance and low oxygen transfer rate.

Keywords: hemicelluloses; chitosan; composite films; oxygen barrier property; food packaging

---

1. Introduction

The utilization of potentially renewable materials is becoming an increasingly acknowledged and promising alternative for future materials products in the sustainable and green society. Over the past decades, the dominating materials of the food packaging are derived from the non-degradable fossil fuels. However, the films produced from fossil fuels have brought much intricate threats to our environment. Meanwhile, the storage volume of fossil fuels was sharply decreased. Therefore, optimized utilization of renewable biomass has attracted more attention in food packaging application [1,2]. Among the biomass polymers, lignocellulosic biomass is a valuable and uniquely sustainable resource because it could be converted into chemicals, polymeric materials and bioproducts.

The lignocellulosic biomass has complex structure consisting of three main components including cellulose, hemicelluloses and lignin. Hemicelluloses are the second abundant polysaccharides in nature. Hemicelluloses demonstrate many valuable properties, such as excellent biodegradability, biodegradability and remarkable film-forming properties [3–5]. Recently, the hemicelluloses based composite films have received increasing concern, especially the application of the films in the food packaging [6]. However, isolated and highly purified hemicelluloses are usually obtained from being extracted by alkaline from the lignocellulosic resources. The alkaline extraction is to add alkaline solution into the lignocellulosic resources and then followed by a series of filtration steps to remove hemicelluloses and some lignin from lignocellulosic resources [7]. In addition, the purification process of hemicelluloses is complex and economically infeasible. Therefore, the forming of hemicelluloses-based films would cost much more energy and time. Wood auto-hydrolysate (WAH) is often removed as the wastewater in the pulping process, which is obtained after hydrothermal treatment the wood chips. In the terms of implementation and commercialization, hemicelluloses-rich wood auto-hydrolysate would be the better alternation to form the films, which could be used in the field of food packaging.

Recently, hemicelluloses-rich WAH is shown to be a feasible resource for the design of films. The dominating polymers of WAH are hemicelluloses, lignin, oligosaccharide and monosaccharide. In the previous work, the films based on the macromolecular hemicelluloses with high purity separated from WAH have some great properties, such as good barriers to oxygen, low cost and easy availability [8]. However, the hemicelluloses based composite films exhibit poor mechanical strength, hygroscopic, poor transparency. These drawbacks of the hemicelluloses based composite films limit the practical applications. To improve the performance of these composite films, plasticizers (such as chitosan [9], carboxymethyl cellulose, [10] and xylitol [11]) are often introduced to improve the mechanical strength of the WAH based composite films. In general, macromolecular hemicelluloses were isolated from the WAH by fractional purification methods such as ethanol precipitation, membrane separation and so forth. This study was to prepare composite films using WAH directly instead of using macromolecular hemicelluloses from the WAH. Therefore, this study aimed at the preparation of the composite films based on WAH and CS with the different ratios in volume. In this work, the components and molecular weight of WAH were determined, the morphology, mechanical properties, thermal stability and water vapor permeability of the composite films were also characterized for the further application.

2. Materials and Methods

2.1. Materials

WAH used in this study, obtained from Eucalyptus wood chips, was provided by Shandong Sun Paper Industry Joint Stock Co., Ltd., Jining, China. Chitosan (CS) was supplied by Sinopharm Chemical Reagent Co. (Shanghai, China), with a medium viscosity of 50–800 mPa·s (CAS 9012-76-4).

2.2. Characterization of WAH

The main hemicelluloses were extracted from WAH by ethanol precipitation. Molecular weight of WAH and the extracted hemicelluloses were measured by Gel permeation chromatography (GPC). [12] The high performance anion exchange chromatography (HPAEC) was applied to determine the sugar components of WAH and the extracted hemicelluloses [12]. The acid insoluble lignin of WAH was analyzed by determining gravimetrically and the acid soluble lignin of WAH was determined by the National Renewable Energy Laboratory (NREL) method [13].

2.3. Preparation of WAH/CS Composite Films

The composite films were prepared from the blended solutions, which consisted of WAH and CS. The forming mechanism of composite films was the result of the Millard reaction, which existed in the carbonyl of WAH and the amino of CS [14,15]. The WAH (2 wt %) was firstly prepared under stirring for 2 h. The CS solution (2 wt %) was prepared under stirring after adding 1% (v/v) acetic acid and the obtained CS solution was centrifuged to remove microbubbles. Then the prepared WAH solution and CS solution were mixed and stirred vigorously for 6 h at room temperature. After absolutely dissolved, each aliquot of 10 mL blended solutions was cast into the 60 mm diameter plastic Petri dish and then all the blended solutions were left to be dried at 50 °C in a vacuum drying chamber. The composite films were obtained after being dried about 5 h and easily peeled from the Petri dishes. As shown in

Table 1, WAH (2 wt %) and were blended with 2 wt % CS in different volume ratios and the forming pathway of the composite films WAH/CS is illustrated in Figure 1.

Table 1. Composite films with different ratios in volume of WAH and chitosan (CS).

Sample Name	WH (2 wt % *v/v*)	Chitosan (2 wt % *v/v*)
$F_{4\text{-}1}$	80	20
$F_{3\text{-}2}$	60	40
$F_{1\text{-}1}$	50	50
$F_{2\text{-}3}$	40	60
$F_{1\text{-}4}$	20	80

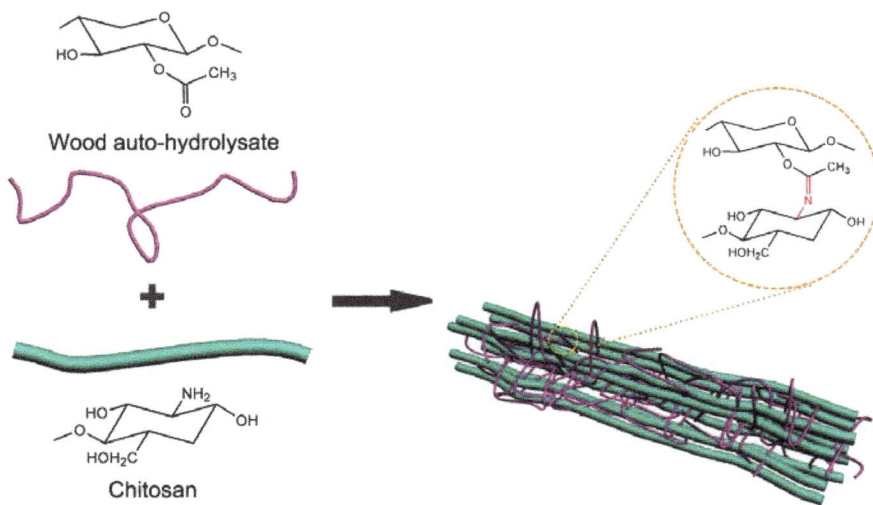

Figure 1. The forming pathway of composite films based wood auto-hydrolysates (WAH) and chitosan (CS).

2.4. Characterization of WAH/CS Composite Films

The surface and cross-section morphology of the composite films based on WAH and CS were analyzed by SEM with the instrument of Hitachi S-3400N II (Hitachi, Tokyo, Japan). Firstly, the composite films were sprayed with gold and sent into the instrument. Then, the SEM images at different magnifications (from $200\times$ to $5000\times$) were obtained. The atomic force microscopy (AFM; Bruker, Germany) images of composite films were used to evaluate the morphology of the surface structure of composite films based on WAH and CS. After gluing the composite films onto metal disks, attaching it to the magnetic sample holder and placing it on the top of scanner tube, the AFM images of composite films were gathered by using a monolithic silicon tip at room temperature. The FT-IR spectra were recorded on a FT-IR Microscope (Thermo Scientific Nicolet In 10, Thermo Electron Scientific Instruments LLC, Madison, WI, USA). The FT-IR spectra of WAH/CS composite films were recorded ranging from 4000 to 650 cm^{-1} at a distinguish ability of wavenumber 4 and 128 cm^{-1} scans.

2.5. Measurement of Thickness

The thickness values of films were measured on a paper thickness gauge (ZH-4, Changchun paper testing machine CO. Ltd., Changchun, China). The indication of the paper thickness gauge furnishes a pinpoint scale with 0.001 mm resolution. The results of all composite films were based on at least 5 sets of data.

2.6. Light Transmittance

The transparency of composite films based on WAH and CS was performed on the UV-Vis spectrophotometer. The cuvettes, the accessory instrument of UV-Vis, were used as the loading gear to load the films WAH-CS. The composite films were cut into be rectangular specimens and then put into the cuvettes for the analysis of the transparencies of the films WAH-CS. The values of transmittance were recorded based on at least 3 sets of data and the corresponding transparencies curves were obtained.

2.7. Tensile Strength Testing

The tensile strength of composite films was measured on an Instron 5566 with Bluehill 2 software. The test was carried out at 50% relative humidity (RH), a stabilized extension rate at 5 mm·min^{-1} and a measure length of 40 mm with a load cell of 100 N volume [16] The composite films were cut into rectangular specimens with the width of 10 mm, afterwards kept in store at room temperature in cabinet containing Mg(NO$_3$)$_2$ solution for at least 3 days. The results of tensile strength were recorded at least 4 specimens.

2.8. Thermal Behavior Analysis

The TGA was carried out on a Mettler Toledo TGA/DSC 851 instrument (Mettler Toledo, Columbus, OH, USA) under a nitrogen atmosphere. The 10 ± 0.5 mg Samples were decomposed on aceramic cup. The weight loss was recorded at the temperature ranging from 40 to 700 °C at a 10 °C·min^{-1} ramp. The samples of 5 ± 0.5 mg were loaded into the sealed aluminum cups with matched punctured lids and heated from 35 to 700 °C at a heating ramp of 10 °C·min^{-1}.

2.9. Oxygen Transfer Rate

According to GB/T1038-2000, the oxygen transfer rate of the WAH/CS composite films were performed on a VAC-V1 differential pressure method of gas permeation apparatus, controlled by the OX2/230 OTR test system. The superficial area of each composite films was 5.0 cm^2 and the OTR tests were carried out at 23 °C for 24 h under the oxygen atmosphere and the relative humidity (RH) was 50%. The thicknesses of WAH/CS composite films were tested by the paper thickness gauge and the display value of the instrument offered a pinpoint scale of 0.001 mm. The results are based on at least 3 specimens.

2.10. Water Vapor Permeability

Water Vapor Permeability of the WAH/CS composite films was determined in accordance with the standard ASTM E 96/E 96M [16]. Each aluminum cup, the loading tools of wet-cup tests, contained 25 g of anhydrous calcium chloride as the desiccant, while the desiccant was dried at 150 °C for 5 h. Then, composite films based on WAH and CS were covered the cups at 23 °C and the cups were put into a cabinet containing water and weighed by a scale of 0.001 g every 1 h. The results are based on at least 3 specimens. The WVP of the composite films were calculated according to the following equation:

$$\text{WVP} = \frac{\text{film thickness (mm)} \times \text{weight augmenter (g)}}{\text{effective area (cm}^2) \times \text{time(s)} \times \Delta P} \tag{1}$$

ΔP is the difference value in water vapor pressure across the composite films (23.76 mmHg).

3. Results and Discussion

3.1. Components of Wood Auto-Hydrolysate

In this work, the components of wood auto-hydrolysate (WAH) were mainly 61.3% hemicelluloses, 10.5% lignin, 12.8% monosaccharide, 13.7% oligosaccharide and 1.65% other insoluble materials of dry WAH. WAH exhibited the molecular weight as follows: M_w of 2300 g·mol^{-1}, M_n of 260 g·mol^{-1} and a polydispersity of 8.8. The hemicelluloses were precipitated by adding three volume of ethanol from WAH. The sugar component of the extracted hemicelluloses is mainly 71.8% xylose, 10.5% glucuronic acid, 7.6% glucose, 7.6% galactose, 1.9% rhamnose and 0.6% arabinose. Based on the sugar composition of the hemicelluloses, the hemicelluloses are mainly composed of glucuronoxylans.

3.2. Structural Analysis of WAH/CS Composite Films

The structural analysis of WAH/CS composite films was characterized by using FT-IR measurement. Figure 2a shows the FT-IR spectra of WAH and CS. The characteristic absorption bands of chitosan observed at 1650 cm^{-1} is assigned to –NH$_2$ [17]. The signals at 1620 cm^{-1} is related to the 4-*O*-methyl-glucuronic acid or glucuronic acid carboxylate of WAH [18,19]. The signal at 1731 cm^{-1} is attributed to C=O stretching of acetyl groups in the WAH. The prominent band at 1035 cm^{-1} is attributed to the C–O–C stretching vibration of glycosidic linkages, which is the representative peak of xylans [20]. The absorption at around 890 cm^{-1} is due to the carbohydrate C–H deformation, which is characteristic β-glycosidic linkage between the sugar units [21]. As shown in Figure 2b, the spectral profiles and peaks of all the bands are extremely similar, indicating that the films prepared from the mixture of WAH and CS in different volume ratios had similar structure. Compared with the Figure 2a,b, the absorption peaks at 1650 cm^{-1} of CS and 1620 cm^{-1} of WAH disappeared and the new bands generated at 1559 cm^{-1} and 1716 cm^{-1}, which suggest that the Millard reaction (C=N double band) occurred between the reducing end of WAH and the amino groups of CS [22].

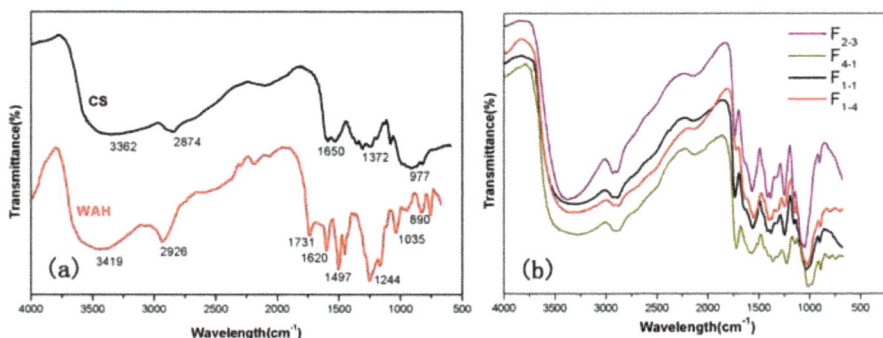

Figure 2. (**a**) Fourier transform-infrared (FT-IR) spectra of WAH and CS, (**b**) FT-IR spectra of represent composite films (F$_{4-1}$, F$_{3-2}$, F$_{1-1}$, F$_{1-4}$).

3.3. Morphology of WAH/CS Composite Films

The homogeneity and topography of WAH/CS composite films were observed by SEM and AFM. The SEM images of the composite films based on WAH and CS are presented in Figure 3. As can be seen from Figure 3a,c,e, the surface of composite films based on WAH and CS are smooth and homogeneous with some irregularities, which are due to the man-made destruction to the films based on WAH and CS. It suggested that WAH and CS were diffused evenly in the composite films. The cross-section images of the films are shown in the Figure 3b,d,e. As can be seen, the cross-sections of composite films based on WAH and CS became rougher when the WAH/CS content ratio changed from 3:2 (F$_{3-2}$)

to 1:1 (F_{1-1}) and to 2:3 (F_{2-3}). It might be due to the increment of the reaction intensities of the Millard reaction between WAH and CS in the composite films F_{1-1}, the network structure of film F_{1-1} were more compact and tighter, thus leading to the rougher cross-section. When the CS content continues to increase, excess CS probably increase the viscosity of the film F_{2-3}, making the cross-section of film F_{2-3} much rougher and obtaining the higher tensile strain and stress strength (Figure 6).

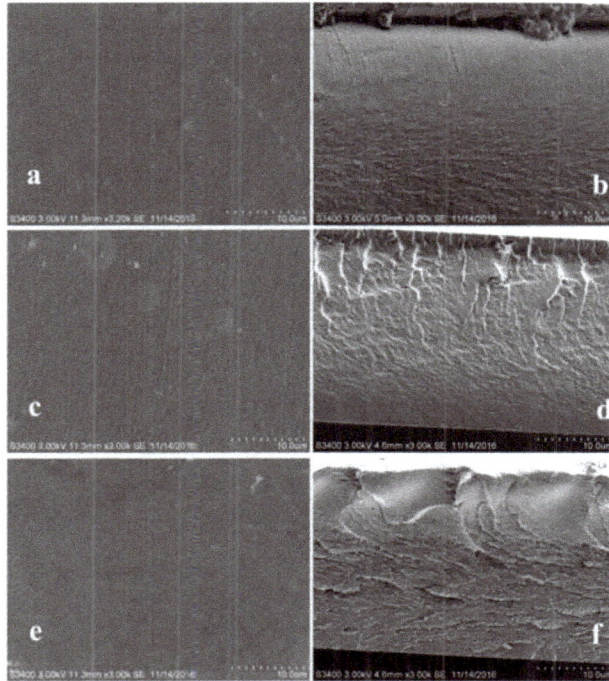

Figure 3. Scanning electron microscope (SEM) images of representative composite films prepared from WAH and CS. (**a,c,e**) Surface of F_{3-2}, F_{1-1}, F_{2-3}; (**b,d,f**) cross-section of F_{3-2}, F_{1-1}, F_{2-3}.

The surface structural analyses of WAH/CS composite films were performed by the AFM, as shown in Figure 4. As can be seen, the surfaces of films were smooth. The root-mean-square (RMS) roughness of the film F_{1-1} was 8.2 nm, which suggested that WAH/CS composite films had smooth surfaces. This is consistent with the SEM results and the smooth surfaces of the films were beneficial to the application of composite films in food packaging materials.

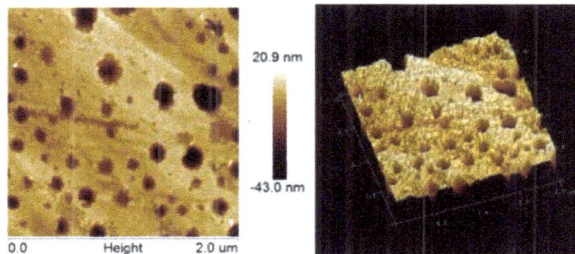

Figure 4. Atomic force microscopy (AFM) images of film, phase image and 3D images of the film F_{1-1} (The scanning scale is 2×2 µm).

3.4. UV-Vis Transmittance of WAH/CS Composite Films

In general, the optical transparencies of composite films reflect the homogeneity of the structure and the miscibility of composite films. The transmittances at wavelength of 200–800 nm and the photograph of composite films are shown in Figure 5a,b respectively. The transmittances of all the composite films under the 800 nm wavelength were above 70%, as shown in Figure 5a, which proved the excellent transparency of WAH/CS composite films. The carboxymethylxylan film were prepared by Alekhina et al. [23], which were highly transparent with a transmittance of 92%. The reason for difference is that the WAH contained some lignin content. In addition, the light transparencies of WAH/CS composite films increased with the increment of wavelength. As can be seen from Figure 5b, composite films based on WAH and CS were at semitransparent and the transparency of the films increased with the increment of WAH/CS content in the films, that is, the relatively high content of WAH is conducive to the transparency of the WAH/CS composite films.

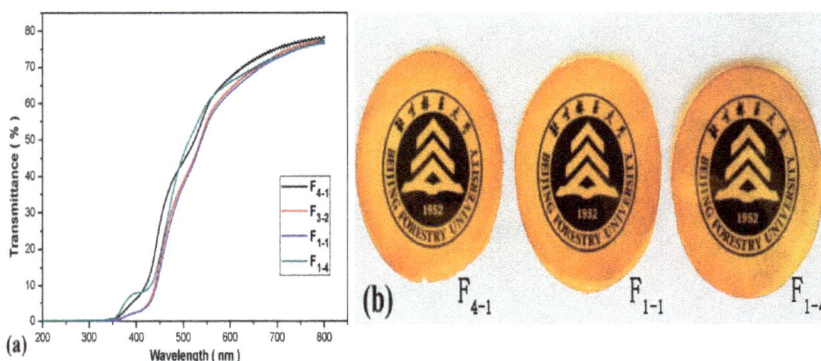

Figure 5. (**a**) UV-Vis transmittance of composite films (F_{4-1}, F_{3-2}, F_{1-1}, F_{1-4}); (**b**) Photograph of composite films (F_{4-1}, F_{1-1}, F_{1-4}).

3.5. Mechanical Properties of WAH/CS Composite Films

In order to ensure the obtained composite films have adequate mechanical properties, the tensile testing is essential to the composite films. The tensile stress, tensile strain at break and thickness of the composite films were summarized in Table 2. The stress-strain curves of composite films are shown in Figure 6. As can be seen, the tensile strengths of F_{3-2}, F_{1-1}, F_{2-3} and F_{1-4} were 28.2, 51.5, 67.5 and 71.2 MPa, respectively. The tensile strength and the tensile strain at break of composite films were improved with the increasing of the CS content. It might be due to the increment of the reaction intensities of WAH and CS with the increment of WAH/CS content ratio from 1:4 to 2:3 to 1:1 and to 3:2. The 100% WH films are very brittle, and so fragile that it cannot be tested; this result is consistent with the xylan-based film reported by Gröndahl et al. [24]. However, the composite film F_{1-4} had an astonishingly higher tensile strength, which is indicated by the tensile strain of 6.1% and stress strength of 71.2 MPa. It might be due to the high viscosity of the unreacted CS which improved the tensile strength of the films. In addition, as can be seen from Table 2, compared with the films reported in literatures [15,25,26], the tensile strength of the WAH/CS films were higher than that of the films based on pure xylan or chitosan. Therefore, the films based on WAH and CS are suitable for the application of food packaging with great mechanical properties.

Table 2. Tensile testing results of the composite films.

Sample	Tensile Strength (MPa)	Tensile Strain at Break (%)	Thickness (μm)
F_{3-2}	28.2 ± 1.3	2.3 ± 0.1	43.1 ± 3.0
F_{1-1}	49.5 ± 1.8	2.5 ± 0.2	42.9 ± 2.0
F_{2-3}	67.5 ± 2.0	3.2 ± 0.2	45.5 ± 3.0
F_{1-4}	71.2 ± 1.5	6.1 ± 0.1	50.5 ± 2.0
Films Reported in Literature			
Major Component (Reference)	Additional Components % (w/w)	Thickness (μm)	Tensile Strength (MPa)	Tensile Strain (%)
Xylan [15]		290–380	1.1–1.4	45.6–56.8
Arabinoxylan [25]	2.7–20 glycerol	22–28	9.7–46.5	5.6–12.1
Chitosan [26]	50–70D-mannan	–	50–60	–

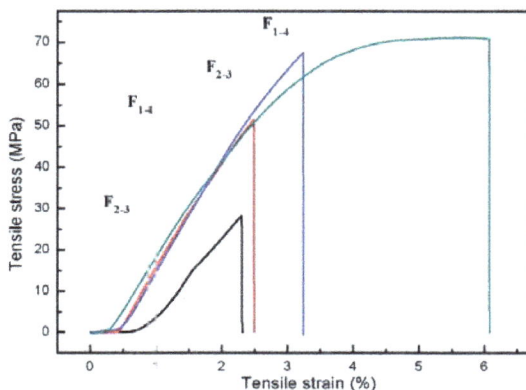

Figure 6. Tensile-strain curves of the composite films.

3.6. Thermal Behavior of WAH/CS Composite Films

The thermal behavior of composite films based on WAH and CS were investigated by thermogravimetric analysis (TGA). In Figure 7a, the initial weight losses of about 6% are ascribed to the evaporation and release of water. All the composite films started to decompose at around 200 °C. And the weight losses of WAH/CS composite films mainly occurred at the temperature range of 200–700 °C, which was due to the degradation of polymers (WAH and CS), such as the glycosidic bonds and C–O band. Additionally, the T_{onset} (the initial degradation temperature), T_{max} (the maximum weight loss temperatures) and the residual values of WAH/CS composite films were determined by the DTG curves and all values are shown in Table 3. In Figure 7b, slight differences in T_{onset} and T_{max} were obviously observed in the DTG curves of the films. As can be seen, the T_{onset} of F_{1-1} and F_{1-4} were 204.2 and 237.6 °C and that of F_{4-1} was 183.4 °C. Therefore, the enhancement of thermal stability may be due to the reaction intensities of WAH and CS increased with the increment of WAH/CS content ratio from 1:4 to 1:1 and to 4:1. It was found that there was a slight shift in T_{max} during the thermal analysis of composite films. The T_{max} of F_{4-1} and F_{1-4} were found at 284.7 and 276.4 °C and that of F_{1-1} was 270.4 °C. In addition, more solid residues were remained in film F_{1-1} than other films at 700 °C (Table 3), which was due to the much stronger interaction between WAH and CS. However, the DTG curve of F_{4-1} had two T_{max} values, which might be due to the excess content of WAH. This result is consistent with the FT-IR results.

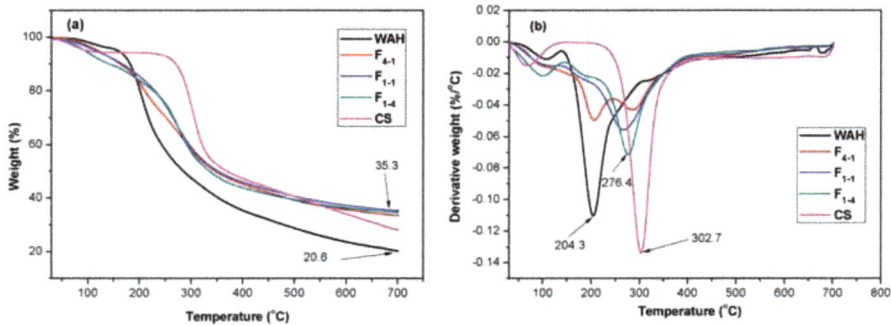

Figure 7. (**a**) The thermogravimetric analysis (TGA) curves of WAH, CS and composite films (F_{4-1}, F_{1-1}, F_{1-4}), (**b**) the DTG curves of WAH, CS and composite films (F_{4-1}, F_{1-1}, F_{1-4}).

Table 3. Thermal characteristics of TGA curves in Figure 7.

Curve	WAH	CS	F_{4-1}	F_{1-1}	F_{1-4}
T_{onset} (°C)	158.1	243.2	183.4	204.2	237.6
T_{max} (°C)	204.3	302.7	284.7	270.4	276.4
Residual (wt %) at 700 °C	20.6	27.9	33.2	35.3	34.8

3.7. Permeability Analysis of WAH/CS Composite Films

Oxygen transfer rate (OTR) and water vapor permeability (WVP) should be as low as possible in order to optimize the applications of composite films in food packaging. The results of OTR and WVP of the films based on WAH and CS are summarized in Table 4. As can be seen, the OTR values of composite films F_{3-2}, F_{1-1}, F_{2-3} and F_{1-4} were 0.34, 0.16, 0.30 and 0.37 $cm^3 \cdot m^{-2} \cdot 24\ h^{-1} \cdot kPa^{-1}$, respectively. The OTR value of F_{1-1} was the lowest among the composite films, which was due to the stronger interactions between WAH and CS. The stronger interactions introduced a barrier of the oxygen molecules. The composite films based on WAH and CS were relatively lower than those of the films in the literatures [27–29]. As reported in literatures, the OTR values of acetylated galactoglucomannans (AcGGM) film and polylactic acid film are 1.28 and 18.65 $cm^3 \cdot m^{-2} \cdot 24 h^{-1} \cdot kPa^{-1}$, respectively. In addition, the standard maximum OTR value of food packaging materials is below 10 $cm^3 \cdot m^{-2} \cdot 24\ h^{-1} \cdot kPa^{-1}$ [29]. As shown in Table 4, the WVP values of the composite films F_{3-2}, F_{1-1}, F_{2-3} and F_{1-4} were 2.42, 2.17, 2.28 and 3.82 ($\times 10^{-10} g \cdot cm \cdot cm^{-2} \cdot s^{-1} \cdot mmHg^{-1}$), respectively. The WVP value of the films F_{1-1} was much lower than those of the film F_{3-2}, F_{2-3} and F_{1-4}, which might be due to the increment of the stronger reaction between WAH and CS. The low WVP value of the composite films is an essential property for the food packaging materials. Therefore, the excellent OTR and WAH made the WAH/CS films more suitable for the application in the food packaging.

Table 4. Oxygen transfer rate (OTR) and water vapor transmission rate (WVP) of the composite films and the films reported in literatures.

Sample	OTR ($cm^3 \cdot m^{-2} \cdot 24\ h^{-1} \cdot kPa^{-1}$)	WVP ($\times 10^{-10}$ $g \cdot cm \cdot cm^{-2} \cdot s^{-1} \cdot mmHg^{-1}$)	Test Area (cm^2)
F_{3-2}	0.34 ± 0.05	2.42 ± 0.33	5.0
F_{1-1}	0.16 ± 0.01	2.17 ± 0.24	5.0
F_{2-3}	0.30 ± 0.06	2.28 ± 0.19	5.0
F_{1-4}	0.29 ± 0.05	3.82 ± 0.36	5.0
Films Reported in Literatures			
Major Component (References)	Additional Components % (*w/w*)	Average Thickness (μm)	OTR ($cm^3 \cdot m^{-2} \cdot 24$ $h^{-1} \cdot kPa^{-1}$)
Arabinoxylan [27]	40 sorbitol	20–50	4.7
Polylactic acid Figurefilm [28]	–	25	18.65
AcGGM [29]	35 CMC	30–60	1.28

4. Conclusions

An easy and rapid way was adopted for preparation the barrier films based on WAH and CS was studied in this works. FT-IR analysis suggested that the obtained composite films was the result of the crosslinking interaction between WAH and CS, which is arose from the Millard reaction of the carbonyl of WAH and the amino of CS. The SEM and AFM images suggested the composite films showed a smooth surface and a dense structure. The physical properties of the composite films with different ratio of WAH and CS were also studied. As the analysis revealed, the tensile strength and oxygen barrier ability of the composite films were improved due to the addition of CS. The films based on WAH and CS showed high tensile strength (71.2 MPa), good thermal stability, high transmittances, low water vapor permeability and excellent oxygen barrier properties (<1 cm$^3 \cdot$m$^{-2} \cdot 24$ h$^{-1} \cdot$kPa^{-1}), these properties are beneficial to constructing packaging materials. Therefore, composite films based on WAH and CS would become attractive in the application of packaging materials in the food packaging. In summary, converting wood auto-hydrolysate into value-added films could lower the production cost, benefit environment and increase revenue for paper making industry.

Author Contributions: Y.-S.N. and J.-D.X. performed the experiments; P.-P.Y., analyzed the data; the paper was written under the direction and supervision of Y.-J.H., J.B., M.-F.L., F.P. and R.-C.S.; J.-D.X. and Y.-S.N. was responsible for writing this work.

Funding: This research was funded by the Fundamental Research Funds for Central Universities (JC2015-03), Beijing Municipal Natural Science Foundation (6182031), Author of National Excellent Doctoral Dissertations of China (201458) and the National Program for Support of Top-notch Young Professionals.

Conflicts of Interest: The authors declare no conflict of interest.

References

1. Ragauskas, A.J.; Williams, C.K.; Davison, B.H.; Britovsek, G.; Cairney, J.; Eckert, C.A.; Frederick, W.J.; Hallett, J.P.; Leak, D.J.; Liotta, C.L. The path forward for biofuels and biomaterials. *Science* **2006**, *311*, 484–489. [CrossRef] [PubMed]
2. Edlund, U.; Ryberg, Y.Z.; Albertsson, A.C. Barrier films from renewable forestry waste. *Biomacromolecules* **2010**, *11*, 2532–2538. [CrossRef] [PubMed]
3. Merdes, F.R.S.; Bastos, M.S.R.; Mendes, L.G.; Silva, A.R.A.; Sousa, F.D.; Monteiro-Moreira, A.C.O.; Cheng, H.N.; Biswas, A.; Moreira, R.A. Preparation and evaluation of hemicellulose films and their blends. *Food Hydrocoll.* **2017**, *70*, 181–190. [CrossRef]
4. Azeredo, H.M.C.; Kontou-Vrettou, C.; Moates, G.K.; Wellner, N.; Cross, K.; Pereira, P.H.F.; Waldron, K.W. Wheat straw hemicellulose films as affected by citric acid. *Food Hydrocoll.* **2015**, *50*, 1–6. [CrossRef]
5. Chen, G.G.; Qi, X.M.; Guan, Y.; Peng, F.; Yao, C.L.; Sun, R.C. High Strength hemicellulose-based nanocomposite film for food packaging applications. *ACS Sustain. Chem. Eng.* **2016**, *4*, 1985–1993. [CrossRef]
6. Hansen, N.M.; Plackett, D. Sustainable films and coatings from hemicelluloses: A review. *Biomacromolecules* **2008**, *9*, 1493–1505. [CrossRef] [PubMed]
7. Peng, F.; Ren, J.L.; Xu, F.; Bian, J.; Peng, F.; Sun, R.C. Fractionation of alkali-solubilized hemicelluloses from delignified *Populus gansuensis*: Structure and properties. *J. Agric. Food Chem.* **2010**, *58*, 5743–5750. [CrossRef] [PubMed]
8. Ibn Yaich, A.; Edlund, U.; Albertsson, A.C. Transfer of biomatrix/wood cell interactions to hemicellulose-based materials to control water interaction. *Chem. Rev.* **2017**, *117*, 8177–8207. [CrossRef] [PubMed]
9. Arnon, H.; Zaitsev, Y.; Porat, R.; Poverenov, E. Effects of carboxymethyl cellulose and chitosan bilayer edible coating on postharvest quality of citrus fruit. *Postharvest Biol. Technol.* **2014**, *87*, 21–26. [CrossRef]
10. Muscat, D.; Adhikari, B.; Adhikari, R.; Chaudhary, D.S. Comparative study of film forming behaviour of low and high amylose starches using glycerol and xylitol as plasticizers. *J. Food Eng.* **2012**, *109*, 189–201. [CrossRef]
11. Ghanbarzadeh, B.; Almasi, H.; Entezami, A.A. Physical properties of edible modified starch/carboxymethyl cellulose films. *Innov. Food Sci. Emerg. Technol.* **2010**, *11*, 697–702. [CrossRef]

12. Peng, F.; Ren, J.L.; Xu, F.; Bian, J.; Peng, P.; Sun, R.C. Comparative study of hemicelluloses obtained by graded ethanol precipitation from sugarcane bagasse. *J. Agric. Food Chem.* **2009**, *57*, 6305–6317. [CrossRef] [PubMed]

13. Sluiter, A.; Hames, B.; Ruiz, R.; Scarlata, C.; Sluiter, J.; Templeton, D.; Crocker, D. Determination of structural carbohydrates and lignin in biomass. *Lab. Anal. Proced.* **2008**, *1617*, 1–16.

14. Luo, Y.; Ling, Y.; Wang, X.; Han, Y.; Zeng, X.; Sun, R.C. Maillard reaction products from chitosan-xylan ionic liquid solution. *Carbohydr. Polym.* **2013**, *98*, 835–841. [CrossRef] [PubMed]

15. Sousa, S.; Ramos, A.; Evtuguin, D.V.; Gamelas, J.A. Xylan and xylan derivatives-their performance in bio-based films and effect of glycerol addition. *Ind. Crop. Prod.* **2016**, *94*, 682–689. [CrossRef]

16. Kumaran, M. Interlaboratory comparison of the ASTM standard test methods for water vapor transmission of materials (E96-95). *J. Test. Eval.* **1998**, *26*, 83–88.

17. Umemura, K.; Kawai, S. Preparation and characterization of Maillard reacted chitosan films with hemicellulose model compounds. *J. Appl. Polym. Sci.* **2008**, *108*, 2481–2487. [CrossRef]

18. Marchessault, R.H.; Liang, C.Y. The infrared spectra of crystalline polysaccharides. VIII. Xylans. *Int. J. Polym. Sci.* **1962**, *59*, 357–378. [CrossRef]

19. Chatjigakisa, A.K.; Pappasa, C.; Proxeniab, N.; Kalantzib, O.; Rodisb, P.; Polissioua, M. FT-IR spectroscopic determination of the degree of esterification of cell wall pectins from stored peaches and correlation to textural changes. *Carbohydr. Polym.* **1998**, *37*, 395–408. [CrossRef]

20. Sun, R.C.; Tomkinson, J. Characterization of hemicelluloses isolated with tetraacetylethylenediamine activated peroxide from ultrasound irradiated and alkali pre-treated wheat straw. *Eur. Polym. J.* **2003**, *39*, 751–759. [CrossRef]

21. Ren, J.L.; Sun, R.C.; Liu, C.F.; Lin, L.; He, B.H. Synthesis and characterization of novel cationic SCB hemicelluloses with a low degree of substitution. *Carbohydr. Polym.* **2007**, *67*, 347–357. [CrossRef]

22. Dash, M.; Chiellini, F.; Ottenbrite, R.M.; Chiellini, E. Chitosan-a versatile semi-synthetic polymer in biomedical applications. *Prog. Polym. Sci.* **2011**, *36*, 981–1014. [CrossRef]

23. Alekhina, M.; Mikkonen, K.S.; Alén, R.; Tenkanen, M.; Sixta, H. Carboxymethylation of alkali extracted xylan for preparation of bio-based packaging films. *Carbohydr. Polym.* **2014**, *100*, 89–96. [CrossRef] [PubMed]

24. Gabrielii, I.; Gatenholm, P.; Glasser, W.G.; Jain, R.K.; Kenne, L. Separation, characterization and hydrogel-formation of hemicellulose from aspen wood. *Carbohydr. Polym.* **2000**, *43*, 367–374. [CrossRef]

25. González-Estrada, R.; Calderón-Santoyo, M.; Carvajal-Millan, E.; Valle, F.D.J.A.; Ragazzo-Sánchez, J.A.; Brown-Bojorquez, F.; Rascón-Chu, A. Covalently cross-linked arabinoxylans films for Debaryomyces hansenii entrapment. *Molecules* **2015**, *20*, 11373–11386. [CrossRef] [PubMed]

26. Sárossy, Z.; Blomfeldt, T.O.J.; Hedenqvist, M.S.; Koch, C.B.; Ray, S.S.; Plackett, D. Composite films of arabinoxylan and fibrous sepiolite: Morphological, mechanical and barrier properties. *ACS Appl. Mater. Interfaces* **2012**, *4*, 3378–3386. [CrossRef] [PubMed]

27. Hettrich, K.; Fischer, S.; Schroder, N.; Engelhardt, J.; Drechsler, U.; Loth, F. Derivatization and characterization of xylan from oat spelts. *Macromol. Symp.* **2005**, *232*, 37–48. [CrossRef]

28. Fukuzumi, H.; Saito, T.; Iwata, T.; Kumamoto, Y.; Isogai, A. Transparent and high gas barrier films of cellulose nanofibers prepared by TEMPO-mediated oxidation. *Biomacromolecules* **2009**, *10*, 162–165. [CrossRef] [PubMed]

29. Hartman, J.; Albertsson, A.C.; Lindblad, M.S.; Sjöberg, J. Oxygen barrier materials from renewable sources: Material properties of softwood hemicellulose-based films. *J. Appl. Polym. Sci.* **2006**, *100*, 2985–2991. [CrossRef]

materials

MDPI

Article

The Evaluation of Physio-Mechanical and Tribological Characterization of Friction Composites Reinforced by Waste Corn Stalk

Yunhai Ma [1,2,3,4], Siyang Wu [1,2,3] ©, Jian Zhuang [1,2,3,*], Jin Tong [1,2,3], Yang Xiao [1,2,3] and Hongyan Qi [1,2,3]

1   State key laboratory of automotive simulation and control, Jilin University, Changchun 130022, China; myh@jlu.edu.cn (Y.M.); siyangwu@outlook.com (S.W.); jtong@jlu.edu.cn (J.T.); xiaoyang_jlu@163.com (Y.X.); m15543610031@163.com (H.Q.)
2   Key Laboratory of Bionic Engineering, Ministry of Education, Jilin University, Changchun 130022, China
3   College of Biological and Agricultural Engineering, Jilin University, Changchun 130022, China
4   State Key Laboratory of Automotive Safety and Energy, Tsinghua University, Beijing 100084, China
*   Correspondence: zhuangjian_2001@163.com; Tel.: +86-187-4302-7198

Received: 3 May 2018; Accepted: 25 May 2018; Published: 27 May 2018

Abstract: This paper addressed the potential use of fibers from waste corn stalk as reinforcing materials in friction composites. The friction composites with different contents of corn stalk fibers were prepared, and their tribological and physio-mechanical behaviors were characterized. It was found that the incorporation of corn stalk fibers had a positive effect on the friction coefficients and wear rates of friction composites. Based on comparisons of the overall performance, FC-6 (containing 6 wt % corn stalk fibers) was selected as the best performing specimen. The fade ratio of specimen FC-6 was 7.8% and its recovery ratio was 106.5%, indicating excellent fade resistance and recovery behaviors. The wear rate of specimen FC-6 was the lowest (0.427×10^{-7} mm^3 (N·mm)$^{-1}$ at 350 °C) among all tested composites. Furthermore, worn surface morphology was characterized by scanning electron microscopy and confocal laser scanning microscopy. The results revealed that the satisfactory wear resistance performances were associated with the secondary plateaus formed on the worn surfaces. This research was contributive to the environmentally-friendly application of waste corn stalk.

Keywords: corn stalk fiber; friction composite; friction and wear; worn surface morphology

---

1. Introduction

Friction composites are commonly used in transmission and brake systems for safe rapid deceleration and immobilization of various vehicles and instruments [1–3]. Friction composites should possess a certain set of outstanding properties, including a moderate friction coefficient, high heat fading resistance and recovery, no or less noise and vibration, and low wear rate under different operating environments [4–7]. For this reason, friction composites normally contain more than ten ingredients, which are separated into four prime classes of reinforcing fibers, friction modifiers, binder resins, and space fillers [8–10]. Among them, reinforcing fibers have a pivotal role to play in deciding the tribological and mechanical properties of friction materials. Ceramic, organic, and metallic fibers are mainly used as substitutes for traditional asbestos fibers in friction composites [11,12].

Among the diverse fibers available for friction composites, natural fibers have drawn much attention as reinforcing materials because of their environmental friendliness, renewability, low density, low costs, excellent acoustic insulating properties as well as their satisfactory mechanical performances [13,14]. In recent years, many studies investigated the influences of natural fibers on the tribological characteristics of friction composites [15]. Chand et al. [16] developed polyester composites reinforced by jute fibers and evaluated the effects of applied load and fiber orientation on friction and sliding wear properties. The results showed that the friction coefficients declined with the rise of applied load and the wear resistance maximized under normal orientation, indicating these reinforced polyester composites may have potential application as friction materials in environmentally-friendly brake pads. Bajpai et al. [17] reported the influence of nettle, grewia optiva and sisal fibers on the wear and frictional behaviors of poly lactic acid (PLA) composites. This study revealed that the addition of these natural fibers remarkably enhanced the wear performance of PLA composites, as the specific wear rate and friction coefficients of the composites were reduced by 10–44% and more than 70%, respectively, in comparison with neat PLA. Nirmal et al. [18] prepared polyester composites reinforced by treated betelnut fibers and studied their mechanical and tribological behaviors under dry/wet sliding conditions. The study suggested that the friction coefficients and average wear rates under the wet sliding condition dropped significantly by about 95% and 54% respectively compared with under dry conditions, and the wear resistance was improved under an anti-parallel orientation to the sliding surface. Fu et al. [19] evaluated the tribological characteristics of phenolic resin-based friction composites containing treated flax fibers under dry contact conditions. The study found that the introduction of flax fibers into resin substrate stabilized the friction coefficients and improved wear resistance, indicating that flax fibers were an ideal substitute for asbestos in brake pads. Given the above advantages and chances of natural fibers, further research is needed to explore and evaluate the tribological behaviors of other types of natural fibers.

Corn is one of the most productive cereals in China, especially in the northeast area. It is estimated that approximately 0.23 billion tons of corn stalks are generated annually as agricultural by-products [20–22]. However, after the harvest, most corn stalks are left on the field or burned, which leads to a waste of resources and environmental degradation [23,24]. Hence, it is essential to seek an effective and environmentally friendly way to improve the utilization value of corn stalks.

This work was aimed to study the feasibility of applying fibers obtained from corn stalks as reinforcement fibers to the manufacture of friction composites. For this purpose, five types of friction composites were fabricated with corn stalk fiber content of 0 wt %, 2 wt %, 4 wt %, 6 wt %, 8 wt % respectively. Then their physio-mechanical and tribological behaviors were characterized and evaluated systematically. Furthermore, the wear mechanisms of the corn stalk fiber-reinforced friction composites were explored and analyzed based on worn surface morphologies.

2. Materials and Methods

2.1. Preparation of Corn Stalk Fibers

The degree of interface adhesion between fibers and matrix affects both the physio-mechanical behaviors of natural fibers-reinforced friction composites and the reinforcing efficiency of the fibers [25,26]. Thus, in the preparation of friction composites reinforced by natural fibers, the indispensable step to improve the fiber-matrix interface bonding is surface modification (e.g., alkali treatment, benzoylation treatment, acetylation treatment, silane treatment and electric discharge treatment) [27–29]. In this study, alkali treatment was used as fiber modification to increase the compatibility with the matrix.

Corn stalks obtained from a local farm in Changchun, China, were naturally air-dried for a few days and then separated into rinds and piths. The rinds were ground into 3–4 mm long fibers, which were surface-treated as described below. In the alkalization, the fibers were immersed in 1% aqueous NaOH solution at 30 °C for about 20 min and rinsed with distilled water until turning pH 7. Finally, excessive solvent and moisture were removed from the corn stalk fibers after treatment in a ZK350S vacuum drying oven (Sanshui, Tianjin, China) at 90 °C for 4 h [30].

2.2. Fabrication of Friction Composites

The detailed compositions of the friction materials (in wt %) are presented in Table 1. The five types of friction composites were numbered as FC-0, FC-2, FC-4, FC-6, FC-8, respectively, according to the content of corn stalk fibers. The friction composites were prepared via compression molding. Firstly the raw materials were mixed thoroughly in a JF805R electrical blender (Wangda, Changchun, China) for 8 min. The uniform mixture was then molded for 30 min at 160 °C under 45 MPa by using a JFY50 hot compression machine (Wangda, Changchun, China). Three intermittent 'breathings' were required in the process of hot pressing to release volatiles. The prepared friction composites were subsequently heat-treated in an oven to remove the remaining stress, which involved three phases (Figure 1): 140 °C × 1 h, 160 °C × 3 h, and 180 °C × 6 h. Finally, the friction composites were air-cooled to room temperature and machined into specimens of 25 × 25 × 6 mm^3.

Table 1. Ingredient ratios of friction composites.

Raw Materials (by wt %)	Specimens				
	FC-0	FC-2	FC-4	FC-6	FC-8
Corn stalk fibers	0	2	4	6	8
Compound mineral fibers	25	24.42	23.84	23.26	22.68
Vermiculite powder	5	4.88	4.76	4.64	4.52
Calcium carbonate	10	9.77	9.54	9.31	9.08
Coke	5	4.88	4.76	4.64	4.52
Graphite	8	7.81	7.62	7.43	7.24
Friction powder	1	0.98	0.96	0.94	0.92
Zirconium silicate	4	3.91	3.82	3.73	3.64
Alumina	6	5.86	5.72	5.58	5.44
Barium sulfate	20	19.54	19.08	18.62	18.16
Zinc stearate	2	1.95	1.90	1.85	1.80
Phenolic resin	14	14	14	14	14

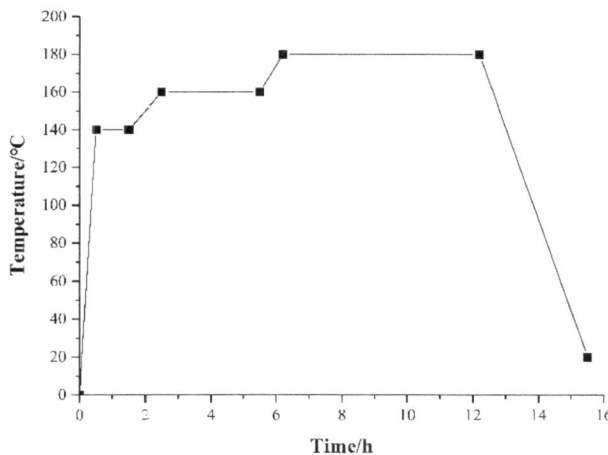

Figure 1. Heat-treatment process of the composites.

2.3. Testing Methods and Equipment

The density of each friction composite was measured on an MP-5002 electronic balance (Junda, Shenzhen, China) following the Archimedes drainage approach. The hardness was tested using an HRSS-150 Rockwell hardness tester (Sannuo, Shenzhen, China) as per the *Test method of Rockwell hardness for friction materials* (GB/T 5766-2007). The impact strength was detected on an XJ-40A impact testing machine (Jianyi, Shanghai, China) based on the *Test Method for Tensile-Impact of Plastics* (GB/T 13525-92). Each specimen was tested in quintuplicate to minimize the error.

Tribological performances of the friction composites were evaluated on a JF150D-II constant-speed friction instrument (Wangda, Changchun, China) as per *Brake Linings for Automobiles* (GB/T 5763-2008), with a schematic diagram showed in Figure 2. The specimens were pressed against the surface of the rotating disk by pressurizing device under a certain load condition. The frictional force and temperature during the test were detected by the tension-compression sensor and temperature sensor respectively. The temperature was controlled and stabilized at the set value through the heating system and cooling system. An HT250 cast iron disk with hardness of 180–220 HB was used as the mating plate. The friction and wear tests consisted of two parts of fade tests and recovery tests. Five parallel tests were carried out to reduce the data scattering.

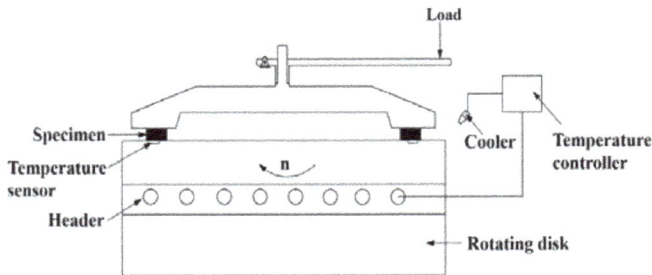

Figure 2. Schematic diagram of friction testing machine.

In the fade tests and recovery tests, the temperature was changed from 100 °C to 350 °C and from 300 °C to 100 °C, respectively, and the disk was rotated 5000 and 7500 revolutions, respectively, before measurement of thickness change and weight loss. Moreover, the rotating speed and contact pressure were set at 480 rpm and 0.98 MPa, respectively in all tests.

The friction coefficient was automatically recorded through the computer attached to the friction tester. The wear rate ΔW was determined by the following equation [31,32]:

$$\Delta W = \frac{1}{2\pi R} \times \frac{A}{N} \times \frac{h_1 - h_2}{f} \tag{1}$$

where R (=150 mm) is the horizontal distance from a friction specimen to the counterpart disk center; A (=625 mm^2) is the area of the friction specimen; N (=5000) is the revolutions of the disk; h_1 and h_2 are the average thicknesses of the friction specimen before and after tests, respectively (mm); f is the mean friction force during tests (N).

The fade ratio F and recovery ratio R were defined by the following equations [33,34]:

$$F = \frac{\mu_{F100°C} - \mu_{F350°C}}{\mu_{F100°C}} \times 100\% \tag{2}$$

$$R = \frac{\mu_{R100°C}}{\mu_{F100°C}} \times 100\% \tag{3}$$

where $\mu_{F100°C}$ and $\mu_{F350°C}$ are the friction coefficients with the temperature rise to 100 °C and 350 °C during fade tests, respectively; $\mu_{R100°C}$ is the friction coefficient with the temperature declined to 100 °C during recovery tests.

After the friction and wear tests, the worn surface morphology of each specimen was observed by an EVO-18 scanning electron microscope (SEM, ZEISS, Jena, Germany) at 20 kV, and 3D profiles and surface roughness were detected on a LEXT OLS3000 confocal laser scanning microscope (CLSM, OLYMPUS, Beijing, China) as per *Geometrical Product Specifications(GPS)—Surface texture—Profile method—Surface roughness—Terminology—Measurement of surface roughness parameters* (GB/T 7220-2004).

3. Results and Discussion

3.1. Surface Morphology of Corn Stalk Fibers

Surface morphology of the raw and treated corn stalk fibers is shown in Figure 3. Clearly, the raw fibers presented smooth outer surfaces with some impurity particles (Figure 3a). After alkali treatment, the surfaces became cleaner and contained a large number of node structures and micropores (Figure 3b), indicating the corn stalk fibers were significantly modified. These changes may be ascribed to the removal of natural and artificial impurities (e.g., lignin, wax, pectin and oils) and the increased amount of exposed cellulose on the fiber surfaces, which could improve the fiber-matrix interfacial adhesion [25,35].

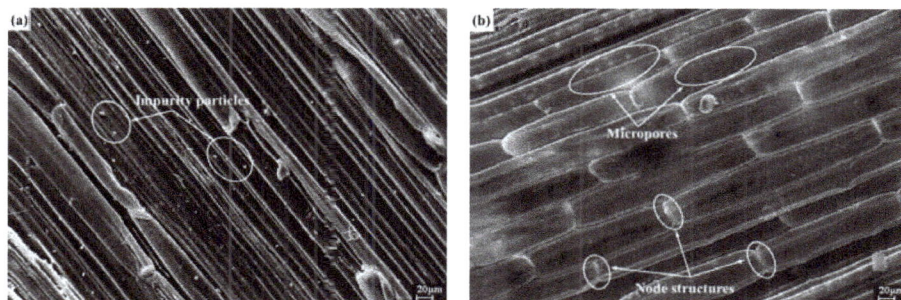

Figure 3. Micrographs of (**a**) raw and (**b**) treated corn stalk fibers.

3.2. Physio-Mechanical Performances

The physio-mechanical performances of friction composites are modestly associated with the reliability and security of vehicle operation. The density, hardness, and impact strength of the composites are summarized in Table 2. The densities of these composites decreased with the increasing content of corn stalk fibers. The density of specimen FC-0 was the largest and that of specimen FC-8 was the smallest among all composites. The hardness of the composites showed a similar tendency with increase in content of corn stalk fibers, as it was maximized in specimen F-0 and minimized in specimen FC-8. However, no obvious variation trend was observed in impact strength. The impact strength of specimen FC-4 was the highest, while specimens FC-2 and FC-8 had the lowest impact strength in all tested composites. This may be because excellent interface adhesion between the reinforcing fibers and the matrix could enhance the impact resistance of friction composites to some degree. It was indicated that the addition of corn stalk fibers can enhance the physio-mechanical properties of the friction composites. This is in agreement with the results of the previous report [36].

Table 2. Physio-mechanical properties of the friction specimens.

Specimens	Density ($g \cdot cm^{-3}$)	Hardness (HRR)	Impact Strength (MPa)
FC-0	2.33	103.6	0.461 ± 0.009
FC-2	2.23	101.4	0.424 ± 0.012
FC-4	2.20	98.9	0.486 ± 0.007
FC-6	2.13	97.2	0.473 ± 0.013
FC-8	2.11	95.8	0.422 ± 0.015

3.3. Friction and Wear Behaviors

Variations in the friction coefficients of the five composites during the fade and recovery tests are presented in Figure 4. The friction coefficient of specimen FC-0 decreased with the temperature rise, while the variant trends of the other four composites were slightly different (Figure 4a). The friction coefficients of the composites incorporated with corn stalk fibers initially increased with the temperature rise from 100 °C to 150 °C and then decreased from 150 °C to 350 °C These changes can be explained by the fact that during the initial phase of the test (100–150 °C), the fibers and some hard particles were exposed to the worn surface due to the removal of the matrix and soft materials, then the wear debris gathered around the nucleation sites formed from protruding fibers, followed by the generation of the third body wear under the action of frictional force and normal pressure, which resulted in the increase of the friction coefficients [2]. When the temperature was higher than 160 °C, the lignin in the fibers began to decompose and the fibers gradually carbonized [37], then some carbon powder appeared on the friction surface, which had a certain lubrication effect, thus leading to the decline of the friction coefficients [36]. Moreover, the sheer strength of friction composites declined with the rise of surface temperature, which also resulted in the decrease of the friction coefficient. Anyway, the friction coefficients at each test temperature were in conformity with Chinese national standards. The friction coefficients of the five friction composites decreased slowly at 250 °C or above, which can be ascribed to the thermal relaxing and degradation of phenolic resins at elevated temperatures during the tests [32].

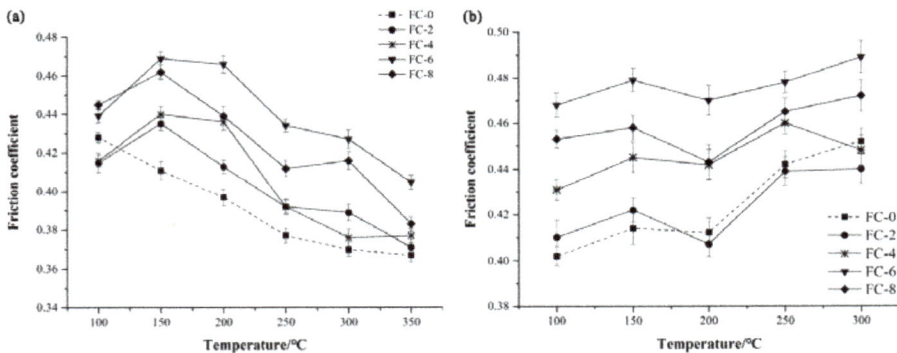

Figure 4. Variation in friction coefficient of the friction specimens: (**a**) fade test and (**b**) recovery test.

In general, the addition of corn stalk fibers improved the friction behaviors at tested conditions barring the 100 °C case. The reason for this phenomenon could be that when the resin matrix was worn off, the reinforcing fibers were exposed to the friction surfaces of the composites and scraped the mating plate, which was transformed to frictional output. This is consistent with the previous report [38]. Among these tested specimens, the specimen FC-6 showed the highest friction coefficient except for the case at 100 °C where the specimen FC-8 showed a little higher friction coefficient.

During the recovery tests (Figure 4b), the friction coefficients of the composites decreased first with the temperature ranging from 300 °C to 200 °C, and then increased from 200 °C to 150 °C, and finally declined with the temperature varying from 150 °C to 100 °C. On the whole, the variation of friction coefficients was relatively stable, and it fluctuated between 0.402 and 0.489. The recovery fluctuation was one of the major affection factors for the performance of automotive braking.

The variations in wear rates of the friction composites with test temperatures are illustrated in Figure 5. It can be seen clearly that the wear rates were dramatically affected by temperature and increased with the temperature rise for all composites. The reason for this behavior may be that the phenolic resin gradually began to soften and decompose as test temperatures rise, causing a decrease in interface binding force between the composite matrix and fillers. As a result, the fillers were loosened and debonded from the matrix, which increased the wear rates of the composites. This is in accordance with the previous study results [39,40].

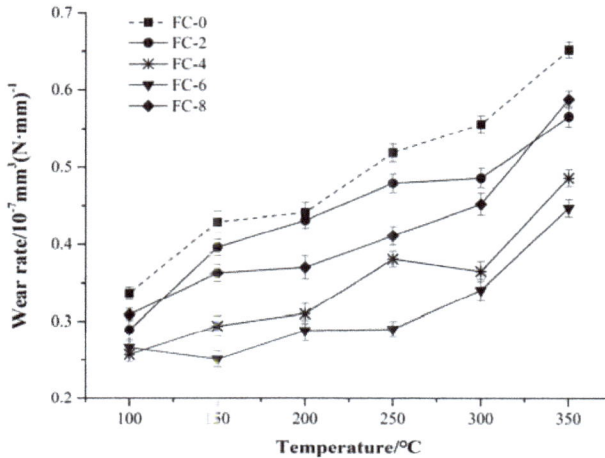

Figure 5. Wear rates of the friction specimens.

Generally, the incorporation of corn stalk fibers enhanced the wear behaviors of the composites, as the wear rates decreased first and then increased with increasing fiber content. It was indicated that there was an optimum fiber content in the formula of friction composites [6]. Among all tested specimens, the wear rate maximized to 0.632×10^{-7} mm^3 (N·mm)$^{-1}$ at 350 °C in specimen FC-0, whereas it minimized to 0.427×10^{-7} mm^3 (N·mm)$^{-1}$ at 350 °C in specimen FC-6 except for the case between 100 °C and 150 °C where the wear rate of specimen FC-4 was a bit lower. This observation suggested that 6 wt % of corn stalk fibers was the optimum dosage for the wear performance of friction composites. A dosage beyond 6 wt % might induce fiber accumulation and uneven distribution in the composite matrix, which would lead to a decline in wear resistance.

3.4. Fade Resistance and Recovery Properties

The fade resistance and recovery behaviors are of critical importance in the performance assessment of friction composites, and they can influence braking reliability and effectiveness during the braking process [33]. The friction coefficients decreased gradually with a temperature rise and recovered after a temperature reduction, which were referred to as fade and recovery phenomena, respectively [34]. Fade ratios and recovery ratios were the main parameters for evaluating the friction stability of friction composites and quantitatively characterizing the fluctuations of the friction coefficients. The fade and recovery behaviors of the five friction composites are illustrated in Figure 6. It can be seen clearly that the composites added with corn stalk fibers showed improved fade and recovery behaviors during the tests. The fade ratios of the friction specimens ranked in the order of FC-0 > FC-8 > FC-2 > FC-4 > FC-6, while the order of recovery ratios was FC-6 > FC-4 > FC-8 > FC-2 > FC-0. In particular, the specimen FC-6 had the fade ratio of 7.8% and recovery ratio of 106.5%, indicating it behaved well in fade resistance and recovery. However, the fade ratio of specimen FC-0 was 14.3% and the recovery ratio was 93.5%, indicating its fade and recovery properties were the worst of the five composites.

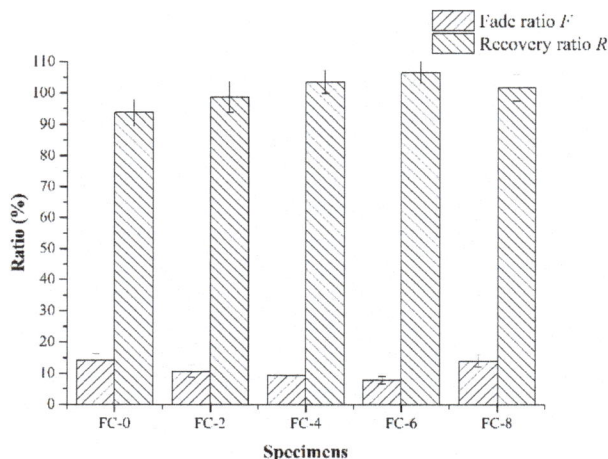

Figure 6. Fade ratios and recovery ratios of the friction composites.

3.5. Analysis of Worn Surface

The tribological performances of friction materials are closely associated with their worn surface morphological properties (e.g., wear debris, primary and secondary plateaus, microcracks and cavities) [31,41,42]. Surface morphology research of friction composites have been reported as an effective tool to interpret the results of tribological behavior analysis and explain the wear mechanisms [43].

In the present study, the worn surface morphology analysis of friction composites was performed by SEM. The typical worn surfaces of the five friction composites at 350 °C are presented in Figure 7a–e. Specifically, the worn surface of the specimen FC-0 was very rough with severe damage and massive destruction (Figure 7a). A number of fine wear debris and hard particles along with large spalling pits presented on the surface, and loose matrix, microcracks, and a lot of grooves were also evident, which corresponded to high wear rate of specimen FC-0. In general, the thermal relaxation and degradation of resin caused matrix loosening and then generated a lot of wear debris and abrasive particles under the action of friction force. Some of these debris and particles were embedded and removed on the worn surface, and then shallow grooves parallel to the sliding direction appeared on the surface of the specimen, which were all typical of abrasive wear. Meanwhile, large flake debris detached from the surface under the sheer force, and spalling pits presented on the matrix surface, which might be the main cause of adhesive wear. Moreover, owing to varying thermal expansion rates in different regions on the friction surface layer, microcracks appeared on the worn surface of the composite under unstable pressure and temperature field, which could be responsible for the fatigue wear [44]. Hence, the main wear mechanisms of the specimen FC-0 were abrasive wear, adhesive wear and fatigue wear.

As illustrated in Figure 7b–e, the worn surfaces of specimens FC-2, FC-4, FC-6 and FC-8 were relatively smooth in comparison with specimen FC-0, indicating that the incorporation of corn stalk fibers prevented the materials from peeling off in large flakes. The micrograph in Figure 7b proved the formation of some wear debris, particles, microcracks and shallow grooves, and meanwhile, some local detached regions also presented on the worn surface, which could account for the high wear rate of specimen FC-2. The surface of specimen FC-8 was covered with fine wear debris, parallel shallow grooves and bare fibers (Figure 7e). This observation could be explained by the fact that the resin adhesiveness to corn stalk fibers decreased with the increasing fiber content, which depressed the fiber-matrix interface bonding strength, and some fibers easily fell off from the matrix under the sheer force and normal pressure. And the broken fibers and hard particles were embedded in the matrix, then scratched and damaged the surface during wear process, leading to the increase of the wear rate [45].

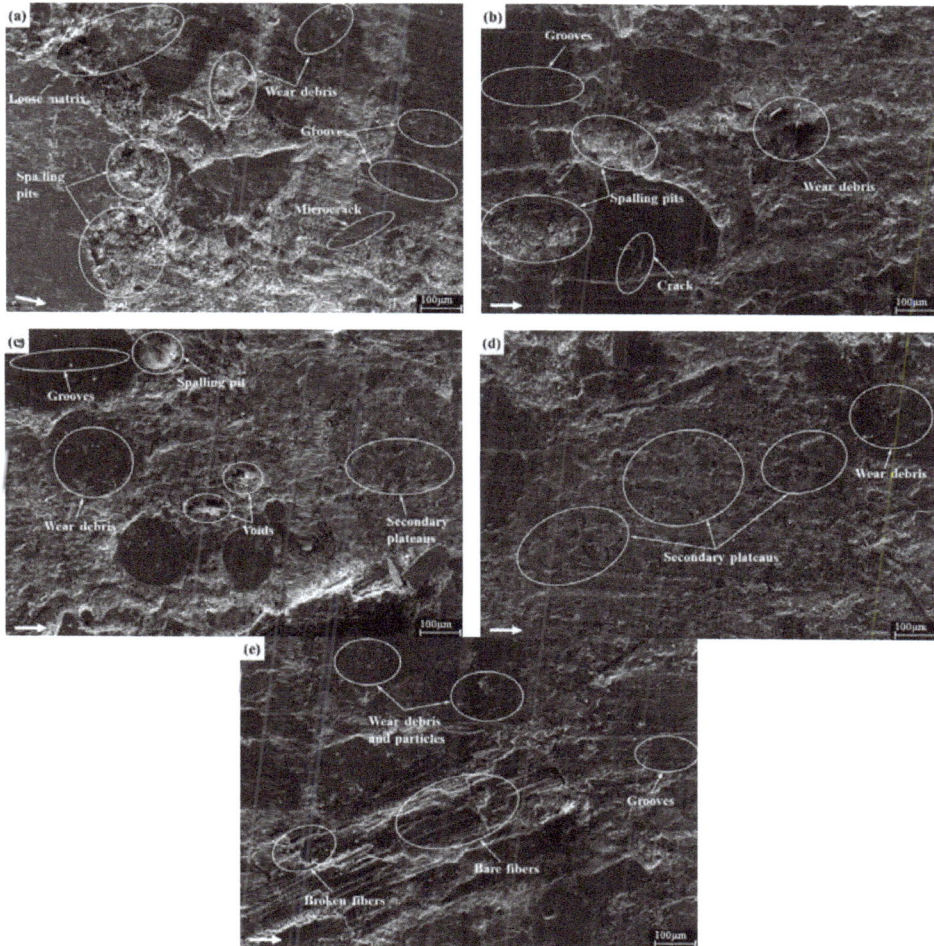

Figure 7. Worn surface morphology of (**a**) FC-0; (**b**) FC-2; (**c**) FC-4; (**d**) FC-6; (**e**) FC-8.

Moreover, as evident from Figure 7c,d, the specimens FC-4 and FC-6 exhibited relatively smooth worn surfaces compared with other composites. Small spalling pits, some fine wear debris and few shallow grooves as well as some apparent voids and secondary plateaus existed on the worn surface of specimen FC-4 (Figure 7c). In general, the formation of voids facilitated the absorption of braking noise to some extent, and meanwhile, some debris and particles were found in the voids, which could contribute to the reduction of the surface damage of specimen FC-4. As shown in Figure 7d, no obvious separation was found at the interface between the resin matrix and fillers, and abundant secondary plateaus presented on the surface of specimen FC-6, which were responsible for its higher wear resistance. During the wear process, the formation and development of secondary contact plateaus were attributed to the compression of wear debris at normal pressure, shear force and friction heat [46].

And the generation of secondary plateaus could induce the formation of friction film on the composite surface, which was correlated with the stable friction coefficient and small wear rate [47,48].

3.6. Analysis of Worn Surface Roughness

The surface roughness of friction composites is significantly related to both friction behavior and wear resistance in a certain manner. An exact analysis of worn surface roughness of the friction composites was carried out by using CLSM, which enabled the three-dimensional reconstruction of surface geometry.

The main surface roughness parameters of the five friction composites, including average roughness (Ra), root-mean-square roughness (Rq), maximum valley depth (Rv) and maximum peak height (Rp), are summarized in Table 3. It can be seen that the values of Ra and Rq were in the order FC-0 > FC-2 > FC-8 > FC-4 > FC-6, which was consistent with the results of the tribological behaviors. As for the Rv and Rp values, no clear trends were observed. The Rv of the specimen FC-4 was the highest, which was attributed to the pull-out of the fibers and the formation of the cavities. The value of Rp was maximized in specimen FC-8. This may be because some wear debris piled up around the fiber ends and were compressed under the normal pressure, then the contact plateaus formed on the worn surface, which resulted in the increase of Rp. This indicated that the specimen FC-6 exhibited the lowest roughness (Ra = 1.746 μm), whereas the specimen FC-0 showed the highest roughness (Ra = 2.786 μm) among all the specimens. These results were consistent with the aforementioned tribological behavior and surface morphology analysis, as well as the reconstructed surface geometry in Figure 8. Under the condition of dry sliding, the larger roughness was mainly ascribed to the serious damage of worn surface which could cause an increase in average roughness. It suggested that the worn surface of specimen FC-6 was much smoother than other friction composites, and the specimen FC-6 possessed higher wear resistance.

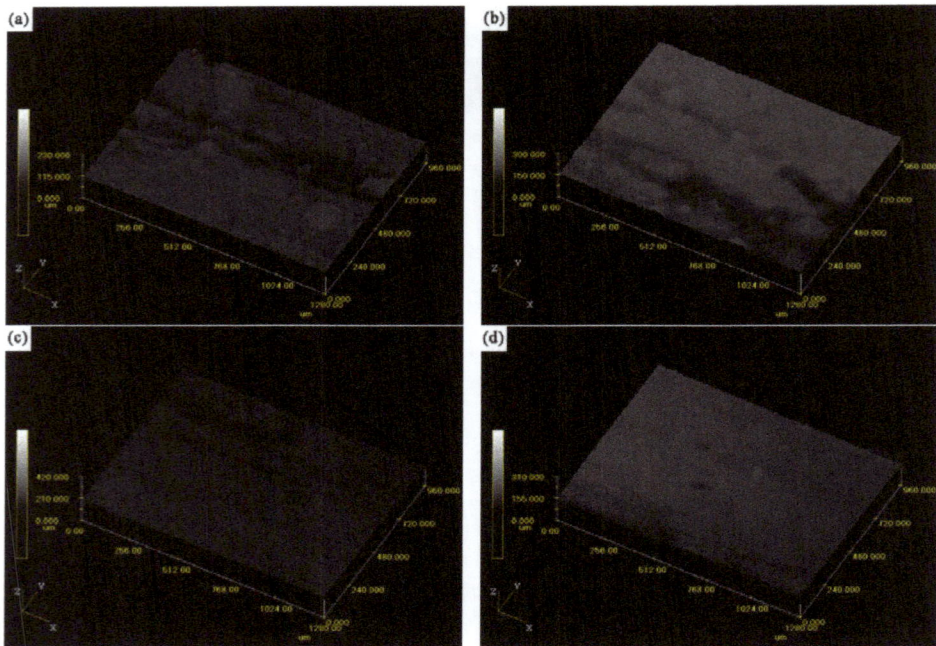

Figure 8. *Cont.*

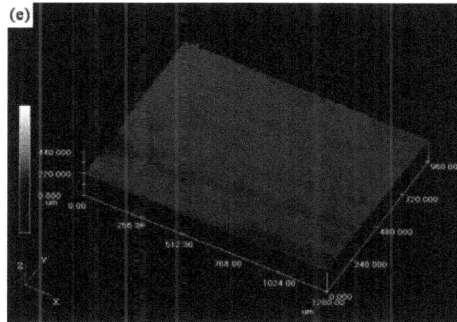

Figure 8. Three-dimensional reconstructions of surface geometry of (**a**) FC-0; (**b**) FC-2; (**c**) FC-4; (**d**) FC-6; (**e**) FC-8.

Table 3. Surface roughness parameters of the friction composites.

Specimens	Average Roughness	Root-Mean-Square Roughness	Maximum Valley Depth	Maximum Peak Height
	Ra (μm)	Rq (μm)	Rv (μm)	Rp (μm)
FC-0	2.786	3.883	71.305	34.012
FC-2	2.506	3.429	38.974	26.658
FC-4	1.838	2.661	75.024	30.059
FC-6	1.746	2.574	42.031	25.824
FC-8	2.407	3.401	48.576	35.839

4. Conclusions

The physio-mechanical and tribological behaviors of friction composites with different relative contents of corn stalk fibers were systematically investigated in the present study. Based on the results, the main conclusions can be summarized as follows:

1. The density and hardness of friction composites decreased with the increasing content of corn stalk fibers. At the same time, the impact strength of specimen FC-4 was the highest in comparison with that of other composites.

2. The friction coefficients of the composites generally increased first and then decreased with the temperature increase. Compared with specimen FC-0, the corn stalk fiber-reinforced friction composites showed higher friction coefficients except for the case at 100 °C.

3. The specimen FC-6 showed a fade ratio of 7.8% and recovery ratio of 106.5%, suggesting superior fade resistance and recovery performances.

4. The wear rates of all composites were significantly influenced by the test temperature and increased with temperature rise. The specimen FC-6 exhibited the lowest wear rate, except for that when the temperatures were about 100–150 °C.

5. The micrographs of worn surface morphology showed that the tribological performances of friction composites were closely associated with the formation of secondary contact plateaus on the surfaces. The specimen FC-6 had a smoother worn surface (Ra = 1.746 μm) than other friction composites and was covered with a great number of secondary plateaus and few shallow grooves, which explained the higher wear resistance.

The physio-mechanical and tribological tests confirmed that corn stalk fibers could be used as reinforcement for friction composites, which is an environmentally-friendly form of utilization of waste corn stalks.

Author Contributions: Y.M. and S.W. conceived and designed the experiments. S.W. and H.Q. performed the experiments. J.Z., J.T. and Y.X. analyzed the data. Y.M., J.Z. and X.Y. discussed the results. Y.M., S.W. and J.T. wrote the paper.

Funding: This study was funded by the State Key Laboratory of Automotive Safety and Energy (No. KF1814), National Natural Science Foundation of China (No. 51475205), Jilin Province Science and Technology Development Plan Item (No. 20170101173JC and 20170204015NY), National Key Research Program of China (No. 2016YFD0701601 and 2017YFD0701103-1), Jilin Provincial Development and Reform Commission (No. 2018C044-3), China-EU H2020 FabSurfWAR project (No. 2016YFE0112100 and 644971), and the 111 Project of China (No. B16020).

Conflicts of Interest: The authors declare no conflict of interest.

References

1. Ingram, M.; Spikes, H.; Noles, J.; Watts, R.F. Contact properties of wet clutch friction material. *Tribol. Int.* **2010**, *43*, 815–821. [CrossRef]
2. Cai, P.; Li, Z.; Wang, T.; Wang, Q. Effect of aspect ratios of aramid fiber on mechanical and tribological behaviors of friction materials. *Tribol. Int.* **2015**, *92*, 109–116. [CrossRef]
3. Wang, Z.H.; Hou, G.H.; Yang, Z.R.; Jiang, Q.; Zhang, F.; Xie, M.H.; Yao, Z.J. Influence of slag weight fraction on mechanical, thermal and tribological properties of polymer based friction materials. *Mater. Des.* **2016**, *90*, 76–83. [CrossRef]
4. Kumar, M.; Bijwe, J. Composite friction materials based on metallic fillers: sensitivity of μ to operating variables. *Tribol. Int.* **2011**, *44*, 106–113. [CrossRef]
5. Kuroe, M.; Tsunoda, T.; Kawano, Y.; Takahashi, A. Application of lignin-modified phenolic resins to brake friction material. *J. Appl. Polym. Sci.* **2013**, *129*, 310–315. [CrossRef]
6. Kim, S.J.; Cho, M.H.; Lim, D.S.; Jang, H. Synergistic effects of aramid pulp and potassium titanate whiskers in the automotive friction material. *Wear* **2001**, *251*, 1484–1491. [CrossRef]
7. Kim, S.J.; Jang, H. Friction and wear of friction materials containing two different phenolic resins reinforced with aramid pulp. *Tribol. Int.* **2000**, *33*, 477–484. [CrossRef]
8. Stachowiak, G.W.; Chan, D. Review of automotive brake friction materials. *Proc. Inst. Mech. Eng. Part D* **2004**, *218*, 953–966.
9. Etemadi, H.; Shojaei, A.; Jahanmard, P. Effect of alumina nanoparticle on the tribological performance of automotive brake friction materials. *J. Reinf. Plast. Compos.* **2014**, *33*, 166–178. [CrossRef]
10. Öztürk, B.; Öztürk, S. Effects of resin type and fiber length on the mechanical and tribological properties of brake friction materials. *Tribol. Lett.* **2011**, *42*, 339–350.
11. Aranganathan, N.; Mahale, V.; Bijwe, J. Effects of aramid fiber concentration on the friction and wear characteristics of non-asbestos organic friction composites using standardized braking tests. *Wear* **2016**, *354*, 69–77. [CrossRef]
12. Jang, H.; Ko, K.; Kim, S.J.; Basch, R.H.; Fash, J.W. The effect of metal fibers on the friction performance of automotive brake friction materials. *Wear* **2004**, *256*, 406–414. [CrossRef]
13. Ramesh, M.; Palanikumar, K.; Reddy, K.H. Plant fibre based bio-composites: Sustainable and renewable green materials. *Renewable Sustainable Energy Rev.* **2017**, *79*, 558–584. [CrossRef]
14. Ray, D.; Sengupta, S.; Sengupta, S.P.; Mohanty, A.K.; Misra, M. A Study of the Mechanical and Fracture Behavior of Jute-Fabric-Reinforced Clay-Modified Thermoplastic Starch-Matrix Composites. *Macromol. Mater. Eng.* **2007**, *292*, 1075–1084. [CrossRef]
15. Ramesh, M.; Palanikumar, K.; Reddy, K.H. Influence of fiber orientation and fiber content on properties of sisal-jute-glass fiber-reinforced polyester composites. *J. Appl. Polym. Sci.* **2016**, *133*, 42968. [CrossRef]
16. Chand, N.; Dwivedi, U.K. Effect of coupling agent on abrasive wear behaviour of chopped jute fibre-reinforced polypropylene composites. *Wear* **2006**, *261*, 1057–1063. [CrossRef]
17. Bajpai, P.K.; Singh, I.; Madaan, J. Tribological behavior of natural fiber reinforced PLA composites. *Wear* **2013**, *297*, 829–840. [CrossRef]
18. Nirmal, U.; Yousif, B.F.; Rilling, D.; Brevern, P.V. Effect of betelnut fibres treatment and contact conditions on adhesive wear and frictional performance of polyester composites. *Wear* **2010**, *268*, 1354–1370. [CrossRef]
19. Fu, Z.; Suo, B.; Yun, R.; Lu, Y.; Wang, H.; Qi, S.; Jiang, S.; Lu, Y.; Matejka, V. Development of eco-friendly brake friction composites containing flax fibers. *J. Reinf. Plast. Compos.* **2012**, *31*, 681–689. [CrossRef]

20. Lu, Z.; Zhao, Z.; Wang, M.; Jia, W. Effects of corn stalk fiber content on properties of biomass brick. *Constr. Build. Mater.* **2016**, *127*, 11–17. [CrossRef]

21. Husseien, M.; Amer, A.A.; El-Maghraby, A.; Hamedallah, N. A comprehensive characterization of corn stalk and study of carbonized corn stalk in dye and gas oil sorption. *J. Anal. Appl. Pyrolysis.* **2009**, *86*, 360–363. [CrossRef]

22. Cai, D.; Li, P.; Luo, Z.; Qin, P.; Chen, C.; Wang, Y.; Wang, Z.; Tan, T. Effect of dilute alkaline pretreatment on the conversion of different parts of corn stalk to fermentable sugars and its application in acetone-butanol-ethanol fermentation. *Bioresour. Technol.* **2016**, *211*, 117–124. [CrossRef] [PubMed]

23. Wang, G.; Chen, C.; Li, J.; Zhou, B.; Xie, M.; Hu, S.; Kawamura, K.; Chen, Y. Molecular composition and size distribution of sugars, sugar-alcohols and carboxylic acids in airborne particles during a severe urban haze event caused by wheat straw burning. *Atmos. Environ.* **2011**, *45*, 2473–2479. [CrossRef]

24. Luo, Z.; Li, P.; Cai, D.; Chen, Q.; Qin, P.; Tan, T.; Cao, H. Comparison of performances of corn fiber plastic composites made from different parts of corn stalk. *Ind. Crops. Prod.* **2017**, *95*, 521–527. [CrossRef]

25. Ghaffar, S.H.; Fan, M.; McVicar, B. Interfacial properties with bonding and failure mechanisms of wheat straw node and internode. *Composites Part A* **2017**, *99*, 102–112. [CrossRef]

26. Ramnath, B.V.; Kokan, S.J.; Raja, R.N.; Sathyanarayanan, R.; Elanchezhian, C.; Prasad, A.R.; Manickavasagam, V.M. Evaluation of mechanical properties of abaca-jute-glass fibre reinforced epoxy composite. *Mater. Des.* **2013**, *51*, 357–366. [CrossRef]

27. Corrales, F.; Vilaseca, F.; Llop, M.; Gironès, J.; Méndez, J.A.; Mutjè, P. Chemical modification of jute fibers for the production of green-composites. *J. Hazard. Mater.* **2007**, *144*, 730–735. [CrossRef] [PubMed]

28. Kabir, M.M.; Wang, H.; Lau, K.T.; Cardona, F. Chemical treatments on plant-based natural fibre reinforced polymer composites: An overview. *Composites Part B* **2012**, *43*, 2883–2892. [CrossRef]

29. Merlini, C.; Soldi, V.; Barra, G.M.O. Influence of fiber surface treatment and length on physico-chemical properties of short random banana fiber-reinforced castor oil polyurethane composites. *Polym. Test.* **2011**, *30*, 833–840. [CrossRef]

30. Goriparthi, B.K.; Suman, K.N.S.; Rao, N.M. Effect of fiber surface treatments on mechanical and abrasive wear performance of polylactide/jute composites. *Composites Part A* **2012**, *43*, 1800–1808. [CrossRef]

31. Ma, Y.; Liu, Y.; Wang, L.; Tong, J.; Zhuang, J.; Jia, H. Performance assessment of hybrid fibers reinforced friction composites under dry sliding conditions. *Tribol. Int.* **2018**, *119*, 262–269. [CrossRef]

32. Ma, Y.; Liu, Y.; Menon, C.; Tong, J. Evaluation of Wear Resistance of Friction Materials Prepared by Granulation. *ACS Appl. Mater. Interfaces* **2015**, *7*, 22814–22820. [CrossRef] [PubMed]

33. Satapathy, B.K.; Bijwe, J. Performance of friction materials based on variation in nature of organic fibres: Part I Fade and recovery behaviour. *Wear* **2004**, *257*, 573–584. [CrossRef]

34. Fu, H.; Fu, L.; Zhang, G.L.; Wang, R.M.; Liao, B.; Sun, B.C. Abrasion mechanism of stainless steel/carbon fiber-reinforced polyether-ether-ketone (PEEK) composites. *J. Mater. Eng. Perform* **2009**, *18*, 973–979. [CrossRef]

35. Rahman, M.M.; Khan, M.A. Surface treatment of coir (Cocos nucifera) fibers and its influence on the fibers' physico-mechanical properties. *Compos. Sci. Technol.* **2007**, *67*, 2369–2376. [CrossRef]

36. Ma, Y.; Liu, Y.; Shang, W.; Gao, Z.; Wang, H.; Guo, L.; Tong, J. Tribological and mechanical properties of pine needle fiber reinforced friction composites under dry sliding conditions. *RSC Adv.* **2014**, *4*, 36777–36783. [CrossRef]

37. Matějka, V.; Fu, Z.; Kukutschová, J. Qi, S.; Jiang, S.; Zhang, X.; Yun, R.; Vaculík, M.; Heliová, M.; Lu, Y. Jute fibers and powderized hazelnut shells as natural fillers in non-asbestos organic non-metallic friction composites. *Mater. Des.* **2013**, *51*, 847–853. [CrossRef]

38. Fei, J.; Luo, W.; Huang, J.F.; Ouyang, H.B.; Xu, Z.W.; Yao, C.Y. Effect of carbon fiber content on the friction and wear performance of paper-based friction materials. *Tribol. Int.* **2015**, *87*, 91–97. [CrossRef]

39. Ji, Z.; Jin, H.; Luo, W.; Cheng, F.; Chen, Y.; Ren, Y.; Wu, Y.; Hou, S. The effect of crystallinity of potassium titanate whisker on the tribological behavior of NAO friction materials. *Tribol. Int.* **2017**, *107*, 213–220. [CrossRef]

40. Cai, P.; Wang, Y.; Wang, T.; Wang, Q. Effect of resins on thermal, mechanical and tribological properties of friction materials. *Tribol. Int.* **2015**, *87*, 1–10. [CrossRef]

41. Silvestre, N. State-of-the-art review on carbon nanotube reinforced metal matrix composites. *Int. J. Compos. Mater.* **2013**, *3*, 28–44.

42. Patnaik, A.; Kumar, M.; Satapathy, B.K.; Tomar, B.S. Performance sensitivity of hybrid phenolic composites in friction braking: effect of ceramic and aramid fibre combination. *Wear* **2010**, *269*, 891–899. [CrossRef]

43. Satapathy, B.K.; Bijwe, J. Fade and recovery behavior of non-asbestos organic (NAO) composite friction materials based on combinations of rock fibers and organic fibers. *J. Reinf. Plast. Compos.* **2005**, *24*, 563–577. [CrossRef]

44. Einset, E.O. Analysis of reactive melt infiltration in the processing of ceramics and ceramic composites. *Chem. Eng. Sci.* **1998**, *53*, 1027–1039. [CrossRef]

45. Straffelini, G.; Maines, L. The relationship between wear of semimetallic friction materials and pearlitic cast iron in dry sliding. *Wear* **2013**, *307*, 75–80. [CrossRef]

46. Kumar, M.; Satapathy, B.K.; Patnaik, A.; Kolluri, D.K.; Tomar, B.S. Hybrid composite friction materials reinforced with combination of potassium titanate whiskers and aramid fibre: assessment of fade and recovery performance. *Tribol. Int.* **2011**, *44*, 359–367. [CrossRef]

47. Kim, S.H.; Jang, H. Friction and vibration of brake friction materials reinforced with chopped glass fibers. *Tribol. Lett.* **2013**, *52*, 341–349. [CrossRef]

48. Prabhu, T.R.; Varma, V.K.; Vedantam, S. Tribological and mechanical behavior of multilayer Cu/SiC+ Gr hybrid composites for brake friction material applications. *Wear* **2014**, *317*, 201–212. [CrossRef]

MDPI

Article

Graphene Oxide Oxygen Content Affects Physical and Biological Properties of Scaffolds Based on Chitosan/Graphene Oxide Conjugates

Iolanda Francolini [1], Elena Perugini [1], Ilaria Silvestro [1], Mariangela Lopreiato [2], Anna Scotto d'Abusco [2], Federica Valentini [3], Ernesto Placidi [4,5], Fabrizio Arciprete [4], Andrea Martinelli [1] and Antonella Piozzi [1,*]

1   Department of Chemistry, Sapienza University of Rome, P.le A. Moro, 5, 00185 Rome, Italy;
    iolanda.francolini@uniroma1.it (I.F.); peruginielena80@gmail.com (E.P.); ilaria.silvestro@uniroma1.it (I.S.);
    andrea.martinelli@uniroma1.it (A.M.)
2   Department of Biochemical Sciences, Sapienza University of Rome, P.le A. Moro, 5, 00185 Rome, Italy;
    mariangela.lopreiato@uniroma1.it (M.L.); anna.scottodabusco@uniroma1.it (A.S.d.)
3   Department of Chemical Science and Technologies, University of Rome Tor Vergata, Via della Ricerca
    Scientifica, 00133 Rome, Italy; federica.valentini@uniroma2.it
4   Department of Physics, University of Rome Tor Vergata, Via della Ricerca Scientifica, 00133 Rome, Italy;
    ernesto.placidi@roma2.infn.it (E.P.); fabrizio.arciprete@roma2.infn.it (F.A.)
5   CNR-ISM, Via Fosso del Cavaliere 100, I-00133 Rome, Italy
*   Correspondence: antonella.piozzi@uniroma1.it; Tel.: +39-06-4991-3692

Received: 13 March 2019; Accepted: 2 April 2019; Published: 8 April 2019

Abstract: Tissue engineering is a highly interdisciplinary field of medicine aiming at regenerating damaged tissues by combining cells with porous scaffolds materials. Scaffolds are templates for tissue regeneration and should ensure suitable cell adhesion and mechanical stability throughout the application period. Chitosan (CS) is a biocompatible polymer highly investigated for scaffold preparation but suffers from poor mechanical strength. In this study, graphene oxide (GO) was conjugated to chitosan at two weight ratios 0.3% and 1%, and the resulting conjugates were used to prepare composite scaffolds with improved mechanical strength. To study the effect of GO oxidation degree on scaffold mechanical and biological properties, GO samples at two different oxygen contents were employed. The obtained GO/CS scaffolds were highly porous and showed good swelling in water, though to a lesser extent than pure CS scaffold. In contrast, GO increased scaffold thermal stability and mechanical strength with respect to pure CS, especially when the GO at low oxygen content was used. The scaffold in vitro cytocompatibility using human primary dermal fibroblasts was also affected by the type of used GO. Specifically, the GO with less content of oxygen provided the scaffold with the best biocompatibility.

Keywords: graphene oxide; chitosan; composites; scaffolds; tissue engineering

---

1. Introduction

Chitosan is a cationic polysaccharide, deriving from chitin deacetylation, which has gained a prominent place in biomedicine for a wide range of applications including drug delivery, wound dressings, bacterial contamination control, fat binding, and tissue engineering [1–5]. The peculiarity of chitosan, compared to other polysaccharides, is that it has been shown to provoke minimal or no foreign-body reaction, including inflammatory response and fibrotic encapsulation when used in hydrogel systems [6,7], polyelectrolyte multilayers [8], biomembranes [9] and as a porous 3-D scaffold [10]. Besides, chitosan has been shown to promote cell adhesion and proliferation in tissue engineering applications, especially when applied for bone tissue regeneration where it showed osteoconductivity and ability to promote osteogenic differentiation [11–16].

The major limitation of chitosan for the repair of bone defects is its low mechanical strength, which precludes pure chitosan scaffolds for load-bearing applications. For such reason, many chitosan composite scaffolds have been lately developed to improve mechanical scaffold properties and bioactivity [16,17]. Main substances used in combination with chitosan to produce scaffolds for bone tissue regeneration are tricalcium phosphate [18], hydroxyapatite [19], silica nanoparticles [20], and, more recently, graphene oxide [21–25]. Particularly, graphene oxide (GO) is obtained by oxidation and exfoliation of graphite [26,27] and consists of a monolayer of sp^2-hybridized carbon atoms arranged in a honeycomb structure decorated with oxygen-containing groups, including hydroxyl, epoxy and carboxylic groups [28]. Such material is considered highly promising for bone tissue engineering because not only it presents high mechanical stiffness and flexibility, but it was also shown to improve osteogenesis and cellular differentiation when combined with other biomaterials [23,29–32]. GO also showed significant antibacterial activity [33] like other 2D-nanomaterials [34]. Therefore, its incorporation into the chitosan matrix opens perspectives for antimicrobial applications of GO-chitosan scaffolds.

A series of different types of GO can be produced, which may vary in terms of layer surface area, structural defects, sp^2/sp^3 ratio and oxygen content. Specifically, the oxygen content, which strongly depends on the GO preparation method and post-oxidation treatments [35,36], can significantly affect GO mechanical properties, conductivity, ability to disperse in water and biocompatibility. Indeed, oxygen groups present on the basal plane and edges of GO enable it to interact with cellular components like proteins, mainly through electrostatic interaction and hydrogen bonds [37–39].

In this study, in order to investigate the effects of GO oxygen content on scaffold mechanical and biological properties, 3D porous scaffolds based on chitosan/graphene oxide conjugates were prepared by employing two types of GO, a GO sample at low oxygen content (commercially available) and a GO sample at high oxygen content obtained by electrochemical exfoliation of graphite. GO was conjugated to chitosan by a carbodiimide-mediated amidation in two concentrations (0.3% and 1%). Then, scaffolds, prepared by either the salt leaching method or freeze-drying, were characterized in terms of water swelling, water retention ability, thermal properties, mechanical response in compression and in vitro cytocompatibility against primary human dermal fibroblasts.

2. Results and Discussion

2.1. GO Characterization

Figure 1a shows the Atomic Force Microscopy (AFM) topography of GO_LiClO$_4$, named GO$_{exfoliated}$. This sample exhibits small surface area (ranging from 0.03 to 0.15 μm^2) and the average height in the range of 3–4 ML (Mono Layer), although also single layer sheets (ca. 0.6 nm in thickness, as shown in Figure 1a) are often observed. Indeed, a single layer of GO on mica typically exhibits a 0.6–1.0 nm height, greater than the graphene thickness.

Figure 1b exhibits the AFM topography of commercial GO product, named GO$_{sigma}$. The latter is characterized by a graphene sheet with well homogeneous in thickness but heterogeneous in area distribution (0.006–0.010 μm^2). In 90% of cases, the sheets are 1–2 layer thick (Figure 1b). Thicker sheets are less frequent but generally wider.

Figure 2 shows the Raman spectrum of GO$_{sigma}$, while Table 1 reports the Raman parameters for both samples. The Raman spectrum of GO$_{exfoliated}$ was already published [40].

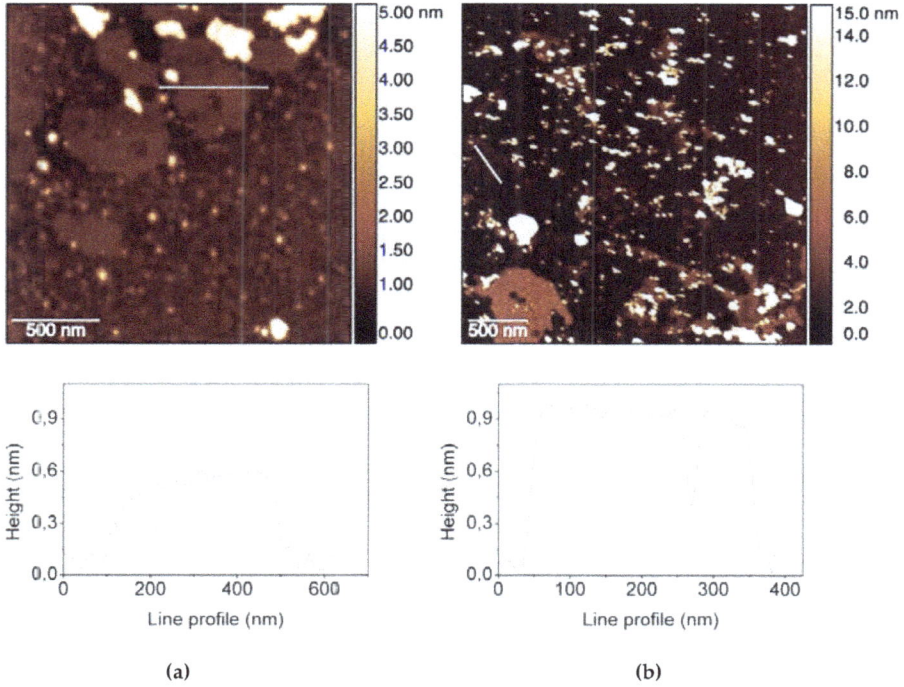

Figure 1. Atomic Force Microscopy (AFM) topographies of: (**a**) GO_LiClO$_4$ sample (GO$_{exfoliated}$); (**b**) commercial GO product (GO$_{sigma}$) and related height profiles.

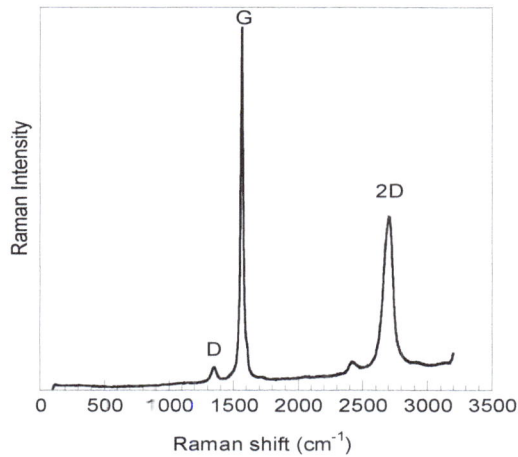

Figure 2. Raman spectrum of GO$_{sigma}$.

Table 1. Raman parameters for GO_{sigma} and $GO_{exfoliated}$.

Sample	Frequency (cm^{-1})	Assigned Bands	I_D/I_G
GO_{sigma}	2706	2D	
	1565	G	0.06
	1350	D	
$GO_{exfoliated}$	2713	2D	
	1582	G	0.27
	1357	D	

In the spectrum, the typical fingerprint of graphene is observable, with G-band at around 1580 cm^{-1}, corresponding to the first-order scattering of the E2g vibration mode, and a 2D band at 2700 cm^{-1}, corresponding to the second-order two phonon mode. Moreover, a D-band is present at 1350 cm^{-1}, reflecting the presence of structural defects (vacancies, edge defects, heteroatoms, etc.) [41]. The intensity ratio of D and G bands (I_D/I_G), reported in Table 1, provides information about structural defects on graphene surface and edges. Between the two samples, GO_{sigma} has the lowest I_D/I_G ratio indicating that this sample has the lowest content of defects.

In agreement with such data, XPS analysis revealed a lower content of oxygenated groups in the GO_{sigma} than the electrochemical synthesized $GO_{exfoliated}$ (Table 2). In Figure 3, the XPS spectrum of commercial GO is reported, while the XPS spectrum of $GO_{exfoliated}$ has already been published [40].

Table 2. Atomic (At) percentages from C1s fit for GO_{sigma} and $GO_{exfoliated}$.

Peak BE (eV)	Species	GO_{sigma} At (%)	$GO_{exfoliated}$ At (%)
283.8	C–C	60.7	50.0
284.7	C–OH	21.7	18.0
285.9	C–O	9.4	16.0
287.0	C=O	8.2	9.0
289.0	C(=O)O	4.3	7.0

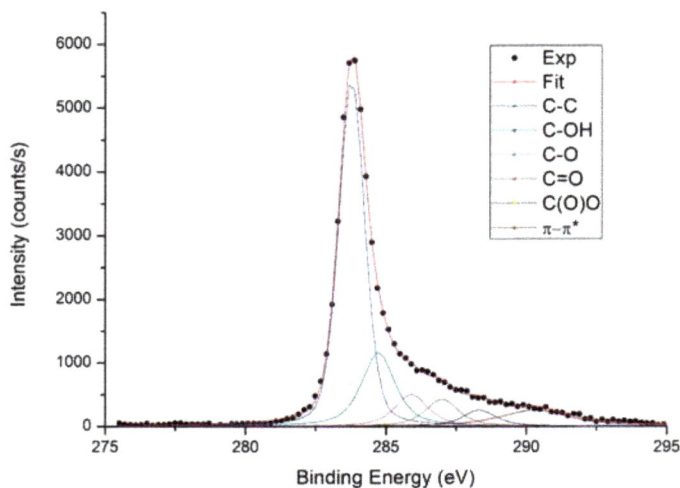

Figure 3. X-ray Photoelectron Spectroscopy (XPS) C1s spectrum of GO_{sigma}.

In Table 2, the deconvolution results of the relative C1s peaks are reported for both samples. The presence of aromatic C=C and aliphatic C–C is demonstrated by the main peak at binding energy 283.8 eV and the π–π* peak at ca. 291 eV attributed π-electrons delocalized in the aromatic network. The C–C peak was fitted with a weighted Voigtian profile instead of a Doniach-Sunjic one to take into account the contribution of C–H group too close to the C–C to be deconvolved singularly. All samples also showed a significant oxidation degree (i.e. functionalization degree), as indicated by the presence of hydroxyl, epoxide, carbonyl, and carboxyl functional groups. Between the two samples, the non-oxidized carbon component was more intense for GO_{sigma}; while $GO_{exfoliated}$ exhibited the highest content of oxygen-containing functional groups, and therefore the highest degree of oxidation/functionalization. This difference in oxidation degree could affect the physico-chemical properties of the resulting chitosan/GO composite materials and, thus their application.

2.2. Preparation and Characterization of GO/CS Composite Scaffolds

In this study, the two characterized GO samples were used in combination with chitosan to prepare composite scaffolds for application in tissue engineering. To achieve intimate contact between the polymeric matrix and the filler, GO was covalently linked to chitosan by an amidation reaction (Figure 4).

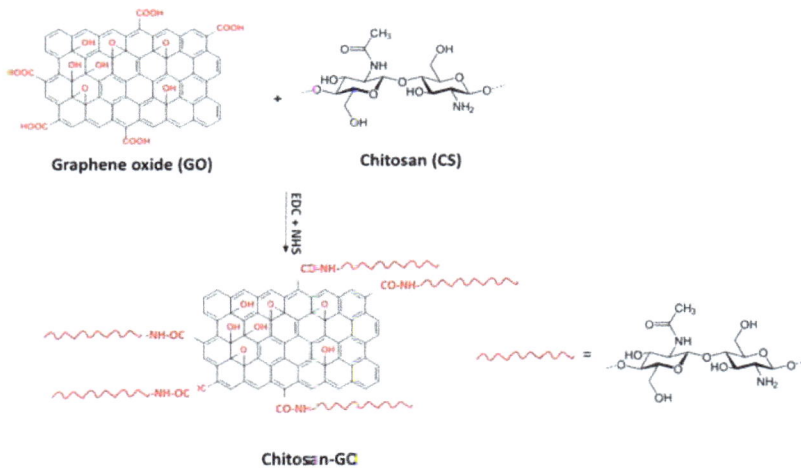

Figure 4. Scheme of Chitosan-GO amidation reaction.

Two GO/CS weight ratios (0.3% and 1%) were used during CS functionalization reaction. These concentrations were chosen as a compromise to obtain scaffolds combining good mechanical and biological properties. Indeed, in the literature, it was found that GO contents higher than 1% in chitosan based-scaffolds could only slightly improve mechanical properties [42] or even worsen them in terms of compressive strength [24]. Also, cell viability could be altered for too high GO content [42].

FTIR IR-ATR analysis confirmed qualitatively the success of amidation reaction in all of the used experimental conditions and for both types of GO. In Figure 5, as an example, the IR-ATR spectra of CS and of the GO-functionalized chitosan obtained with the GO_{sigma}—at a 0.3% GO/CS weight ratio—are reported. The spectra of GO_{sigma} and $GO_{exfoliated}$ were also reported for comparison. In both spectra, the absorption peaks of oxygenated groups are present (3427 cm^{-1} stretching O–H; 1738 stretching C=O; 1050–1280 cm^{-1} stretching C–O–C). As for the amidation reaction, the peaks at 1644 cm^{-1} and at 1560 cm^{-1}, attributed to stretching of C=O amide of CS and bending N–H of CS primary amine respectively, are present in the spectrum of the GO-functionalized chitosan but in a different ratio with respect to pure chitosan. Specifically, in all of the chitosan derivatives, an increase in the intensity ratio A_{1644}/A_{1560} was observed (Table 3), which is presumably related to a decrease in the number of NH$_2$ (decrease in the intensity of the peak at 1560 cm^{-1}) as a consequence of amidation with GO. However, physical interactions between chitosan and GO cannot be excluded, and they can occur to some extent [43].

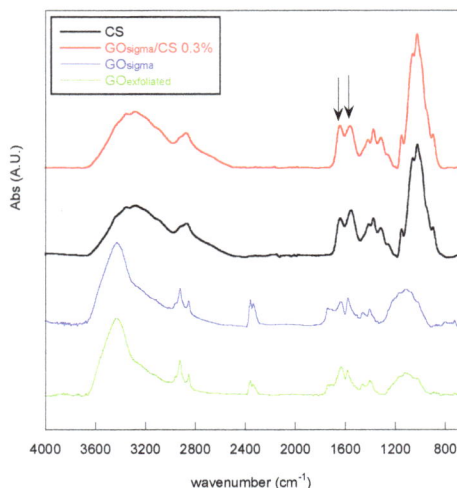

Figure 5. FTIR spectra of GO$_{sigma}$, GO$_{exfoliated}$, chitosan and of the GO-functionalized chitosan obtained with GO$_{sigma}$, and a GO/CS weight ratio of 0.3%.

Table 3. Intensity ratio of peaks at 1644 cm^{-1} and 1560 cm^{-1} (A_{1644}/A_{1560}) for the GO/CS samples.

Sample	CS Conc. (%, w/v)	A_{1644}/A_{1560}
CS	–	0.82
GO$_{sigma}$/CS 0.3%	1	0.93
GO$_{sigma}$/CS 1%	1	1.02
GO$_{exfoliated}$/CS 0.3%	1	0.88
GO$_{exfoliated}$/CS 1%	1	0.91

The GO-functionalized chitosan derivatives were then used to prepare porous scaffolds by the freeze-drying method or the salt leaching method. In Table 4, the types of prepared scaffolds and corresponding acronyms were reported.

Table 4. Types of prepared scaffolds and corresponding acronyms.

Acronym	Method for Scaffold Preparation	GO/CS % (w/w)
CS$_{SL}$	SL *	–
GO$_{sigma}$/CS$_{SL}$ 0.3%	SL	0.3
GO$_{sigma}$/CS$_{SL}$ 1%	SL	1
GO$_{sigma}$/CS$_{FD}$ 1%	FD *	1
GO$_{exfoliated}$/CS$_{SL}$ 0.3%	SL	0.3
GO$_{exfoliated}$/CS$_{SL}$ 1%	SL	1
GO$_{exfoliated}$/CS$_{FD}$ 1%	FD	1

* SL = salt leaching method; FD = Freeze-drying method.

In Figure 6, the porosity determined with either the gravimetric method or the liquid displacement method for all the prepared scaffolds is reported.

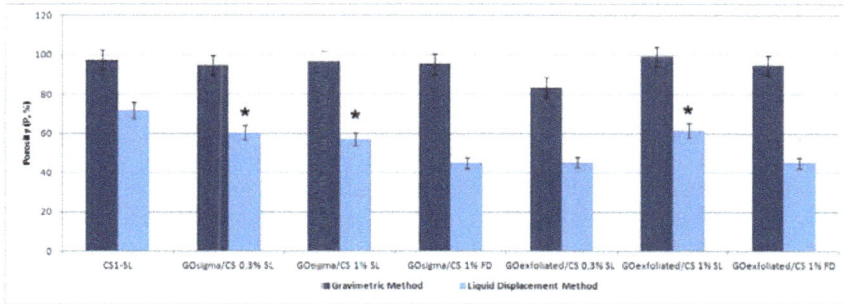

Figure 6. Porosity (%) of all the prepared scaffolds. The symbol (*) indicates samples with the best interconnected porosity.

As it can be observed, the porosity values determined by the gravimetric method were quite high (an average of 95% for all of the systems) and always greater than those determined by the liquid displacement method. That is because the gravimetric method provides information about the total scaffold porosity while the second method only about the pore fraction accessible to the liquid (ethanol), that is the fraction of interconnected pores. Some differences were observed in terms of such fraction among the various scaffolds.

Specifically, for both types of GC, the salt leaching permitted to obtain scaffold with a greater fraction of interconnected pores compared to freeze-drying. The GO content (0.3% or 1%) only slightly affected pore interconnection in the case of $GO_{exfoliated}$.

Field Emission Scanning Electron Microscope (FESEM) observations confirmed the higher level of porosity of the scaffolds prepared with the salt leaching method compared to the freeze-drying (Figure 7).

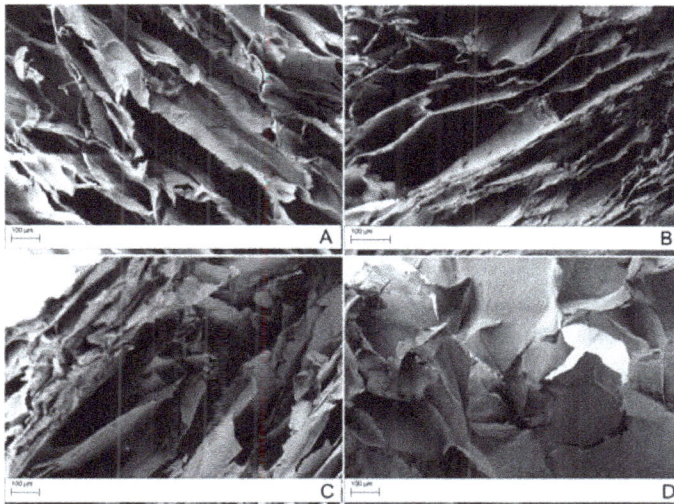

Figure 7. Field Emission Scanning Electron Microscope (FESEM) micrographs of scaffolds obtained with 1% CS solution: (**A**) CS; (**B**) GC_{sigma}/CS_{SL} 1%; (**C**) $GO_{exfoliated}/CS_{SL}$ 1%; (**D**) GO_{sigma}/CS_{FD} 1%.

Water swelling and water retention efficiencies are important properties for a scaffold since a low swelling degree may hamper nutrient diffusion through the scaffold while a high swelling degree may compromise scaffold integrity over usage [44]. As shown in Figure 8A, all of the scaffolds absorbed water very quickly, reaching the equilibrium in approximately 20 minutes. Composite scaffolds showed a lower swelling degree than pure CS scaffold, with the only exception of $GO_{exfoliated}/CS_{SL}$ 0.3%. For a fixed GO concentration (0.3% or 1%), the $GO_{exfoliated}$ make the scaffolds more hydrophilic than the GO_{sigma}, due to its high content of oxygenated groups. For a fixed type of GO ($GO_{exfoliated}$ or GO_{sigma}), instead, an increase in GO concentration decreased scaffold hydrophilicity because of the hydrophobicity of the aromatic C=C of the graphene plane. As reported in Table 5, the highest equilibrium swelling ratio was shown by the $GO_{exfoliated}/CS_{SL}$ 0.3% sample. A reduced degree of swelling of the composite scaffolds compared to CS could contribute to improving scaffold mechanical stability while the high initial rate of swelling could ensure a suitable supply of nutrients to cells seeded in the scaffold [45].

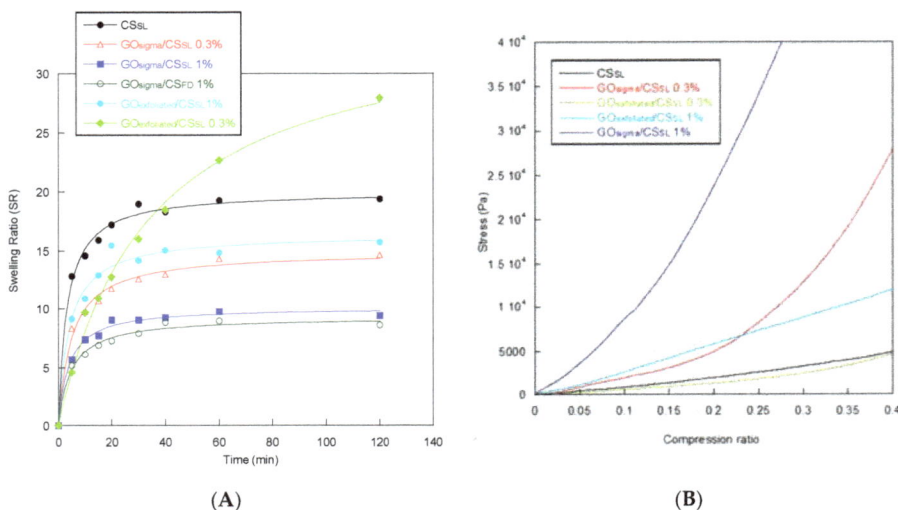

(A)   (B)

Figure 8. Swelling ratio (**A**) and mechanical behavior in compression tests (**B**) of the scaffolds.

Table 5. Equilibrium swelling ratio (SW), water retention (WR), degradation temperature (T_d) and glass transition temperature (T_g) of chitosan and GO/CS composite scaffolds.

Sample	Equilibrium Swelling Ratio (SR)	Water Retention (WR)	T_d (°C)	T_g (°C)	Compressive Modulus (KPa)
CS_{SL}	19 ± 2	10.1 ± 0.5	280	72	12 ± 3
GO_{sigma}/CS_{SL} 0.3%	14 ± 1	7.1 ± 0.5	296	84	22 ± 2
GO_{sigma}/CS_{SL} 1%	9 ± 2	5.9 ± 0.5	304	82	108 ± 5
GO_{sigma}/CS_{FD} 1%	8 ± 1	4.2 ± 0.5	ND *	ND	ND
$GO_{exfoliated}/CS_{SL}$ 0.3%	28 ± 4	7.6 ± 0.5	283	86	10 ± 3
$GO_{exfoliated}/CS_{SL}$ 1%	16 ± 3	5.5 ± 0.5	286	84	31 ± 4
$GO_{exfoliated}/CS_{FD}$ 1%	10 ± 2	4.0 ± 0.5	ND	ND	ND

* ND = Not determined.

As far as the water retention efficiency is concerned, the composite GO/CS scaffolds had sufficiently high water retention (WR) values ranging from 4 of the scaffolds obtained by freeze drying to 8 of GO$_{exfoliated}$/CS$_{SL}$ 0.3%, this latter being the sample also showing the highest swelling in water (Table 5). WR decreased with increasing GO content presumably because a higher CS functionalization resulted in a decreased number of CS amino groups available for interaction with water. In agreement with such a hypothesis, the scaffold made of pure CS showed the highest WR value (Table 5).

Thermal properties of the GO/CS composite scaffolds were studied by thermogravimetric analysis and differential scanning calorimetry. In Table 5, the degradation temperature (T_d) and glass transition temperature (T_g) of chitosan and GO/CS composite scaffolds are reported. As it can be observed, the composite scaffolds showed T_d values higher than that of CS, suggesting the formation of a crosslinked structure where presumably GO forms bridges among the polymer chains. Such network seemed to be stronger with increasing the GO content. The values of the glass transition temperature of the GO/CS composite scaffolds (Table 5) are also coherent with the formation of a cross-linked structure induced by CS amidation with GO. Indeed, all of the composite scaffolds showed a T_g higher than CS indicating a reduced polymer chain mobility.

Finally, the mechanical behavior in compression of the scaffolds was studied. In Figure 8B the stress versus the compression ratio is reported for CS and composite scaffolds while in Table 5 the Compressive Modulus for all of the scaffolds is reported.

The composite scaffolds showed Compressive Modulus values greater than CS$_{SL}$ scaffold, especially at 1% GO content. The only exception was GO$_{exfoliated}$/CS$_{SL}$ 0.3% that had a Compressive Modulus similar to CS$_{SL}$. At a fixed GO content, either 0.3% or 1%, the scaffolds obtained with the GO$_{sigma}$ were stiffer than those obtained with GO$_{exfoliated}$, suggesting a better dispersion of GO$_{sigma}$ in the CS structure. Presumably, the average size of GO$_{sigma}$ sheets smaller than GO$_{exfoliated}$, as resulted from AFM observations, contributed to the observed better dispersion and enhanced mechanical properties of the GO$_{sigma}$/CS scaffolds [46,47]. Aggregation phenomenon in GO$_{exfoliated}$ could also be related to the formation of hydrogen bonds among oxygenated functional groups present in the sheet basal planes, mediated by water molecules entrapped within the interlayer cavities. Indeed, it is known that a hydrogen bond network is present among GO sheets and water molecules even after prolonged drying [48–50]. The scaffold's mechanical properties could also be affected by the extent of the amidation reaction. In our case, even if the amidation degree was not quantified, it can be hypothesized that GO$_{exfoliated}$ reacted more efficiently with CS leading to a higher GO incorporation for a fixed GO/CS weight ratio. That could contribute to increasing GO aggregation phenomena in GO$_{exfoliated}$/CS scaffolds with a consequent decrease in the scaffold's mechanical properties, as found by Sivashankari et al. in scaffolds of hydroxypropyl chitosan-graft-graphene oxide [24].

2.3. Assessment of Cell Viability in the Scaffolds

The biocompatibility of the scaffolds was evaluated on human dermal fibroblasts by a mitochondrial activity-based assay that uses the tetrazolium dye [3-(4,5-dimethylthiazol-2-yl)-5-(3-carboxymethoxyphenyl)-2-(4-sulfophenyl)-2H-tetrazolium] MTS. Figure 9 exhibits the viability of the cells in the presence of scaffolds after 48 h of culture. As can be seen, pure chitosan showed good biocompatibility with cell viability of ca 75% compared to control. In contrast, pristine GO$_{exfoliated}$ and GO$_{sigma}$ samples were found to be toxic for the cells, especially GO$_{exfoliated}$ having the highest oxygen content.

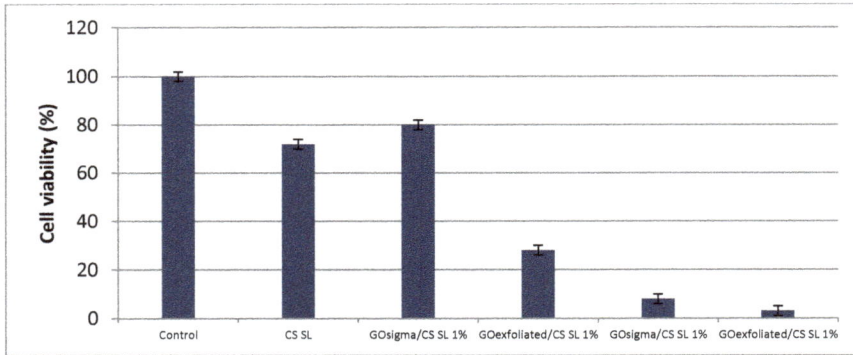

Figure 9. Cell viability on CS and composite scaffolds.

Literature data on GO cytotoxicity are controversial since several factors can affect the cytocompatibility of such material including concentration, size, shape and oxygen content [51]. A greater hemolytic activity was shown by GO at high oxygen content compared to graphene sheets [51]. Similarly, GO was shown to generate more reactive oxygen species (ROS) than reduced GO materials (lower oxygen content) when tested against murine lung epithelial cells [52]. However, the toxicity of GO can be drastically reduced by coating it with biocompatible polymers like polyvinylpyrrolidone [53], collagen [54] and chitosan [25,51].

Also in our case, biocompatibility of composite scaffolds was affected by the type of used GO. Specifically, the GO at low oxygen content (GO_{sigma}) increased biocompatibility of CS scaffold; cell viability being ca. 80% compared to 75% of pure CS (Figure 9). That was not true for $GO_{exfoliated}$, for which a reduction of scaffold biocompatibility was observed following GO incorporation. Presumably, in the case of $GO_{exfoliated}$, GO aggregates present within the composite structure were not adequately shielded by CS to avoid interaction with cells. That would be in agreement with the observed lower mechanical strength of the $GO_{exfoliated}$/CS scaffolds compared with the GO_{sigma}/CS series.

In order to study possible cell morphology changes induced by contact with GO/CS scaffold, in Figure 10, as an example, the optical images of cells cultured on 96 well plates compared to those remaining in the plate after contact with the scaffold GO_{sigma}/CS_{SL} 1% are reported.

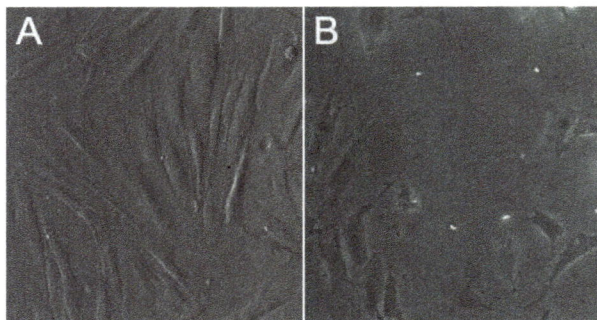

Figure 10. Optical images of fibroblast cells grown in the absence (**A**) or in the presence (**B**) of the scaffold GO_{sigma}/CS_{SL} 1%.

As it can be observed, cells grown in contact with the scaffold did not present signs of damage (Figure 10B). Only a slight morphology change was observed with a less elongated and more enlarged cell shape (Figure 10B) with respect to control (Figure 10A). That finding is doubtless related to the low number of cells remained in the culture plate due to penetration of seeded cells in the scaffold. Indeed, such a low number of cells could distribute in a larger area compared to control.

3. Materials and Methods

3.1. Graphene Oxide

Two types of graphene oxide (GO) were used in this study: a) Graphene oxide from Sigma-Aldrich (Merck KGaA, Darmstadt, Germany; 4–10% edge-oxidized, with 15–20 number of layers, 1.8 g/cm^3 bulk density), named GO$_{sigma}$; b) Graphene oxide prepared by electrochemical exfoliation of graphite using the salt KClO$_4$ in the electrolytic solution, as previously described [40]. This sample was named GO$_{exfoliated}$.

3.2. Characterization of Graphene Oxide Samples

The morphology of the GO sheets was evaluated by Atomic Force Microscopy (AFM) using a Veeco AFM Multimode™ (Veeco, Plainview, NY, USA) equipped with a Nanoscope IIIa controller. For the analysis, a drop (10 μl) of a GO dispersed in deionized water (0.01 mg·mL^{-1}) was layered onto a clean silicon wafer with negligible roughness. All images were obtained in tapping mode acquiring topography, amplitude and phase data, by using a Rectangular Tip Etched Silicon Probe (RTESP, Bruker, Billerica, MA, USA; nominal parameters r = 8 mm, f = 300 kHz, k = 40 N/m) and with a 512 × 512 pixels resolution. The software Gwyddion 2.31 (Version 2.31, Gwyddion, Brno, Czech Republic, http://gwyddion.net/) was used to correct images by polynomial background filters and to calculate average thickness and dimensions of GO sheets.

The presence of functional groups on GO edges and basal planes was investigated by Fourier transform infrared spectroscopy (FTIR) and X-ray Photoelectron Spectroscopy (XPS). FTIR spectra were acquired in transmission by a Nicolet 6700 FTIR (Thermo Fisher Scientific, Waltham, MA, USA), by co-adding 100 scans at a resolution of 2 cm^{-1}. GO powder (ca. 1 mg) was pelleted in 150 mg of KBr using a Specac manual hydraulic press, by applying a pressure of 2 tons for 5 min. XPS was performed by an Omicron DAR 400 Al/Mg Kα non-monochromatized X-ray source (Scienta Omicron GmbH, Taunusstein, Germany), and a VG-CLAM2 electron spectrometer (Thermo Fisher Scientific, Waltham, MA, USA). For the analysis, GO was dispersed in ethanol to a 1 mg·mL^{-1} content and deposited onto a silicon wafer.

The carbon structure of GO sheets was analyzed by Raman Spectroscopy. Raman spectrometer XY Dilor (HORIBA Jobin Yvon GmbH, Bensheim, Germany), recording the spectrum from 1200 to 2900 cm^{-1}, with a resolution of 2 cm^{-1}, using an excitation wavelength of 514.5 nm. The power of the laser beam was 3.5 mW, focused on the sample by using a 100× objective and performing 10 repetitions of 60 s, for each measurement.

3.3. Functionalization of Chitosan with GO

GO was covalently linked to chitosan (CS, Sigma Aldrich, Merck KGaA, Darmstadt, Germany; medium molecular weight, 75–85% deacetylated) by amidation between GO carboxylic groups and CS amino groups (Figure 4). First, chitosan was dissolved in 1% acetic acid aqueous solution at 1% (w/v) concentration. CS solution was then dialyzed in water (membrane cutoff = 3.5 KDa) to remove acetic acid and low molecular weight by-products.

The determined amount of GO was suspended in water and exfoliated by sonication for 4 h at 40°C. Then, GO carboxylic groups were activated by 1-Ethyl-3-(3-dimethylaminopropyl) carbodiimide (EDC Sigma Aldrich, Merck KGaA, Darmstadt, Germany) and N-Hydroxysuccinimide (NHS, Sigma Aldrich, Merck KGaA, Darmstadt, Germany), added in amounts such to achieve a 0.1 M concentration of each. After 2 h of activation at room temperature, chitosan solution at 1% (w/v) concentration was added to the GO suspension such to have a GO/CS weight ratio of either 0.3% or 1%. The amidation was carried out under stirring at room temperature for 24 h.

Following reaction, CS/GO suspensions were centrifuged at 3500 RPM for 10 min to eliminate the unreacted GO; then the supernatant was recovered and dried under vacuum. The obtained polymer samples were named GO_X/CS Y% where X was the type of employed GO (sigma or exfoliated), and Y was the GO/CS weight ratio used for CS amidation (0.3% or 1%).

The amidation reaction was followed by FTIR analysis. Spectra were acquired in attenuated total reflection (ATR) by a Nicolet 6700 (Thermo Fisher Scientific, Waltham, MA, USA) equipped with a Golden Gate single reflection diamond ATR accessory at a resolution of 2 cm^{-1} and co-adding 100 scans.

3.4. Preparation of GO/CS Composite Scaffolds

Porous GO/CS composite scaffolds were prepared by employing two methods, the freeze-drying (FD) and the salt leaching (SL). In the first method, after the reaction, the solution of GO-functionalized chitosan was poured into a steel mold with a square base and frozen in liquid nitrogen. Then, the frozen polymer was removed from the mold and lyophilized for 1 day. In the second method, a porogen, sodium acetate (100–200 μm), was added in the solution of GO-functionalized chitosan. After stirring for 1 h, the solution of the GO-functionalized chitosan containing the porogen was poured into a steel mold, frozen and lyophilized. Then, the lyophilized polymer was immersed first in ethanol/water solutions (96%, 80%, 60%, 40% v/v), 2 h for each concentration, and then in water for 48 h to remove the salt and form a porous structure. Finally, the wet porous scaffold was dried by lyophilization. A scaffold of pure chitosan was also prepared by the salt leaching method employing a 1% CS solution.

The obtained scaffolds were named as follow: GO_X/CS_M Y% where X was the type of employed GO (sigma or exfoliated), M was the method used to prepare the scaffold (FD or SL), and Y was the GO/CS weight ratio used for CS amidation (0.3% or 1%). In Table 4, all the prepared samples with the corresponding acronyms are reported.

3.5. Characterization of GO/CS Composite Scaffolds

The porosity of GO/CS composite scaffolds was measured by the gravimetric method [55] and the liquid displacement method [56].

The gravimetric method permits to evaluate the scaffold porosity (P) by determining the bulk and true density of the scaffold as shown in the equations below:

$$\rho_s = \frac{m}{V} \tag{1}$$

$$P(\%) = \left(1 - \frac{\rho_s}{\rho_c}\right) \cdot 100 \tag{2}$$

where ρ_s is the apparent density of the scaffold, m is the weight of the scaffold, V is the volume of the scaffold, and ρ_m is the density of the material used to prepare the scaffold, in our case chitosan. Chitosan was considered to have a density of 1.41 g/cm^3.

As for the liquid displacement method, ethanol was used as the displacement liquid because it penetrated easily into the pores and, being a non-solvent of chitosan, did not induce shrinkage or swelling of the scaffold. A scaffold sample with the initial weight W_0 and initial volume V_0 was immersed for 30 min in a cylinder containing a known volume of ethanol (V_1). Then, the scaffold was removed and weighed (W_1). Sample porosity (P) was calculated as follow:

$$P(\%) = \left(\frac{W_1 - W_0}{\rho_{EtOH} \cdot V_0}\right) \cdot 100 \tag{3}$$

where ρ_{EtOH} is the density of ethanol (0.806 g/cm^3 a 20 °C).

The microstructure of the scaffolds were observed by field emission scanning electron microscope (FESEM, AURIGA Carl Zeiss AG, Oberkochen, Germany). For the analysis, the scaffolds were fractured in liquid nitrogen, then fixed on stubs and gold sputter before observation.

Water uptake of scaffolds was determined at room temperature by immersing the scaffolds in water for increasing times. At determined intervals, scaffolds were removed from water and weighed, after removal of the excess of solvent using filter paper. The analysis was repeated until constant weight (equilibrium swelling weight, W') was reached. The swelling ratio, SR, was calculated by applying the following equation:

$$SR = \left(\frac{W_t - W_0}{W_0} \right) \tag{4}$$

where W_t was the weight of the sample after swelling at the time t and W_0 was the initial weight of the film. Five parallel swelling experiments were performed for each sample and data were reported as average value ± standard deviation.

The water retention efficiency was determined by transferring the swollen scaffold, with the maximum swelling (W'), into a centrifuge tube having a filter paper at the bottom. Then the sample was centrifuged at 500 rpm for 3 min and immediately weighed (W_f). Water retention (WR) of the scaffold was calculated as follows:

$$WR = \frac{W' - W_f}{W'} \tag{5}$$

Differential scanning calorimetry (DSC) was performed from -100 to $+150\ ^\circ$C under N_2 flux by using a Mettler TA-3000 DSC apparatus (Mettler Toledo, Columbus, OH, USA). The scan rate used for the experiments was $10\ ^\circ$C·min^{-1} and the sample weight of 6–7 mg. Thermo-gravimetric analysis (TGA) was carried out employing a Mettler TG 50 thermobalance (Mettler Toledo, Columbus, Ohio, Stati Uniti) at a heating rate of $10\ ^\circ$C·min^{-1} under N_2 flow in the temperature range 25–600 $^\circ$C.

The compressive strength of scaffolds was determined by an ISTRON 4502 instrument (INSTRON, Norwood, MA, USA). Measurements were performed on parallelepipeds 10 mm height with a square base (5 mm × 5 mm), obtained by using a proper steel mold. Particularly, solutions of GO-functionalized chitosan were poured into the mold and frozen in liquid nitrogen. Then, the frozen samples were removed from the mold and lyophilized for 1 day. When a porogen was used, the lyophilized samples were immersed in water/ethanol solutions to remove the porogen (see Section 3.4) and lyophilized again. The crosshead speed of the Instron tester was set at 1 mm/min, and load was applied until 40% reduction in specimen height. Five parallel samples were tested for every scaffold, and mechanical properties were reported as average value ± standard deviation.

3.6. Assessment of Cell Viability in Scaffolds by MTS Assay

Cell compatibility and cytotoxicity were analyzed by culturing human primary dermal fibroblasts in the presence of the scaffolds. Cells were obtained from young adult male patients complaining of phimosis, full ethical consent was obtained from all donors and the Research Ethics Committee, Sapienza University of Roma, approved the study. Scaffolds were set down in 96 well tissue culture plate and conditioned in Dulbecco's Modified Eagle's Medium DMEM without red phenol supplemented with L-glutamine, penicillin/streptomycin, Na-pyruvate, non-essential amino acids, plus 10% Fetal Bovine Serum (FBS) for 2 hours at 37 °C, in 95% humidity and 5% CO_2 atmosphere. Equal numbers (8×10^3) of cells were seeded in each well containing the scaffolds and allowed to proliferate for 48h. Cellular viability/proliferation was quantified by measuring the mitochondrial dehydrogenase activity using tetrazolium dye MTS [3-(4,5-dimethylthiazol-2-yl)-5-(3-carboxymethoxyphenyl)-2-(4-sulfophenyl)-2H-tetrazolium] (Promega Corporation, Madison, WI, USA) based colorimetric assay, according to the manufacturer's instructions. Briefly, after 48 h, 20% (v/v) of MTS dye was added in the culture media and cells were cultured for 4 h to allow the formation of soluble formazan crystals by viable cells. Spectrophotometric absorbance was measured at 490 nm using a multi-plate reader Appliskan (Thermo Fisher, Waltham,

MA, USA). Cells cultured in the absence of the scaffolds were taken as a control. In order to analyze the cytotoxicity of pristine GO, human primary fibroblasts (8×10^3) were cultured in the presence of 0.1 µg/µl (final concentration) powder of both $GO_{exfoliated}$ and GO_{sigma} and then the cells were analyzed as described above with MTS dye.

4. Conclusions

This study confirms the ability of graphene oxide to act as reinforcing filler for chitosan scaffolds. Findings suggest that the oxygen content of GO affects the final properties of GO/CS composite scaffolds. Specifically, high oxygen content in GO can promote aggregation of GO sheets in the chitosan matrix and, hence, reduce its reinforcement effect. Additionally, high content of functional groups in GO have a negative effect on material biocompatibility, presumably because they could promote the generation of reactive oxygen species as reported in the literature. In contrast, the conjugation of chitosan with a GO sample at low oxygen content resulted in scaffolds with improved compression modulus and biocompatibility compared to pristine CS. Overall, GO_{sigma}/CS scaffold at 1% GO content showed good potentiality for application in tissue engineering.

Author Contributions: Conceptualization, I.F. and A.P.; Methodology I.F., A.S.d, A.M. and A.P.; Formal Analysis, E.P. (Elena Perugini) I.S., M.L., E.P. (Ernesto Placidi), F.A. and F.V., Investigation, E.P. (Elena Perugini), I.S. and M.P.; Data Curation, A.M. and I.F.; Writing—Original Draft Preparation, I.F.; Writing—Review & Editing, A.P., F.V. and A.S.d.; Funding Acquisition, A.P.

Funding: The work was funded by Sapienza University of Rome, through a grant to A.P.

Conflicts of Interest: The authors declare no conflict of interest.

References

1. Ramya, R.; Venkatesan, J.; Kim, S.K.; Sudha, P.N. Biomedical applications of chitosan: An overview. *J. Biomim. Biomater. Tissue Eng.* **2012**, *2*, 100–111. [CrossRef]
2. Al-Jbour, N.D.; Beg, M.D.; Gimbun, J.; Alam, A.K.M.M. An overview of chitosan nanofibers and their applications in drug delivery process. *Curr. Drug Deliv.* **2019**. [CrossRef] [PubMed]
3. Cuzzucoli Crucitti, V.; Migneco, L.M.; Piozzi, A.; Taresco, V.; Garnett, M.; Argent, R.H.; Francolini, I. Intermolecular interaction and solid state characterization of abietic acid/chitosan solid dispersions possessing antimicrobial and antioxidant properties. *Eur. J. Pharm. Biopharm.* **2018**, *125*, 114–123. [CrossRef] [PubMed]
4. Amato, A.; Migneco, L.M.; Martinelli, A.; Pietrelli, L.; Piozzi, A.; Francolini, I. Antimicrobial activity of catechol functionalized-chitosan versus *Staphylococcus epidermidis*. *Carbohydr. Polym.* **2018**, *179*, 273–281. [CrossRef]
5. Singh, R.; Shitiz, K.; Singh, A. Chitin and chitosan: Biopolymers for wound management. *Int. Wound J.* **2017**, *14*, 1276–1289. [CrossRef]
6. Moura, M.J.; Brochado, J.; Gil, M.H.; Figueiredo, M.M. In situ forming chitosan hydrogels: Preliminary evaluation of the in vivo inflammatory response. *Mater. Sci. Eng. C Mater. Biol. Appl.* **2017**, *75*, 279–285. [CrossRef] [PubMed]
7. Iglesias, D.; Bosi, S.; Melchionna, M.; Da Ros, T.; Marchesan, S. The Glitter of carbon nanostructures in hybrid/composite hydrogels for medicinal use. *Curr. Top. Med. Chem.* **2016**, *16*, 1976–1989. [CrossRef]
8. Zhou, G.; Niepel, M.S.; Saretia, S.; Groth, T. Reducing the inflammatory responses of biomaterials by surface modification with glycosaminoglycan multilayers. *J. Biomed. Mater. Res. A* **2016**, *104*, 493–502. [CrossRef]
9. Moraes, P.C.; Marques, I.C.S.; Basso, F.G.; Rossetto, H.L.; Pires-de-Souza, F.C.P.; Costa, C.A.S.; Garcia, L.D.F.R. Repair of bone defects with chitosan-collagen biomembrane and scaffold containing calcium aluminate cement. *Braz. Dent. J.* **2017**, *28*, 287–295. [CrossRef] [PubMed]
10. Haifei, S.; Xingang, W.; Shoucheng, W.; Zhengwei, M.; Chuangang, Y.; Chunmao, H. The effect of collagen-chitosan porous scaffold thickness on dermal regeneration in a one-stage grafting procedure. *J. Mech. Behav. Biomed. Mater.* **2014**, *29*, 114–125. [CrossRef]
11. Chen, C.K.; Chang, N.J.; Wu, Y.T.; Fu, E.; Shen, E.C.; Feng, C.W.; Wen, Z.H. Bone formation using cross-linked chitosan scaffolds in rat calvarial defects. *Implant Dent.* **2018**, *27*, 15–21. [CrossRef]

12. Balagangadharan, K.; Dhivya, S.; Selvamurugan, N. Chitosan based nanofibers in bone tissue engineering. *Int. J. Biol. Macromol.* **2017**, *104*, 1372–1382. [CrossRef]

13. LogithKumar, R.; KeshavNarayan, A.; Dhivya, S.; Chawla, A.; Saravanan, S.; Selvamurugan, N. A review of chitosan and its derivatives in bone tissue engineering. *Carbohydr. Polym.* **2016**, *151*, 172–188. [CrossRef]

14. Yang, X.; Chen, X.; Wang, H. Acceleration of osteogenic differentiation of preosteoblastic cells by chitosan containing nanofibrous scaffolds. *Biomacromolecules* **2009**, *10*, 2772–2778. [CrossRef]

15. Di Martino, A.; Sittinger, M.; Risoud, M.V. Chitosan: A versatile biopolymer for orthopaedic tissue-engineering. *Biomaterials* **2005**, *26*, 5983–5990. [CrossRef]

16. Palma, P.J.; Ramos, J.C.; Martins, J.B.; Diogenes, A.; Figueiredo, M.H.; Ferreira, P.; Viegas, C.; Santos, J.M. Histologic evaluation of regenerative endodontic procedures with the use of chitosan scaffolds in immature dog teeth with apical periodontitis. *J. Endod.* **2017**, *43*, 1279–1287. [CrossRef]

17. Stepniewski, M.; Martynkiewicz, J.; Gosk, J. Chitosan and its composites: Properties for use in bone substitution. *Polim. Med.* **2017**, *47*, 49–53. [CrossRef]

18. Bi, L.; Cheng, W.; Fan, H.; Pei, G. Reconstruction of goat tibial defects using an injectable tricalcium phosphate/chitosan in combination with autologous platelet-rich plasma. *Biomaterials* **2010**, *31*, 3201–3211. [CrossRef]

19. Zhou, D.; Qi, C.; Chen, Y.X.; Zhu, Y.J.; Sun, T.W.; Chen, F.; Zhang, C.Q. Comparative study of porous hydroxyapatite/chitosan and whitlockite/chitosan scaffolds for bone regeneration in calvarial defects. *Int. J. Nanomed.* **2017**, *12*, 2673–2687. [CrossRef]

20. Keller, L.; Regiel-Futyra, A.; Gimeno, M.; Eap, S.; Mendoza, G.; Andreu, V.; Wagner, Q.; Kyzioł, A.; Sebastian, V.; Stochel, G.; et al. Chitosan-based nanocomposites for the repair of bone defects. *Nanomedicine* **2017**, *13*, 2231–2240. [CrossRef]

21. Shamekhi, M.A.; Mirzadeh, H.; Mahdavi, H.; Rabiee, A.; Mohebbi-Kalhori, D.; Baghaban Eslaminejad, M. Graphene oxide containing chitosan scaffolds for cartilage tissue engineering. *Int. J. Biol. Macromol.* **2019**, *127*, 396–405. [CrossRef]

22. Valencia, C.; Valencia, C.H.; Zuluaga, F.; Valencia, M.E.; Mina, J.H.; Grande-Tovar, C.D. Synthesis and application of scaffolds of chitosan-graphene oxide by the freeze-drying method for tissue regeneration. *Molecules* **2018**, *23*, 2651. [CrossRef]

23. Hermenean, A.; Codreanu, A.; Herman, H.; Balta, C.; Rosu, M.; Mihali, C.V.; Ivan, A.; Dinescu, S.; Ionita, M.; Costache, M. Chitosan-graphene oxide 3D scaffolds as promising tools for bone regeneration in critical-size mouse calvarial defects. *Sci. Rep.* **2017**, *7*, 16641. [CrossRef]

24. Sivashankari, P.R.; Moorthi, A.; Abudhahir, K.M.; Prabaharan, M. Preparation and characterization of three-dimensional scaffolds based on hydroxypropyl chitosan-graft-graphene oxide. *Int. J. Biol. Macromol.* **2018**, *110*, 522–530. [CrossRef]

25. Dinescu, S.; Ionita, M.; Pandele, A.M.; Galateanu, B.; Iovu, H.; Ardelean, A.; Costache, M.; Hermenean, A. In vitro cytocompatibility evaluation of chitosan/graphene oxide 3D scaffold composites designed for bone tissue engineering. *Biomed. Mater. Eng.* **2014**, *24*, 2249–2256.

26. Hummers, W.; Offeman, R. Preparation of graphitic oxide. *J. Am. Chem. Soc.* **1958**, *80*, 1339. [CrossRef]

27. Park, S.; Ruoff, R.S. Chemical methods for the production of graphenes. *Nat. Nanotechnol.* **2009**, *4*, 217–224. [CrossRef]

28. Lerf, A.; He, H.; Forster, M.; Klinowski, J. Structure of graphite oxide revisited. *J. Phys. Chem. B* **1998**, *102*, 4477–4482. [CrossRef]

29. Wu, C.; Xia, L.; Han, P.; Xu, M.; Fang, B.; Wang, J.; Chang, J.; Xiao, Y. Graphene oxide modified beta-tricalcium phosphate bioceramics stimulate in vitro and in vivo osteogenesis. *Carbon* **2015**, *93*, 116–129. [CrossRef]

30. Lee, J.H.; Shin, Y.C.; Lee, S.M.; Jin, O.; Kang, S.H.; Hong, S.W.; Jeong, C.M.; Huh, J.B.; Han, D.W. Enhanced osteogenesis by reduced graphene oxide/hydroxyapatite nanocomposites. *Sci. Rep.* **2015**, *5*, 18833. [CrossRef]

31. La, W.G.; Jin, M.; Park, S.; Yoon, H.H.; Jeong, G.J.; Bhang, S.H.; Park, H.; Char, K.; Kim, B.S. Delivery of bone morphogenetic protein-2 and substance P using graphene oxide for bone regeneration. *Int. J. Nanomed.* **2014**, *9*, 107–116.

32. Depan, D.; Girase, B.; Shah, J.S.; Misra, R.D. Structure-process-property relationship of the polar graphene oxide-mediated cellular response and stimulated growth of osteoblasts on hybrid chitosan network structure nanocomposite scaffolds. *Acta Biomater.* **2011**, *7*, 3432–3445. [CrossRef]

33. Liu, Y.; Wen, J.; Gao, Y.; Li, T.; Wang, H.; Yan, H.; Niu, B.; Guo, R. Antibacterial graphene oxide coatings on polymer substrate. *Appl. Surf. Sci.* **2018**, *436*, 624–630. [CrossRef]

34. Pandit, S.; Karunakaran, S.; Boda, S.K.; Basu, B.; De, M. High Antibacterial activity of functionalized chemically exfoliated MoS₂. *ACS Appl. Mater. Interfaces* **2016**, *8*, 31567–31573. [CrossRef]

35. Krishnamoorthy, K.; Veerapandian, M.; Yun, K.; Kim, S.-J. The chemical and structural analysis of graphene oxide with different degrees of oxidation. *Carbon* **2013**, *53*, 38–49. [CrossRef]

36. Mattevi, C.; Eda, G.; Agnoli, S.; Miller, S.; Andre Mkhoyan, K.; Celik, O.; Mastrogiovanni, D.; Granozzi, G.; Garfunkel, E.; Chhowalla, M. Evolution of electrical, chemical, and structural properties of transparent and conducting chemically derived graphene thin films. *Adv. Funct. Mater.* **2009**, *19*, 2577–2583. [CrossRef]

37. Feng, R.; Yu, Y.; Shen, C.; Jiao, Y.; Zhou, C. Impact of graphene oxide on the structure and function of important multiple blood components by a dose-dependent pattern. *J. Biomed. Mater. Res. A* **2015**, *103*, 2006–2014. [CrossRef]

38. Kucki, M.; Rupper, P.; Sarrieu, C.; Melucci, M.; Treossi, E.; Schwarz, A.; León, V.; Kraegeloh, A.; Flahaut, E.; Vázquez, E.; et al. Interaction of graphene-related materials with human intestinal cells: An in vitro approach. *Nanoscale* **2016**, *8*, 8749–8760. [CrossRef]

39. Wei, X.Q.; Hao, L.Y.; Shao, X.R.; Zhang, Q.; Jia, X.Q.; Zhang, Z.R.; Lin, Y.F.; Peng, Q. Insight into the interaction of graphene oxide with serum proteins and the impact of the degree of reduction and concentration. *ACS Appl. Mater. Interfaces* **2015**, *7*, 13367–13374. [CrossRef]

40. Costa de Oliveira, M.A.; Mecheri, B.; D'Epifanio, A.; Placidi, E.; Arciprete, F.; Valentini, F.; Perandini, A.; Valentini, V.; Licoccia, S. Graphene oxide nanoplatforms to enhance catalytic performance of iron phthalocyanine for oxygen reduction reaction in bioelectrochemical systems. *J. Power Sources* **2017**, *356*, 381–388. [CrossRef]

41. Ferrari, A.C. Raman spectroscopy of graphene and graphite: Disorder, electrone phonon coupling, doping and non adiabatic effects. *Solid State Commun.* **2007**, *143*, 47–57. [CrossRef]

42. Mazaheri, M.; Akhavan, O.; Simchi, A. Flexible bactericidal graphene oxide-chitosan layers for stem cell proliferation. *Appl. Surf. Sci.* **2014**, *301*, 456–462. [CrossRef]

43. Pakulski, D.; Czepa, W.; Witomska, S.; Aliprandi, A.; Pawluć, P.; Patroniak, V.; Ciesielski, A.; Samorì, P. Graphene oxide-branched polyethylenimine foams for efficient removal of toxic cations from water. *J. Mater. Chem. A* **2018**, *6*, 9384–9390. [CrossRef]

44. Kavya, K.C.; Jayakumar, R.; Nair, S.; Chennazhi, K.P. Fabrication and characterization of chitosan/gelatin/nSiO2 composite scaffold for bone tissue engineering. *Int. J. Biol. Macromol.* **2013**, *59*, 255–263. [CrossRef]

45. Unnithan, A.R.; Park, C.H.; Kim, C.S. Nanoengineered bioactive 3D composite scaffold: A unique combination of graphene oxide and nanotopography for tissue engineering applications. *Compos. Part B* **2016**, *90*, 503–511. [CrossRef]

46. Kundie, F.; Azhari, C.H.; Muchtar, A.; Ahmad, Z.A. Effects of filler size on the mechanical properties of polymer-filled dental composites: A review of recent developments. *J. Phys. Sci.* **2018**, *29*, 141–165. [CrossRef]

47. Kim, J.; Kim, S.W.; Yun, H.; Kim, B.J. Impact of size control of graphene oxide nanosheets for enhancing electrical and mechanical properties of carbon nanotube–polymer composites. *RSC Adv.* **2017**, *7*, 30221–30228. [CrossRef]

48. Medhekar, N.V.; Ramasubramaniam, A.; Ruoff, R.S.; Shenoy, V.B. Hydrogen bond networks in graphene oxide composite paper: Structure and mechanical properties. *ACS Nano* **2010**, *4*, 2300–2306. [CrossRef]

49. Compton, O.C.; Cranford, S.W.; Putz, K.W.; An, Z.; Brinson, L.C.; Buehler, M.J.; Nguyen, S.T. Tuning the mechanical properties of graphene oxide paper and its associated polymer nanocomposites by controlling cooperative intersheet hydrogen bonding. *ACS Nano* **2012**, *6*, 2008–2019. [CrossRef]

50. Dikin, D.A.; Stankovich, S.; Zimney, E.J.; Piner, R.D.; Dommett, G.H.B.; Evmenenko, G.; Nguyen, S.T.; Ruoff, R.S. Preparation and Characterization of Graphene Oxide Paper. *Nature* **2007**, *448*, 457–460. [CrossRef]

51. Liao, K.H.; Lin, Y.S.; Macosko, C.W.; Haynes, C.L. Cytotoxicity of graphene oxide and graphene in human erythrocytes and skin fibroblasts. *ACS Appl. Mater. Interfaces* **2011**, *3*, 2607–2615. [CrossRef]

52. Bengtson, S.; Kling, K.; Madsen, A.M.; Noergaard, A.W.; Jacobsen, N.R.; Clausen, P.A.; Alonso, B.; Pesquera, A.; Zurutuza, A.; Ramos, R.; et al. No cytotoxicity or genotoxicity of graphene and graphene oxide in murine lung epithelial FE1 cells in vitro. *Environ. Mol. Mutagen.* **2016**, *57*, 469–482. [CrossRef]

53. Zhi, X.; Fang, H.; Bao, C.; Shen, G.; Zhang, J.; Wang, K.; Guo, S.; Wan, T.; Cui, D. The immunotoxicity of graphene oxides and the effect of PVP-coating. *Biomaterials* **2013**, *34*, 5254–5261. [CrossRef]
54. De Marco, P.; Zara, S.; De Colli, M.; Radunovic, M.; Lazović, V.; Ettorre, V.; Di Crescenzo, A.; Piattelli, A.; Cataldi, A.; Fontana, A. Graphene oxide improves the biocompatibility of collagen membranes in an in vitro model of human primary gingival fibroblasts. *Biomed. Mater.* **2017**, *12*, 055005. [CrossRef]
55. Loh, Q.L.; Choong, C. Three-dimensional scaffolds for tissue engineering applications: Role of porosity and pore size. *Tissue Eng. Part B Rev.* **2013**, *19*, 485–502. [CrossRef]
56. Han, J.; Zhou, Z.; Yin, R.; Yang, D.; Nie, J. Alginate-chitosan/hydroxyapatite polyelectrolyte complex porous scaffolds: Preparation and characterization. *Int. J. Biol. Macromol.* **2010**, *46*, 199–205. [CrossRef]

materials

MDPI

Review

Materials for the Spine: Anatomy, Problems, and Solutions

Brody A. Frost [1], Sandra Camarero-Espinosa [2,*] and E. Johan Foster [1,*]

[1]   Department of Materials Science and Engineering, Macromolecules Innovation Institute, Virginia Tech, Blacksburg, VA 24061, USA; bfrost12@vt.edu
[2]   Complex Tissue Regeneration Department, MERLN Institute for Technology-inspired Regenerative Medicine, Maastricht University, P.O. Box 616, 6200MD Maastricht, The Netherlands
*   Correspondence: s.camarero-espinosa@maastrichtuniversity.nl (S.C.-E.); johanf@vt.edu (E.J.F.)

Received: 29 November 2018; Accepted: 5 January 2019; Published: 14 January 2019

Abstract: Disc degeneration affects 12% to 35% of a given population, based on genetics, age, gender, and other environmental factors, and usually occurs in the lumbar spine due to heavier loads and more strenuous motions. Degeneration of the extracellular matrix (ECM) within reduces mechanical integrity, shock absorption, and swelling capabilities of the intervertebral disc. When severe enough, the disc can bulge and eventually herniate, leading to pressure build up on the spinal cord. This can cause immense lower back pain in individuals, leading to total medical costs exceeding $100 billion. Current treatment options include both invasive and noninvasive methods, with spinal fusion surgery and total disc replacement (TDR) being the most common invasive procedures. Although these treatments cause pain relief for the majority of patients, multiple challenges arise for each. Therefore, newer tissue engineering methods are being researched to solve the ever-growing problem. This review spans the anatomy of the spine, with an emphasis on the functions and biological aspects of the intervertebral discs, as well as the problems, associated solutions, and future research in the field.

Keywords: spinal anatomy; intervertebral disc; degenerative disc disease; herniated disc; spinal fusion; total disc replacement; tissue engineering

---

1. Human Spinal Anatomy

The spine, or vertebral column, is a bony structure that houses the spinal cord and extends the length of the back, connecting the head to the pelvis [1].

The most important function of the spine is to protect the spinal cord, which is the nerve supply for the entire body originating in the brain [1]. Along with this major function, others include supporting the mass of the body, withstanding external forces, and allowing for mobility and flexibility while dissipating energy and protecting against impact. The spine is connected to the muscles and ligaments of the trunk for postural control and spinal stability [2]. It can be separated into five distinct sections, the cervical spine, the thoracic spine, the lumbar spine, the sacrum, and the coccyx, all of which are comprised of independent bony vertebrae and intervertebral discs [3], Figure 1. To describe the differences between the spinal column sections, each one has been further discussed.

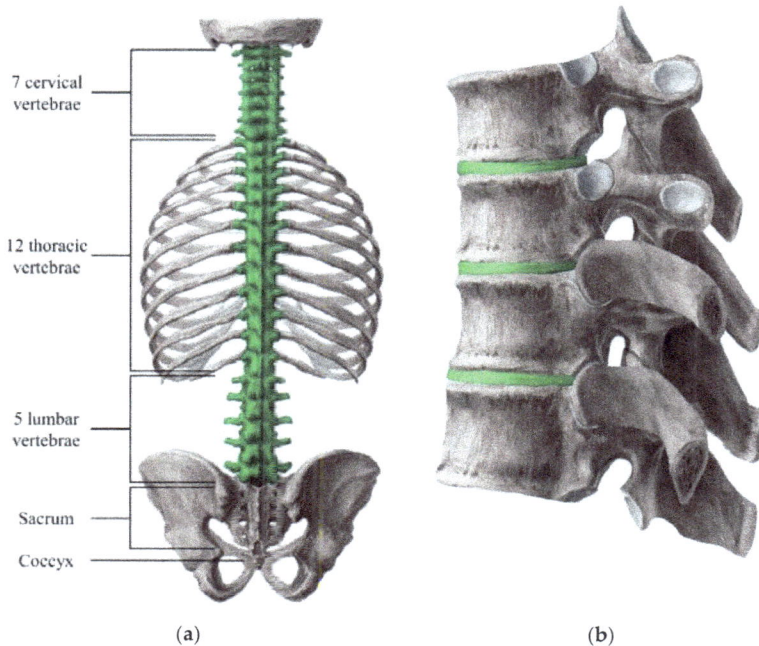

Figure 1. Overview of the vertebral column with each specific section labeled for clarification (**a**). The green highlighted section refers to the part of the spine that contain individual vertebrae, as well as intervertebral discs (IVD). The structure of the vertebrae and IVD (green highlighted) have been added for better visualization (**b**) [4]

1.1. Cervical Spine

The cervical section of the spine consists of seven vertebrae (C1–C7) and six intervertebral discs, and extends from the base of the skull to the top of the trunk, where the thoracic vertebrae and rib cage start [3] Figure 1. The cervical spine's major functions include supporting and cushioning loads to the head/neck while allowing for rotation, and protecting the spinal cord extending from the brain [5].

Of these seven vertebrae, the atlas (C1) and the axis (C2) are among the most important for rotation and movement of the head [6]. The atlas is the only cervical vertebra that does not contain a vertebral body, but instead has a more ring-like structure for cradling the skull at the occipital bone, creating the atlanto-occipital joint. This joint in particular makes up for about 50% of the head's flexion and extension range of motion [5–7]. The axis contains a large bony protrusion (the odontoid process) that extends from the body superiorly, into a facet on the ring-shaped atlas, forming the atlanto-axial joint [5,6]. This connection allows the head and atlas to rotate from side to side as one unit, and accounts for about 50% of the neck's rotation, as well as having the function of transferring the weight of the head through the rest of the cervical spine [5–7]. The rest of the vertebrae (C3–C7), have significantly reduced mobility, however are mainly used as support for the weight bearing of the head and other loads applied onto the neck.

The cervical spine protects both the efferent and afferent nerves that stem from the spinal cord and, if damaged, can lead to dramatic effects on the nervous system eventually affecting the patient's daily activity, and even causing a potential paralysis [8]. The cushioning and support of loads by the intervertebral discs are crucial to the longevity of vertebrae, and therefore, the nerves, since they run through the same joint separation [9]. However, because of the extensive movement that occurs in the cervical spine, the intervertebral discs go through drastic changes in stresses and strains causing them to be much more susceptible to injury, which can cause damage to or impingements on these nerves [9]. This can lead to feelings of weakness, numbness, tingling, and potentially loss of feeling.

1.2. Thoracic Spine

The thoracic section of the spine consists of twelve vertebrae (T1–T12) and twelve intervertebral discs, and extends from the bottom of the cervical spine to the beginning of the lumbar spine [3], Figure 1. The thoracic spine's major functions include heavy load bearing and protection of the spinal cord, supporting posture and stability throughout the trunk, and connection of the rib cage that houses and protects vital organs, such as the heart and lungs [10].

This connection poses a significant decrease in mobility, as compared to the cervical spine section, and a greater stability and support of the entire trunk, usually leading to fewer cases of disc degeneration [10,11]. The vertebrae that make up the thoracic spine have body sizes (thickness, width, and depth) that drastically increases descending from T1 to T12, corresponding to an increased load bearing that is transferred from the vertebra above [12]. All other features stay relatively the same throughout, except for the T11 and T12 vertebrae, in which no ribs are connected. Along with this change towards the end of the thoracic spine, the T12 plays an interfacial role and has distinct thoracic characteristics superiorly and lumbar characteristics inferiorly for articulation with the L1 vertebra, allowing rotational movements with T11 while disallowing movements with L1 [12].

The thoracic spine contains nerves that are much less specialized per vertebrae like that of the cervical and lumbar spine, however they are no less important. The afferent and efferent nerves that stem from the spinal cord in this section power the muscles that lie around (major back, chest, and abdominal muscles) and between (intercostal muscles) the ribs [13]. The sympathetic nervous system, which stems from the entire thoracic spine and top two lumbar vertebrae and help power the intercostal muscles, is necessary for vital involuntary functions such as increasing heart rate, increasing blood pressure, controlling breathing rate, regulating body temperature, air passage dilation, decreasing gastric secretions, bladder function (bladder muscle relaxation, and storage of urine), and sexual function [13]. The thoracic spine and sacrum are the only sections of the spinal cord that these involuntary nervous systems stem from, and if impinged, can cause similar problems as discussed for the cervical spine. As mentioned previously, with these nerves passing through the same proximity as the intervertebral discs, cushioning of loads and proper weight dissipation is crucial for disc health and nerve protection, although the structural support of the ribcage makes damage to these discs much less prevalent [11].

1.3. Lumbar Spine

The lumbar section of the spine consists of five vertebrae (L1–L5) and five intervertebral discs, and extends from the bottom of the thoracic spine to the beginning of the sacrum, which attaches the spine to the pelvis [3], Figure 1. The lumbar spine's major functions include heavy load bearing and protection of the spinal cord during locomotion and bending/torsion of the trunk, providing maximum stability while maintain crucial mobility of the trunk about the hips/pelvis [14].

This particular section of the spine needs to be the most resilient due to the vital functions it provides. Not only does it need to support all of the transferred weight from the previous spinal sections (virtually the entire human body), but it also needs to be able to retain its mobility under these strenuous conditions. The lumbar spine, from bending over to standing straight, can go through more than a 50° range for the average person (± 28.0° from 0° bend) [15]. As well as bending motion, rotation becomes a big factor, with each normal lumbar segment having the ability to undergo up to 7°–7.5° of rotation [16]. When weight is added to these conditions, such as bending over to pick up a backpack or a weight from the floor, an immense amount of stress and strain is induced into the lumbar spine [17]. Because of this, the vertebrae and intervertebral discs in the lumbar spine are the greatest in thickness, width, and depth [18]. The L1 vertebra starts out with a thickness, width, and depth greater than any of the cervical or thoracic vertebrae, and the trend only continues as the lumbar spine continues to descend to the L5 vertebra [18]. Although the vertebrae increase in size as the lumbar spine descends, none of the vertebrae themselves are specialized in any way like the aforementioned atlas and axis of the cervical spine. The L5 vertebra is not much different to the others other than in size, but since it is the most inferior vertebra in the spine, it takes more load bearing responsibility than any other vertebra in the spine making it a necessity to be the biggest and strongest [19,20].

The lumbar spine contains afferent and efferent nerves that are much more similar to those of the cervical spine, in that each one that comes out of the different levels have very specialized functions, which if damaged, can hinder an individual's daily life and potentially leave them paralyzed from the waist down [21,22]. These nerves control mainly the front of the lower extremities, and when impinged can lead to loss of feeling, mobility, weakness, isolated lower back pain, and extending leg pain [23]. With all of the load bearing, torsion, and bending, these nerves tend to have the most significant chance to be impinged or damaged (roughly 95% in individuals aged 25–55 years, further discussed in Section 3.4), compared to any other spinal section [22,24].

1.4. Sacrum

The sacrum consists of five fused vertebrae (S1–S5) that connect to the pelvis at the sacro-iliac joint, and acts as the only skeletal connection between the trunk and the lower body [3]. While in adolescence, the sacrum remains unfused as an individual grows into adulthood, the sacrum begins to fuse together. The fusion of the sacrum tends to begin with the lateral elements fusing around puberty, and the vertebral bodies fusing at about 17 or 18 years of age, becoming fully fused by 23 years of age [3], Figure 1. The sacrum has few active roles in the body, however one of those roles are incredibly vital, being the bridge between the hips with the rest of the spine [25].

Although the sacrum has no intervertebral discs, it does have very important afferent and efferent nerves that stem from the spinal cord, going through the entire lower extremity. The most important and commonly injured of these nerves travels through the L5/S1 space, which is more commonly known as the sciatic nerve. When this nerve is damaged or impinged it leads to pain and numbness down the legs hindering much of an individual's way of life [26].

1.5. Coccyx

The coccyx consists of three to five fused vertebrae depending on the individual (four is most common) that are connected to the bottom of the sacrum, and is usually referred to as the tail bone [3], Figure 1. The coccyx's major functions include acting as an attachment site for pelvic tendons, ligaments, and muscles, mainly those of which make up the pelvic floor, and supporting and stabilizing the body while in a sitting position [27].

The coccyx has no intervertebral discs nor do any nerves pass through it, therefore it is insignificant with regards to disc degeneration and disc damage.

2. Intervertebral Discs

Every vertebra in the cervical (excluding the C1 and C2 vertebrae), thoracic, and lumbar spines is separated by intervertebral discs, each named for the two vertebrae they sit between (e.g., C6–C7, T7–T8, and L4–L5, also sometimes denoted as L4/L5). These discs make up about 20–30% of the total length of the spine, and have incredibly important functions including load cushioning, reducing stress caused by impact (shock absorber), weight dispersion, allowing for movement of individual vertebrae, and allowing for the passage of nutrients and fluid to the spine and spinal cord [28]. Although each disc grants almost identical functions to the spine, based on their location, their structure and mechanical properties change to adapt to the different loads, stresses, and strains produced [29]. For example, as the expected weight-bearing role of each disc increases, descending from the base of the skull along the length of the spine, the transverse cross-sectional area of the discs also increases. The pressure exerted on the discs however, does not increase to the same extent due to the fact that the cross-sectional area increases in the inferior direction [29].

Along with the changes in the cross-sectional areas of the discs, the height (thickness) of each disc changes throughout the spine as well. The cervical and lumbar spines have been shown to have much thicker discs than that of the thoracic spine, most likely being adapted to the higher range of motion expected from these sections, for both flexion-extension and torsion [29]. All cross-sectional areas and thicknesses for the continuation of this review will be associated with the transverse plane and disc height, respectively.

On a smaller scale, the three components that form the disc, the annulus fibrosus, the nucleus pulposus, and the vertebral endplates, (further discussed in Section 2.2), change throughout the spinal sections as well [28]. For example, as the discs increase in thickness, the length of reinforcing fibers of the annulus fibrosus increase as well. This change allows for a decrease in fiber strain caused by a given movement for thicker discs compared to thinner discs [29]. Although there is a general trend between the structural and mechanical properties of the intervertebral discs and the spinal sections they belong to, each individual disc of the same section have their differences as well.

2.1. Classification of Intervertebral Discs

2.1.1. Cervical Discs

The cervical spine consists of six intervertebral discs (C2/C3–C7/T1), with the absence of a disc between the atlas (C1) and the axis (C2) [3]. These discs are smaller in cross-sectional area than any of the other discs in the spine, due to the load bearing role of the cervical spine being much less than that in any other section, therefore decreasing the need for load distribution [29]. The average cross-sectional areas and thicknesses taken from 70 cervical discs range from 190–440 mm^2 and 3.5 to 4.5 mm, respectively, shown in Table 1 [29,30].

Table 1. Average dimensions of the intervertebral discs in the cervical, thoracic, and lumbar spine [29–33].

Cervical IVD Dimensions	C2/C3	C3/C4	C4/C5	C5/C6	C6/C7	C7/T1
Area (mm^2)	190 ± 10	280 ± 40	240 ± 20	300 ± 30	460 ± 5	440 ± 5
Thickness (mm)	3.51 ± 0.71	3.74 ± 0.36	4.07 ± 0.36	4.45 ± 0.21	4.11 ± 0.28	4.50 ± 0.53
Thoracic IVD Dimensions	T1/T2	T2/T3	T3/T4	T4/T5	T5/T6	T6/T7
Area (mm^2)	510 ± 50	490 ± 5	485 ± 5	450 ± 40	605 ± 20	750 ± 10
Thickness (mm)	4.40 ± 0.65	3.50 ± 0.69	3.30 ± 0.50	3.20 ± 0.47	3.50 ± 0.47	4.10 ± 0.47
Thoracic IVD Dimensions	T7/T8	T8/T9	T9/T10	T10/T11	T11/T12	T12/L1
Area (mm^2)	710 ± 30	900 ± 10	840 ± 30	1080 ± 20	1170 ± 30	1190 ± 40
Thickness (mm)	3.90 ± 0.72	5.30 ± 0.80	4.80 ± 1.07	6.50 ± 0.97	5.40 ± 0.95	6.8 ± 0.21
Lumbar IVD Dimensions	L1/L2	L2/L3	L3/L4	L4/L5	L5/S1	
Area (mm^2)	1400 ± 20	1640 ± 50	1690 ± 40	1660 ± 30	1680 ± 30	
Thickness (mm)	7.65 ± 0.57	8.90 ± 0.25	9.25 ± 0.29	9.90 ± 0.49	9.35 ± 1.06	

(From References [29–33]). Cervical disc thicknesses were taken from 19 Chinese cadaveric humans of no specified age or gender. Standard deviations were estimated from graphical error bars. Thoracic disc thicknesses were taken from 15 healthy female and male cadaveric humans with average ages of 58.67 ± 10.74 years and 56.20 ± 11.65 years, respectively. Lumbar disc thicknesses were taken from 607 female and 633 male human spines with age ranges from 20–92 years and 20–87 years, respectively. Standard deviations were estimated from graphical error bars. All of the cross-sectional areas were taken from 4 full human cadaver spines of the following demographics: Male of 73 years, female of 86 years, female of 85 years, and female of 80 years.

In adults, the maximum flexion and extension of the cervical spine occurs around the C5/C6 disc, therefore its thickness is representative of such and will be, on average, thicker than the others. The cervical discs also show a maximum thickness in the anterior section and a minimum height in the posterior section, giving it a natural convex curvature [30]. Because of the mobility of the cervical spine, its discs have a significantly higher risk of damage from bending and torsion, making it the second most common spinal section for disc injury [34].

2.1.2. Thoracic Discs

The thoracic spine consists of twelve intervertebral discs (T1/T2–T12/L1) [3]. These discs are greater in cross-sectional area than the cervical discs, however are still less than that of the lumbar discs. This is due to the amount of extra load transferred to the thoracic spine from the vertebrae above, therefore increasing the need for greater load distribution [29]. The average cross-sectional areas and thicknesses taken from 72 thoracic discs range from 500–1200 mm^2 and 4.4 to 6.8 mm, respectively, shown in Table 1 [31,32].

Although the thoracic discs are greater in cross-sectional area than the cervical discs, they are still thinner in comparison. This is because the thoracic spine does not go through as much flexion/extension and rotation as the other sections of the spine, mainly due to the attachment of the rib cage [29]. The majority of the thoracic discs also show a greater height in the anterior section as opposed to the posterior section (exception of T4/T5, T5/T6, and T10/T11), like that of the cervical discs, however, the difference is not to the same extent as the other sections of the spine [31,32].

Because of the lack of mobility throughout the thoracic spine, its discs tend to have very little torsional stress, giving them a very low chance to become injured from degradation. However, if a high impact is sustained in the thoracic spine, there is a possibility of disc damage, although it is much more common for one of the vertebra to fracture before damage to the disc occurs [35].

2.1.3. Lumbar Discs

The lumbar spine consists of five intervertebral discs (L1/L2–L5/S1) [3]. These discs have the greatest cross-sectional area out of all of the spinal sections, with L2/L3–L5/S1 being virtually equal. This is because the lumbar discs need to withstand the greatest amount of load without building up too much pressure and failing [29]. The average cross-sectional areas taken from roughly 1200 lumbar discs range from 1400–1700 mm^2 and 7.6 to 9.4 mm, respectively, shown in Table 1 [32,33].

Like the cervical spine, the lumbar spine goes through a large amount of flexion/extension and torsion causing a high stress and strain on the discs. Due to these factors, they are the thickest discs and they have the largest surface area [32]. The lumbar discs also have a high ratio of anterior disc thickness to posterior disc thickness, the greatest being the L5/S1 disc, causing the lumbar spine's natural convex curvature similar to the cervical spine [32,33]. Because of the mobility of the lumbar spine and the high loads applied to it, sometimes being in the order of thousands of newtons, its discs have a significantly higher chance of becoming damaged from bending and torsion, making it the most common spinal section for disc injury [36].

2.2. Intervertebral Disc Physiology

Each intervertebral disc is a complex structure comprised of three main components, a thick outer ring of fibrous cartilage called the annulus fibrosus, a more gelatinous core called the nucleus pulposus, and the cartilage vertebral endplates. All together, they bring structural and mechanical integrity to the organ. These components combine to give the necessary structural and mechanical properties to the intervertebral discs as a whole (further discussed in Sections 2.2.1–2.2.3) [37], Figure 2.

The intervertebral discs are the among the largest avascular tissues within the body, due to the lack of vessel penetration throughout the internal sections. Therefore, a flow of nutrients occurs via diffusion from the pre-disc vessels that reach into the outer most layers of the disc [37]. The increase in vascularization into the inner parts of the discs are contributed to their degeneration (further discussed in Sections 2.2.4 and 3). To better understand the functions and properties of each component, they will be further described in detail.

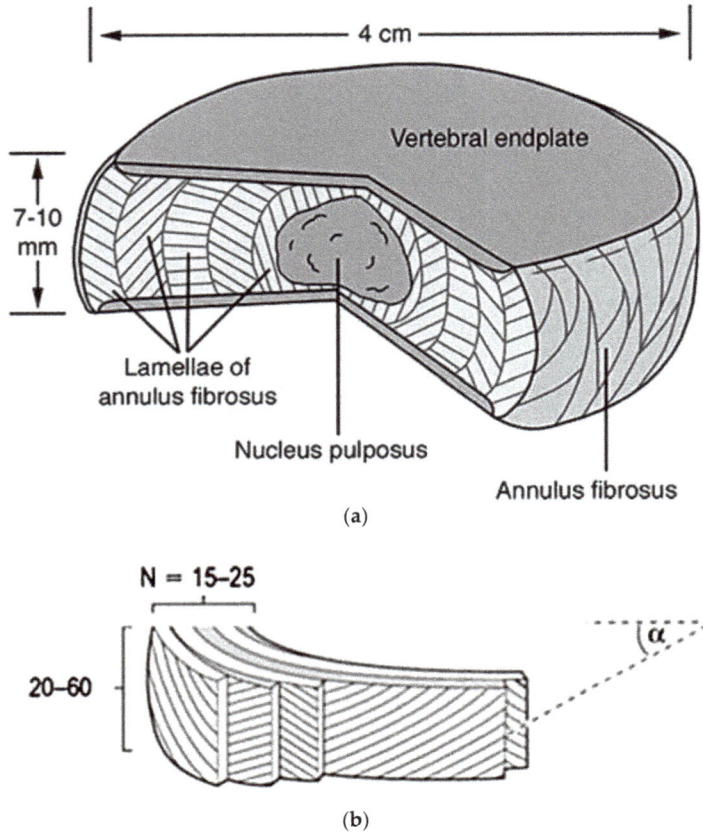

Figure 2. Pictured (**a**) is a cut out portion of a normal disc depicting the nucleus pulposus, vertebral endplates, and annulus fibrosus. The chosen intervertebral disc is 4 cm wide and 7–10 mm thick [37]. Depicted in the lower image (**b**) is a diagram showing the detailed structure of the annulus fibrosus, with its 15–25 lamellae comprised of 20–60 collagen fiber bundles. Also shown, is the angle α, which correlates to the directionality of the fibers' bundles in relation to the vertebrae [38].

2.2.1. Annulus Fibrosus

The annulus fibrosus is a fibrocartilaginous tissue that is structured as concentric rings, or lamellae, surrounding the nucleus pulposus (Figure 2), and is referred to as having two main sections, the inner and outer annulus fibrosus. Both of these sections are composed of mostly water (70–78% inner and 55–65% outer wet weight), collagens (type I and type II collagen, 25–40% inner and 60–70% outer dry weight), proteoglycans (11–20% inner and 5–8% outer dry weight), and other minor proteins building-up the extracellular matrix (ECM). The composition of the ECM varies gradually with increasing radial distance from the nucleus, mainly the type of collagen (having more collagen type I as the distance increases) and decrease of proteoglycans [37,39,40]. These ECM components help create the more rigid structure of the annulus fibrosus necessary to withstand the loads and strains applied.

The annulus fibrosus accounts for a multi-layered structure with alternating collagen fiber angles (varying in degrees throughout the lamella) that help creating a structurally stable material, housing the nucleus pulposus, keeping it under pressure and from impinging on the spine, and enabling the disc to withstand complex loads with its inhomogeneous, anisotropic, and nonlinear mechanical behaviors [41].

(1) Composition

The annulus fibrosus is a unicellular tissue comprising of annulus fibrosus cells embedded in an ECM composed mainly of collagen types I and II, and proteoglycans, which are responsible for the high load-bearing properties of the tissue [41]. Collagens play structural roles, contributing to the mechanical properties, tissue organization, and shape of the annulus fibrosus. Many different isoforms of collagen exist, more than 28 of which have already been identified. It is one of the most abundant ECM proteins in the body, and can take varying structures such as fibrils, short-helix or globular structures [42]. The annulus fibrosus contains only fibril forming collagen, collagen type I and type II, which form the fibrocartilage of the lamellae, Table 2. The collagen types I and II replace one another in a smooth gradient, transitioning from 100% type I in the furthest outer lamella, to 100% type II in the furthest inner lamella [40]. However, based on discs of different individuals, some might include minute amounts of the opposing collagen in the inner and outer lamellae. Not only does the type of collagen change as radial distance increases, but the concentration of collagen as well, increasing from inner annulus to outer annulus [40]. This creates a smooth transition zone between the softer nucleus pulposus and the stronger outer annulus fibrosus [43].

Table 2. Types of collagen found in lamellae of the annulus fibrosus [42–44].

Collagen Type	Structure	Genes	Alpha Chains	% Collagen Distribution
Collagen I	Large diameter, 67-nm banded fibrils	COL1A1 COL1A2	$\alpha1(I)$ $\alpha2(I)$	Increases from $0 \rightarrow 100$ from inner to outer regions
Collagen II	67-nm banded fibrils	COL2A1	$\alpha1(II)$	Decreases from $100 \rightarrow 0$ from inner to outer regions

All collagen consists of a triple helix structure comprised of three polypeptide chains [45]. These polypeptide chains, called alpha (α) chains (procollagens), further diversify the collagen family by creating several molecular isoforms for the same collagen, as well as hybrid isoforms comprised of two different collagen types. The size of these α chains can vary from 662 to 3152 amino acids for humans, and can either be identical to form homotrimers or different to form heterotrimers [42]. Collagen type I is considered a heterotrimer consisting of $\alpha1(I)$ and $\alpha2(I)$, while collagen type II is considered a homotrimer consisting of only $\alpha1(II)$, both of which are found in the annulus fibrosus.

After the transcription and translation of the procollagen α chains, four distinct stages occur for the assembly of collagen fibrils. The first stage is transportation of the α chains into the rough endoplasmic reticulum, where they are modified to form the triple-helical procollagen. The second stage is the modification of the procollagen in the Golgi apparatus and its packaging into secretory vesicles. The third stage is the formation of the collagen molecule in the extracellular space by cleavage of the procollagen. The final stage is the crosslinking between the collagen molecules to stabilize the supramolecular collagen structure [45], Figure 3. These collagen fibrils are vital to the structure, strength, and flexibility of the fibrocartilage in the annulus fibrosus lamellae.

Proteoglycans are glycosylated proteins which have covalently attached highly anionic glycosaminoglycans (GAGs). Major GAGs include heparin sulphate, chondroitin sulphate, dermatan sulphate, hyaluronan, and keratin sulphate [45]. They are less abundant glycoproteins found in the annulus fibrosus ECM, and instead of being predominantly fibrillar in structure, like collagen, they form higher ordered brush-like ECM structures around cells. The main proteoglycans present in the annulus fibrosus are aggrecan and versican, which promote hydration and mechanical strength within the tissue. The keratin sulphate and chondroitin sulphate attached to their protein cores provide the ability to aggregate to hyaluronic acid, resulting in substantial osmotic swelling pressure crucial for the biomechanical properties of the tissue [45,46]. To clarify, their major biological function is to bind water to provide hydration and swelling pressure to the tissue, giving it compressive resistance. More specifically, the negative charges of the sulfated and carboxylated GAGs help trap water within the brushes, generating large drag forces when a load is applied to the tissue, as well as creating osmotic pressure for added resistance [46]. Inverse of the collagen, the proteoglycan concentration has an increasing gradient from outer annulus to inner annulus, or transition zone [40]. Other proteoglycans present in smaller amount on the ECM are the small leucine-rich repeat proteoglycans (SLRPs), such as decorin and byglycan, which are implicated in fibrillary collagen assembly.

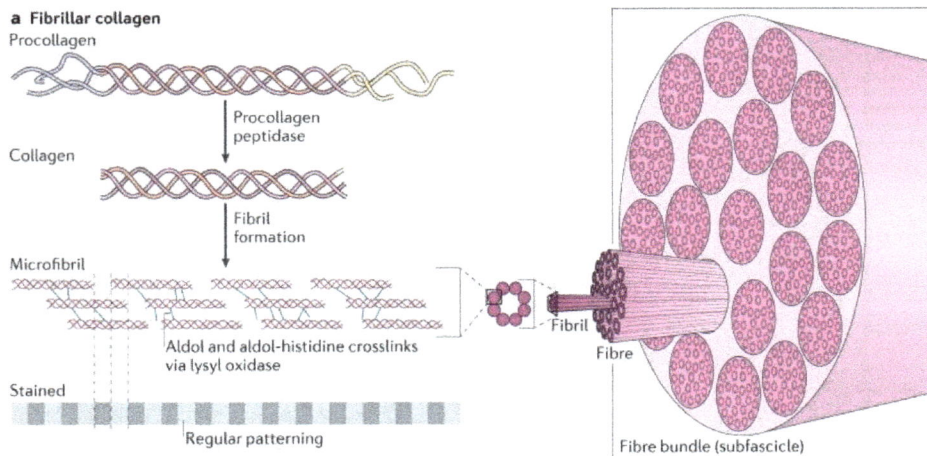

Figure 3. Construction of fibrillary collagen as described above [45].

(2) Structure

The annulus fibrosus has a unique structure consisting of anywhere from 15 to 25 distinct layers (lamellae), depending on the circumferential location, the spine level, and the individual's age, with the thickness of these individual lamellae varying both circumferentially and radially, increasing as age increases [47]. Each adjacent lamella is held together by discrete collagenous bridging structures comprised of type VI collagen, and aggrecan and versican, which are orientated radially to wrap around individual collagen fibers and prevent severe delamination [48,49]. Based on the location of the disc, the amount of collagen fibril bundles in each lamella can vary from 20 to 60 bundles over the total height of the disc, with an average inter-bundle spacing of 0.22 mm and bundle thickness of roughly 10 microns [47], Figure 2. These bundles sit at different angles ranging anywhere from 55° to 20°, alternating direction every other layer, and have a planar zig-zag (crimped) structure. This allows them to be stretched and extend more as the crimps straighten out, resulting in the rotational and flexion/extension mobility of the spine [48,50]. Although the components within the annulus fibrosus are relatively the same, as previously stated, the organization of components such as microfibrils, collagen fibers, and elastin fibers differ with respect to the outer and inner annulus fibrosus [51]. This gives rise to different mechanical properties throughout the structure, detailed in Table 3 and Figure 4 [50,52,53].

The annulus fibrosus' unique structure helps give it its mechanical functions of containing the radial bulge of the nucleus, enabling a uniform distribution and transfer of compressive loads between vertebral bodies, and to distend and rotate, allowing and facilitating joint mobility [40].

(3) Mechanical Properties

Like the collagen and proteoglycan concentration, the mechanical properties of the annulus fibrosus differ with an increase in radial distance, usually becoming stronger and stiffer towards the outer annulus. These mechanical properties are highly anisotropic and nonlinear in uniaxial tension, compression, and shear, and have a high tensile modulus in the circumferential direction [52].

In particular, the tensile properties of the lamella show drastic differences depending on the tested samples and the orientation at which they are tested. When testing parallel to the alignment of the collagen fiber bundles as opposed to perpendicular, the strength and modulus increases due to the strength and reinforcement given by the fibers, and the same correlation can be found when testing the outer lamellae as opposed to the inner lamellae, Table 3 [50,52–55].

Table 3. Mechanical properties of the annulus fibrosus and nucleus pulposus [50,52–55].

Tensile Properties of the Annulus Fibrosus						
Sample	Sample Specification	Ultimate Stress, MPa	Elastic Modulus, MPa	Yield Strain, %	Ultimate Strain, %	Stiffness, N/m
Bulk Annulus	Outer, A	3.9 ± 1.8	16.4 ± 7.0	20–30 *	65 ± 16	5.7 ± 3.4
	Outer, P	8.6 ± 4.3	61.8 ± 23.2	20–30 *	34 ± 11	5.7 ± 3.4
	Inner	0.9	–	20–30 *	33	1.2 ± 1.1
Single Lamella	Parallel	–	80–120	–	–	–
	Perpendicular	–	0.22	–	–	–

Compressive Properties of the Annulus Fibrosus			
Section	Swell Pressure, (P_{sw}), MPa	Modulus, (H_A), MPa	Permeability, (k), ($\times 10^{-15}$ m^4/N-s)
Anterior	0.11 ± 0.05	0.36 ± 0.15	0.26 ± 0.12
Posterior	0.14 ± 0.06	0.40 ± 0.18	0.23 ± 0.09
Outer	0.11 ± 0.07	0.44 ± 0.21	0.25 ± 0.11
Middle	0.14 ± 0.04	0.42 ± 0.10	0.22 ± 0.06
Inner	0.12 ± 0.04	0.27 ± 0.11	0.27 ± 0.13

Compressive Properties of the Nucleus Pulposus			
Sample	Swell Pressure, (P_{sw}), MPa	Modulus, (H_A), MPa	Permeability, (k), ($\times 10^{-16}$ m^4/N-s)
Nucleus Pulposus	0.138	1.0	9.0

A/P, anterior/posterior section of the annulus. Parallel/Perpendicular, alignment of testing in relation to the fiber orientation. * Only one value was ascertained for entirety of the annulus fibrosus. Tensile properties for the bulk annulus fibrosus were taken from 7 cadaveric human lumbar spines. Tensile properties for the single lamella were taken from 8 male and 3 female cadaveric human lumbar spines with an average age of 57.9 ± 15.4 years. The spines were harvested within 24 h of death. Compressive properties of the annulus fibrosus were taken from cadaveric humans of no specified age or gender. Compressive properties of the nucleus pulposus were taken from 10 IRB-approved cadaveric human lumbar spines with ages between 19–80 years (average of 57.5 years) and of no specified gender.

Although the elastic modulus of the lamella differs by a factor of roughly 500, with respect to fiber orientation, when tested as a whole, the tensile elastic modulus instead hovers around 18–45 MPa [52,53]. As the stress induced on the annulus fibrosus increases, the rigidity of the system increases. This mechanical behavior is the result of the un-crimping of the collagen fibers that leads to the stiffening of the intervertebral disc tissue for larger strains. Not only does the stiffness relate to amount of strain on the annulus fibrosus, but also the load rate of the induced stress [54].

The annulus fibrosus is the only section of the disc that undergoes tensile stress, and it is usually due to these stresses that the collagen fibrils breakdown and deteriorate, making its unique tensile properties a focus when studying disc degeneration. However, while tensile properties are important for the understanding of how much stress and strain the annulus fibrosus can withstand, the injuries sustained are rarely due to a single impact, but more often the cyclic loading or wear and tear of the spine that causes deterioration of the collagen fibrils [50,53]. Therefore, cyclic loading tests are crucial for the understanding of the annulus fibrosus' mechanical integrity and resiliency of the tissue. For example, both the anterior and posterior sections of a healthy annulus fibrosus have been shown to withstand more than 10,000 applied cycles with a stress magnitude of 45% or less of its ultimate tensile strength [50].

Although not as important for the annulus fibrosus as it is for the nucleus pulposus, compressive stresses and strains still occur on the lamellae, Table 3. However, they have very little effect on the degradation of the annulus fibrosus. Most often only the swell pressure (P_{sw}), modulus (H_A), and permeability (k) are characterized [55].

2.2.2. Nucleus Pulposus

The nucleus pulposus resides in the middle of the disc surrounded by the annulus fibrosus, which keeps it from leaking into the spinal canal. It consists of randomly organized collagen type II fibers (15–20% dry weight) and radially arranged elastin fibers, housed in a proteoglycan hydrogel (50% dry weight), with chondrocyte-like cells interspersed at a low density of approximately 5000/mm^3 [37,56]. The nucleus is an incompressible structure that it is made up of about 80–90% water, which helps it carry out its vital roles in the intervertebral disc of compressive load dispersion, compressive shock absorption, and keeping the inside of the disc swollen for necessary internal pressure [57].

(1) Composition

There are four main components found in the nucleus pulposus; collagen type II fibrils and elastin fibers (roughly 150 micrometers in length), proteoglycans, and chondrocyte-like cells. Each play a vital role in the performance and health of the nucleus pulposus, providing it with the necessary mechanical properties to serve its functions [58]. For a description of collagen and proteoglycan formation and structure, the reader is referred to Section 2.2.1, (1).

Unlike the annulus fibrosus, the collagen in the nucleus forms a loose network, which is joined by the network of elastin fibers. The elastin fibers are necessary for maintaining collagen organization and recovery of the disc size and shape after the disc deforms under various loads. It accomplishes this with its unique structure of microfibrils forming a meshwork around a central elastin core, Figure 4. These microfibrils are structural elements of the nucleus' ECM, and have been found distributed in connective and elastic tissues such as blood vessels, ligament, and lung [51].

The microfibrils play vital roles in the properties of the elastic fibers, such as conferring mechanical stability and limited elasticity to tissues, contributing to growth factor regulation, and playing a role in tissue development and homeostasis. Microfibrils are made up of a multicomponent system, consisting of a glycoprotein fibrillin core (three known types), microfibril associated proteins (MFAPs), and microfibril associated glycoproteins (MAGPs). The MFAPs and MAGPs, as well as a few other peripheral molecules, contribute to link microfibrils to elastin, to other ECM components, and to cells [59].

In the nucleus pulposus, the chondrocyte-like cells act as metabolically active cells that synthesize and turnover a large volume of ECM components, mainly collagen and proteoglycans [60]. They produce and maintain the ECM with the presence of Golgi cisternae and well-developed endoplasmic reticulum, and are able to withstand very high compressive loads and help with the movement of water and ions within the matrix [61]. They also maintain tissue homeostasis, play a role in the physio-chemical properties of cartilage-specific macromolecules, and prevent degenerative diseases like degenerative disc disease and osteoarthritis. However, with age these cells start to become necrotic, increasing from about 2% at birth to 50% in most adults. This can lead to cartilage/collagen degradation, abnormal bone growth formation on the vertebrae (osteophyte) where bone on bone friction occurs, and stiffening of joints [58,60].

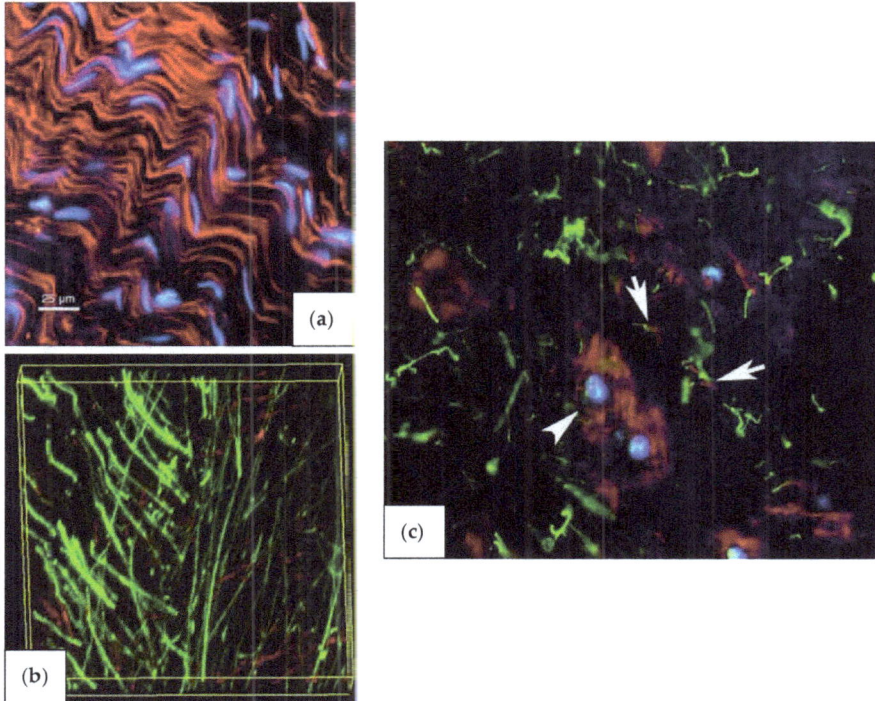

Figure 4. Fluorescence microscopic images of stained components in the outer annulus fibrosus (**a**), inner annulus fibrosus (**b**), and nucleus pulposus (**c**). The microfibrils in relation to cell distribution (blue) and collagen fiber organization (red) indicates the organization of the microfibrils within the ECM of the outer annulus fibrosus. Opposite however, the microfibrils (red) and elastin fibers (green) in the inner annulus fibrosus do not demonstrate any organization or co-localization to any great degree within the ECM. These two distinct characteristics of organization give rise to the varying mechanical properties of each, Section 2.2.1, (3). The microfibrils (red) show a tendency to hover/organize around the nucleus pulposus cells (blue), while the elastin fibers (green) have a tendency to stay dispersed through the entire ECM [51].

(2) Structure

The nucleus pulposus is a soft, gelatinous mass that is irregularly ovoid and is found under pressure in the center of the disc. Because it is mostly water (between 80–90%), it does not have a definite structure or form, but like a liquid, takes the shape of wherever it is confined [62]. From birth to adolescence, the nucleus pulposus is a semi-fluid mucoid mass formed by proliferation and degeneration of embryological notochord cells with a few scattered chondrocytes and collagen fibers. As age increases into adulthood, the notochord cells completely degenerate and become replaced by chondrocyte-like cells, which deposit a specialized ECM to provide the nucleus tissue with its structure and mechanical properties. Also with age, the nucleus becomes less fluid-like and more cartilaginous as the collagen fibrils start to crosslink together forming fibers like the collagen type II fibers of the annulus [63].

(3) Mechanical Properties

Being a virtually incompressible liquid, the nucleus pulposus does not endure any tensile stresses or strains, and the loads it can withstand in compression are largely due to the force that the annulus fibrosus can resist radially. The natural swell pressure of the nucleus at rest is 0.138 MPa, which is correlated to the water uptake and retention during resting periods. However, as compressive forces are introduced to the nucleus, the swell pressure increases to withstand the loads within the confined space of the annulus fibrosus [52]. When testing for compressive properties, the nucleus is confined so that accurate measurements can be taken, Table 3. Confining the nucleus during testing allows for a more accurate resemblance of the resistance towards outward deformation controlled by the annulus fibrosus, as well as keeping the nucleus from being infinitely compressed, since it is a virtually incompressible liquid.

During everyday activities, the lumbar compressive forces can fluctuate between 800 N and 3000 N. This causes the nucleus to become pressurized up to 0.4 MPa while lying down, 1.5 MPa while standing or sitting, and up to 2.3 MPa while actively lifting, however these stresses can vary slightly due to the different dimensional areas of the disc [64]. Although the mechanical testing of the nucleus pulposus is not quite as extensive as that of the annulus fibrosus, it does not make it any less important to the structural and mechanical properties of the disc as a whole.

2.2.3. Vertebral Endplates

The vertebral endplates are situated on the top and bottom of each intervertebral disc, and are comprised of hyaline cartilage [65]. Their main function is to function as an interface between the dense, harder cortical bone shell of the vertebrae and the annulus and nucleus via mechanical interlocking, and to keep the nucleus pressurized and from bulging into the soft, spongy/cancellous trabecular bone center of the vertebrae, Figure 5. The vertebral endplates are the strongest part of the intervertebral disc, and usually fail after the vertebral body has already fractured [38].

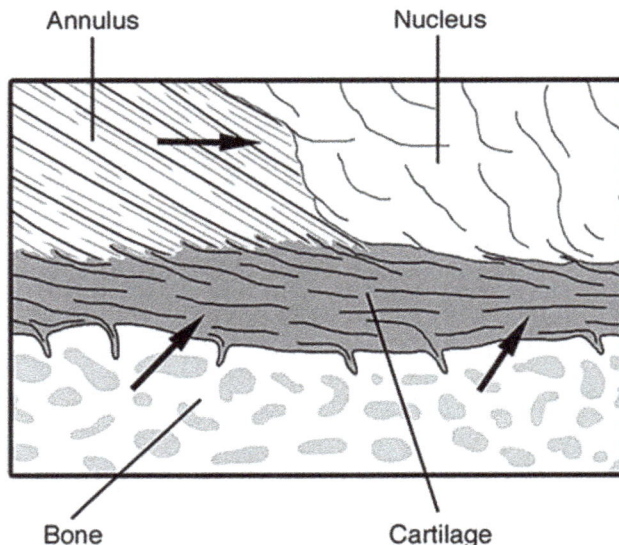

Figure 5. The connection of the hyaline cartilage vertebral endplate to the perforated cortical bone of the vertebral body and collagen fibers of the annulus and nucleus. The arrows in the figure refer to the direction of nutrients and blood flow through the different components of the disc, mainly coming from the bone through the vertebral endplates [37].

The vertebral endplates also have the unique role of acting as the main transport for nutrients in and out of the disc. This provides the nucleus and annulus with the cells and other required components that keep the disc alive, and from degenerating [64].

(1) Composition

The vertebral endplates are composed of an osseous and a cartilaginous component. The hyaline cartilage within differs from the articular cartilage of the joins on its structure. While both are composed of chondrocytes, proteoglycans and a string collagenous network, the former is not connected to the underlying bone [65]. The hyaline cartilage of the vertebral endplates maintains very similar macromolecules in their ECM as that of the nucleus pulposus, however the ratios of proteoglycan to collagen content differs drastically. The typical ratio of glycosaminoglycan to collagen in the endplates is roughly 2:1, providing to the tissue with higher mechanical properties than the nucleus pulposus with a ratio of 27:1 [66]. Also, distinctively different from the annulus fibrosus' fibrocartilage which contain large collagen fiber bundles, the endplates have fine collagen fibers similar to the nucleus, but they are closely packed together. The hyaline cartilage in the endplates are made up of multiple types of collagen. Collagen Type II is the main collagenous component on the endplates. Collagens are often employed as a measure of the degeneration state (hypertrophy of chondrocytes and ossification) of the endplate, being the downregulation of collagen II and upregulation of collagen X the most characteristic markers. [65]. The other collagens, Type I, III, V, VI, IX, and XI are present in small amounts, and only contribute to a minor portion of the cartilage with the main functions of forming and stabilizing the collagen Type II fibril network [67–69].

All of the collagen structures and cellular make-up are the same for the hyaline cartilage as previously discussed in the annulus fibrosus (Section 2.2.1, (1)).

(2) Structure

Two major structures can be distinguished in the vertebral endplates, the collagen fibers of the cartilaginous section (roughly 0.1 to 0.2 mm thick) that connect to the annulus fibrosus and the bony layer of the vertebral section (roughly 0.2 to 0.8 mm thick) that connect to the vertebrae. For the cartilaginous section, the proteoglycan hydrogel-enveloped collagen fibers run horizontal and parallel to the vertebral bodies, however the fibers then continue into the annulus fibrosus at an angle parallel to the currently residing fibers [37]. The integration between the collagen fibers in the nucleus and the endplates is more convoluted. For the vertebral section, the bony component of the endplate is a porous layer of fused trabecular bone with osteocytes embedded within saucer-shaped lamellar packets, resembling the structure of the vertebral cortex [64].

The most important structural features of the endplate biomechanical functions are the thickness, porosity, and curvature. For example, thick, dense endplates with a high degree of curvature are stronger than thin, porous, and flat endplates [64]. They are typically less than 1.0 mm thick, and cover the entire surface area of the top and bottom of the intervertebral disc. The thickness across the width of the disc is not uniform, varying considerably, while tending to be the thinnest in the central region adjacent to the nucleus [65]. The density tends to increase towards the vertebral periphery where the subchondral bone growth starts, however porosity can increase up to 50–130% with aging and disc degeneration. Due to the variations throughout the structure of the vertebral endplate, its mechanical properties vary as well [70,71].

(3) Mechanical Properties

The mechanical properties of the vertebral endplates vary with the region on which the endplate is tested, as well as the region of the spine from which they are extracted. The central area of the endplates tends to be the weakest, and increases in strength and stiffness radially towards the outer annulus [70,71]. When tested in different sections of the spine, the endplates show a significant increase in strength and stiffness from superior to inferior sections of the spine. Not only do the properties change between spinal sections, but also within the same section, such as the stiffness and strength increasing as the lumbar spine descends (L1/L2–L5/S1) [70]. Due to the unique structure of the endplates, they are able to withstand high loads, outlasting the vertebral body more often than not. The failure of the vertebral endplate tends to occur at around 10.2 kN, however the failure of the vertebral body, usually due to fracture, occurs around 4.2 kN in individuals 60 years of age or older, and around 7.6 kN in individuals 40 years or younger [47,72]. Not only do the endplates have great strength, but they also possess great stiffness (1965 ± 804 N/mm) that allow it to be semi-flexible during the loads put onto the spine. This helps the nucleus move and cushion loads more readily inside of the disc, while also protecting the endplates from tensile damage, of which they are most likely to fail [64,73].

2.2.4. Blood Vessels and Nerve Supply

Because the intervertebral disc is one of the most avascular tissues in the human body, in a healthy adult, it tends to have very few microvessels. However, during early stages of skeletal development, blood and lymph vessels are present throughout the majority of the disc with the exception of the nucleus. With maturation of the skeleton, blood and lymph vessels found within the disc start to decrease and migrate towards the outer parts of the annulus fibrosus. These blood vessels extend through the cartilaginous endplates into the inner and outer annulus and slightly into the nucleus up to 12 months of age. However, as age increases past 12 months into skeletal maturity (around 20 years of age), the blood vessels start to recede from the nucleus and inner annulus, until they only remain in the outer annulus and endplates, Figure 6 [37,74].

Given the size of the tissue, once the blood vessels retract from the disc in adulthood, the discs rely on diffusion through the endplates and annulus for the nutritional supply of the disc cells [75]. This reduced nutrient supply is thought to contribute to the degeneration of the discs and to be responsible of the lower regenerative potential of the tissue during aging, giving a reason for the low structural and functional restoration properties of the tissue during aging [75].

The intervertebral discs are innervated organs with some of the most important nerves residing in the cervical and lumbar spine. Recurrent sinuvertebral nerves innervate the posterior and some of the posterolateral aspects of the disc, and the posterior longitudinal ligament, branching off the dorsal root ganglion extending from the spinal cord. The other posterolateral aspects receive branches from the adjacent ventral primary rami and from the grey rami communicants [76]. Lateral aspects of the disc receive other branches from the rami communicantes, some of which cross the intervertebral disc and are embedded within the surrounding connective tissue of the disc, such as the origin of the psoas for the lumbar spine. Lastly, the anterior aspects along with the anterior longitudinal ligament are innervated by recurrent branches of rami communicantes, Figure 7 [76].

Opposite of blood vessels, in a healthy young adult, the sensory nerve endings of the disc can be found on the superficial layers of the annulus and in the outer third of the annulus, only extending about 3 mm into the disc [77]. With age and degeneration, the nerves tend to creep into the inner parts of the disc by means of neoinnervation, arising from granulation tissue growing in the disc. This can cause innervation of the middle and inner annulus, and potentially of the nucleus pulposus. As innervation progresses, significant problems with regards to lower back pain can arise from the amount of pressure being induced onto the discs, and therefore pressure onto the nerves [77,78].

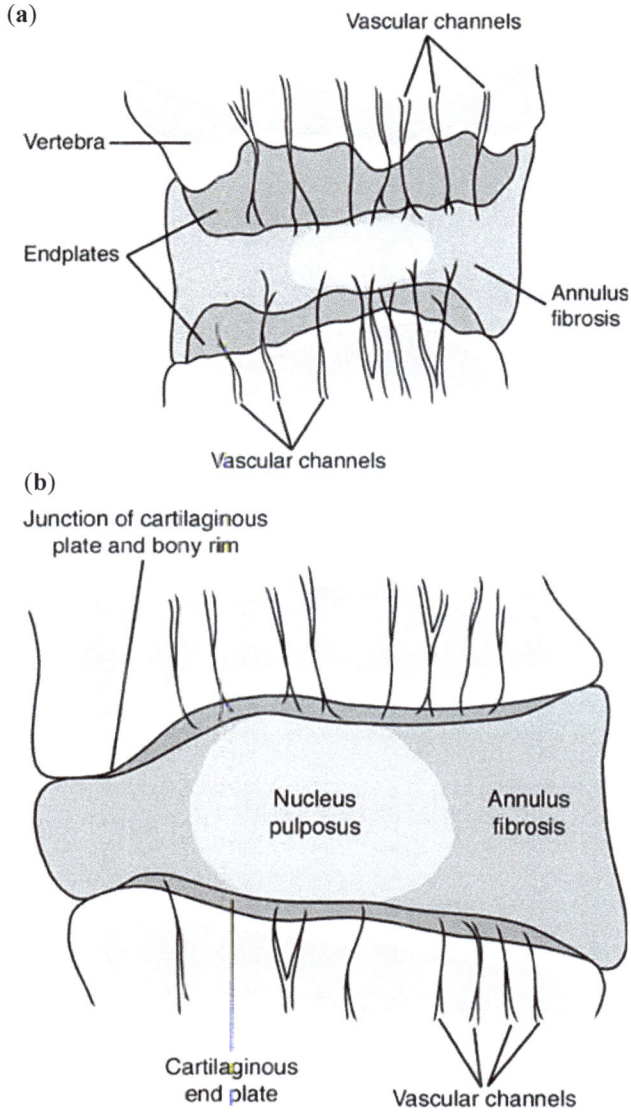

Figure 6. (**a**) Schematic representation of the multiple longer and thicker vascular channels throughout the intervertebral disc on a 10-month old female; while (**b**) represents the vascular channels throughout the disc of a 50-year old adult, showing the retraction and thinning of the channels [37].

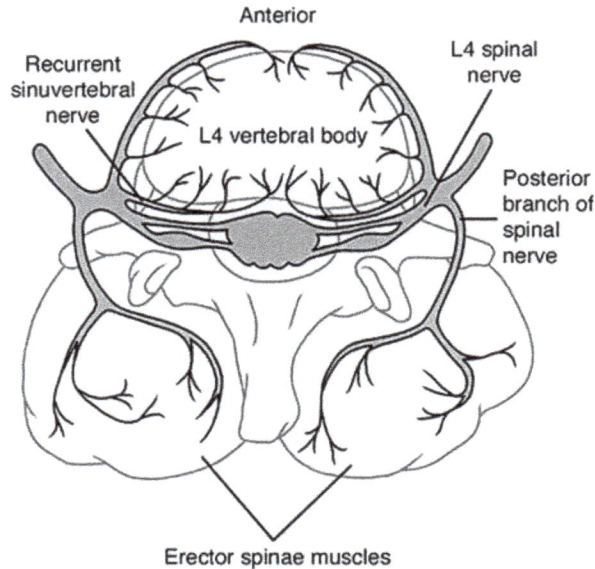

Figure 7. The innervation of a healthy intervertebral disc, showing the sinuvertebral nerves and rami communicantes extending into the vertebral foramen and the outer annulus of the disc [37].

3. Spinal Degeneration and Lower Back Pain

Back pain is a major health problem in Western industrialized societies, inflicting suffering and distress on a large number of patients, especially those of old age, increasing with the increased aged population. The effects of this problem are vast, with a study in the year 2000 in the UK showing prevalence rates ranging from 12% to 35%, and around 10% of sufferers becoming chronically disabled [79]. With total costs, including direct medical costs, insurance, lost production, and disability benefits, reaching into the billions of dollars, an enormous economic burden is placed on society [79]. In the United States alone, costs associated with lower back pain exceeds $100 billion per year, two-thirds resulting from lost wages and reduced productivity [80]. Among the other third are direct costs for medical treatments of back pain diagnoses, estimated at $34 billion out of the total $47 billion for all treatments for pain diagnoses in 2010. These costs include office-based visits, hospital outpatients, emergency services, hospital inpatients, and prescription drugs [81]. This back pain is strongly associated with disc degeneration and injury, the majority of the time occurring in the lumbar spine due to the increased stresses, strains, and torsion compared to other sections, and the thoracic spine being the least affected [11].

Intervertebral discs can degenerate due to injury or due wear and tear, as a result of the stress and strain to which the tissue is exposed to on a daily basis. However, intervertebral discs are among the most avascular tissues in the human body and together with the low proliferative potential of cells within, being almost quiescent, results in a tissue that is unable to adequately self-regenerate [82]. Multiple factors promote the degeneration of the tissue other than just wear and tear, such as genetic predisposition, impaired metabolite transport, altered levels of enzyme activity, cell senescence and death, changes in matrix macromolecules and water content, osteoarthritis, structural failure, and neurovascular ingrowth. Although genetic inheritance is the greatest risk factor, it does not cause discs to degenerate by itself, but instead increases their susceptibility to environmental factors such as high and repetitive mechanical loading and smoking cigarettes [83].

3.1. Degenerative Disc Disease

Degenerative disc disease is defined by the degeneration of intervertebral discs due to aging and other environmental factors, with genetic inheritance playing a significant role in the rate of degradation. Approximately 50–70% of the variability in disc degeneration is caused by an individual's genetic inheritance [83,84]. The inherited genes associated with disc degeneration include those for collagen type I and IX (COL1A1, and COL9A2 and COL9A3, respectively), aggrecan, vitamin D receptor, matrix metallopeptidase-3 (MMP3), and cartilage intermediate layer protein (CILP). The strength of musculoskeletal tissue, like that of intervertebral discs, is affected by the composition of the ECM, such as the strength of the collagen fibrils throughout the annulus fibrosus, which is regulated by the aforementioned genes (and others) [85]. Although an unfavorable genetic inheritance is present at birth, disc degeneration only becomes prevalent and common in the individual's 40's, and usually only in the lower lumbar spine [83,84]. Some individuals however, can become inflicted by this disease much earlier than the norm, depending on both the severity of their genetic deficiencies and lifestyles.

Degeneration of intervertebral discs can occur at faster rates than for other tissues and is sometimes presented on individuals as young as 11–16 years of age, usually found in the lumbar section [79]. Degenerative disc disease affects about 20% of people in their teens, showing mild signs of degeneration before their second decade of life. However, because the discs have yet to undergo progressive innervation, most cannot feel the pain and disabilities associated with degeneration until it propagates through to the later years of life. Therefore, this disease increases drastically with age, causing the discs of around 10% of 50-year-old population and 60% of 70-year-old population to become severely degenerated, significantly hindering daily activities [79].

Degenerative disc disease can affect the tissue in many ways, causing it to undergo striking alterations in volume, shape, structure, and composition that result on a decreased motion and an altered biomechanical properties of the nucleus pulposus and annulus fibrosus tissues, thus altering the mechanics of the spine [34]. Both the nucleus pulposus and annulus fibrosus experience changes individually, mainly in the ECM composition and structure. Consequentially, due to the compositional changes on the discs ECM, such as collagen, proteoglycan, and water content, the major structural properties become hindered as well. The main structural effects tend to be the loss of swelling ability, and therefore volume of the nucleus, and tears or fissures forming in the annulus [86]. When these fissures are formed in the annulus, there is also frequently a cleft formation of some sort, particularly in the nucleus, and the morphology becomes more and more disorganized, Figure 8. The vertebral endplates also go through some deformation and changes, such as an increase of porosity from 50 to 130%, the natural curvature becoming less apparent and flattening out, and a significant decrease in the thickness by roughly 20 to 50% [64,71]. These changes make the vertebral endplate much more likely to fracture under the stresses of the spine and tensile stresses induced by the nucleus.

Along with major structural changes, many biochemical changes occur throughout the disc as well. With age and degeneration, comes an increased incidence in these changes, including cell proliferation and death, mucous degeneration, decrease in proteoglycan content, increase in collagen fibril cross-linking (mainly nucleus), granular changes, and concentric tears in the annulus [79]. Innervation and vascularization of the disc are thought to cause the increase in cell proliferation in the nucleus, which leads to the formation of clusters of living, necrotic, and apoptotic cells. The appearance of these apoptotic and necrotic cells can promote cell death in the healthy living cells. Unfortunately, these mechanisms tend to be very common with age, with more than 50% of cells in adult discs being necrotic [79].

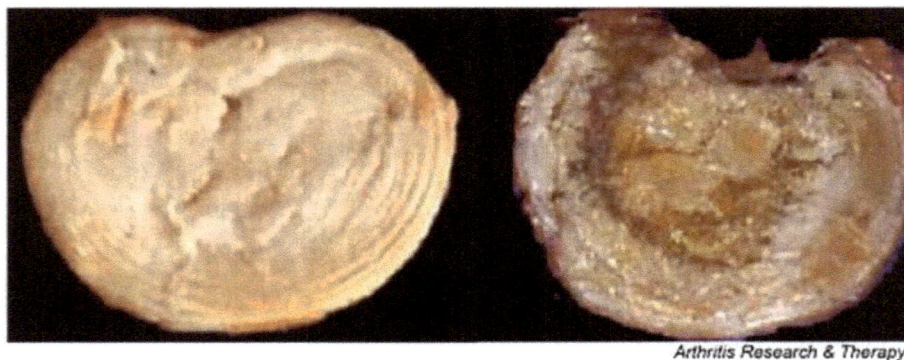

Arthritis Research & Therapy

Figure 8. A healthy, normal intervertebral disc on the left, shows a distinct difference between the swollen, softer looking nucleus and the ringed annulus. However, during growth and skeletal maturation, the boundary between these components becomes less obvious, and with the nucleus generally becoming more fibrotic and less gel-like, like the highly degenerate disc on the right [79].

As degeneration progresses, compositional and structural changes to the discs become more and more apparent. The status of the degeneration is commonly studied via Magnetic Resonance Imaging (MRI) and evaluated with the Magnetic Resonance Classification System with rankings from Grade I to Grade V [87]. The ranks are based on disc structure, signal intensity, distinction between the nucleus and annulus, and the height of the disc, Table 4.

Table 4. Distinction between different grades of disc degeneration based on magnetic resonance imaging (MRI) scans [87].

Grade	Structure	Distinction of Nucleus and Annulus	Signal Intensity	Height of Intervertebral Disc
I	Homogenous, bright white	Clear	Hyperintense, isointense to cerebrospinal fluid	Normal
II	Inhomogeneous with or without horizontal bands	Clear	Hyperintense, isointense to cerebrospinal fluid	Normal
III	Inhomogeneous, gray	Unclear	Intermediate	Normal to slightly decreased
IV	Inhomogeneous, gray to black	Lost	Intermediate to hypointense	Normal to moderately decreased
V	Inhomogeneous, black	Lost	Hypointense	Collapsed disc space

Although the grading scale has shifted from the previous radiographic imaging systems, which focuses on the antero-posterior abnormalities of the discs, distinguishing among bulging, protrusion, and extrusion, (Grade I through Grade III respectively), the MRI images used for the Magnetic Resonance Classification System still show the symptoms of all three past grades, Figure 9. It can be seen that Grade II–III shows a slight bulging of the nucleus (more prominent in Grade III), Grade IV shows the beginning stages of protrusion of the disc, and Grade V shows a fully blown-out disc in which the entire nucleus has been extruded into the spinal canal [87].

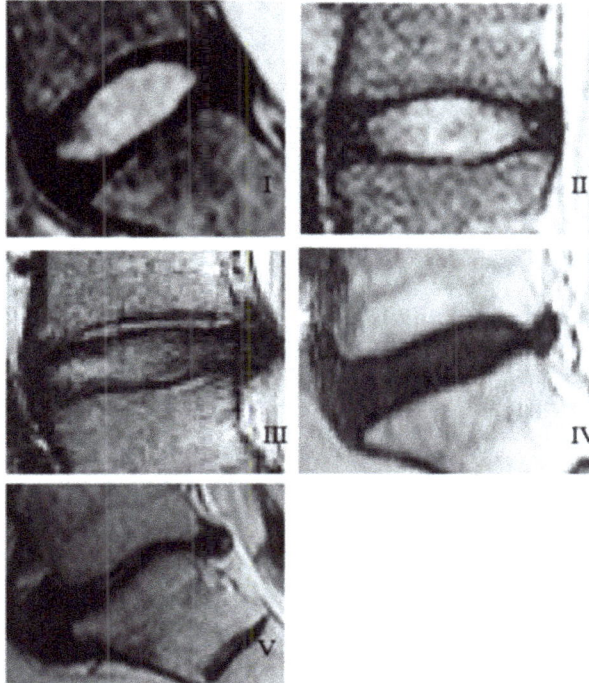

Figure 9. MRI scans showing the different grades of disc degeneration based on the Pfirrmann grading system, (**I–V**) referring to Grades (**I–V**): I is representative of Grade (**I**) degeneration, (**II**) is representative of Grade (**II**) degeneration, (**III**) is representative of Grade (**III**) degeneration, (**IV**) is representative of Grade (**IV**) degeneration, and (**V**) is representative of Grade (**V**) degeneration [88].

3.2. Osteoarthritis

Although not as common of a cause for disc degeneration as degenerative disc disease, osteoarthritis can have a significant impact on the structural changes of the intervertebral discs, causing major problems at long term. Osteoarthritis is a degenerative disorder of the articular cartilage affecting over 30% of the population above the age of 65 and is associated with hypertrophic changes of the tissue affecting the facet joints and vertebrae of the spine, especially the lumbar spine [89,90]. Many risk factors can affect the probability as well as severity of osteoarthritis including genetic inheritance, female gender, past physical trauma, increased age, and obesity. Symptoms usually include joint pain that increases with movement, trouble or disability with activities of daily living, and lower back pain associated with narrowing disc space. With the current U.S. population living longer and becoming more obese, osteoarthritis has become more common than it ever has before, affecting an estimated 27 million adults in the U.S. [89,91].

Peripheral joints such as hips, knees and hands, were most commonly thought of with regards to osteoarthritis, with prevalence in the spine often being ignored. However, the prevalence of disabilities and functional distress caused to the spine by osteoarthritis are actually quite high. In the lumbar spine, it is a very common condition, with a prevalence range of roughly 40–85% based on age, weight, and other factors. The spinal degeneration process has been partly linked to both osteoarthritis and changes in facet joint structure. Osteoarthritis leads to the narrowing of disc spacing from the formation of vertebral osteophytes introducing increased pressure to the disc. Being comprised of the same type of cartilage as appendicular joints, facet joints have similar pathological degenerative processes, such as crystal deposition within the cartilage, degradation from high impact and torsional loads, and joint instability, which all can cause additional stress to the discs [91]. Both the intervertebral discs and facet joints play vital roles in the motion of the spine, especially in the cervical and lumbar spines, therefore when they are heavily affected by osteoarthritis, the mobility of the spine can decrease significantly, and pain can ensue from even the slightest of movements.

Three main components are observed with regards to osteoarthritis in spine, referred to as the "three joint complex". These components include the structure of vertebral osteophytes, facet joint osteoarthritis, and disc space narrowing. With the amount of nerve supply running through all of these spinal structures, lower back pain can be generated by any of them [91]. With further progression of disc degeneration in the spine, the facet joints as well as vertebrae further degenerate, due to disc space narrowing, which in turn puts even more stresses onto the intervertebral discs. Facet joint osteoarthritis is a multifactorial process that is highly affected by disc degeneration, leading to greater loads and motions endured by the joints [92]. This, consequently, leads to the breakdown of the layer of hyaline cartilage between the two subchondral bones, creating friction and grinding between them, and finally abnormal bone growth and pressure. However, facet joint osteoarthritis can still occur in the absence of disc degeneration, in which case it causes more stress and motion on the intervertebral disc leading to quicker degeneration [93].

Changes in the structure of the vertebral osteophytes on the shape of formation of bony outgrowths which arise from the periosteum at the junction of the bone and cartilage, lead to disc space narrowing, Figure 10. Although it is highly correlated to disc degeneration, like that of the osteoarthritis in the facet joints, osteophyte formation in the vertebral column can occur without any signs of cartilage damage, implying that with the general aging process, they may form in an otherwise healthy joint [91]. In this case, the vertebral osteophytes can cause extra stresses on the discs, mainly in the annulus fibrosus, potentially weakening it for further degeneration, damage, and tears/fissures [91,94].

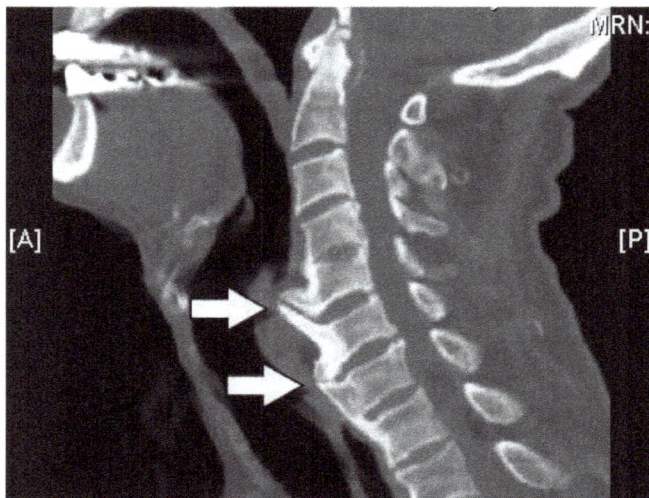

Figure 10. Sagittal computerized axial tomography (CT scan) image of the cervical spine showing large anterior osteophytes (indicted by the arrows) extending from C5 to C7, which affect the intervertebral disc space [95].

Osteoarthritis, along with the aforementioned degenerative disc disease and mechanical loading factors endured by the spine, can cause severe lower back pain because of the potential impingement and injury that can happen to the spinal cord in a couple ways such as bulging discs, disc prolapse and protrusion, and finally disc herniation/rupture and extrusion [94].

3.3. Bulging Disc

Bulging discs are considered the starting stage for problems with impingement to the spine and are generally associated with fatigue failure from mechanical loading and disc degeneration of Grade 0 (negligible degeneration), Grade I, and Grade II [96]. In the early stages of disc degeneration, when the annulus fibrosus starts to dry out and become more fibrous, the amount of mechanical strain it can take decreases. With high compressive loads that are put onto the discs that require the nucleus to push out causing pressure to the annulus, this can cause problems such as small tears part way through the lamellae. When some of these lamellae tear, usually in the posterior section of the disc, the pressure from the nucleus can make the discs bulge outwards due to the lack of support from the annulus, Figure 11 [97].

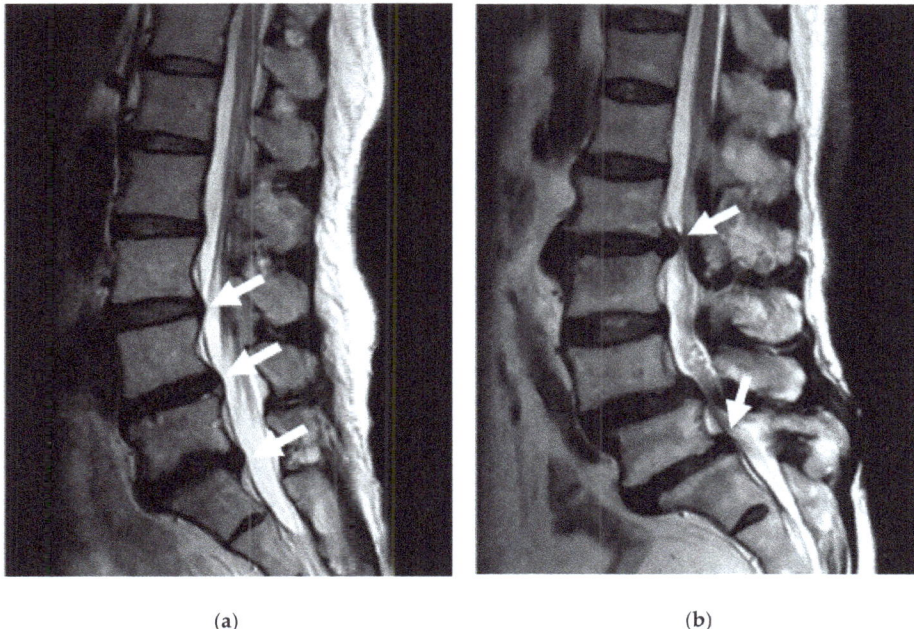

(a)   (b)

Figure 11. MRI image showing a slight bulge of the annulus into the spinal canal without severe impingement (**a**). MRI image showing a full lumbar disc herniation with substantial spinal stenosis and nerve-root compression (**b**) [97].

When the disc bulges into the spinal canal, it can put pressure onto the spinal cord and other spinal nerves, one of the most prominent being the sciatic nerve, causing pain and sometimes even numbness [22]. Although the pain from these bulging discs is bearable, if left untreated, they can lead to even more severe problems such as disc herniation.

3.4. Disc Herniation (Prolapse/Rupture)

Disc herniation, also referred to as disc prolapse, rupture, and extrusion, occurs in later stages of disc degeneration, Grades III–V, and is brought about by increased mechanical loading and fatigue of the annulus that has typically already started to bulge [96]. As the annulus becomes more and more fibrous with degeneration, there is an increase in tears through the lamellae due to the forces of the nucleus. When the tears penetrate all the way through the annulus, the nucleus starts to push out and leak into the spinal canal, Figure 11 [98]. Unlike bulging discs, because the nucleus actually leaks into the spinal canal, it tends to have much more significant impacts on an individual's life due to the severe impingement on nerves of the spinal cord, causing pain, numbness, tingling, and weakness [99].

The most common area for disc herniation is in the lumbar spine, particularly in the lower lumbar, with roughly 56% of herniations occurring in the L4/L5 disc and roughly 41% occurring in the L5/S1 disc [99]. Both of these disc herniations can play significant roles in the quality of an individual's life, since they both are involved with the sciatic nerve. The sciatic nerve, as mentioned in the above anatomy, runs all the way from the lower spine down through the back of the leg. When impinged, this can cause severe problems with motions such as standing up from a seated position, walking, bending over, and twisting of the upper body, and can cause pain, numbness, weakness and general discomfort throughout the entire low extremity. With disc herniation, surgery is very often required to fix it, however with a bulging disc or other lower back pain, some other less invasive procedures exist [100].

4. Current Treatment Techniques

Depending on the severity of disc degeneration, and whether or not a disc is bulging or herniated, there are multiple treatment options, both invasive (surgical) and noninvasive (nonsurgical). The most common treatments include physical therapy, epidural injections, and medications for noninvasive, and radiofrequency ablation, spinal fusion surgery, synthetic total disc replacements, and annulus fibrosus repair for invasive. Although pain and disability are usually relieved for a period of time, the effectiveness of these treatments are less than ideal, due to certain problems associated with each, further discussed below. Therefore, along with the invasive and noninvasive options, other less-traditional treatments are being researched such as the use of stem cells, growth factors, and gene therapy with the theoretical potential to prevent, slow, or even reverse disc degeneration, as well as tissue engineered scaffolds in order to completely replace degenerated discs [101].

4.1. Nonsurgical Treatments

4.1.1. Physical Therapy

With disc degeneration, comes lack of support and stability of the spine due to the decreasing biomechanical functions of the intervertebral disc. In order to regain this loss of function, the muscles surrounding the spine and supporting spinal loads must increase in strength and stability, therefore decreasing the need for intervertebral disc support for the spine. A solution to this problem is physical/functional therapy, of which benefits include increased strength, flexibility, and range of motion [102]. Improving motion in a joint is one of the optimal ways to relieve pain. This can be accomplished by stretching and flexibility exercises which improve mobility in the joints and muscles of the spine and extremities. The next is increasing strength with exercises for the trunk muscles, providing greater support for the spinal joints, and arm and leg muscles, reducing the workload required by the spinal joints. Aerobic exercising has also been shown to relieve lower back pain by promoting a healthy body weight and improving overall strength and mobility [102]. Other therapies include deep tissue massaging, posture and movement education for daily life (functional therapy), and special treatments such as ice, electrical stimulation, traction, and ultrasound. Ultrasound treatment, in particular, has been shown to significantly improve lower back pain for individuals suffering from degenerative and even prolapsed discs, although it is only a temporary solution [103]. Physical therapy does not reverse the age-related disc degenerative changes, however, healing should be promoted by stimulating cells, boosting metabolite transport, and preventing adhesions and re-injury, which in turn will relieve pain caused by degenerative disc disease [104].

4.1.2. Epidural Steroid Injections

Epidural steroid injections are one of the most common injections for relief of pain, by reducing inflammation caused by degenerative disc disease. The injections consist of cortisone, which has anti-inflammatory properties reducing and further preventing additional inflammation, combined with a local anesthetic, which offers immediate short-term pain relief. Both of these components help to turn off the inflammatory chemicals produced by the body's immune system that can lead to future flare-ups [105]. It is injected into the epidural space that surrounds the membrane covering the spine and nerve roots. Because it is administered so close to the area of pain, this treatment tends to have better effects and outcomes than that of oral and topical medications, however it can only be performed three times a year due to the negative side effects of the steroids in the body and the effects only last 1–2 months. Also, it does not reverse the changes of degenerative disc disease already caused by aging, with over two-thirds of patients undergoing an additional invasive treatment within two years of the epidural injections [106].

4.1.3. Medications

For low to moderate lower back pain caused by degeneration of the discs and spine, oral and topical medications can be prescribed. These medications include over-the-counter acetaminophen (Tylenol) and non-steroidal anti-inflammatory drugs (NSAIDs), anti-depressants, skeletal muscle relaxants, neuropathic agents, opioids (narcotics), and prescription NSAIDs, each having individual and unique benefits depending on the severity and type of pain [107].

The acetaminophen and NSAIDs are usually taken for very low, dull chronic pain. Acetaminophen such as Tylenol is used to essentially block the brain's pain receptors, while NSAIDs such as ibuprofen, naproxen, or aspirin are used to reduce inflammation. The NSAIDs however, need to be taken on a daily basis because they work to build up an anti-inflammatory effect in the immune system [108]. This means that only taking them when pain is present does not work to limit inflammation as well as taking them regularly. Tricyclic anti-depressants are usually given for chronic lower back pain as well. These anti-depressants work similarly to acetaminophen, blocking pain messages on their way to the brain. They also help to increase the body's production of endorphins, a natural painkiller, and help individuals sleep better, allowing the body to regenerate and recover [107,108]. Skeletal muscle relaxants, such as tizanidine and cyclobenzaprine, are needed for individuals who have acute back pain due to muscle spasms. When their muscles spasm, they put additional stresses onto the discs and spinal nerves causing intense pain through the spine. Neuropathic agents, such as Neurontin and Lyrica, are used when the nerves of the spine are impinged due to a bulging or herniated discs. These medications allow for the specific targeting of nerves to block signals sent to the brain in order to prevent pain. Opioids (narcotics), such as Vicodin and Percocet, are used in extreme cases of spinal pain given their addictive qualities. They work by attaching to receptors in the brain, similar to acetaminophen, however with much higher strength and effect, tending to cause side effects such as slow breathing, general calmness/drowsiness, and an anti-depressant effect. Prescription NSAIDs work exactly the same as over-the-counter NSAIDs, however they tend to work better given their increased strength and potency [107,108].

4.2. Surgical Treatments

4.2.1. Radiofrequency Ablation

Radiofrequency ablation is a technique that uses heat put through the tip of a needle, either by continuous or pulsed radiofrequency, to denervate an injured disc causing pain to an individual. Nerves of which can be denervated to help with low back pain are the facet nerves, sympathetic nerves, communicating rami, and nerve branches in the disc itself. After anesthesia is administered to the procedure site, a needle or electrode is inserted into the disc or near the small nerve branch, under X-ray, fluoroscopy, computerized axial tomography, or magnetic resonance guidance [109,110].

When in the right position, the tip of the needle or electrode is heated up to the point in which it causes damage or heat lesions to the nerves, destroying them to the point that back pain is relieved. Pain can be relieved usually for 6 to 12 months, and in some cases can last for a few years. It is one of the less invasive operations, and therefore is considered an outpatient surgery, in which the patient is put under local anesthesia and can go home that day without being hospitalized [109,110]. This procedure is usually recommended for patients who have already undergone procedures such as epidural steroid injections, facet joint injections, sympathetic nerve blocks, or other nerve blocks with pain relief lasting shorter than desired. The average cost of this procedure ranges anywhere from $2000 to $5000 based on practitioner, amount of nerves destroyed, and location of spine. If, however degenerative disc disease becomes too severe, this method will not be suitable for long term, and other surgeries or total disc replacements will have to be considered.

4.2.2. Spinal Fusion Surgery

Spinal fusion surgery has been widely accepted as a useful treatment option for correcting severe disc degeneration disease, however its efficacy and success remain controversial. Multiple approaches for this procedure can be taken such as posterolateral fusion, anterior lumbar interbody fusion, posterior lumbar interbody fusion, and lateral lumbar interbody fusion, each being a minimally invasive technique to lumbar spinal fusion [101]. For this treatment, the damaged disc is completely removed from the spine and replaced with either an osteoconductive-filled titanium cage or a hydroxyapatite bone graft extender that sits in between the two vertebrae [111,112]. Titanium plates are then attached to the vertebrae above and below the titanium cage, using titanium pedicle screws as fasteners, to offer additional support to the spine after surgery, Figure 12. This allows for stability of the spine and correct anatomic alignment of the spinal segments by sharing the loads acting on the spine, until the point in which solid biological fusion occurs into a single bone [113]. This is important because if the adjacent segment motion is altered, it can lead to further degeneration of additional discs and motion segments [101]. Once this occurs, the patient can opt to have the plates and screws removed via another surgery.

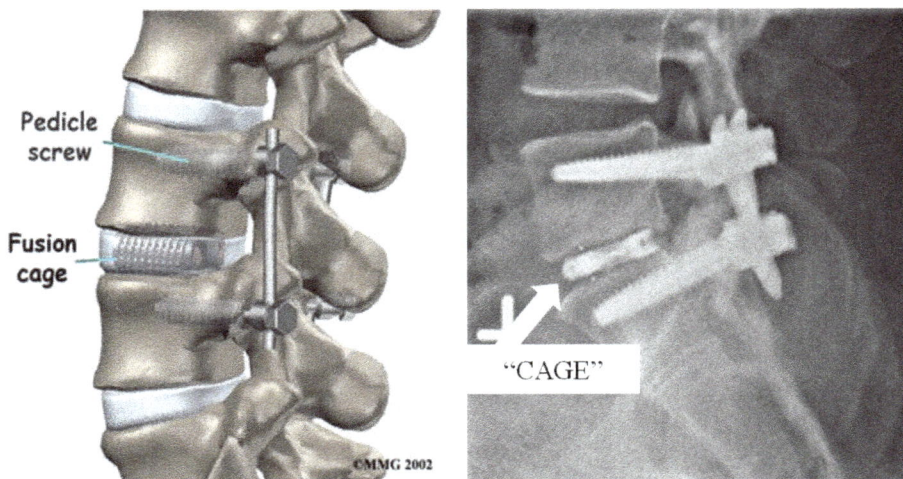

Figure 12. Example image of spinal fusion surgery using titanium cages loaded with hydroxyapatites and pedicle screws and rods to keep stability and anatomic alignment in spinal segment [114].

Although spinal fusion surgery tends to alleviate discogenic pain associated with degenerative changes, due to eliminating motion between certain vertebrae, some other problems can arise that could potentially be more detrimental in the long run. When two vertebrae are fused together, there becomes no load absorbing center, which severely limits shock absorption and increases loads and stresses on surrounding tissues and discs, as well as limiting mobility [101,113]. This gives way to additional intervertebral disc degeneration in the adjacent levels, which will then potentially need to be fused as well. However, since the lumbar is the main contributor to the mobility of the spine, preserving that mobility is vital to everyday activity. For this reason, most doctors refuse to fuse more than three levels of the spine together so to not hinder the movements of everyday life and cause more problems than leaving the damaged disc in the spine [115]. It is estimated that over 137,000 cervical and 162,000 lumbar spinal fusion surgeries are performed every year in the United States alone, totaling over 325,000 fusions, each costing over $34,000 for the average hospital bill, excluding professional fees and equipment fees [116,117]. In the last few years however, interest in total disc replacement instead of spinal fusion surgery has grown due to their ability to retain motion of the lumbar motion segments [116].

4.2.3. Total Disc Replacement

Total disc replacements (TDR) is a treatment option that consists of the removal of the degenerative native disc and replacing it with a synthetic implant. This option offers the mobility that is required for the lumbar section that spinal fusion surgery does not, however, they are still not as mainstream as fusion surgery [116]. In order for a TDR to be considered effective, the implant must fulfill four main requirements: (1) a solid, nondestructive interface with the adjacent vertebral bodies; (2) provide mobility to mimic the range of motion of the natural disc; (3) resist wear and tear in the body to reduce debris contamination in the body; (4) have the ability to absorb shock and distribute loads evenly and effectively [118]. In all of these requirements, the lumbar spine TDR must perform at a more demanding level than that of the cervical spine due to the extra loads it must bear. Therefore, fabrication of TDRs for the lumbar spine have proven to be much more difficult when compared to those for the cervical spine. Lumbar TDR can be classified according to their configuration, materials, bearing type, and regulatory status, Table 5. The configurations of the TDR devices are designed to maximize the range of motion within the realm of natural disc mobility and permit the most freedom. Each configuration of TDR is dependent upon the type of modules involved in the working disc, therefore current designs are built around a bearing for maximum mobility [118]. The bearing systems used includes one-piece (1P), Metal-on-Metal (MoM), or Metal-on-Polymer (MoP), with MoM and MoP bearings using a ball and socket design to allow for motion in all directions. Only two lumbar disc prostheses have currently been approved for use by the Food and Drug Administration (FDA), the Charite® from DePuy Spine and the Prodisc® L from DePuy Synthes, although many more are becoming prevalent through trial testing such as Maverick™, Kineflex®, Freedom®, and Mobidisc® [118].

Although there are a lot of different TDR options, each has their disadvantages, with only the two previously mentioned even being FDA approved. Ball and socket bearing systems give way to the possibility of hypermobility within the motion segment, greater amounts of debris from wear, and stress concentration within bearing itself, which causes higher stresses to act on the vertebrae. It has also been shown that these systems show no elastic shock absorption properties, even between MoM and ultra-high molecular weight polyethylene (UHMWPE) cores (MoP) [119]. The one-piece bearing systems were designed to potentially counteract the above flaws by adequately mimicking the natural disc behavior; reducing the number of surfaces on which wear can occur, reducing the hypermobility of the joint, and distributing load and absorbing shock [118,119]. The flaws with the one-piece systems, however, are that the elastomer core used suffers greater chance of material tears either within the material or at the adhesion interface between the different materials. They experience short fatigue life and are still recent designs, needing further evaluation of wear and corrosion resistance [118]. Creep deformations and hysteresis properties of the elastomeric material may be limiting factors as well [119]. Each TDR system experiences failure through two mechanisms of degradation of the implant, wear and corrosion. These degradations are to be expected with articulating bearings in harsh environments, however act more heavily on some materials as opposed to others, Table 6.

Table 5. Summary of current total disc replacement (TDR) classification, materials, bearing type, and regulatory status [118].

Device	Classification	Biomaterials	Bearing Design	Examples of Manufacturer
CHARITE	MoP	CoCr-UHMWPE	Mobile	DePuy Spine
Prodisc-L	MoP	CoCr-UHMWPE	Fixed	DePuy Synthes
Activ-L	MoP	CoCr-UHMWPE	Mobile	Aesculap
Mobidisc	MoP	CoCr-UHMWPE	Mobile	LDR Medical
Baguera	MoP	DLC coated Ti-UHMWPE	Fixed	Spineart
NuBlac	PoP	PEEK-PEEK	Fixed	Pioneer
Maverick	MoM	CoCr-CoCr	Fixed	Medtronic
Kineflex	MoM	CoCr-CoCr	Mobile	SpinalMotion
Flexicore	MoM	CoCr-CoCr	Constrained	Stryker
XL-TDR	MoM	CoCr-CoCr	Fixed	NuVasive
CAdisc-L	1P	PU-PC graduated modulus	1P	Rainier Technology
Freedom	1P	Ti plates; silicone PU-PC core	1P	Axiomed
eDisc	1P	Ti plates; elastomer core	1P	Theken
Physio-L	1P	Ti plates; elastomer core	1P	NexGen Spine
M6-L	1P	Ti plates; PU-PC core with UHMWPE fiber encapsulation	1P	Spinal Kinetics
LP-ESP (elastic spine pad)	1P	Ti endplates; PU-PC coated silicone gel with microvoids	1P	FH Orthopedics

CoCr—Cobalt-chromium alloy. UHMWPE—Ultra-high molecular weight polyethylene. DLC—Diamond-like carbon. Ti—Titanium. PEEK—Polyether ether ketone. PU-PC—Polyurethane-polycarbonate elastomer.

Table 6. Common problems of different implant materials and their effects leading to failure [118].

Bearing Type	Material	Problems	Effects
Ball and Socket	CoCr	Reactive wear ions and fibrous particles	Metal sensitivity reactions, Inflammation, Osteolysis
		Metallosis	–
		No shock absorption	Compressive stresses on vertebral bodies
	UHMWPE	Large wear volume and wear debris	Bone resorption, Osteolysis
		Plastic deformation	–
		Increased range of motion (hypermobility)	Facet and ligament loading
		No shock absorption	Compressive stresses on vertebral bodies
	PEEK	Prosthesis migration	Biomechanical incompatibility, Stress on remaining annulus, Total rejection of device
		Endplate reaction	Severe biological rejection
1P	PUPC	More studies necessary	

Note: The effects stated are correlated to the problems directly next to it.

When using MoM devices, the degradation due to wear is minimal when compared to MoP devices and PEEK-on-PEEK devices (PoP), however the toxicity introduced to the body is relatively the same. Although the volume of wear particles might be smaller, the CoCr wear particles are chemically reactive within the body causing corrosion, tribocorrosion, and toxic and biological responses, such as metallosis, biological reactions, osteolysis, and inflammation. When MoP devices wear, the particles produced tend to be fine, needle and fiber-shaped particles which are less chemically reactive than the metal particles although bigger in size. The PoP devices shows properties of resisting expulsion of nucleus particles, and superior fatigue resistance and wear resistance, however severe biological reactions occur causing device rejection and migration of device into surrounding muscle tissue [118]. Each of these systems have their benefits and disadvantages when compared to each other, however when compared to spinal fusion surgery shows great advantages in the range of mobility. If a disc has undergone some degeneration, but is not yet to the point of spinal fusion or total disc replacement, other actions can be taken such as annulus fibrosus repair.

4.2.4. Repair of Annulus Fibrosus

The annulus fibrosus is involved in almost any pathological condition of the degenerating spine, therefore when its function becomes impaired, plays a fundamental role in two specific clinical situations. It acts as the main source of discogenic low back pain, and as the origin of disc herniation due to its insufficiency caused by degenerative disc disease. As previously discussed, when small fissures occur in the annulus, a repair process takes place in which granulation tissue is formed along with neovascularization and concomitant ingrowth of nerve fibers. This causes chronic discogenic pain throughout the disc due to the pressure being sustained by the nerves. Annulus fibrosus repair is the procedure to fix those tears before the disc herniates, and is usually performed in relatively young patients with very minor degenerative changes. Efficient annulus repair could significantly limit the need for future surgeries in certain cases in which there is potential of disc herniation, however no herniation has currently occurred. When the ruptures are treated, the focus is on improving cell-biomaterial interaction, using an initial implant to provide immediate closure of the tear and maintain mechanical properties of the disc, while the cellular component starts the regenerative process within the disc. This process, however, is not complete or satisfactory when it comes to being a permanent solution, but instead is a preventive measure for disc prolapse [120]. The most straight-forward solution is suturing the annulus tear shut, helping give the disc a stronger tendency to heal itself. However, its sole purpose is the containment of the nucleus pulposus and does not compensate for the loss of the annulus nor reverse the biomechanical changes [121]. One way to adjust for the lack of compensation could be the addition of growth factors in order to enhance the regenerative process of the annulus tissue [122]. Vadala et al. studied the potential of Transforming Growth Factor-β (TGF-β) loaded microfibrous poly(L-lactide) scaffold in vitro. The biological evaluation of the scaffolds was performed using bovine annulus fibrosus cells that were cultured on the scaffold for up to three weeks [122]. These electrospun scaffolds allowed for the closure of the defect site while releasing the TGF-β, inducing an anabolic stimulus on the annulus cells, mimicking the ECM environment of the tissue [122]. The scaffolds, together with the TGF-β release, promoted rapid cell growth compared to the control, resulting in the deposition of significantly greater amounts of GAGs and total collagen within the annulus tissue, as well as a higher neo-ECM thickness [122]. Another method studied by Cruz et al., focuses on the repair of annulus fibrosus defects through a cell-seeded adhesive biomaterial, further detailed in Section 5 [123].

Annulus fibrosus repair gives great advantages to those who have yet to have a full disc herniation, as well as those only experiencing minimal degeneration, giving them the opportunity to forgo the potential chance for surgery in the future. It should be noted however, that this treatment option is not a cure for degenerative disc disease, but a preventative measure taken to increase the longevity of the native disc, potentially permanently, depending on an individual's particular life style.

5. Tissue Engineering and Regeneration Strategies

With all of these options facing difficult challenges, tissue engineering and regenerative strategies stand out as potential solutions. These include some form of gene therapy, regeneration strategies via delivery of bioactive molecules, e.g., growth factors, or a material scaffold with or without cells. Gene therapy and regeneration with growth factors, cells, or enzymes, such as ADAMTS5, have been researched in rats for early stage trials, showing greater GAGs and total collagen deposition for the TGF-β1 treated animals, and successful suppression of the degradation of the nucleus pulposus for ADAMTS5 treated ones [122,124]. Along with these, a similar approach was studied by Guterl et al. to target caspase 3 in rabbit models, in order to disrupt the execution of apoptosis [125]. A direct injection of Alexa Fluor 555-caspase 3 small interfering RNA (siRNA) into the rabbit intervertebral disc was used to determine the effect on suppression of degenerative changes within the disc. Compared to the caspase 3 siRNA control, the Alexa Fluor 555-caspase 3 siRNA resulted in a significant decrease in serum-starved apoptotic cells, as well as a significant suppression of the degenerative changes to the disc [125].

In more current regeneration efforts, FDA-cleared Phase III adult stem cells were used in a test study to treat chronic lower back pain associated with degenerative disc disease. The use of mesenchymal precursor cells directly injected into the lumbar disc will hopefully show some ability to regenerate lost tissue of the disc [126]. Another system studied by Alini et al. uses implanted intervertebral disc cells in a scaffold of collagen and hyaluronan, or entrapped into a chitosan gel, with either fetal calf serum (FCS) or growth factors (TGF-β1, bFGF, and IGF-1) to modulate ECM synthesis [124]. The FCS and TGF-β1 were able to induce proteoglycan synthesis, while the presence of bFGF and IGF-1 reduced proteoglycan synthesis. However, the IGF-1 was shown to stimulate cell division by the greatest extent [124]. By day 20 of the culture, in FCS and the varying growth factors, not only did the matrices contain aggrecan, but also other small leucine-rich repeat proteoglycan found in the normal disc and both collagen type I and II [124]. Although all proteoglycans found in a normal disc were synthesized, the construct was not able to retain the majority of its proteoglycans, resulting in the inability to withstand the compressive loads normally subjected to an intervertebral disc [124]. Another future direction is to look more into the regeneration and repair of the annulus tissues as opposed to the nucleus tissues. Efforts for novel therapies have mainly been directed towards nucleus tissue regeneration and replacement, however a main challenge is the development of strategies and techniques that deal with the degenerated annulus, preferably in a combined approach with the nucleus [121]. A recent study performed by Cruz et al. shows the possibility to help repair damaged annulus fibrosus tissue through a cell-seeded adhesive biomaterial [123]. Multiple genipin-crosslinked fibrin adhesive cell carriers were developed with varying genipin to fibrin ratios, to determine the optimal composition for mimicking natural annulus fibrosus tissue. Among the adhesive cell carrier, were encapsulated bovine annulus fibrosus cells to show the feasibility of cell delivery to the injured tissue [123]. The cell-seeded adhesive demonstrated shear and compressive properties matching those of the annulus fibrosus tissue, while significantly improving failure strength in situ. As well, the adhesive showed increased cell viability and GAGs production [123]. These efforts could propel the future of annulus repair, offering successful preventative methods, as opposed to perpetuating the need for herniation surgeries and issues associated with them.

In the last few years, tissue engineered scaffolds for total intervertebral disc replacement have risen to the forefront of current biomaterial literature and research in order to address the challenges mentioned above [123,127–144]. Scaffolds have been fabricated using both natural and synthetic materials, as well as most containing embedded cells for natural tissue growth and integration [127]. Two recent studies by Iu et al. and Yang et al. focus on total intervertebral disc replacement using a hierarchically organized annulus fibrosus and a hydrogel-like nucleus pulposus, however differ with respect to materials, procedures, and objectives [129,130]. Iu et al. focuses on an in vitro generated intervertebral disc with the ability of tissue integration between the fabricated annulus and nucleus to better mimic the natural disc [129]. The annulus fibrosus was created using six lamellae comprised of aligned nanofibrous polycarbonate urethane scaffolds cultured with annulus fibrosus cells in a Dulbecco's modified Eagle's medium (DMEM) with 20% fetal bovine serum (FBS) for three weeks to produce an integrated type I collagen-rich ECM [129]. This surrounded the nucleus pulposus tissue comprised of a type II collagen- and aggrecan-rich ECM hydrogel which was cultured with nucleus pulposus cells in DMEM with 20% FBS solution for four weeks [129]. Both tissues were then combined and co-cultured to create the full intervertebral disc model with integration between the tissues [129]. It was shown that this system successfully integrated the annulus fibrosus lamellae not only to the nucleus tissue, but also to each other, allowing for interlamellar connectivity. When biologically and mechanically tested both in vitro and in vivo, the tissue engineered intervertebral disc showed no inflammatory reaction and was able to stand up to the interlamellar and annulus-nucleus interface shear forces experienced by the disc in the spine [129]. The in vitro studies demonstrated the possibility to create an intervertebral disc with mechanically stable tissue integration that was able to grow similar ECMs as the natural tissue. The in vivo studies demonstrated the ability of the engineered nucleus pulposus to form tissue in vivo, as well as test the disc's ability to develop intradiscal swelling pressure under load [129]. However, this experiment was evaluated using a bovine caudal spine rather than a human spine, resulting in a smaller intervertebral disc model, and would therefore need to pursue further research to evaluate the scalability and suitability of the system for human biological disc replacement [129]. Yang et al., on the other hand, focused on creating a total intervertebral disc replacement that can integrate natural tissue in vivo, and demonstrates excellent hydrophilicity and functional performance [130]. The hydrophilicity property of this scaffold is highly important for not only the swelling properties of the disc for mechanical stability, but also diffusion of nutrients through the disc for cell viability. The annulus fibrosus mimicry was fabricated using electrospun polycaprolactone/poly(D,L-lactide-*co*-glycolide)/collagen type I nanofibers to create a hierarchically organized, concentric ring-aligned structure, and the nucleus pulposus mimicry was fabricated using an alginate hydrogel [130]. Both components were cultured for 3 days in a DMEM/F12, 10% FBS, and 1% penicillin-streptomycin with a seeded cell density of 2500 cells/cm, and tested for biocompatibility and mechanical integrity before being implanted into rat caudal spine models [130]. In vivo, the replacement discs demonstrated excellent hydrophilicity, mimicking the highly hydrated native tissue, as well as shape maintenance, integration with surrounding natural tissue, acceptable mechanical support, and flexibility [130]. This study shows the potential of scaffold materials as intervertebral disc tissue engineering and regeneration platforms in vivo. However, like the previous study, the fabricated disc was smaller than that of a human, since it was synthesized for a rat model, resulting in the need for much lower mechanical properties for the materials. Therefore, further research would be needed to scale this approach to human trials [130].

Bhunia et al. studied a method for correcting degenerative disc disease that lies between annulus fibrosus repair and total intervertebral disc replacement [131]. Bhunia et al. focused on the recapitulation of form and function of the intervertebral disc through a silk protein-based multilayered, disc-like angle-ply annulus fibrosus scaffold comprising of multiple concentric lamellae. The scaffold was fabricated to resemble the hierarchical structure of the natural tissue, which was verified through electron microscopy [131]. These "biodiscs" demonstrated mechanical properties similar to those of the native tissue, as well as support of human mesenchymal stem cell proliferation and differentiation, and deposition of a sufficient amount of ECM after 14 days of culture. A section of the biodisc was implanted subcutaneously in a mice model and retrieved after one and four weeks of implantation, showing negligible immune response [131]. However, the proposed system lacked the ability to replace the entirety of the intervertebral disc, leading to the need for further research with the addition of an implantable nucleus. A recent study by Ghorbani et al. shows a promising method for nucleus pulposus replacement utilizing an injectable hydrogel [132]. The hydrogel was comprised of chitosan-β-glycerophosphate-hyaluronic acid, chondroitin-6-sulfate, type II collagen, gelatin, and fibroin silk, in order to replicate the complexity of the natural nucleus pulposus ECM. The synthesized nucleus demonstrated ideal hydrophilicity, stability, and strength when subjected to loads, with the storage modulus remaining nearly constant over a wide range of strain. In vitro tests were conducted using MTT and trypan blue to quantify and qualify cell growth and cytotoxicity, revealing the hydrogel to be cytocompatible with good cell attachment and growth [132]. Like the study performed by Bhunia et al., the solution only focuses on nucleus replacement/regeneration, therefore further research is needed with the combination of a tissue engineered annulus fibrosus.

As pointed out earlier, many of these studies use relatively weak electrospun scaffolds or combinations thereof with hydrogels that lead to mechanical properties that are on the range of rat native IVD but that are far from recapitulating those of human IVD. Recent studies have focused their attention on the development of sophisticated scaffolds and materials that can better mimic the outstanding mechanical properties of human IVD (Table 7). Novel materials and composites on the form of hydrogels have been investigated for the replacement of the nucleus pulposus. These include interpenetrating networks based on dextran, gelatin and poly (ethylene glycol) [133]; cross-linked collagen-II, aggrecan and hyaluronan [134]; and silk-fibrin and hyaluronic acid composite hydrogels [135], among others. For the annulus fibrosus scaffolds on the shape of fibrous matts or polymer films are generally preferred, mimicking the structure of the native tissue. To this end, novel materials have also been investigated such as nanocellulose reinforced gellan-gum hydrogels [136]; electrospun aligned polyurethane scaffolds or poly(trimethylene carbonate) structures prepared by lithography and covered with a polyester urethane membrane [137], among others [138]. However, these studies report on the fabrication of individual tissues rather than the recapitulation of the whole organ, what makes difficult the extrapolation of these results to a more complete and practicable approach.

Hu et al. recently reported on the fabrication of 3D printed scaffolds based on the combination of poly (lactic acid) (PLA) and gellan gum-poly (ethylene glycol) diacrylate (GG-PEGDA) double network hydrogel [139]. This combination allowed fine tuning of the mechanical properties of the overall organ by changing the infill patterns and the density of the PLA framework. Initial studies with in-situ bioprinted human mesenchymal stem cells (hMSCs) show good cell viability and spreading within the constructs. Although this first study shows an interesting 3D printing approach, the final scaffold represents a rather homogeneous construct and not the clearly compartmented native organ. Yang et al. overcame this issue by designing and fabricating a triphasic scaffolds that aimed at recapitulating the three main structures of native IVD, the nucleus pulposus and the inner and outer rings of the annulus fibrosus, while targeting the mechanical properties of human IVD [140]. The authors used a chitosan hydrogel to mimic the inner nucleus pulposus that was then surrounded by a poly(butylene succinate-co-terephthalate) (PBST) fiber film and a poly(ether ether ketone) (PEEK) ring to mimic the inner and outer annulus fibrosus, respectively. This multi-layered structure was seeded with porcine IVD cells and used on an in vivo porcine spine model. After 4 and 8 weeks of implantation the cell-scaffold construct retained its original height and showed a histological gross appearance that resembled that of the native tissue. Moreover, the compressive Young's modulus of the construct was 58.4 ± 12.9 MPa, similar to that measured for the native tissue (71.5 ± 18.2 MPa) [140]. Under a similar concept, Choy et al. studied the potential of biphasic scaffolds for full IVD tissue engineering. They developed a collagen and GAG hydrogel core that was encapsulated on a multiple lamella of photochemically cross-linked collagen membranes, mimicking the nucleus pulposus and the annulus fibrosus, respectively [141]. These constructs were capable of recovering up to 87% their original size after compression and showed a dynamic mechanical stiffness similar of that of the native rabbit IVD. Although this studied showed great promise in terms of mechanical properties and shape recovery of the constructs, a detailed biological study is still missing [141].

Other studies, such as those by Hudson et al., have focused more on the adaptation of tissue engineered intervertebral discs when exposed to certain environments and conditions [142,143]. In both studies, the intervertebral disc was fabricated by floating an injection molded alginate hydrogel nucleus pulposus in a collagen type I annulus fibrosus that contracted around the nucleus given ample time. This study showed that hypoxic expansion of human mesenchymal stem cells enhances the maturation of the tissue engineered disc, as opposed to normoxic environments [142]. Hypoxic conditions, which correlated to 1 to 5% oxygen content, resulted in an increase in ECM production, as well as driving chondrogenesis of the embedded stem cells, when compared to normoxic conditions (21% oxygen). Also, the hypoxic discs were stiffened up to 141%, and showed an increase in GAGs and collagen content within the nucleus, compared to normoxic [142]. The results obtained in this study show the benefit of hypoxic maturation of stem cells within the tissue engineered disc before implantation, however, to fully grasp the effectiveness of this scaffold, in vivo tests will need to be performed. Another study focused on the potential of dynamic unconfined compressive loading on the tissue regeneration/deposition rate [143]. Each tissue engineered disc was subjected to mechanical stimulation from a strain amplitude range of 1–10% for two weeks with a cycle of one hour on, one hour off, one hour on. The discs were then evaluated for biochemical and mechanical properties, which showed an increase in GAGs and hydroxyproline content, and equilibrium and instantaneous modulus for both the nucleus and annulus [143]. These results suggest that dynamic loading increases the functionality of the tissue engineered disc, with each section experiencing region dependent responses, which could be used to expedite maturation for implantation. Although promising, further research would need to be performed in vivo as well as on a larger scale models, bearing a closer resemblance to the natural intervertebral disc [143].

Altogether, these studies show the need of further development and study of materials and scaffolds fabrication techniques for the regeneration of full IVD. The outstanding mechanical properties and complexity of the multi-phasic structure of the native organ will require the development of also complex systems that can recapitulate these features.

Table 7. Materials, scaffold architecture, mechanical properties, and cell types used in tissue engineering approaches for IVD.

Targeted Tissue	Material	Structure	Mechanical Properties	Cells	Comments	Reference
Total IVD	AF: Poly caprolactone urethane NP: Collagen II and aggrecan	AF: Nanofibrous, aligned NP: Hydrogel	Compressive modulus of 17.2 ± 7.5 kPa	AF and NP cells	Integration between the two compartments. Tested in vitro and in vivo in a bovine model	[129]
Total IVD	AF: Polycaprolactone/poly(D,L-lactide-co-glycolide)/collagen type I NP: Alginate	AF: Electrospun nanofibers to create a concentric ring-structure NP: Hydrogel	Tensile Young's modulus of 380 MPa	Rat AF and NP cells	Integration with host tissue and between compartment in in vivo rat caudal spine model	[130]
AF	Silk	Concentric layers of lamella sheets on an angle-ply construct	499.18 ± 86.45 kPa	Porcine AF cells and human MSCs	Subcutaneous implantation in rat showed negligible immune response	[131]
NP	chitosan-β-glycerophosphate-hyaluronic acid, chondroitin-6-sulfate, type II collagen, gelatin, and fibroin silk	Hydrogel	≈50 Pa	Rabbit NP cells	Preliminary study with in vitro cell compatibility assays	[132]
IVD	PLA and GG_PEGDA	3D printed	Compressive Young's modulus of ≈400 MPa	hMSCs	Preliminary study on cell viability	[133]
IVD	NP: Chitosan; inner AF: PBST and outer AF: PEEK	NP: hydrogel and AF fiber film and ring	Compressive Young's modulus of 58.4 ± 12.9 MPa	Porcine IVD cells	In vivo implantation on a porcine spine model	[134]
NP	Dextran, gelatin and poly (ethylene glycol);	Hydrogel	Compressive Young's modulus of 15.86 ± 1.7 kPa	Porcine NP cells	In vivo subcutaneous implantation in Lewis rats	[135]
NP	Cross-linked collagen-II, aggrecan and hyaluronan	Hydrogel	Storage modulus of ≈1.25 kPa	Bovine NP cells	7 days in vitro cell culture studies	[136]
NP	Silk-fibrin and hyaluronic acid composite hydrogels	Hydrogel	Compressive modulus of ≈5–7 kPa	Human primary chondrocytes	Full in vitro study with up to 4 weeks cell culture	[137]
AF	Nanocellulose reinforced gellan-gum hydrogels	Hydrogel	Compressive modulus of ≈45–55 kPa	Bovine AF cells	Preliminary in vitro studies	[138]
AF	Electrospun aligned polyurethane scaffolds	Fibrous scaffold	N/A	Rabbit AF derived progenitor cells	7 days in vitro cultures	[139]
AF	poly(trimethylene carbonate) and polyester urethane	Fibrous scaffold	Yield strength of 4.9 ± 1.4 MPa	Human MSCs	In vitro bovine caudal spine organ culture model with or without dynamic load.	[140]

NP: nucleus pulposus and AF: annulus fibrosus.

6. Conclusions

Although great strides have been made in the field of degenerative disc disease, there is still a lot more progress to be made, given the challenges faced with every treatment option currently available. In early stages of degenerative disc disease, noninvasive treatments or treatments such as radiofrequency ablation and annulus fibrosus repair can be of great help, however they only mitigate the symptoms instead of the actual cause. Noninvasive treatments face the challenges of only dealing with some of the symptoms of the pain rather than dealing with the actual degeneration of the discs, therefore allowing the discs to continue to degrade to the future point of needing invasive treatments. Radiofrequency, although good for reducing pain, has the challenge of only lasting short term, a few months to a year in most cases. Also, it is an expensive procedure that has to be repeated every six months [109]. Annulus repair seems to be a better option for young adults with degeneration to the point just before herniation to significantly reduce the need for future surgery, but faces challenges of, again, only fixing the symptoms of the main problem as well as not being to mend any biological changes/losses within the annulus [120]. When disc degeneration gets even worse, greater procedures need to take place, such as spinal fusion surgery and TDR. Spinal fusion surgery is, as of today, the most common life-long solution to severe disc degeneration, however it is struck with multiple challenges such as significantly limiting mobility and adding additional stresses to the adjacent motion segments potentially causing greater degeneration in other intervertebral discs [113]. TDR has been shown to help retain the mobility that spinal fusion cannot, but can sometimes lead to hypermobility of the joint, can wear and corrode causing a biological reaction in the body, and more often than not, does not distribute load nor absorb shock, but rather transfers it directly into the adjacent vertebrae [118]. These challenges have led to vast research in the field of tissue engineering for disc degeneration. Even though scaffolds for disc regeneration are taking strides in the right direction, many still remain in a premature state. Each have their own benefits, but also complications, including scalability, tunability, tissue integration, or optimal mechanical properties. Therefore, the gap between the translation of this research to the clinic still remains fairly large with many hurdles to overcome, leading to the need for future research [127,144]. With degenerative disc disease posing such a large problem for individuals and society, and no current ideal treatment options that come without complications, there is a welcoming for future research in the field of tissue engineered biomaterials for the solution of total intervertebral disc replacement [128–144].

Author Contributions: The following contributions for this review are associated with the authors listed: Writing-Original Draft Preparation, B.F.; Writing-Review & Editing, S.C.E. and E.J.F.

Funding: This research received no external funding.

Acknowledgments: The authors would like to thank Priya Venkatraman for her comments and input on this review.

Conflicts of Interest: The authors declare no conflict of interest.

References

1. Vertebral Column. In *Encyclopaedia Britannica*; Britannica, T.E.o.E. (Ed.) Encyclopaedia Britannica, Inc.: Chicago, IL, USA, 2014.
2. Kibler, W.B.; Press, J.; Sciascia, A. The role of core stability in athletic function. *Sports Med.* **2006**, *36*, 189–198. [CrossRef] [PubMed]
3. Agur, A.M.R.; Dalley, A.F. *Grant's Atlas of Anatomy*, 12th ed.; Lipincott Williams & Wilkins: Pennsylvania, PA, USA, 2009; p. 841.
4. Vertebral Column. Available online: https://www.kenhub.com/en/start/c/vertebral-column (accessed on 14 April 2016).
5. Bogduk, N.; Mercer, S. Biomechanics of the cervical spine. I: Normal kinematics. *Clin. Biomech.* **2000**, *15*, 633–648. [CrossRef]

6.   Swartz, E.E.; Floyd, R.T.; Cendoma, M. Cervical spine functional anatomy and the biomechanics of injury due to compressive loading. *J. Athl. Train.* **2005**, *40*, 155–161. [PubMed]

7.   Panjabi, M.M.; Crisco, J.J.; Vasavada, A.; Oda, T.; Cholewicki, J.; Nibu, K.; Shin, E. Mechanical properties of the human cervical spine as shown by three-dimensional load-displacement curves. *Spine* **2001**, *26*, 2692–2700. [CrossRef]

8.   Caridi, J.M.; Pumberger, M.; Hughes, A.P. Cervical radiculopathy: A review. *HSS J.* **2011**, *7*, 265–272. [CrossRef] [PubMed]

9.   Yeung, J.T.; Johnson, J.I.; Karim, A.S. Cervical disc herniation presenting with neck pain and contralateral symptoms: A case report. *J. Med. Case Rep.* **2012**, *6*, 166. [CrossRef] [PubMed]

10. Edmondston, S.J.; Singer, K.P. Thoracic spine: Anatomical and biomechanical considerations for manual therapy. *Man. Ther.* **1997**, *2*, 132–143. [CrossRef]

11. Son, E.S.; Lee, S.H.; Park, S.Y.; Kim, K.T.; Kang, C.H.; Cho, S.W. Surgical treatment of t1-2 disc herniation with t1 radiculopathy: A case report with review of the literature. *Asian Spine J.* **2012**, *6*, 199–202. [CrossRef]

12. Goh, S.; Tan, C.; Price, R.I.; Edmondston, S.J.; Song, S.; Davis, S.; Singer, K.P. Influence of age and gender on thoracic vertebral body shape and disc degeneration: An MR investigation of 169 cases. *J. Anat.* **2000**, *197 Pt 4*, 647–657. [CrossRef]

13. Cervero, F.; Tattersall, J.E. Somatic and visceral sensory integration in the thoracic spinal cord. *Prog. Brain Res.* **1986**, *67*, 189–205.

14. Boszczyk, B.M.; Boszczyk, A.A.; Putz, R. Comparative and functional anatomy of the mammalian lumbar spine. *Anat. Rec.* **2001**, *264*, 157–168. [CrossRef]

15. Troup, J.D.; Hood, C.A.; Chapman, A.E. Measurements of the sagittal mobility of the lumbar spine and hips. *Ann. Phys. Med.* **1968**, *9*, 308–321. [CrossRef]

16. Haughton, V.M.; Rogers, B.; Meyerand, M.E.; Resnick, D.K. Measuring the axial rotation of lumbar vertebrae in vivo with MR imaging. *Am. J. Neuroradiol.* **2002**, *23*, 1110–1116.

17. Granhed, H.; Jonson, R.; Hansson, T. The loads on the lumbar spine during extreme weight lifting. *Spine* **1987**, *12*, 146–149. [CrossRef]

18. Tan, S.H.; Teo, E.C.; Chua, H.C. Quantitative three-dimensional anatomy of cervical, thoracic and lumbar vertebrae of Chinese Singaporeans. *Eur. Spine J.* **2004**, *13*, 137–146. [CrossRef]

19. Crawford, R.P.; Cann, C.E.; Keaveny, T.M. Finite element models predict in vitro vertebral body compressive strength better than quantitative computed tomography. *Bone* **2003**, *33*, 744–750. [CrossRef]

20. Shah, J.S.; Hampson, W.G.; Jayson, M.I. The distribution of surface strain in the cadaveric lumbar spine. *J. Bone Jt. Surg.* **1978**, *60*, 246–251. [CrossRef]

21. Bogduk, N. The innervation of the lumbar spine. *Spine* **1983**, *8*, 286–293. [CrossRef]

22. Luoma, K.; Riihimaki, H.; Luukkonen, R.; Raininko, R.; Viikari-Juntura, E.; Lamminen, A. Low back pain in relation to lumbar disc degeneration. *Spine* **2000**, *25*, 487–492. [CrossRef]

23. Nygaard, O.P.; Mellgren, S.I. The function of sensory nerve fibers in lumbar radiculopathy—Use of quantitative sensory testing in the exploration of different populations of nerve fibers and dermatomes. *Spine* **1998**, *23*, 348–352. [CrossRef]

24. Takahashi, I.; Kikuchi, S.; Sato, K.; Sato, N. Mechanical load of the lumbar spine during forward bending motion of the trunk—A biomechanical study. *Spine* **2006**, *31*, 18–23. [CrossRef]

25. Bogduk, N. *Clinical and Radiological Anatomy of the Lumbar Spine*, 5th ed.; Elsevier Health Sciences: Philadelphia, PA, USA, 2012.

26. Koes, B.W.; van Tulder, M.W.; Peul, W.C. Diagnosis and treatment of sciatica. *BMJ* **2007**, *334*, 1313–1317. [CrossRef]

27. Lirette, L.S.; Chaiban, G.; Tolba, R.; Eissa, H. Coccydynia: An overview of the anatomy, etiology, and treatment of coccyx pain. *Ochsner J.* **2014**, *14*, 84–87.

28. Humzah, M.D.; Soames, R.W. Human intervertebral disc: Structure and function. *Anat. Rec.* **1988**, *220*, 337–356. [CrossRef]

29. Pooni, J.S.; Hukins, D.W.; Harris, P.F.; Hilton, R.C.; Davies, K.E. Comparison of the structure of human intervertebral discs in the cervical, thoracic and lumbar regions of the spine. *Surg. Radiol. Anat.* **1986**, *8*, 175–182. [CrossRef]

30. Mahendra, K.A.; Rajani, J.A.; Shailendra, J.S.; Narsinh, H.G. Morphometric study of the cervical intervertebral disc. *Int. J. Anat. Phys. Biochem.* **2015**, *2*, 22–26.

31. Kunkel, M.E.; Herkommer, A.; Reinehr, M.; Bockers, T.M.; Wilke, H.J. Morphometric analysis of the relationships between intervertebral disc and vertebral body heights: An anatomical and radiographic study of the human thoracic spine. *J. Anat.* **2011**, *219*, 375–387. [CrossRef]

32. Shao, Z.; Rompe, G.; Schiltenwolf, M. Radiographic changes in the lumbar intervertebral discs and lumbar vertebrae with age. *Spine* **2002**, *27*, 263–268. [CrossRef]

33. Twomey, L.; Taylor, J. Age changes in lumbar intervertebral discs. *Acta Orthop. Scand.* **1985**, *56*, 496–499. [CrossRef]

34. Davis, H. Increasing Rates of Cervical and Lumbar Spine Surgery in the United-States, 1979–1990. *Spine* **1994**, *19*, 1117–1124. [CrossRef]

35. Williams, M.P.; Cherryman, G.R.; Husband, J.E. Significance of thoracic disc herniation demonstrated by MR imaging. *J. Comput. Assist. Tomogr.* **1989**, *13*, 211–214. [CrossRef] [PubMed]

36. Adams, M.A.; Roughley, P.J. What is intervertebral disc degeneration, and what causes it? *Spine* **2006**, *31*, 2151–2161. [CrossRef] [PubMed]

37. Raj P.P. Intervertebral disc: Anatomy-physiology-pathophysiology-treatment. *Pain Pract.* **2008**, *8*, 18–44. [CrossRef] [PubMed]

38. Adams, M.A. Intervertebral Disc Tissues. In *Mechanical Properties of Aging Soft Tissues*; Derby, B., Akhtar, R., Eds.; Springer International Publishing: Cham, Switzerland, 2015; pp. 7–35.

39. Galante, J.O. Tensile properties of the human lumbar annulus fibrosus. *Acta Orthop. Scand.* **1967**, *38* (Suppl. 100), 1–91. [CrossRef]

40. Guerin, H.L.; Elliott, D.M. Quantifying the contributions of structure to annulus fibrosus mechanical function using a nonlinear, anisotropic, hyperelastic model. *J. Orthop. Res.* **2007**, *25*, 508–516. [CrossRef] [PubMed]

41. Smith, L.J.; Fazzalari, N.L. The elastic fibre network of the human lumbar anulus fibrosus: Architecture, mechanical function and potential role in the progression of intervertebral disc degeneration. *Eur. Spine J.* **2009**, *18*, 439–448. [CrossRef]

42. Ricard-Blum, S. The Collagen Family. *CSH Perspect. Biol.* **2011**, *3*, a004978. [CrossRef]

43. Eyre, D.R.; Muir, H. Types I and II collagens in intervertebral disc. Interchanging radial distributions in annulus fibrosus. *Biochem. J.* **1976**, *157*, 267–270. [CrossRef]

44. Kielty, C.M.; Grant, M.E. The Collagen Family: Structure, Assembly, and Organization in the Extracellular Matrix. In *Connective Tissue and Its Heritable Disorders: Molecular, Genetic, and Medical Aspects*, 2nd ed.; Royce, P.M., Steinmann, B., Eds.; Wiley-Liss: Hoboken, NJ, USA, 2003; pp. 159–221.

45. Mouw, J.K.; Ou, G.; Weaver, V.M. Extracellular matrix assembly: A multiscale deconstruction. *Nat. Rev. Mol. Cell Biol.* **2014**, *15*, 771–785. [CrossRef]

46. Yanagishita, M. Function of proteoglycans in the extracellular matrix. *Pathol. Int.* **1993**, *43*, 283–293. [CrossRef]

47. Marchand, F.; Ahmed, A.M. Investigation of the laminate structure of lumbar disc anulus fibrosus. *Spine* **1990**, *15*, 402–410. [CrossRef] [PubMed]

48. Melrose, J.; Smith, S.M.; Appleyard, R.C.; Little, C.B. Aggrecan, versican and type VI collagen are components of annular translamellar crossbridges in the intervertebral disc. *Eur. Spine J.* **2008**, *17*, 314–324. [CrossRef] [PubMed]

49. Hickey, D.S.; Hukins, D.W.L. Relation between the Structure of the Annulus Fibrosus and the Function and Failure of the Intervertebral-Disk. *Spine* **1980**, *5*, 106–116. [CrossRef] [PubMed]

50. Green, T.P.; Adams, M.A.; Dolan, P. Tensile properties of the annulus fibrosus II. Ultimate tensile strength and fatigue life. *Eur. Spine J.* **1993**, *2*, 209–214. [CrossRef]

51. Yu, J.; Tirlapur, U.; Fairbank, J.; Handford, P.; Roberts, S.; Winlove, C.P.; Cui, Z.; Urban, J. Microfibrils, elastin fibres and collagen fibres in the human intervertebral disc and bovine tail disc. *J. Anat.* **2007**, *210*, 460–471. [CrossRef] [PubMed]

52. Nerurkar, N.L.; Elliott, D.M.; Mauck, R.L. Mechanical design criteria for intervertebral disc tissue engineering. *J. Biomech.* **2010**, *43*, 1017–1030. [CrossRef] [PubMed]

53. O'Connell, G.D.; Sen, S.; Elliott, D.M. Human annulus fibrosus material properties from biaxial testing and constitutive modeling are altered with degeneration. *Biomech. Model. Mechan.* **2012**, *11*, 493–503. [CrossRef]

54. Ambard, D.; Cherblanc, F. Mechanical behavior of annulus fibrosus: A microstructural model of fibers reorientation. *Ann. Biomed. Eng.* **2009**, *37*, 2256–2265. [CrossRef] [PubMed]

55. Best, B.A.; Guilak, F.; Setton, L.A.; Zhu, W.B.; Saednejad, F.; Ratcliffe, A.; Weidenbaum, M.; Mow, V.C. Compressive Mechanical-Properties of the Human Anulus Fibrosus and Their Relationship to Biochemical-Composition. *Spine* **1994**, *19*, 212–221. [CrossRef]

56. Perie, D.S.; MacLean, J.J.; Owen, J.P.; Iatridis, J.C. Correlating material properties with tissue composition in enzymatically digested bovine annulus fibrosus and nucleus pulposus tissue. *Ann. Biomed. Eng.* **2006**, *34*, 769–777. [CrossRef]

57. Iatridis, J.C.; Setton, L.A.; Weidenbaum, M.; Mow, V.C. Alterations in the mechanical behavior of the human lumbar nucleus pulposus with degeneration and aging. *J. Orthop. Res.* **1997**, *15*, 318–322. [CrossRef] [PubMed]

58. Trout, J.J.; Buckwalter, J.A.; Moore, K.C. Ultrastructure of the human intervertebral disc: II. Cells of the nucleus pulposus. *Anat. Rec.* **1982**, *204*, 307–314. [CrossRef] [PubMed]

59. Bonetti, M.I. Microfibrils: A cornerstone of extracellular matrix and a key to understand Marfan syndrome. *Ital. J. Anat. Embryol.* **2009**, *114*, 201–224.

60. Akkiraju, H.; Nohe, A. Role of Chondrocytes in Cartilage Formation, Progression of Osteoarthritis and Cartilage Regeneration. *J. Dev. Biol.* **2015**, *3*, 177–192. [CrossRef]

61. Muir, H. The chondrocyte, architect of cartilage. Biomechanics, structure, function and molecular biology of cartilage matrix macromolecules. *Bioessays* **1995**, *17*, 1039–1048. [CrossRef] [PubMed]

62. Calve, J.; Galland, M. The intervertebral nucleus pulposus—Its anatomy, its physiology, its pathology. *J. Bone Jt. Surg.* **1930**, *12*, 555–578.

63. Keyes, D.C.; Compere, E.L. The normal and pathological physiology of the nucleus pulposus of the intervertebral disc—An anatomical, clinical, and experimental study. *J. Bone Jt. Surg.* **1932**, *14*, 897–938.

64. Lotz, J.C.; Fields, A.J.; Liebenberg, E.C. The Role of the Vertebral End Plate in Low Back Pain. *Glob. Spine J.* **2013**, *3*, 153–163. [CrossRef]

65. Moore, R.J. The vertebral endplate: Disc degeneration, disc regeneration. *Eur. Spine J.* **2006**, *15*, S333–S337. [CrossRef]

66. Mwale, F.; Roughley, P.; Antoniou, J. Distinction between the extracellular matrix of the nucleus pulposus and hyaline cartilage: A requisite for tissue engineering of intervertebral disc. *Eur. Cell Mater.* **2004**, *8*, 58–63. [CrossRef]

67. Sophia Fox, A.J.; Bedi, A.; Rodeo, S.A. The basic science of articular cartilage: Structure, composition, and function. *Sports Health* **2009**, *1*, 461–468. [CrossRef] [PubMed]

68. Miller, E.J.; Rhodes, R.K. Preparation and characterization of the different types of collagen. *Methods Enzym.* **1982**, *82 Pt A*, 33–64.

69. Kuhn, K.; Schmid, T.M.; Linsenmayer, T.F.; Rest, M.; Mayne, R. *Structure and Function of Collagen Types*, 1st ed.; Mayne, R., Burgeson, R.E., Eds.; Academic Press: New York, NY, USA, 1987; pp. 1–281.

70. Grant, J.P.; Oxland, T.R.; Dvorak, M.F. Mapping the structural properties of the lumbosacral vertebral endplates. *Spine* **2001**, *26*, 889–896. [CrossRef] [PubMed]

71. Rodriguez, A.G.; Rodriguez-Soto, A.E.; Burghardt, A.J.; Berven, S.; Majumdar, S.; Lotz, J.C. Morphology of the human vertebral endplate. *J. Orthop. Res.* **2012**, *30*, 280–287. [CrossRef] [PubMed]

72. Herkowitz, H.N.; Spine, I.S.S.L. *The Lumbar Spine*; Lippincott Williams & Wilkins: Philadelphia, PA, USA, 2004; pp. 1–943.

73. Nekkanty, S.; Yerramshetty, J.; Kim, D.G.; Zauel, R.; Johnson, E.; Cody, D.D.; Yeni, Y.N. Stiffness of the endplate boundary layer and endplate surface topography are associated with brittleness of human whole vertebral bodies. *Bone* **2010**, *47*, 783–789. [CrossRef] [PubMed]

74. Rudert, M.; Tillmann, B. Lymph and Blood-Supply of the Human Intervertebral Disc—Cadaver Study of Correlations to Discitis. *Acta Orthop. Scand.* **1993**, *64*, 37–40. [CrossRef] [PubMed]

75. Nerlich, A.G.; Schaaf, R.; Walchli, B.; Boos, N. Temporo-spatial distribution of blood vessels in human lumbar intervertebral discs. *Eur. Spine J.* **2007**, *16*, 547–555. [CrossRef]

76. Bogduk, N.; Tynan, W.; Wilson, A.S. The Nerve Supply to the Human Lumbar Intervertebral Disks. *J. Anat.* **1981**, *132*, 39–56.
77. Edgar, M.A. The nerve supply of the lumbar intervertebral disc. *J. Bone Jt. Surg. Br.* **2007**, *89*, 1135–1139. [CrossRef]
78. Freemont, A.J.; Peacock, T.E.; Goupille, P.; Hoyland, J.A.; OBrien, J.; Jayson, M.I.V. Nerve ingrowth into diseased intervertebral disc in chronic back pain. *Lancet* **1997**, *350*, 178–181. [CrossRef]
79. Urban, J.P.G.; Roberts, S. Degeneration of the intervertebral disc. *Arthritis Res.* **2003**, *5*, 120–130. [CrossRef]
80. Gaskin, D.J.; Richard, P. *Relieving Pain in America: A Blueprint for Transforming Prevention, Care, Education, and Research*; National Academies Press: Washington, DC, USA, 2011.
81. Crow, W.T.; Willis, D.R. Estimating cost of care for patients with acute low back pain: A retrospective review of patient records. *J. Am. Osteopath. Assoc.* **2009**, *109*, 229–233. [PubMed]
82. Roberts, S.; Evans, H.; Trivedi, J.; Menage, J. Histology and pathology of the human intervertebral disc. *J. Bone Jt. Surg. Am.* **2006**, *88* (Suppl. 2), 10–14.
83. Battie, M.C.; Videman, T.; Levalahti, E.; Gill, K.; Kaprio, J. Genetic and environmental effects on disc degeneration by phenotype and spinal level: A multivariate twin study. *Spine* **2008**, *33*, 2801–2808. [CrossRef] [PubMed]
84. Buckwalter, J.A. Aging and degeneration of the human intervertebral disc. *Spine* **1995**, *20*, 1307–1314. [CrossRef] [PubMed]
85. Chan, D.; Song, Y.; Sham, P.; Cheung, K.M. Genetics of disc degeneration. *Eur. Spine J.* **2006**, *15* (Suppl. 3), S317–S325. [CrossRef]
86. Inoue, N.; Espinoza Orias, A.A. Biomechanics of intervertebral disk degeneration. *Orthop. Clin. North. Am.* **2011**, *42*, 487–499. [CrossRef]
87. Pfirrmann, C.W.A.; Metzdorf, A.; Zanetti, M.; Hodler, J.; Boos, N. Magnetic Resonance Classification of Lumbar Intervertebral Disc Degeneration. *Spine* **2001**, *26*, 1873–1878. [CrossRef]
88. Radek, M.; Pacholczyk-Sienicka, B.; Jankowski, S.; Albrecht, L.; Grodzka, M.; Depta, A.; Radek, A. Assessing the correlation between the degree of disc degeneration on the Pfirrmann scale and the metabolites identified in HR-MAS NMR spectroscopy. *Magn. Reson. Imaging* **2016**, *34*, 376–380. [CrossRef]
89. Sinusas, K. Osteoarthritis: Diagnosis and treatment. *Am. Fam. Physician* **2012**, *85*, 49–56.
90. Maetzel, A.; Li, L.C.; Pencharz, J.; Tomlinson, G.; Bombardier, C.; The Community Hypertension and Arthritis Project Study Team. The economic burden associated with osteoarthritis, rheumatoid arthritis, and hypertension: A comparative study. *Ann. Rheum. Dis.* **2004**, *63*, 395–401. [CrossRef]
91. Goode, A.P.; Carey, T.S.; Jordan, J.M. Low Back Pain and Lumbar Spine Osteoarthritis: How Are They Related? *Curr. Rheumatol. Rep.* **2013**, *15*, 305. [CrossRef] [PubMed]
92. Dunlop, R.B.; Adams, M.A.; Hutton, W.C. Disc space narrowing and the lumbar facet joints. *J. Bone Jt. Surg. Br.* **1984**, *66*, 706–710. [CrossRef]
93. Fujiwara, A.; Tamai, K.; An, H.S.; Kurihashi, A.; Lim, T.H.; Yoshida, H.; Saotome, K. The relationship between disc degeneration, facet joint osteoarthritis, and stability of the degenerative lumbar spine. *J. Spinal Disord.* **2000**, *13*, 444–450. [CrossRef] [PubMed]
94. Fujiwara, A.; Lim, T.H.; An, H.S.; Tanaka, N.; Jeon, C.H.; Andersson, G.B.J.; Haughton, V.M. The effect of disc degeneration and facet joint osteoarthritis on the segmental flexibility of the lumbar spine. *Spine* **2000**, *25*, 3036–3044. [CrossRef] [PubMed]
95. Horkoff, M.; Maloon, S. Dysphagia secondary to esophageal compression by cervical osteophytes: A case report. *BCMJ* **2014**, *56*, 442–444.
96. Milette, P.C.; Fontaine, S.; Lepanto, L.; Cardinal, E.; Breton, G. Differentiating lumbar disc protrusions, disc bulges, and discs with normal contour but abnormal signal intensity. Magnetic resonance imaging with discographic correlations. *Spine* **1999**, *24*, 44–53. [CrossRef] [PubMed]
97. Rim, D.C. Quantitative Pfirrmann Disc Degeneration Grading System to Overcome the Limitation of Pfirrmann Disc Degeneration Grade. *Korean J. Spine* **2016**, *13*, 1–8. [CrossRef]
98. Adams, M.A.; Hutton, W.C. Gradual disc prolapse. *Spine* **1985**, *10*, 524–531. [CrossRef] [PubMed]
99. Kortelainen, P.; Puranen, J.; Koivisto, E.; Lahde, S. Symptoms and Signs of Sciatica and Their Relation to the Localization of the Lumbar-Disk Herniation. *Spine* **1985**, *10*, 88–92. [CrossRef]
100. Frymoyer, J.W. Back Pain and Sciatica. *N. Engl. J. Med.* **1988**, *318*, 291–300. [CrossRef] [PubMed]

101. Taher, F.; Essig, D.; Lebl, D.R.; Hughes, A.P.; Sama, A.A.; Cammisa, F.P.; Girardi, F.P. Lumbar Degenerative Disc Disease: Current and Future Concepts of Diagnosis and Management. *Adv. Orthop.* **2012**, *2012*, 970752. [CrossRef] [PubMed]

102. Physical Therapist's Guide to Degenerative Disc Disease. Available online: http://www.moveforwardpt.com/symptomsconditionsdetail.aspx?cid=514086b4-1272-4584-8742-ec6d2aa8f8cb (accessed on 8 March 2017).

103. Nwuga, V.C. Ultrasound in treatment of back pain resulting from prolapsed intervertebral disc. *Arch. Phys. Med. Rehabil.* **1983**, *64*, 88–89. [PubMed]

104. Adams, M.A.; Stefanakis, M.; Dolan, P. Healing of a painful intervertebral disc should not be confused with reversing disc degeneration: Implications for physical therapies for discogenic back pain. *Clin. Biomech.* **2010**, *25*, 961–971. [CrossRef] [PubMed]

105. Will Steroid Injections Help My Degenerative Disc Disease? Available online: http://www.arksurgicalhospital.com/will-steroid-injections-help-my-degenerative-disc-disease/ (accessed on 8 March 2017).

106. Buttermann, G.R. The effect of spinal steroid injections for degenerative disc disease. *Spine J.* **2004**, *4*, 495–505. [CrossRef] [PubMed]

107. Chou, R.; Huffman, L.H. Medications for acute and chronic low back pain: A review of the evidence for an American Pain Society/American College of Physicians clinical practice guideline. *Ann. Int. Med.* **2007**, *147*, 505–514. [CrossRef] [PubMed]

108. Drugs, Medications, and Spinal Injections for Degenerative Disc Disease. Available online: https://www.spineuniverse.com/conditions/degenerative-disc/drugs-medications-spinal-injections-degenerative-disc-disease (accessed on 8 March 2017).

109. Sluijter, M.E.; Cosman, E.R. Method and Apparatus for Heating an Intervertebral Disc for Relief of Back Pain. U.S. Patent 5433739A, 18 July 1995.

110. Sluijter, M.E.; Cosman, E.R. Thermal Denervation of an Intervertebral Disc for Relief of Back Pain. U.S. Patent 5571147A, 5 November 1996.

111. Sulaiman, S.B.; Keong, T.K.; Cheng, C.H.; Saim, A.B.; Idrus, R.B. Tricalcium phosphate/hydroxyapatite (TCP-HA) bone scaffold as potential candidate for the formation of tissue engineered bone. *Indian J. Med. Res.* **2013**, *137*, 1093–1101.

112. Spivak, J.M.; Hasharoni, A. Use of hydroxyapatite in spine surgery. *Eur. Spine J.* **2001**, *10* (Suppl. 2), S197–S204. [CrossRef]

113. Nouh, M.R. Spinal fusion-hardware construct: Basic concepts and imaging review. *World J. Radiol.* **2012**, *4*, 193–207. [CrossRef] [PubMed]

114. Confusion about Spinal Fusion. Available online: https://www.spineuniverse.com/treatments/surgery/lumbar/confusion-about-spinal-fusion (accessed on 22 March 2017).

115. Multilevel Spinal Fusion for Low Back Pain. Available online: https://www.spine-health.com/treatment/spinal-fusion/multilevel-spinal-fusion-low-back-pain (accessed on 21 March 2017).

116. Quirno, M.; Goldstein, J.A.; Bendo, J.A.; Kim, Y.; Spivak, J.M. The Incidence of Potential Candidates for Total Disc Replacement among Lumbar and Cervical Fusion Patient Populations. *Asian Spine J.* **2011**, *5*, 213–219. [CrossRef] [PubMed]

117. Deyo, R.A.; Nachemson, A.; Mirza, S.K. Spinal-fusion surgery—The case for restraint. *N. Engl. J. Med.* **2004**, *350*, 722–726. [CrossRef] [PubMed]

118. Reeks, J.; Liang, H. Materials and Their Failure Mechanisms in Total Disc Replacement. *Lubricants* **2015**, *3*, 346–364. [CrossRef]

119. Serhan, H.; Mhatre, D.; Defossez, H.; Bono, C.M. Motion-preserving technologies for degenerative lumbar spine: The past, present, and future horizons. *Int. J. Spine Surg.* **2011**, *5*, 75–89. [CrossRef] [PubMed]

120. Guterl, C.C.; See, E.Y.; Blanquer, S.B.; Pandit, A.; Ferguson, S.J.; Benneker, L.M.; Grijpma, D.W.; Sakai, D.; Eglin, D.; Alini, M.; et al. Challenges and strategies in the repair of ruptured annulus fibrosus. *Eur. Cell Mater.* **2013**, *25*, 1–21. [CrossRef]

121. Bron, J.L.; Helder, M.N.; Meisel, H.J.; Van Royen, B.J.; Smit, T.H. Repair, regenerative and supportive therapies of the annulus fibrosus: Achievements and challenges. *Eur. Spine J.* **2009**, *18*, 301–313. [CrossRef]

122. Vadala, G.; Mozetic, P.; Rainer, A.; Centola, M.; Loppini, M.; Trombetta, M.; Denaro, V. Bioactive electrospun scaffold for annulus fibrosus repair and regeneration. *Eur. Spine J.* **2012**, *21* (Suppl. 1), S20–S26. [CrossRef]

123. Cruz, M.A.; Hom, W.W.; DiStefano, T.J.; Merrill, R.; Torre, O.M.; Lin, H.A.; Hecht, A.C.; Illien-Junger, S.; Iatridis, J.C. Cell-Seeded Adhesive Biomaterial for Repair of Annulus Fibrosus Defects in Intervertebral Discs. *Tissue Eng. Part A* **2018**, *24*, 187–198. [CrossRef]

124. Alini, M.; Roughley, P.J.; Antoniou, J.; Stoll, T.; Aebi, M. A biological approach to treating disc degeneration: Not for today, but maybe for tomorrow. *Eur. Spine J.* **2002**, *11* (Suppl. 2), S215–S220.

125. Sudo, H.; Minami, A. Caspase 3 as a therapeutic target for regulation of intervertebral disc degeneration in rabbits. *Arthritis Rheum.* **2011**, *63*, 1648–1657. [CrossRef] [PubMed]

126. Sakai, D.; Mochida, J.; Iwashina, T.; Hiyama, A.; Omi, H.; Imai, M.; Nakai, T.; Ando, K.; Hotta, T. Regenerative effects of transplanting mesenchymal stem cells embedded in atelocollagen to the degenerated intervertebral disc. *Biomaterials* **2006**, *27*, 335–345. [CrossRef] [PubMed]

127. D'Este, M.; Eglin, D.; Alini, M. Lessons to be learned and future directions for intervertebral disc biomaterials. *Acta Biomater.* **2018**, *78*, 13–22. [CrossRef] [PubMed]

128. Bowles, R.D.; Setton, L.A. Biomaterials for intervertebral disc regeneration and repair. *Biomaterials* **2017**, *129*, 54–67. [CrossRef] [PubMed]

129. Iu, J.; Massicotte, E.; Li, S.Q.; Hurtig, M.B.; Toyserkani, E.; Santerre, J.P.; Kandel, R.A. In Vitro Generated Intervertebral Discs: Toward Engineering Tissue Integration. *Tissue Eng. Part A* **2017**, *23*, 1001–1010 [CrossRef] [PubMed]

130. Yang, J.C.; Yang, X.L.; Wang, L.; Zhang, W.; Yu, W.B.; Wang, N.X.; Peng, B.A.; Zheng, W.F.; Yang, G.; Jiang, X.Y. Biomimetic nanofibers can construct effective tissue-engineered intervertebral discs for therapeutic implantation. *Nanoscale* **2017**, *9*, 13095–13103. [CrossRef]

131. Bhunia, B.K.; Kaplan, D.L.; Mandal, B.B. Silk-based multilayered angle-ply annulus fibrosus construct to recapitulate form and function of the intervertebral disc. *Proc. Natl. Acad. Sci. USA* **2018**, *115*, 477–482. [CrossRef]

132. Ghorbani, M.; Ai, J.; Nourani, M.R.; Azami, M.; Beni, B.H.; Asadpour, S.; Bordbar, S. Injectable natural polymer compound for tissue engineering of intervertebral disc: In vitro study. *Mater. Sci. Eng. C Mater.* **2017**, *80*, 502–508. [CrossRef]

133. Gan, Y.; Li, P.; Wang, L.; Mo, X.; Song, L.; Xu, Y.; Zhao, C.; Ouyang, B.; Tu, B.; Luo, L.; et al. An interpenetrating network-strengthened and toughened hydrogel that supports cell-based nucleus pulposus regeneration. *Biomaterials* **2017**, *136*, 12–28. [CrossRef]

134. Halloran, D.O.; Grad, S.; Stoddart, M.; Dockery, P.; Alini, M.; Pandit, A.S. An injectable cross-linked scaffold for nucleus pulposus regeneration. *Biomaterials* **2008**, *29*, 438–447. [CrossRef]

135. Park, S.-H.; Cho, H.; Gil, E.S.; Mandal, B.B.; Min, B.-H.; Kaplan, D.L. Silk-Fibrin/Hyaluronic Acid Composite Gels for Nucleus Pulposus Tissue Regeneration. *Tissue Eng. Part A* **2011**, *17*, 2999–3009. [CrossRef] [PubMed]

136. Pereira, D.R.; Silva-Correia, J.; Oliveira, J.M.; Reis, R.L.; Pandit, A.; Biggs, M.J. Nanocellulose reinforced gellan-gum hydrogels as potential biological substitutes for annulus fibrosus tissue regeneration. *Nanomed. Nanotechol.* **2018**, *14*, 897–908. [CrossRef] [PubMed]

137. Liu, C.; Zhu, C.; Li, J.; Zhou, P.; Chen, M.; Yang, H.; Li, B. The effect of the fibre orientation of electrospun scaffolds on the matrix production of rabbit annulus fibrosus-derived stem cells. *Bone Res.* **2015**, *3*, 15012. [CrossRef] [PubMed]

138. Pirvu, T.; Blanquer, S.B.G.; Benneker, L.M.; Grijpma, D.W.; Richards, R.G.; Alini, M.; Eglin, D.; Grad, S.; Li, Z. A combined biomaterial and cellular approach for annulus fibrosus rupture repair. *Biomaterials* **2015**, *42*, 11–19. [CrossRef] [PubMed]

139. Hu, D.; Wu, D.; Huang, L.; Jiao, Y.; Li, L.; Lu, L.; Zhou, C. 3D bioprinting of cell-laden scaffolds for intervertebral disc regeneration. *Mater. Lett.* **2018**, *223*, 219–222. [CrossRef]

140. Yang, F.; Xiao, D.; Zhao, Q.; Chen, Z.; Liu, K.; Chen, S.; Sun, X.; Yue, Q.; Zhang, R.; Feng, G. Fabrication of a novel whole tissue-engineered intervertebral disc for intervertebral disc regeneration in the porcine lumbar spine. *RSC Adv.* **2018**, *8*, 39013–39021. [CrossRef]

141. Choy, A.T.H.; Chan, B.P. A Structurally and Functionally Biomimetic Biphasic Scaffold for Intervertebral Disc Tissue Engineering. *PLoS One* **2015**, *10*, e0131827. [CrossRef]

142. Hudson, K.D.; Bonassar, L.J. Hypoxic Expansion of Human Mesenchymal Stem Cells Enhances Three-Dimensional Maturation of Tissue-Engineered Intervertebral Discs. *Tissue Eng. Part A* **2017**, *23*, 293–300. [CrossRef]

143. Hudson, K.D.; Mozia, R.I.; Bonassar, L.J. Dose-Dependent Response of Tissue-Engineered Intervertebral Discs to Dynamic Unconfined Compressive Loading. *Tissue Eng. Part A* **2015**, *21*, 564–572. [CrossRef]
144. Buckley, C.T.; Hoyland, J.A.; Fujii, K.; Pandit, A.; Iatridis, J.C.; Grad, S. Critical aspects and challenges for intervertebral disc repair and regeneration—Harnessing advances in tissue engineering. *JOR Spine* **2018**, *1*, e1029. [CrossRef]

materials

MDPI

Review

Developing a New Generation of Therapeutic Dental Polymers to Inhibit Oral Biofilms and Protect Teeth

Ke Zhang [1,2], Bashayer Baras [2], Christopher D. Lynch [3], Michael D. Weir [2], Mary Anne S. Melo [2], Yuncong Li [4], Mark A. Reynolds [2], Yuxing Bai [1,*], Lin Wang [2,5,*], Suping Wang [2,6,*] and Hockin H. K. Xu [2,7,8]

1   Department of Orthodontics, School of Stomatology, Capital Medical University, Beijing 100069, China; tuzizhangke@163.com
2   Department of Advanced Oral Sciences and Therapeutics, University of Maryland Dental School, Baltimore, MD 21201, USA; bbaras@umaryland.edu (B.B.); mweir@umaryland.edu (M.D.W.); mmelo@umaryland.edu (M.A.S.M.); mreynolds@umaryland.edu (M.A.R.); hxu2@umaryland.edu (H.H.K.X.)
3   Restorative Dentistry, University Dental School and Hospital, University College Cork, Wilton T12 E8YV, Ireland; chris.lynch@ucc.ie
4   Clinical Research Center of Shaanxi Province for Dental and Maxillofacial Diseases, Key Laboratory of Shaanxi Province for Craniofacial Precision Medicine Research, College of Stomatology, Xi'an Jiaotong University, Xi'an 710004, China; yuncong1982@mail.xjtu.edu.cn
5   Department of Oral Implantology, School of Dentistry, Jilin University, Changchun 130012, China
6   Department of Operative Dentistry and Endodontics & Stomatology Center, The First Affiliated Medical School of Zhengzhou University, Zhengzhou 450052, China
7   Center for Stem Cell Biology & Regenerative Medicine, University of Maryland School of Medicine, Baltimore, MD 21201, USA
8   University of Maryland Marlene and Stewart Greenebaum Cancer Center, University of Maryland School of Medicine, Baltimore, MD 21201, USA
*   Correspondence: byuxing@263.net (Y.B.); dentistwanglin@126.com (L.W.); wangsupingdent@163.com (S.W.); Tel.: +86-10-5709-9004 (Y.B.); +86-431-8879-6025 (L.W.); +86-371-6691-3114 (S.W.)

Received: 29 August 2018; Accepted: 14 September 2018; Published: 17 September 2018

Abstract: Polymeric tooth-colored restorations are increasingly popular in dentistry. However, restoration failures remain a major challenge, and more than 50% of all operative work was devoted to removing and replacing the failed restorations. This is a heavy burden, with the expense for restoring dental cavities in the U.S. exceeding $46 billion annually. In addition, the need is increasing dramatically as the population ages with increasing tooth retention in seniors. Traditional materials for cavity restorations are usually bioinert and replace the decayed tooth volumes. This article reviews cutting-edge research on the synthesis and evaluation of a new generation of bioactive dental polymers that not only restore the decayed tooth structures, but also have therapeutic functions. These materials include polymeric composites and bonding agents for tooth cavity restorations that inhibit saliva-based microcosm biofilms, bioactive resins for tooth root caries treatments, polymers that can suppress periodontal pathogens, and root canal sealers that can kill endodontic biofilms. These novel compositions substantially inhibit biofilm growth, greatly reduce acid production and polysaccharide synthesis of biofilms, and reduce biofilm colony-forming units by three to four orders of magnitude. This new class of bioactive and therapeutic polymeric materials is promising to inhibit tooth decay, suppress recurrent caries, control oral biofilms and acid production, protect the periodontium, and heal endodontic infections.

Keywords: polymeric composites; bonding agents; antibacterial; oral biofilms; periodontal pathogens; caries inhibition

1. Introduction

Tooth caries is a widespread problem in the world. More than half of all dental restorations fail within 10 years, and recurrent (secondary) caries is a main reason for failures [1–3]. Replacing the failed restorations accounts for 50–70% of all tooth cavity restorations performed [4]. This represents a large economic burden; for example, the annual expense for restoring tooth cavities in the U.S. was $46 billion in 2005 [5]. In addition, the expense is rapidly climbing because of an aging population with longer life expectancies, and seniors are retaining more and more of their natural teeth [6]. Tooth-colored polymeric composites and bonding agents are the primary materials for restoring tooth cavities [3,7–12]. This is because advances in polymer chemistry and filler particle compositions have enhanced the composite restoration properties [13–18]. However, one key disadvantage is that polymeric composite materials tend to accumulate more oral biofilms than other dental materials such as metals and ceramics [19]. Oral biofilms ferment carbohydrates and produce acids that can lead to dental caries [20,21]. Therefore, researchers have devoted effort to synthesizing new antibacterial polymers for dental applications [22–27]. In general, antibacterial dental resins and composites can be divided into two classes. Class 1 uses polymerizable quaternary ammonium methacrylates (QAMs) where the antibacterial agent is bonded to be part of the polymer network. Class 2 uses filler particles with antibacterial activities that are filled into the polymer matrix. For class 1, QAMs were developed and incorporated into dental polymeric materials [22,23]. The first such material, 12-methacryloyloxydodecyl-pyridinium bromide (MDPB), was copolymerized with dental polymers and provided potent anti-biofilm effects [22,23]. Since then, other antibacterial resins were also synthesized and exhibited the capability to hinder bacterial growth and biofilm formation [25,26,28–36]. For Class 2, antibacterial fillers such as silver, zinc oxide and bioglass particles were mixed into polymer matrices, in which the antibacterial effect was achieved by the release of the agents [37–43]. While some studies reported sustainable long-term release of ions to exert antibacterial effects [44], other studies showed that the release and antibacterial efficacy decreased with increasing time [42,43]. Controlled long-term release of antibacterial agents has great potential for dental applications to combat caries and oral pathogens, especially via the use of nanotechnology and recharge and re-release mechanisms. This article focuses on Class 1 and reviews innovative developments in QAM-containing dental polymers and their exciting potential in restorative, preventive, root caries, periodontal, and endodontic applications.

2. Antibacterial Polymeric Dental Composites

Novel antibacterial polymeric composites were synthesized with functions to reduce oral biofilm acids and dental caries formation [22,23]. Antibacterial monomer MDPB was copolymerized into a resin composite which substantially reduced the glucan synthesis by *Streptococcus mutans* (*S. mutans*), a major cariogenic species, on the composite surface [45,46]. This was achieved without negatively influencing the composite's mechanical properties and degree of polymerization conversion. A separate study synthesized polymeric composites with antibacterial and fluoride-releasing properties, which caused a large decrease in *S. mutans* biofilm formation [26]. Another study synthesized novel nanoparticles of quaternary ammonium polyethylenimine (QPEI) and incorporated them into a polymeric composite [47]. The QPEI composite resulted in a strong anti-biofilm activity in human participants in vivo against oral salivary bacteria [47]. In another study, researchers developed a furanone-containing composite with antibacterial functions, achieving a 16%–68% decrease in the viability of *S. mutans* grown on the composite surface [48].

Recently, a new class of QAMs with the alkyl chain length (CL) from 3 to 18 were developed and mixed into dental polymers to develop composites [36]. The QAMs were developed using a Menschutkin method in which a tertiary amine had reaction with an organo-halide [25,49]. Five QAMs with different CL values of 3 to 18 were produced. To fabricate a composite, the model polymer matrix was made of bisphenol A glycidyl dimethacrylate (BisGMA) and triethylene glycol dimethacrylate (TEGDMA) (Esstech, Essington, PA) which were mixed at 1:1 by weight, although the method was also applicable to other polymer matrices as well. To render the BisGMA-TEGDMA resin light-curable, camphorquinone (0.2%) and ethyl 4-*N*,*N*-dimethylaminobenzoate (0.8%) were added. This polymer matrix was denoted BT. To develop the composite, a filler level of 50% mass fraction of silanated glass filler particles (barium boroaluminosilicate glass, median particle size = 1.4 µm, Caulk/Dentsply, Milford, DE, USA) were incorporated for improving the mechanical properties to enable the composite in load-bearing restorations [36]. In addition, nanoparticles of amorphous calcium phosphate (NACP) were also mixed into the composite at a 20% mass fraction for the releases of calcium and phosphate ions and remineralization properties. Each QAM with each CL was incorporated into the composite at a 3% by weight [36]. The flexural strength and elastic modulus of the composite indicated that adding 3% QAM did not negatively compromise the mechanical properties (Figure 1A). All the QAM composites possessed mechanical properties similar to those of the composite without QAM and a commercial control composite without antibacterial properties [36].

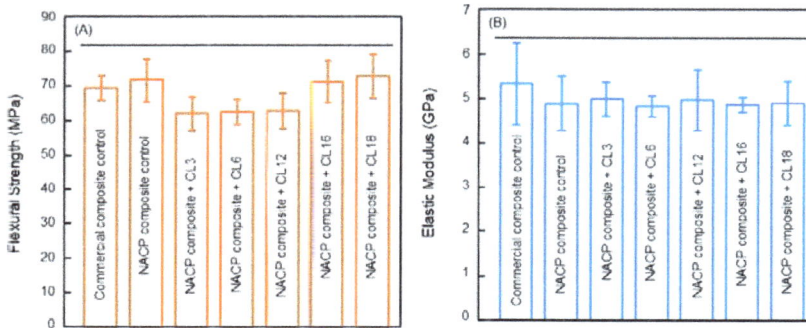

Figure 1. Mechanical properties of dental polymeric composites: (**A**) flexural strength, and (**B**) elastic modulus (mean ± SD; *n* = 6). Adding QAMs with amine alkyl chain length (CL) from 3 to 18 produced no significant loss in strength and elastic modulus. All QAM composites had mechanical properties similar to control composites without QAM. Horizontal line indicates *p* > 0.1. Adapted from [36], with permission from © 2015 Springer Nature.

To test the antibacterial properties, saliva from human donors was used as an inoculum to obtain oral biofilms consisting of organisms from the mouth. This enabled the use of a dental plaque microcosm biofilm model [36]. Live/dead staining assay of two-day biofilms grown on the composite surface showed that increasing the CL of the QAM in the polymeric composite strengthened the antibacterial potency, which was the greatest at CL16. Raising the CL further to 18 reduced the antibacterial activity, compared to that of CL16. This was consistent with the lactic acid results from the biofilms on the surfaces of the composites (Figure 2) [36]. The two-day microcosm biofilms grown on the two control composite surfaces yielded the greatest amounts of lactic acid. Raising the CL from 3 to 16 substantially reduced the lactic acid production, reaching the minimum acid at CL16. Therefore, CL16 appeared to possess the strongest antibacterial activity among the groups tested. For the composite with CL16, the acid production of the adherent biofilms was reduced by an order of magnitude when compared with control composites. This acid reduction could contribute to reducing tooth mineral dissolution and caries occurrence [36].

Regarding the antibacterial mechanism, the QAM-incorporated polymer composite had quaternary amine N$^+$ with positive charges which could interact with the cell membrane of the bacteria having negative charges. This could disrupt the membrane and cause cytoplasmic leakage, leading to bacterial death [30]. Other possible antibacterial mechanisms include preventing material transports across the bacterial cell membrane, interfering with signaling pathways or adhesive molecules at the bacterial wall, etc. It was suggested that quaternary ammonium materials with relatively long chains would be particularly effective with insertion into the bacterial membrane, thus inducing physical disruption to compromise the bacteria [22,23,30]. Indeed, a previous report on antibacterial glass ionomer materials showed greater antibacterial potency by using longer chain lengths [50]. Another study on three-dimensional biofilms also demonstrated that the oral biofilm thickness and the mass of biofilms were substantially reduced when the alkyl chain was raised from 3 to 16 [51]. These findings are in agreement with Figure 2 showing an increasing antibacterial potency for composites with increasing CL from 3 to 16, with CL16 being the most potent [36]. However, when CL was further increased to 18, the anti-biofilm potency was reduced. A possible explanation may be that when the alkyl chain becomes excessively long, the alkyl chain may be bent or curled. This would then contribute to the partial covering of the positively-charged quaternary ammonium groups, thus to some extent blocking the interactions electrostatically with the bacterial cells, and yielding a decrease in the anti-biofilm efficacy [34,36]. Another possible reason is that increasing the chain length leads to a larger thermal fluctuation amplitude that reduces the probability of these molecules penetrating into the outer bacterial membrane. Further study is needed to determine and understand the relationship between the quaternary amine chain length and the antibacterial potency. Meanwhile, tailoring and tuning of the polymeric compositions are needed to optimize the anti-biofilm, acid reduction, and mechanical and physical properties of dental composites.

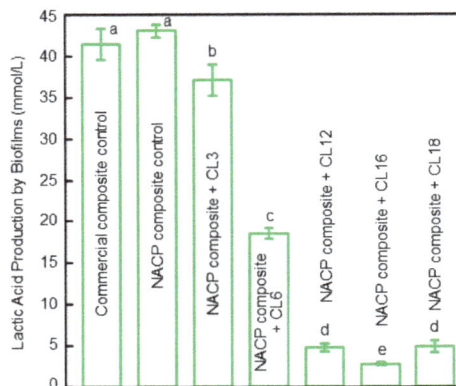

Figure 2. Lactic acid production by two-day dental plaque microcosm biofilms on the composites vs. QAM amine alkyl chain length (CL) (mean $\pm$ SD; $n = 6$). The polymeric composite using CL16 had the strongest anti-biofilm activity. Values indicated by dissimilar letters are statistically significantly different from each other ($p < 0.05$). Adapted from [36], with permission from © 2015 Springer Nature.

3. Antibacterial Dental Bonding Agents

Bonding agents are used clinically to adhere the restoration to enamel and dentin, enabling the restoration to sustain repeated chewing forces in the oral environment without detachment. However, the weakest link of the restoration is the bonded composite-tooth interface, and its failure is the primary reason for the failure of the entire restoration. Therefore, extensive efforts were made to enhance the dentin bond strength and investigate the mechanisms of the tooth-restoration bond [7,52]. Studies indicated that it would be advantageous for the bonding agent to be antibacterial in order to suppress biofilm acids and avoid caries formation at the tooth–composite margins [22,23,28,29,31]. Studies suggested that antibacterial adhesives could help eradicate the residual bacteria inside the tooth cavity, as well kill the invading bacteria due to marginal leakage, which otherwise would allow the oral bacteria to invade into the tooth-restoration margins [28,29]. Indeed, previous studies demonstrated that dental adhesives with MDPB incorporation were able to kill *S. mutans* growth [23,53]. In addition, methacryloxyl ethyl cetyl dimethyl ammonium chloride (DMAE-CB) was synthesized and incorporated into adhesive to inhibit bacterial growth [54]. Furthermore, antibacterial primer containing MDPB was also developed, which demonstrated strong antibacterial functions [53].

In addition to MDPB, chlorhexidine (CHX) particles were mixed into a primer to obtain antibacterial properties [55]. Besides modifying commercial bonding agents with antibacterial agents, novel experimental bonding agents with antibacterial functions were also developed.

More recently, a therapeutic adhesive was synthesized that contained three agents: a QAM named dimethylaminododecyl methacrylate (DMADDM) with antibacterial activity, nanoparticles of silver (NAg), and NACP for remineralization [56]. This bonding agent showed a long-term durability in dentin bond strength. There was no reduction in dentin bond strength from one day to six months of immersion in water, while the commercial control bonding agent lost approximately one-third of its dentin bond strength at six months (Figure 3) [56]. Although many dental adhesives show satisfactory dentin bond strength in the short term, the long-term durability and stability of the resin–dentin interface remain a big challenge [57,58]. The resin–dentin bond strength demonstrated progressive decreases with increasing time in aging [59–62]. The reason for this decrease was attributed to the hydrolysis and enzymatic degradation of the exposed collagen and the adhesive resin, leading to the degradation of the hybrid layer at the dentin–adhesive interface [60]. There was water sorption in the aqueous oral environment because of the polar ether-linkages and the hydroxyl groups in the adhesive resin [61]. This could cause hydrolysis especially for the relatively more hydrophilic components in the resin [60,63]. Furthermore, the bacterial enzymes and the matrix metalloproteinases (MMPs) in the host tissues likely contributed significantly to the hybrid layer degradation [64]. During the dentin bonding, the MMPs were released and activated, which in turn could break down the collagen fibrils which became unprotected in the hybrid layer [64–67]. Such a damage of the collagen would in turn further increase the water sorption content, thus producing even more collagen degradation and causing deterioration in the dentin bonded interface [58]. In previous studies, CHX was shown to possess capabilities to inhibit the MMPs and suppress the enzymes [68]. Indeed, CHX was shown to nearly completely inhibit the collagen degradation of the demineralized dentin [69,70]. However, CHX can be dissolved in and cannot be co-polymerized with the resin, and, therefore, would be released in a relatively short amount of time, thus losing its long-term anti-MMP efficacy [24]. In contrast, DMADDM in Figure 3 was co-polymerized and immobilized in the polymer structure, and would not be leached out to diminish its effect over time, and hence could provide long-term MMP-inhibition [56]. Its durable anti-MMP effect likely contributed to maintaining the dentin bond strength without any decrease from one day to six months of water-aging treatment [56].

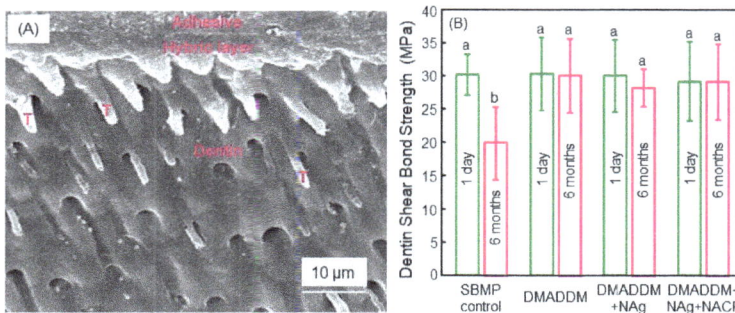

Figure 3. Dentin bonding. (**A**) Typical SEM image of the dentin–adhesive bonded region after one day of immersion. "T" indicates resin tags. (A) is for the DMADDM + Nag + NACP group; similar features were found in other groups. (**B**) Dentin bond strengths measured in shear (mean ± SD; $n = 10$). Values indicated by dissimilar letters are significantly different from each other ($p < 0.05$). Water-aging for six months caused a decrease of 35% in dentin bond strength for the commercial control bonding agent. In sharp contrast, the novel bioactive bonding agents containing DMADDM, NAg, and NACP showed no decrease in bond strength from one day to six months of water-aging. Adapted from [56], with permission from © 2013 Elsevier.

In addition, the bonding agent with DMADDM, NAg, and NACP incorporation possessed a strong antibacterial function with no decrease in the antibacterial potency from one day to six months of water-aging (Figure 4) [56]. This is consistent with the antibacterial agent being copolymerized and covalently bonded with the polymer network. This long-term antibacterial activity is beneficial considering that recurrent caries at the tooth-restoration margins is the primary cause for failures. By suppressing biofilm growth and reducing acids and enzymes, the antibacterial bonding agent could help suppress secondary dental caries. Furthermore, when the clinical requirements prevented the complete removal of the caries tissues such as avoiding the perforation of the pulp, as well as in minimal intervention dentistry [71], greater amounts of carious tissues were left in the tooth cavity. These carious tissues contained numerous residual bacteria inside the dentinal tubules in the prepared tooth cavity. The unpolymerized primer with the DMADDM anti-biofilm monomer, once applied to tooth cavity, would have direct contact with the tooth structure when flowing into the dentinal tubules, thereby eradicating the residual bacteria in the tubules. Then, upon polymerization, the adhesive resin at the margin would be in contact with the new invading bacteria, thus inhibiting their growth into the microgaps at the tooth-restoration interfaces [72]. While the six-month water-aging study indicated that the DMADDM copolymerization and covalent bonding with the polymer network enabled a long-term antibacterial activity, further longer-term study lasting for more than two years is needed on dentin bond strength, biofilm response, and caries prevention at the margins.

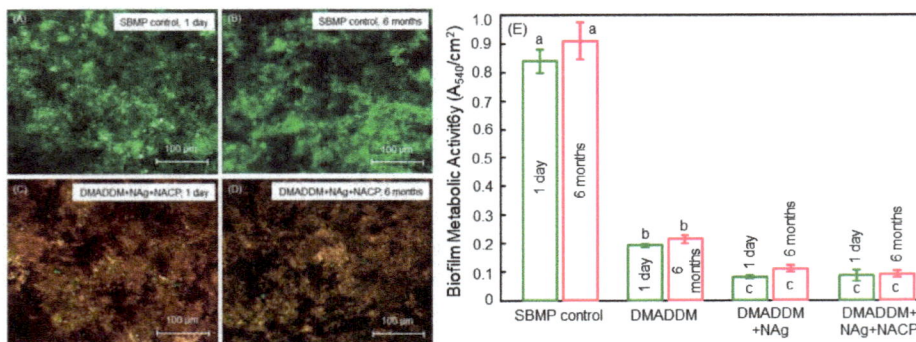

Figure 4. Anti-biofilm bonding resin after water-aging for 6 months. (**A,B**) Typical confocal laser scanning microscopy live/dead images for SBMP control; (**C,D**) DMADDM + Nag + NACP, at one day and six months, respectively. DMADDM and DMADDM + NACP groups had features similar to (**C,D**) (not shown). The novel bioactive bonding agent had primarily compromised bacteria. SBMP control had mostly live bacteria. (**E**) Metabolic activity (mean ± SD; *n* = 6). The potent antibacterial function remained after water-aging for six months. Adapted from Ref. [56], with permission from © 2013 Elsevier.

4. Antibacterial Composite for Tooth Root Caries Treatments

Senior people generally show greater risks of forming tooth root caries due to gingival recession and less saliva flow [73]. Periodontitis can lead to gingival recession, which in turn leads to more and more root surfaces to be exposed to the oral environment. Reduced saliva leads to more plaque buildup and less remineralization by saliva. These factors contribute to an increased risk of root caries. Root caries can be treated with Class V restorations. However, these restorations often have margins that are subgingival, which can provide pockets for bacterial growth that are difficult to clean, thus gradually leading to the loss of the periodontal attachment of the tooth. Indeed, it is a well-established knowledge that microbial biofilms are the primary etiological factor that causes periodontitis [74]. There are three primary species that are most often found in subgingival plaque from the periodontitis and periimplantitis areas: They are *Porphyromonas gingivalis* (*P. gingivalis*), *Prevotella intermedia* (*P. intermedia*), and *Aggregatibacter actinomycetemcomitans* (*A. actinomycetemcomitans*) [75]. In the periodontal pockets, these bacteria can generate virulence factors which lead to the gradual loss of the alveolar bone and the bone in periapical regions [75]. In areas with progressing periodontitis, *P. gingivalis* can serve as a keystone pathogen and as a portion of the climax group in the periodontal biofilms [76]. Being able to use estrogen and progesterone as an essential source of nutrients instead of using vitamin K, the *P. intermedia* species is connected with pregnancy gingivitis and periodontitis [77]. The third species, *A. actinomycetemcomitans*, is related to localized aggressive periodontitis. In addition to these three species, the fourth species, *Prevotella nigrescens* (*P. nigrescens*), is related to both healthy and diseased periodontium, and is biochemically comparable to *P. intermedia* [78]. In addition, *Fusobacterium nucleatum* (*F. nucleatum*), the fifth species, is linked to greater probing depth and periodontal ligament reduction [79]. Moreover, *F. nucleatum* can also promote the invasion of *P. gingivalis* into the gingival epithelial and aortic endothelial cells [80]. Last, *Enterococcus faecalis* (*E. faecalis*), the sixth species, is mainly considered an endodontic pathogen; however, it is also discovered in biofilms in the regions and in the saliva of patients who have chronic types of periodontal infections [81].

Therefore, these six species were selected in a recent study [82]. That study reported a novel polymeric composite for Class-V tooth cavity restorations with therapeutic functions to combat the six types of pathogens related to the start and the exacerbation of periodontitis [82]. The polymer matrix consisted of ethoxylated bisphenol A dimethacrylate (EBPADMA) and pyromellitic glycerol dimethacrylate (PMGDM) at 1:1 mass ratio (referred to as EBPM). Dimethylaminohexadecyl methacrylate (DMAHDM) was added at 3% mass fraction into the composite. Disks of the polymeric composite were transferred to a new 24-well plate. Each type of bacteria was inoculated in 1.5 mL of medium at 10^7 CFU/mL concentration in each well and cultured for 24 h. Then, the biofilm-disk constructs were transferred to new 24-well plates. New medium was added and the samples were cultured for another 24 h, thus totaling two days of culture to grow biofilms on the polymer surface [82]. Figure 5 shows the biofilm biomass after curing for two days which was measured using the absorbance values tested at OD_{600nm} [82]. The commercial control composite and the EBPM composite with 0% DMAHDM had similar biomass values. The EBPM composite with 3% DMAHDM had much less biofilm biomass. Therefore, the DMAHDM composite diminished the biomass of the biofilms for all six types of periodontitis-related pathogens [82].

In another study, protein-repellent agent 2-methacryloyloxyethyl phosphorylcholine (MPC) and antibacterial agent DMAHDM were combined in the polymeric composite to inhibit periodontal pathogens [83]. Figure 6 shows the polysaccharide amounts produced by the biofilms on the composites with: (A) *P. gingivalis*, (B) *P. intermedia*, (C) *A. actinomycetemcomitans*, and (D) *F. nucleatum* [83]. Biofilms on the commercial control composite and EBPM control composite produced similar quantities of polysaccharide. In contrast, biofilms on the EBPM + 3DMAHDM + 3MPC composite produced much less polysaccharide. Hence, the composite EBPM + 3DMAHDM + 3MPC could suppress periodontal pathogens and their production of the extracellular matrix [83]. Furthermore, the addition of MPC and DMAHDM into the polymeric composite did not adversely affect the mechanical properties. In addition, the use of dual agents of MPC + DMAHDM exerted a substantially more potent anti-biofilm activity, than using MPC or DMAHDM alone, against periodontal pathogens [83]. Therefore, the polymeric composite containing 3% DMAHDM and 3% MPC appeared to be the optimal composition. It showed a high potential for applications in Class-V tooth cavity restorations to inhibit periodontal biofilms, by reducing biofilm CFU by four orders of magnitude for all the types of periodontitis-related pathogens examined in that study [83].

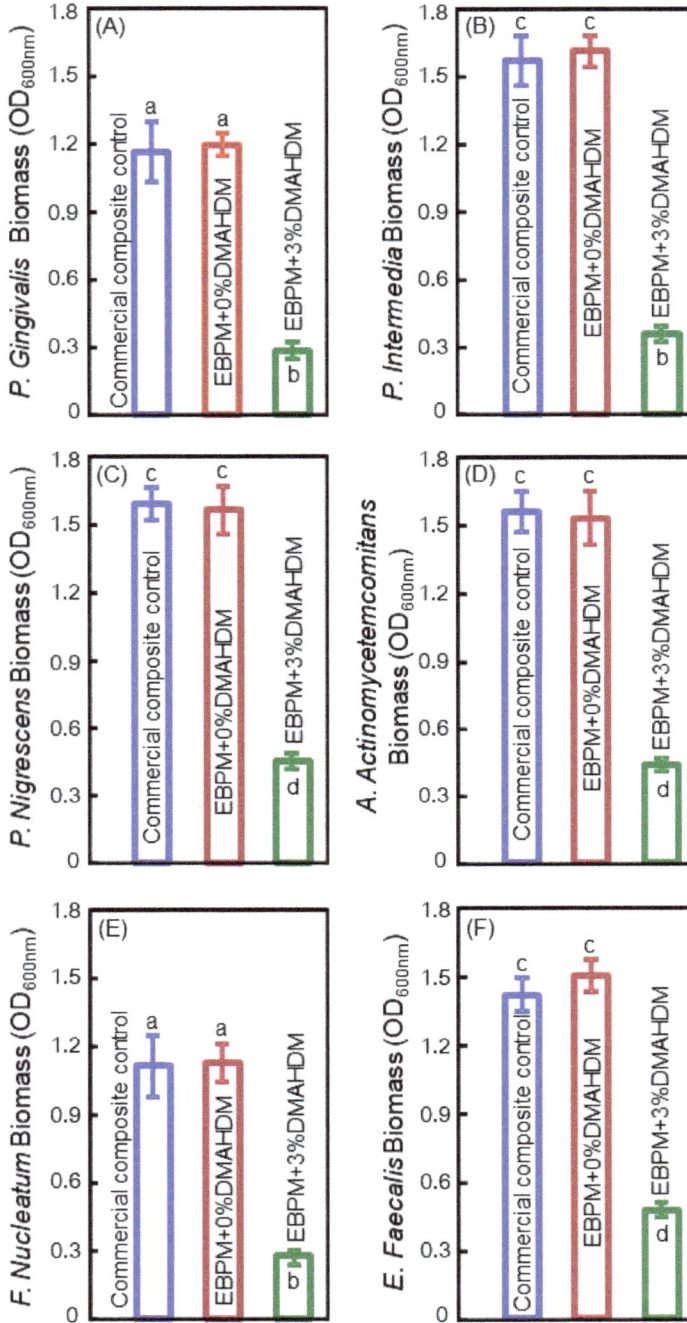

Figure 5. The biomass results of two-day biofilms grown on composites, evaluated using crystal violet assay and spectrophotometric optical density (OD600nm) (mean ± SD; $n = 6$): (**A**) *P. gingivalis*, (**B**) *P. intermedia*, (**C**) *P. nigrescens*, (**D**) *A. actinomycetemcomitans*, (**E**) *F. nucleatum* and (**F**) *E. faecalis*. The biofilm biomass on composites with DMAHDM was substantially reduced, as compared to that on composite without DMAHDM. Bars indicated by dissimilar letters are significantly different from each other ($p < 0.05$). Adapted from [82], with permission from © 2016 Elsevier.

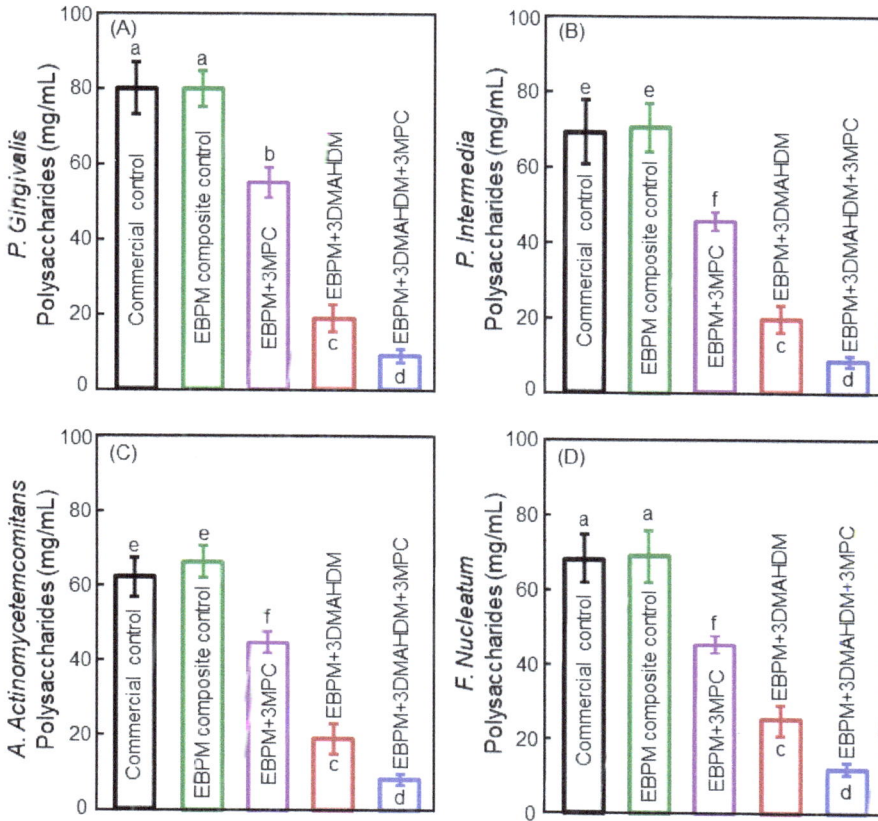

Figure 6. Polysaccharide results in biofilms grown for two days on polymeric composites: (**A**) *P. gingivalis*, (**B**) *P. intermedia*, (**C**) *A. actinomycetemcomitans*, and (**D**) *F. nucleatum* (mean ± SD; $n = 6$). Values indicated by dissimilar letters are significantly different ($p < 0.05$). Adapted from [83], with permission from © 2016 Elsevier.

5. Antibacterial Bonding Agents Inhibiting Periodontal Pathogens

Three bioactive agents (NACP for remineralization, MPC for protein-repellency, and DMAHDM for anti-biofilm activity) were combined into a polymeric bonding agent to suppress periodontal pathogens [84]. The adhesive contained PMGDM, EBPADMA, 2-hydroxyethyl methacrylate (HEMA) and BisGMA at 45/40/10/5 mass ratio (referred to as PEHB). The dentin shear bond strength results showed that adding 30% NACP into the adhesive did not compromise the dentin bond strength, compared to the control without NACP. In addition, incorporation of 5% DMAHDM + 5% MPC into both the primer and the adhesive did not negatively influence the dentin bond strength, compared to PEHB-NACP group without DMAHDM and MPC [84]. However, the incorporation of 5% DMAHDM + 7.5% MPC did lower the bond strengths. Therefore, a mass fraction of 30% NACP was incorporated into the adhesive, and mass fractions of 5% DMAHDM + 5% MPC were incorporated into both primer and adhesive [84].

Figure 7 shows the (A) metabolic activity, (B) polysaccharide, and (C) biofilm colony-forming units (CFU) for the multispecies periodontal biofilms [84]. The commercial bonding agent control and the PEHB-NACP without DMAHDM and MPC had similar CFU counts, indicating that NACP had little anti-biofilm activity. In contrast, DMAHDM or MPC each alone substantially decreased the biofilm CFU than those of the controls. Furthermore, the incorporation of 5% DMAHDM + 5% MPC resulted in the lowest metabolic activity, polysaccharide, and biofilm CFU counts. The CFU of the periodontal biofilm grown on the PEHB + 5DMAHDM + 5MPC polymer was three orders of magnitude less than that grown on the PEHB control polymer [84].

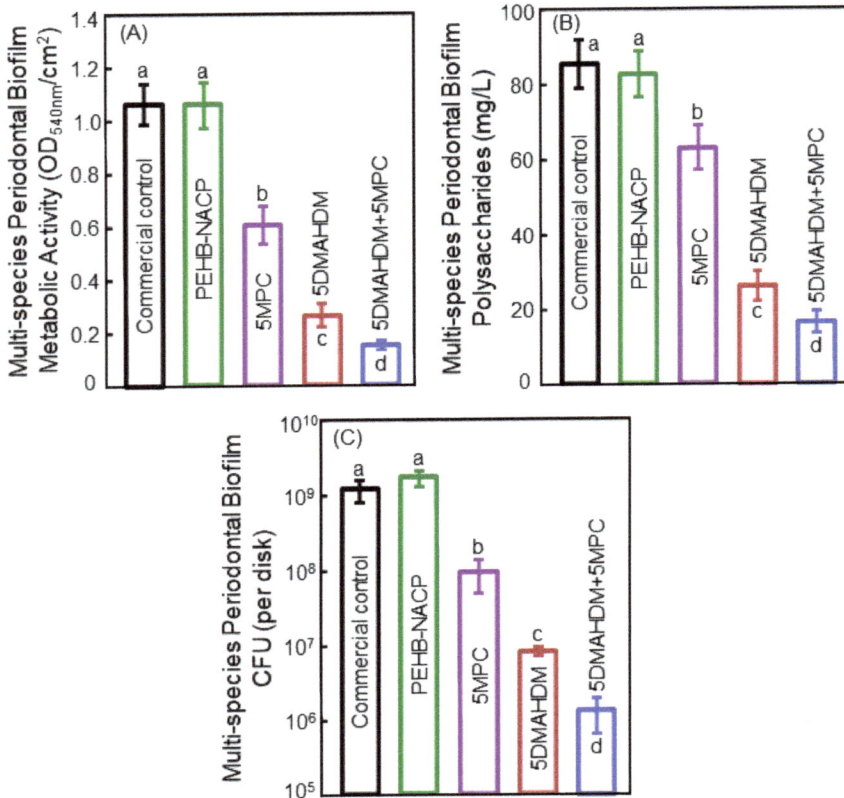

Figure 7. Multi-species periodontal biofilm: (**A**) metabolic activities, (**B**) polysaccharides production, and (**C**) CFU values on different dental bonding adhesive disks (mean ± SD; $n = 6$). The y-axis has the log scale in plot (**C**). Values indicated by dissimilar letters are significantly different from each other ($p < 0.05$). Adapted from [84], with permission from © 2017 RSC Advance.

On a clean polymer surface in the oral environment, the saliva-derived proteins deposit on the polymer first, and then bacteria start to attach to the polymer. Salivary protein adsorption on the surface is required and is a prerequisite for oral bacteria adherence on the polymer surface [85]. This mechanism indicates that developing a protein-repellent polymer can greatly reduce biofilm growth on the polymer restoration in the oral environment. MPC is a methacrylate with phospholipid polar group in the side chain, with the capability to reduce protein adsorption and bacterial adhesion [34]. The mechanism for the protein-repellency was attributed to that in the hydrated MPC polymer, a large amount of free water exists around the phosphorylcholine groups which could detach the proteins [86]. Adding 5% of MPC into the bonding agent decreased the amount of protein adsorption by more than an order of magnitude [84]. In addition, combining the MPC with DMAHDM incorporation produced the strongest suppression of periodontal biofilms. The periodontal multi-species biofilm CFU was approximately 10^9 counts on the control adhesive polymer. The CFU was decreased to 10^8 counts by the use of MPC. The CFU was lowered to 10^7 counts with the use of DMAHDM. In contrast, the CFU was reduced to only 10^6 counts when both MPC and DMAHDM were used together in the bonding agent [84]. This synergistic reduction in biofilm growth on polymer surfaces was related to the mode of action, which was contact-inhibition [22,23]. When the cell membrane of the bacteria with negative charges contacts the quaternary amine N^+ with positive charges on the polymer, the membrane could be disrupted, thus causing cytoplasmic leakage [30,47]. This mechanism of contact-inhibition implied that, when the polymer surface was covered by the salivary protein pellicles, the polymer surface was separated from the overlaying biofilm. This reduced the extent of contact, and hence the contact-inhibition efficacy was decreased. Therefore, because of the protein-repellency of the MPC, it helped diminish protein coverage on the polymer surface, thus exposing more polymer surface with quaternary amine N^+ sites, thereby promoting the contact-inhibition ability. Therefore, the dual use of DMAHDM and MPC in the dental polymer could work synergistically to maximize the periodontal bacteria inhibition capability [84].

6. Antibacterial Polymeric Endodontic Sealers

Endodontic treatment is needed to eradicate bacterial infection in the tooth root canal, to avoid the microorganisms from harming the periapical healing and causing apical lesions [87]. Clinically, the anatomic complexity of the tooth root canal renders the complete debridement of bacteria practically impossible [88]. Such persistence of bacteria in the tooth root canal often results in post-treatment diseases [89]. One promising approach to address this challenge is the development of antibacterial root canal sealers with the capability to kill endodontic pathogens.

A recent study developed a bioactive endodontic sealer with a good sealing ability in bonding to root dentin, indicated by a push-out strength being similar to those of commercial control without bioactive properties [90]. The push-out bond strength results to root wall dentin are shown in Figure 8A. The addition of 5% DMAHDM and 3% MPC into both the primer and the sealer paste did not adversely influence the dentin bond strength. However, when 5% DMAHDM and 4.5% MPC were incorporated together into the sealer, the push-out strength decreased. Hence, the composition of 5% DMAHDM and 3% MPC was determined to be optimal and was employed for the endodontic sealer and the primer [90]. Figure 8B shows the CFU results of the multispecies endodontic biofilms grown for 14 days on polymer samples. The commercial control group and the PEHB control polymer had similar CFU results. The addition of either DMAHDM or MPC alone reduced the endodontic biofilm CFU. The bioactive endodontic sealer containing 5% DMAHDM and 3% MPC had the lowest biofilm CFU. The 14-day endodontic biofilm CFU on the PEHB + NACP + 5DMAHDM + 3MPC polymer samples was three orders of magnitude less than that on the PEHB+NACP control polymer samples [90].

In a study on three-dimensional (3D) biofilms grown on dental polymer surfaces, the percentage of live bacteria was determined as a function of the location of the 2D cross-section inside the 3D biofilm at various distances from the polymer surface [91]. Near the surface of the polymer which contained DMAHDM, there were more dead bacteria in the biofilm. In the 3D biofilm away from the polymer surface, the percentage of live bacteria increased, likely due to a decrease in the contact-inhibition efficacy [91]. These results are consistent with the results of the DMAHDM-containing endodontic sealer, which achieved a greater reduction in the biofilm CFU at 3 days, compared to the reduction at 14 days [90]. The reason for the more killing effect at 3 days, and less killing effect at 14 days, was likely related to the contact-killing mechanism. The compromised bacteria on the polymer surface acted as a bridge for the further adherence and growth of bacteria, and the next layer of bacteria were away from the polymer surface with a reduced extent of contact-inhibition [92]. Therefore, the contact-inhibition mode of action would indicate that the antibacterial activity against the 14-day biofilms would be decreased because of a lack of direct contact when the microbes lived in the 3D biofilm structure away from the polymer surface. Therefore, the 14-day biofilm model represented a rigorous test of antibacterial activity. The fact that the PEHB + NACP + 5DMAHDM + 3MPC polymer was able to successfully kill and reduce the 14-day biofilm CFU by three orders of magnitude (Figure 8) indicates a novel bioactive endodontic sealer with an extremely potent anti-biofilm function [90]. Novel dental biomaterial development has the potential to bring tremendous benefits to treatment efficacy and improve the quality of life [7–17,93,94]. Further investigation is needed to achieve the long-lasting biofilm-eradication, therapeutic effects, and tooth-protection via the new bioactive dental polymeric materials using clinically-relevant experiments in the oral environment of human participants.

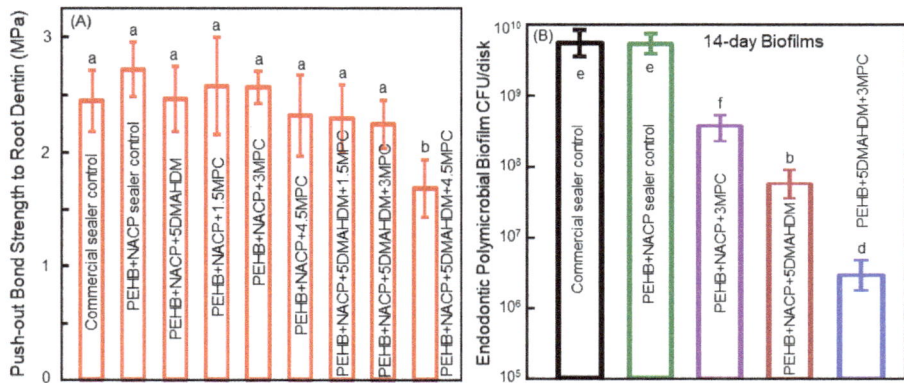

Figure 8. Polymeric endodontic sealers. (**A**) The push-out bond strength values to tooth root dentin (mean $\pm$ SD; $n = 6$). All the groups had similar strengths, except PEHB+NACP+5DMAHDM+4.5MPC which had a lower strength ($p < 0.05$). (**B**) The CFU of endodontic biofilm on the endodontic sealer grown for 14 days (mean $\pm$ SD; $n = 6$). In each plot, values with dissimilar letters are significantly different from each other ($p < 0.05$). Adapted from [90], with permission from © 2017 Elsevier.

7. Conclusions

Currently available dental polymeric composites and bonding agents for tooth cavity restorations are usually bioinert. Since oral bacteria and biofilms play an important role in dental caries and oral infections, a new generation of dental polymeric materials are being developed that are bioactive and possess therapeutic effects including antibacterial, acid-reduction, protein-repellent, and remineralization capabilities. This article reviewed cutting-edge research on the development and properties of novel antibacterial dental polymeric composites, antibacterial bonding agents, bioactive root caries composites for senior patients, adhesives that can suppress periodontal pathogens, and antibacterial and protein-repellent endodontic sealers that can kill endodontic pathogens. Substantial reductions in oral biofilm metabolic activity, acid production, biomass, and polysaccharide synthesis were achieved with the tailored polymeric compositions. Biofilm CFU counts were reduced by three to four orders of magnitude. One advantage of QAM-containing polymers is that the antibacterial agent is co-polymerized and covalently bonded with the polymer, and hence it has long-term antibacterial function that is not leached out and lost over time. The disadvantage is that QAM-polymers rely on the contact-inhibition mechanism, with reduced antibacterial efficacy when the polymer surface is covered by a layer of salivary proteins. As alluded in the Introduction, one potential future development would be to combine strategies from Class 1 and Class 2 so that the dental polymer would possess long-term contact-inhibition as well as the release of antibacterial agents to inhibit bacteria away from the polymer surface throughout the three-dimensional biofilm. The advances in the anti-biofilm properties and therapeutic capabilities of the new generation of dental polymeric materials are expected to bring significant benefits to a wide range of restorative and preventive dental applications.

Author Contributions: K.Z. performed experiments that yielded Figures 1–4, and helped in writing and discussions. B.B. contributed in data collection, data analysis and data interpretation, and discussions on endodontic sealers. C.D.L. contributed to the conception and discussions on the development of dental resin composites and bonding agents, clinical need in the field, and future trend in materials development. M.D.W. served as the lab manager and contributed in materials development, polymer synthesis, development of antibacterial monomers, and fabrication of nano silver. M.A.S.M. contributed in conception, biofilm experiments, saliva collection, and protein repellent tests. Y.L. contributed in performing experiments, data collection, data analysis and data interpretation. M.A.R. contributed in the conception, discussions on the clinical relevance in developing novel compositions of bioactive dental materials, as well as the need in periodontal treatments, and the formulations of therapeutic materials to combat periodontal pathogens. Y.B. served as the co-supervisor of K.Z. and contributed the conception, provided discussions to H.H.K.X. on antibacterial studies and how to suppress caries development. L.W. performed experiments that yielded Figures 5–8, and contributed in writing and data analyses and interpretations. S.W. helped perform antibacterial and biofilm experiments, contributed in data collection, data analysis and data interpretation, and helped H.H.K.X. in writing, literature search, and putting together the references. H.H.K.X. contributed in the conception and design of the work, supervised the studies, and wrote the manuscript. All authors approve the final version to be published.

Funding: This work was supported by NIH R01 DE17974 (H.X.), National Science Foundation of China NSFC 81400487 (L.W.), Jilin Youth Fund of Science and Technology 20150520043JH (L.W.), NSFC 81400540 (K.Z.), University of Maryland School of Dentistry bridging fund (H.X.) and University of Maryland seed grant (H.X.).

Acknowledgments: We thank Satoshi Imazato, Lei Cheng, Ning Zhang, Xianju Xie, Fang Li, Xuedong Zhou, Joseph M. Antonucci, Nancy J. Lin, Sheng Lin-Gibson, Jirun Sun, and Ashraf F. Fouad for discussions and help.

Conflicts of Interest: The authors declare no conflict of interest.

References

1. Sakaguchi, R.L. Review of the current status and challenges for dental posterior restorative composites: clinical, chemistry, and physical behavior considerations. *Dent. Mater.* **2005**, *21*, 3–6. [CrossRef] [PubMed]
2. Mjor, I.A.; Toffeneti, F. Secondary caries: a literature review with caries reports. *Quintessence Int.* **2000**, *31*, 165–179. [PubMed]
3. Ferracane, J.L. Resin composite-state of the art. *Dent. Mater.* **2011**, *27*, 29–38. [CrossRef] [PubMed]
4. Frost, P.M. An audit on the placement and replacement of restorations in a general dental practice. *Prim. Dent. Care* **2002**, *9*, 31–36. [CrossRef] [PubMed]
5. Beazoglou, T.; Eklund, S.; Heffley, D.; Meiers, J.; Brown, L.J.; Bailit, H. Economic impact of regulating the use of amalgam restorations. *Public Health Rep.* **2007**, *122*, 657–663. [CrossRef] [PubMed]
6. Saunders, R.H.; Meyerowitz, C. Dental caries in older adults. *Dent. Clin. N. Am.* **2005**, *49*, 293–308. [CrossRef] [PubMed]
7. Breschi, L.; Mazzoni, A.; Ruggeri, A.; Cadenaro, M.; Di Lenarda, R.; De Stefano Dorigo, E. Dental adhesion review: aging and stability of the bonded interface. *Dent. Mater.* **2008**, *24*, 90–101. [CrossRef] [PubMed]
8. Spencer, P.; Ye, Q.; Park, J.G.; Topp, E.M.; Misra, A.; Marangos, O.; Wang, Y.; Bohaty, B.S.; Singh, V.; Sene, F.; Eslick, J.; Camarda, K.; Katz, J.L. Adhesive/dentin interface: The weak link in the composite restoration. *Annals Biomed. Eng.* **2010**, *38*, 1989–2003. [CrossRef] [PubMed]
9. Milward, P.J.; Adusei, G.O.; Lynch, C.D. Improving some selected properties of dental polyacid-modified composite resins. *Dent. Mater.* **2011**, *27*, 997–1002. [CrossRef] [PubMed]
10. Farrugia, C.; Camilleri, J. Antimicrobial properties of conventional restorative filling materials and advances in antimicrobial properties of composite resins and glass ionomer cements—A literature review. *Dent. Mater.* **2015**, *31*, 89–99. [CrossRef] [PubMed]
11. Ilie, N.; Hilton, T.J.; Heintze, S.D.; Hickel, R.; Watts, D.C.; Silikas, N.; Stansbury, J.W.; Cadenaro, M.; Ferracane, J.L. Academy of dental materials guidance—Resin composites: Part I—Mechanical properties. *Dent. Mater.* **2017**, *33*, 880–894. [CrossRef] [PubMed]
12. Maas, M.S.; Alania, Y.; Natale, L.C.; Rodrigues, M.C.; Watts, D.C.; Braga, R.R. Trends in restorative composites research: what is in the future? *Braz. Oral. Res.* **2017**, *31*, 55. [CrossRef] [PubMed]
13. Xu, X.; Ling, L.; Wang, R.; Burgess, J.O. Formation and characterization of a novel fluoride-releasing dental composite. *Dent. Mater.* **2006**, *22*, 1014–1023. [CrossRef] [PubMed]
14. Ferracane, J.L. Placing dental composites—A stressful experience. *Oper. Dent.* **2008**, *333*, 247–257. [CrossRef] [PubMed]
15. Wei, Y.J.; Silikas, N.; Zhang, Z.T.; Watts, D.C. Hygroscopic dimensional changes of self-adhering and new resin-matrix composites during water sorption/desorption cycles. *Dent. Mater.* **2011**, *27*, 259–266. [CrossRef] [PubMed]

16. Huang, S.; Podgórski, M.; Zhang, X.; Sinha, J.; Claudino, M.; Stansbury, J.W.; Bowman, C.N. Dental restorative materials based on thiol-michael photopolymerization. *J. Dent. Res.* **2018**, *97*, 530–536. [CrossRef] [PubMed]

17. Vallittu, P.K.; Boccaccini, A.R.; Hupa, L.; Watts, D.C. Bioactive dental materials—Do they exist and what does bioactivity mean? *Dent. Mater.* **2018**, *34*, 693–694. [CrossRef] [PubMed]

18. Kitagawa, H.; Miki-Oka, S.; Mayanagi, G.; Abiko, Y.; Takahashi, N.; Imazato, S. Inhibitory effect of resin composite containing S-PRG filler on *Streptococcus mutans* glucose metabolism. *J. Dent.* **2018**, *70*, 92–96. [CrossRef] [PubMed]

19. Beyth, N.; Domb, A.J.; Weiss, E.I. An in vitro quantitative antibacterial analysis of amalgam and composite resins. *J. Dent.* **2007**, *35*, 201–206. [CrossRef] [PubMed]

20. Totiam, P.; Gonzalez-Cabezas, C.; Fontana, M.R.; Zero, D.T. A new in vitro model to study the relationship of gap size and secondary caries. *Caries Res.* **2007**, *41*, 467–473. [CrossRef] [PubMed]

21. Cenci, M.S.; Pereira-Cenci, T.; Cury, J.A.; ten Cate, J.M. Relationship between gap size and dentine secondary caries formation assessed in a microcosm biofilm model. *Caries Res.* **2009**, *43*, 97–102. [CrossRef] [PubMed]

22. Imazato, S. Review: Antibacterial properties of resin composites and dentin bonding systems. *Dent. Mater.* **2003**, *19*, 449–457. [CrossRef]

23. Imazato, S. Bioactive restorative materials with antibacterial effects: new dimension of innovation in restorative dentistry. *Dent. Mater. J.* **2009**, *28*, 11–19. [CrossRef] [PubMed]

24. Tezvergil-Mutluay, A.; Agee, K.A.; Uchiyama, T.; Imazato, S.; Mutluay, M.M.; Cadenaro, M.; Breschi, L.; Nishitani, Y.; Tay, F.R.; Pashley, D.H. The inhibitory effects of quaternary ammonium methacrylates on soluble and matrix-bound MMPs. *J. Dent. Res.* **2011**, *90*, 535–540. [CrossRef] [PubMed]

25. Antonucci, J.M.; Zeiger, D.N.; Tang, K.; Lin-Gibson, S.; Fowler, B.O.; Lin, N.J. Synthesis and characterization of dimethacrylates containing quaternary ammonium functionalities for dental applications. *Dent. Mater.* **2012**, *28*, 219–228. [CrossRef] [PubMed]

26. Xu, X.; Wang, Y.; Liao, S.; Wen, Z.T.; Fan, Y. Synthesis and characterization of antibacterial dental monomers and composites. *J. Biomed. Mater. Res.* **2012**, *100B*, 1151–1162. [CrossRef] [PubMed]

27. Cheng, L.; Weir, M.D.; Zhang, K.; Wu, E.; Xu, S.M.; Zhou, X.D.; Xu, H.H.K. Dental plaque microcosm biofilm behavior on calcium phosphate nanocomposite with quaternary ammonium. *Dent. Mater.* **2012**, *28*, 853–862. [CrossRef] [PubMed]

28. Imazato, S.; Kinomoto, Y.; Tarumi, H.; Ebisu, S.; Tay, F.R. Antibacterial activity and bonding characteristics of an adhesive resin containing antibacterial monomer MDPB. *Dent. Mater.* **2003**, *19*, 313–319. [CrossRef]

29. Imazato, S.; Tay, F.R.; Kaneshiro, A.V.; Takahashi, Y.; Ebisu, S. An in vivo evaluation of bonding ability of comprehensive antibacterial adhesive system incorporating MDPB. *Dent. Mater.* **2007**, *23*, 170–176. [CrossRef] [PubMed]

30. Beyth, N.; Yudovin-Farber, I.; Bahir, R.; Domb, A.J.; Weiss, E.I. Antibacterial activity of dental composites containing quaternary ammonium polyethylenimine nanoparticles against Streptococcus mutans. *Biomaterials* **2006**, *27*, 3995–4002. [CrossRef] [PubMed]

31. Namba, N.; Yoshida, Y.; Nagaoka, N.; Takashima, S.; Matsuura-Yoshimoto, K.; Maeda, H.; Van Meerbeek, B.; Suzuki, K.; Takashiba, S. Antibacterial effect of bactericide immobilized in resin matrix. *Dent. Mater.* **2009**, *25*, 424–430. [CrossRef] [PubMed]

32. Li, F.; Weir, M.D.; Xu, H.K. Effects of quaternary ammonium chain length on antibacterial bonding agents. *J. Dent. Res.* **2013**, *92*, 932–938. [CrossRef] [PubMed]

33. Li, F.; Weir, M.D.; Fouad, A.F.; Xu, H.H. Effect of salivary pellicle on antibacterial activity of novel antibacterial dental adhesives using a dental plaque microcosm biofilm model. *Dent. Mater.* **2014**, *30*, 182–191. [CrossRef] [PubMed]

34. Zhang, N.; Ma, J.; Melo, M.A.S.; Weir, M.D.; Bai, Y.X.; Xu, H.H.K. Protein-repellent and antibacterial dental composite to inhibit biofilms and caries. *J. Dent.* **2015**, *43*, 225–234. [CrossRef] [PubMed]

35. Cheng, L.; Zhang, K.; Weir, M.D.; Melo, M.A.S.; Zhou, X.D.; Xu, H.H.K. Nanotechnology strategies for antibacterial and remineralizing composites and adhesives to tackle dental caries. *Nanomedicine* **2015**, *10*, 627–641. [CrossRef] [PubMed]

36. Zhang, K.; Cheng, L.; Weir, M.D.; Bai, Y.X.; Xu, H.H.K. Effects of quaternary ammonium chain length on antibacterial and remineralizing calcium phosphate nanocomposite. *Int. J. Oral Sci.* **2016**, *8*, 45–53. [CrossRef] [PubMed]

37. Kasraei, S.L.; Sami, L.; Hendi, S.; Alikhani, M.-Y.; Rezaei-Soufi, L.; Khamverdi, Z. Antibacterial properties of composite resins incorporating silver and zinc oxide nanoparticles on Streptococcus mutans and Lactobacillus. *Restor. Dent. Endod.* **2014** *39*, 109–114. [CrossRef] [PubMed]

38. Stencel, R.; Kasperski, J.; Pakiela, W.; Mertas, A.; Bobela, E.; Barszczewska-Rybarek, I.; Chladek, G. Properties of experiment dental composite containing antibacterial silver-releasing filler. *Materials* **2018**, *11*, 1031. [CrossRef] [PubMed]

39. Khvostenko, D.; Hilton, T.J.; Ferracane, J.L.; Mitchell, J.C.; Kruzic, J.J. Bioactive glass fillers reduce bacterial penetration into marginal gaps for composite restorations. *Dent. Mater.* **2016**, *32*, 73–81. [CrossRef] [PubMed]

40. Al-Eesa, N.A.; Wong, F.S.L.; Johal, A.; Hill, R.G. Fluoride containing bioactive glass composite for orthodontic adhesive–ion release properties. *Dent. Mater.* **2017**, *33*, 1324–1329. [CrossRef] [PubMed]

41. Cheng, L.; Weir, M.D.; Xu, H.H.K.; Antonucci, J.M.; Lin, N.J.; Lin-Gibson, S.; Xu, S.M.; Zhou, X. Effect of amorphous calcium phosphate and silver nanocomposites on dental plaque microcosm biofilms. *J. Biomed. Mater. Res. B Appl. Biomater.* **2012**, *100*, 1378–1386. [CrossRef] [PubMed]

42. Tavassoli Hojati, S.; Alaghemand, H.; Hamze, F.; Ahmadian Babaki, F.; Rajad-Nia, R.; Rezvni, M.B.; Kaviani, M.; Atai, M. Antibacterial, physical and mechanical properties of flowable resin composites containing zinc oxide nanoparticles. *Dent. Mater.* **2013**, *29*, 495–505. [CrossRef] [PubMed]

43. Aydin Sevinc, B.; Hanley, L. Antibacterial activity of dental composites containing zinc oxide nanoparticles. *J. Biomed. Mater. Res. B Appl. Biomater.* **2010**, *94*, 22–31. [CrossRef] [PubMed]

44. Cheng, L.; Zhang, K.; Zhou, C.C.; Weir, M.D.; Zhou, X.D.; Xu, H.H. One-year water-ageing of calcium phosphate composite containing nano-silver and quaternary ammonium to inhibit biofilms. *Int. J. Oral Sci.* **2016**, *8*, 172–181. [CrossRef] [PubMed]

45. Imazato, S.; Torii, M.; Tsuchitani, Y.; McCabe, J.F.; Russell, R.R. Incorporation of bacterial inhibitor into resin composite. *J. Dent. Res.* **1994**, *73*, 1437–1443. [CrossRef] [PubMed]

46. Lin, J.; Qiu, S.; Lewis, K.; Klibanov, A.M. Bactericidal properties of flat surfaces and nanoparticles derivatized with alkylated polyethylenimines. *Biotechnol. Prog.* **2002**, *18*, 1082–1086. [CrossRef] [PubMed]

47. Beyh, N.; Yudovin-Farber, I.; Perez-Davidi, M.; Domb, A.J.; Weiss, E.I. Polyethyleneimine nanoparticles incorporated into resin composite cause cell death and trigger biofilm stress in vivo. *Proc. Natl. Acad. Sci. USA* **2010**, *107*, 22038–22043. [CrossRef] [PubMed]

48. Weng, Y.; Howard, L.; Guo, X.; Chong, V.J.; Gregory, R.L.; Xie, D. A novel antibacterial resin composite for improved dental restoratives. *J. Mater. Sci. Mater. Med.* **2012**, *23*, 1553–1561. [CrossRef] [PubMed]

49. Cheng, L.; Zhang, K.; Melo, M.A.; Weir, M.D.; Zhou, X.; Xu, H.H. Anti-biofilm dentin primer with quaternary ammonium and silver nanoparticles. *J. Dent. Res.* **2012**, *91*, 598–604. [CrossRef] [PubMed]

50. Xie, D.; Weng, Y.; Guo, X.; Zhao, J.; Gregory, R.L.; Zheng, C. Preparation and evaluation of a novel glass-ionomer cement with antibacterial functions. *Dent. Mater.* **2011**, *27*, 487–496. [CrossRef] [PubMed]

51. Zhou, H.; Weir, M.D.; Antonucci, J.M.; Antonucci, J.M.; Schumacher, G.E.; Zhou, X.D.; Xu, H.H. Evaluation of three-dimensional biofilms on antibacterial bonding agents containing novel quaternary ammonium methacrylates. *Int. J. Oral Sci.* **2014**, *6*, 77–86. [CrossRef] [PubMed]

52. Pashley, D.H.; Tay, F.R.; Breschi, L.; Tjaderhane, L.; Carvalho, R.M.; Carrilho, M.; Tezvergil-Mutluay, A. State of the art etch-and-rinse adhesives. *Dent. Mater.* **2011**, *27*, 1–16. [CrossRef] [PubMed]

53. Imazato, S.; Kuramoto, A.; Takahashi, Y.; Ebisu, S.; Peters, M.C. In vitro antibacterial effects of the dentin primer of Clearfil Protect Bond. *Dent. Mater.* **2006**, *22*, 527–532. [CrossRef] [PubMed]

54. Li, F.; Chen, J.; Chai, Z.; Zhang, L.; Xiao, Y.; Fang, M.; Ma, S. Effects of a dental adhesive incorporating antibacterial monomer on the growth, adherence and membrane integrity of Streptococcus mutans. *J. Dent.* **2009** *37*, 289–296. [CrossRef] [PubMed]

55. Hiraishi, N.; Yiu, C.K.; King, N.M.; Tay, F.R. Effect of chlorhexidine incorporation into a self-etching primer on dentine bond strength of a luting cement. *J. Dent.* **2010**, *38*, 496–502. [CrossRef] [PubMed]

56. Zhang, K.; Cheng, L.; Wu, E.J.; Weir, M.D.; Bai, Y.; Xu, H.H.K. Effect of water-aging on dentin bond strength and anti-biofilm activity of bonding agent containing antibacterial monomer dimethylaminododecyl methacrylate. *J. Dent.* **2013**, *41*, 504–513. [CrossRef] [PubMed]

57. Burrow, M.F.; Tyas, M.J. Clinical evaluation of three adhesive systems for the restoration of non-carious cervical lesions. *Oper. Dent.* **2007**, *32*, 11–15. [CrossRef] [PubMed]

58. Manuja, N.; NAgpal, R.; Pandit, I.K. Dental adhesion: mechanism, techniques and durability. *J. Clin. Pediatr. Dent.* **2012**, *36*, 223–234. [CrossRef] [PubMed]

59. Kitasako, Y.; Burrow, M.F.; Katahira, N.; Nikaido, T.; Tagami, J. Shear bond strengths of three resin cements to dentine over 3 years in vitro. *J. Dent.* **2001**, *29*, 139–144. [CrossRef]

60. De Munck, J.; Van Landuyt, K.; Peumans, M.; Poitevin, A.; Lambrechts, P.; Braem, M.; Van Meerbeek, B. A critical review of the durability of adhesion to tooth tissue: methods and results. *J. Dent. Res.* **2005**, *84*, 118–132. [CrossRef] [PubMed]

61. Koshiro, K.; Inoue, S.; Sano, H.; De Munck, J.; Van Meerbeek, B. In vivo degradation of resin–dentin bonds produced by a self-etch and an etch-and-rinse adhesive. *Eur. J. Oral Sci.* **2005**, *113*, 341–348. [CrossRef] [PubMed]

62. Sideridou, I.; Tserki, V.; Papanastasiou, G. Study of water sorption, solubility and modulus of elasticity of light-cured dimethacrylate-based dental resins. *Biomaterials* **2003**, *24*, 655–665. [CrossRef]

63. Tay, F.R.; Pashley, D.H. Water treeing—A potential mechanism for degradation of dentin adhesives. *Am. J. Dent.* **2003**, *16*, 6–12. [PubMed]

64. Pashley, D.H.; Tay, F.R.; Yiu, C.; Hashimoto, M.; Breschi, L.; Carvalho, R.M.; Ito, S. Collagen degradation by host-derived enzymes during aging. *J. Dent. Res.* **2004**, *83*, 216–221. [CrossRef] [PubMed]

65. Hashimoto, M.; Ohno, H.; Sano, H.; Tay, F.R.; Kaga, M.; Kudou, Y.; Oguchi, H.; Araki, Y.; Kubota, M. Micromorphological changes in resin–dentin bonds after 1 year of water storage. *J. Biomed. Mater. Res.* **2002**, *63*, 306–311. [CrossRef] [PubMed]

66. Wang, Y.; Spencer, P. Hybridization efficiency of the adhesive/dentin interface with wet bonding. *J. Dent. Res.* **2003**, *82*, 141–145. [CrossRef] [PubMed]

67. Nishitani, Y.; Yoshiyama, M.; Wadgaonkar, B.; Breschi, L.; Mannello, F.; Mazzoni, A.; Carvalho, R.M.; Tjäderhane, L.; Tay, F.R.; Pashley, D.H. Activation of gelatinolytic/collagenolytic activity in dentin by self-etching adhesives. *Eur. J. Oral Sci.* **2006**, *114*, 160–166. [CrossRef] [PubMed]

68. Gendron, R.; Grenier, D.; Sorsa, T.; Mayrand, D. Inhibition of the activities of matrix metalloproteinases 2, 8, and 9 by chlorhexidine. *Clin. Diagn. Lab. Immunol.* **1999**, *6*, 437–439. [PubMed]

69. Breschi, L.; Mazzoni, A.; Nato, F.; Carrilho, M.; Visintini, E.; Tjaderhane, L.; Ruggeri, A.J.R.; Tay, F.R.; Dorigo Ede, S.; Pashley, D.H. Chlorhexidine stabilizes the adhesive interface: A 2-year in vitro study. *Dent. Mater.* **2010**, *26*, 320–325. [CrossRef] [PubMed]

70. Osorio, R.; Yamauti, M.; Osorio, E.; Ruiz-Requena, M.E.; Pashley, D.; Tay, F.; Toledano, M. Effect of dentin etching and chlorhexidine application on metalloproteinase-mediated collagen degradation. *Eur. J. Oral Sci.* **2011**, *119*, 79–85. [CrossRef] [PubMed]

71. Lynch, C.D.; Frazier, K.B.; McConnell, R.J.; Blum, I.R.; Wilson, N.H. Minimally invasive management of dental caries: contemporary teaching of posterior resin-based composite placement in U.S. and Canadian dental schools. *J. Am. Dent. Assoc.* **2011**, *142*, 612–620. [CrossRef] [PubMed]

72. Loguercio, A.D.; Reis, A.; Bortoli, G.; Patzlaft, R.; Kenshima, S.; Rodrigues Filho, L.E.; Accorinte, M.d.e.L.; van Dijken, J.W. Influence of adhesive systems on interfacial dentin gap formation in vitro. *Oper. Dent.* **2006**, *31*, 431–441. [CrossRef] [PubMed]

73. Fure, S. Ten-year incidence of tooth loss and dental caries in elderly Swedish individuals. *Caries Res.* **2003**, *37*, 462–469. [CrossRef] [PubMed]

74. Ravald, N.; Johansson, C.S. Tooth loss in periodontally treated patients. A long-term study of periodontal disease and root caries. *J. Clin. Periodontol.* **2012**, *39*, 73–79. [CrossRef] [PubMed]

75. Kumar, P.S.; Griffen, A.L.; Moeschberger, M.L.; Leys, E.J. Identification of candidate periodontal pathogens and beneficial species by quantitative 16S clonal analysis. *J. Clin. Microbiol.* **2005**, *43*, 3944–3955. [CrossRef] [PubMed]

76. Darveau, R.; Hajishengallis, G.; Curtis, M. Porphyromonas gingivalis as a potential community activist for disease. *J. Dent. Res.* **2012**, *91*, 816–820. [CrossRef] [PubMed]

77. Fteita, D.; Könönen, E.; Söderling, E.; Gürsoy, U.K. Effect of estradiol on planktonic growth, coaggregation, and biofilm formation of the Prevotella intermedia group bacteria. *Anaerobe* **2014**, *27*, 7–13. [CrossRef] [PubMed]

78. Charalampakis, G.; Leonhardt, Å.; Rabe, P.; Dahlén, G. Clinical and microbiological characteristics of peri-implantitis cases: a retrospective multicentre study. *Clin. Oral Implants Res.* **2012**, *23*, 1045–1054. [CrossRef] [PubMed]

79. Signat, B.; Roques, C.; Poulet, P.; Duffaut, D. Role of Fusobacterium nucleatum in periodontal health and disease. *Curr. Issues Mol. Biol.* **2011**, *13*, 25–36. [PubMed]

80. Saito, A.; Inagaki, S.; Kimizuka, R.; Okuda, K.; Hosaka, Y.; Nakagawa, T.; Ishihara, K. Fusobacterium nucleatum enhances invasion of human gingival epithelial and aortic endothelial cells by Porphyromonas gingivalis. *FEMS Immunol. Med. Microbiol.* **2008**, *54*, 349–355. [CrossRef] [PubMed]

81. Souto, R.; Colombo, A.P.V. Prevalence of Enterococcus faecalis in subgingival biofilm and saliva of subjects with chronic periodontal infection. *Arch. Oral Biol.* **2008**, *53*, 155–160. [CrossRef] [PubMed]

82. Wang, L.; Melo, M.A.S.; Weir, M.D.; Xie, X.J.; Reynolds, M.A.; Xu, H.H.K. Novel bioactive nanocomposite for Class-V restorations to inhibit periodontitis-related pathogens. *Dent. Mater.* **2016**, *32*, 351–361. [CrossRef] [PubMed]

83. Wang, L.; Xie, X.J.; Imazato, S.; Weir, M.D.; Reynolds, M.; Xu, H.H.K. Protein-repellent and antibacterial nanocomposite for Class-V restorations to inhibit periodontitis-related pathogens. *Mater. Sci. Eng. C Mater. Biol. Appl.* **2016**, *67*, 702–710. [CrossRef] [PubMed]

84. Wang, L.; Li, C.Y.; Weir, M.D.; Zhang, K.; Zhou, Y.M.; Xu, H.H.K.; Reynolds, M.A. Novel multifunctional dental bonding agent for class-V restorations to inhibit periodontal biofilms. *RSC Adv.* **2017**, *7*, 29004–29014. [CrossRef] [PubMed]

85. Periasamy, S.; Kolenbrander, P.E. Mutualistic biofilm communities develop with Porphyromonas gingivalis and initial, early, and late colonizers of enamel. *J. Bacteriol.* **2009**, *191*, 6804–6811. [CrossRef] [PubMed]

86. Park, J.H.; Lee, J.K.; Um, H.S.; Chang, B.S.; Lee, S.Y. A periodontitis-associated multispecies model of an oral biofilm. *J. Periodontal Implant Sci.* **2014**, *44*, 79–84. [CrossRef] [PubMed]

87. Nair, P. On the causes of persistent apical periodontitis: a review. *Int. Endod. J.* **2006**, *39*, 249–281. [CrossRef] [PubMed]

88. Siqueira, J.F.; Araújo, M.C.; Garcia, P.F.; Fraga, R.C.; Dantas, C.J.S. Histological evaluation of the effectiveness of five instrumentation techniques for cleaning the apical third of root canals. *J. Endod.* **1997**, *23*, 499–502. [CrossRef]

89. Stuart, C.D.; Schwartz, S.A.; Beeson, T.J.; Owatz, C.B. Enterococcus faecalis: its role in root canal treatment failure and current concepts in retreatment. *J. Endod.* **2006**, *32*, 93–98. [CrossRef] [PubMed]

90. Wang, L.; Xie, X.J.; Li, C.Y.; Liu, H.B.; Zhang, K.; Zhou, Y.M.; Chang, X.F.; Xu, H.H.K. Novel bioactive root canal sealer to inhibit endodontic multispecies biofilms with remineralizing calcium phosphate ions. *J. Dent.* **2017**, *60*, 25–35. [CrossRef] [PubMed]

91. Zhou, H.; Liu, H.; Weir, M.D.; Reynolds, M.A.; Zhang, K.; Xu, H.H. Three-dimensional biofilm properties on dental bonding agent with varying quaternary ammonium charge densities. *J. Dent.* **2016**, *53*, 73–81. [CrossRef] [PubMed]

92. Webb, J.S.; Thompson, L.S.; James, S.; Charlton, T.; Tolker-Nielsen, T.; Koch, B.; Givskov, M.; Kjelleberg, S. Cell death in Pseudomonas aeruginosa biofilm development. *J. Bacteriol.* **2003**, *185*, 4585–4592. [CrossRef] [PubMed]

93. Ferracane, J.L.; Giannobile, W.V. Novel biomaterials and technologies for the dental, oral, and craniofacial structures. *J. Dent. Res.* **2014**, *93*, 1185–1186. [CrossRef] [PubMed]

94. Price, R.B.; Ferracane, J.L. Enhancing the value of dental biomaterials research: Reducing the noise. *J. Dent. Res.* **2018**, *97*, 481–482. [CrossRef] [PubMed]

![materials logo] *materials*

MDPI

Article

Surface-Attached Poly(oxanorbornene) Hydrogels with Antimicrobial and Protein-Repellent Moieties: The Quest for Simultaneous Dual Activity

Monika Kurowska, Vania Tanda Widyaya, Ali Al-Ahmad and Karen Lienkamp *[ID]

Freiburg Center for Interactive Materials and Bioinspired Technologies (FIT) and Department of Microsystems Engineering (IMTEK), Albert-Ludwigs-Universität, Georges-Köhler-Allee 105, 79110 Freiburg, Germany; monika.kurowska@imtek.uni-freiburg.de (M.K.); vania.widyaya@imtek.uni-freiburg.de (V.T.W.); ali.al-ahmad@uniklinik-freiburg.de (A.A.-A.)
* Correspondence: lienkamp@imtek.uni-freiburg.de; Tel.: +49 761 203 95090

Received: 23 July 2018; Accepted: 9 August 2018; Published: 11 August 2018

Abstract: By copolymerizing an amphiphilic oxanorbornene monomer bearing N- tert-butyloxycarbonyl (Boc) protected cationic groups with an oxanorbornene-functionalized poly(ethylene glycol) (PEG) macromonomer, bifunctional comb copolymers were obtained. Varying the comonomer ratios led to copolymers with PEG contents between 5–25 mol %. These polymers were simultaneously surface-immobilized on benzophenone-bearing substrates and cross-linked with pentaerythritoltetrakis(3-mercaptopropionate). They were then immersed into HCl to remove the Boc groups. The thus obtained surface-attached polymer hydrogels (called SMAMP*-*co*-PEG) were simultaneously antimicrobial and protein-repellent. Physical characterization data showed that the substrates used were homogeneously covered with the SMAMP*-*co*-PEG polymer, and that the PEG moieties tended to segregate to the polymer–air interface. Thus, with increasing PEG content, the interface became increasingly hydrophilic and protein-repellent, as demonstrated by a protein adhesion assay. With 25 mol % PEG, near-quantitative protein-adhesion was observed. The antimicrobial activity of the SMAMP*-*co*-PEG polymers originates from the electrostatic interaction of the cationic groups with the negatively charged cell envelope of the bacteria. However, the SMAMP*-*co*-PEG surfaces were only fully active against *E. coli*, while their activity against *S. aureus* was already compromised by as little as 5 mol % (18.8 mass %) PEG. The long PEG chains seem to prevent the close interaction of bacteria with the surface, and also might reduce the surface charge density.

Keywords: antimicrobial polymer; coatings; hydrogel; protein-repellent polymer; surface-attached polymer network

1. Introduction

Bacterial infections related to the use of medical devices is one of the main causes of nosocomial infections [1], and might cost the lives of up to 10 million people worldwide every year by 2050 if current trends continue [2]. Thus, numerous strategies to prevent biofilm formation have been reported in the past two decades [3–19]. One prominent strategy is to coat materials with polycationic surfaces which then kill bacteria upon contact due to their interaction with the bacteria's negatively charged cell membranes [9,20,21]. The exact mechanism of this interaction is still unclear, however it seems as if charge density plays an important role to either damage the bacterial membrane or to pin the bacteria to the surface and thereby prevent their proliferation [21–23]. However, it was also shown that cationic polymer surfaces get contaminated by the debris of dead bacteria, and therefore they do not prevent bacterial biofilm formation on the long run [24,25]. A number of studies have addressed this problem by combining antimicrobial and protein-repellent moieties in one material. This has been nicely reviewed by Chen [26]. For example, Paris et al. grafted the antimicrobial peptide nisin onto surface-immobilized anti-adhesive anionic hyaluronic acid [27]. Yang et al. coupled antibacterial chitosan to protein-repellent poly(2-hydroxyethyl methacrylate) polymer brushes [28]. Ye et al. presented a membrane that was surface-functionalized with block copolymer brushes containing non-fouling polyzwitterionic moieties and antimicrobial quaternary ammonium parts [29]. In these and other examples, some degree of simultaneous dual antimicrobial activity and protein-repellency was obtained, yet they also demonstrate that the integration of anti-adhesive and antibacterial components into one surface coating without loss of one of the two functionalities remains a challenge [30]. We also previously reported the integration of polymeric synthetic mimics of antimicrobial peptides (SMAMPs) and protein-repellent polyzwitterions, either side-by-side using micro-/nanostructuring approaches [31,32], or hierarchically using a grafting-onto approach [31]. However, these were either obtained via complicated, multi-step procedures (which is undesirable for technical applications) [31,32], and/or the surface modification could not be sufficiently controlled, so that no clear structure–property relationships were obtained [31].

We here present a bifunctional comb copolymer consisting of antimicrobial, polycationic SMAMP moieties and protein-repellent poly(ethylene glycol) (PEG) side chains (Scheme 1). SMAMPs are well known for their high antimicrobial activity combined with good cell compatibility [33], yet they are also protein-adhesive due to their cationic nature [25]. PEG, on the other hand, is well known for its protein-repellency [17,34]. The aim of this study was to investigate the structure–property relationships of such SMAMP-PEG copolymers, and to see whether simultaneously bifunctional materials could be obtained using this synthetic platform. Importantly, the PEG moieties were obtained through a macromonomer, so that the resulting material had a hierarchical structure, with PEG extending over the SMAMP moieties. The question was how much PEG was needed to sufficiently shield the SMAMP moieties from proteins, yet still retain the antimicrobial activity. For this, copolymers with varying SMAMP and PEG content (5, 10, and 25 mol % PEG, named SMAMP*-co-X%PEG, where X refers to the PEG content) were synthesized and surface-immobilized, and their physical properties and bioactivity were evaluated and compared.

2. Experiment

2.1. Materials

All chemicals were obtained as reagent grade from Sigma-Aldrich (Taufkirchen, Germany), Carl Roth (Karlsruhe, Germany), Fluka (Taufkirchen, Germany), or Alfa Aeser (Karlsruhe, Germany), and used as received. High performance liquid chromatography (HPLC) grade solvents were purchased dry from Carl Roth (Karlsruhe, Germany), and used as received. Dichloromethane (DCM) was freshly distilled over P_2O_5 before use.

2.2. Instrumentation.

Gel permeation chromatography (GPC, in THF, calibrated with poly(styrene) standards) was performed on a PSS SDV column (PSS, Mainz, Germany). NMR spectra were recorded on a Bruker 250 MHz spectrometer (Bruker, Madison, WI, USA). MALDI-TOF mass spectra were measured on Bruker Autoflex III TOF/TOF mass spectrometer (Bruker, Billerica, MA, USA) equipped with a 200 Hz beam laser. The measurement was performed by using trans-2-[3-(4-tert-butylphenyl)-2-methyl-2-propenylidene]malononitrile (DCTB) as matrix and Na^+ as ionization agent in $CHCl_3$. Measurements were done at the Institute for Macromolecular Chemistry of the University of Freiburg. The thickness of the dry polymer layers on silicon wafers was measured with the auto-nulling imaging ellipsometer Nanofilm EP3 (Nanofilm Technologie GmbH, Göttingen, Germany), which was equipped with a 532 nm solid-state laser. For each sample, the average value from three different positions was taken. The irradiation of samples with UV light was conducted using a BIO-LINK Box (Vilber Lourmat

GmbH, Eberhardzell, Germany) with different wavelengths (254 and 365 nm). Attenuated total reflection Fourier transform infrared spectroscopy (ATR-FTIR) spectra were recorded from 4000 to 400 cm^{-1} with a Bio-Rad Excalibur spectrometer (Bio-Rad, München, Germany), using a spectrum of the blank double side polished silicon wafer as background. Double side polished silicon wafers were used as substrates for the FTIR experiments. Contact angles were measured on an OCA 20 system (Dataphysics GmbH, Filderstadt, Germany). The average value of the contact angles was calculated from four measurements per sample. The topography of the surfaces was imaged with a Dimension Icon atomic force microscope (AFM) (Bruker). Commercial Bruker ScanAsyst Air cantilevers (length: 115 μm; width: 25 μm; spring constant: 0.4 Nm^{-1}; resonance frequency: 70 kHz) were used. All AFM images were recorded in the ScanAsyst mode in air, respectively. The obtained images were analyzed and processed with the software 'Nanoscope Analysis 9.1'. For each sample, the root mean square (RMS) average roughness from three images of an area of 5 × 5 μm^2 at different positions was taken. Photoelectron spectroscopy (XPS) data was obtained from on a Perkin Elmer PHI 5600 ESCA System (PerkinElmer, Waltham, MA, USA). The X-ray source was a Mg anode with an energy of 1253.6 eV. The aperture size was 400 μm, the angle was 45°. The typical measurement size is 10 μm^2. Samples were measured at room temperature. Surface plasmon resonance spectroscopy (SPR) experiments were performed on a RT2005 RES-TEC device in Kretschmann configuration from Res-Tec, Framersheim, Germany. Excitation was done with a He-Ne-Laser with λ = 632.8 nm. SPR substrates were homemade (LaSFN9 glass from Hellma GmbH, Müllheim, Germany; coated with 1 nm Cr and 50 nm Au at the Cleanroom Service Center (RSC) of the Department of Microsystems Engineering, University of Freiburg, using the device CS 730 S (Von Ardenne, Dresden, Germany). The set-up and measuring procedures of the kinetics experiments and the full angular reflectivity scans have been reported previously [25]. Experiments to test the antimicrobial activity of the polymer networks were performed using a previously described spray assay, which is a modification of the Japanese Industrial Standard JIS Z 2801:2000 "Antibacterial Products Test for Anti-bacterial Activity and Efficacy", and was reported previously [35]. *S. aureus* (ATCC29523) and *E. coli* (ATCC25922) were used.

2.3. Synthesis

Monomers—The synthesis and characterization of the monomers P and M was described previously [36,37].

Copolymerization—Polymerizations were performed under nitrogen using standard Schlenk techniques. The respective amounts of propyl SMAMP and PEG monomers were dissolved in anhydrous THF (5 mL for SMAMP-*co*-5%PEG and SMAMP-*co*-10%PEG copolymers, and 6 mL for SMAMP-*co*-25%PEG copolymer). The Grubbs third generation catalyst was dissolved separately in 2 mL anhydrous THF and added to the monomers solution in one shot. After 40 min stirring, the polymerization was quenched by adding 1 mL (750 mg, 10 mmol) ethyl vinyl ether. The mixture was stirred for 30 min. The solvent was removed under reduced pressure and the crude polymer was purified by precipitation into a cold diethyl ether/n-hexane mixture, yielding white solid. Yield: 80%. The reagent amounts for each copolymer are included in Table 1. The SMAMP reference polymer was synthesized as reported previously [33].

Table 1. Reagent amounts for the synthesis of the SMAMP-*co*-PEG polymers.

Polymer	Monomer M		Monomer P		Catalyst G3		Solvent
	n/mmol	m/mg	n/mmol	m/mg	n/mmol	m/mg	mL
SMAMP*-*co*-5%PEG	0.07	116	1.36	500	5.7·10^{-3}	4.2	7
SMAMP*-*co*-10%PEG	0.15	244	1.36	500	6.5·10^{-3}	4.7	7
SMAMP*-*co*-25%PEG	0.45	733	1.36	500	9.5·10^{-3}	6.9	8

Surface anchor groups and surface functionalization—The molecule used to covalently bind the SMAMP-*co*-PEG copolymers to the Si surfaces was a benzophenone-functionalized triethoxysilane (3EBP); a lipoic acid disulfide (LS-BP) was used for the gold substrates for the surface plasmon resonance spectroscopy (SPR) experiments. LS-BP and 3EBP-silane were synthesized as described previously; the surfaces were also functionalized as described previously [25,33].

Surface-attached polymer networks. A stock solution (Solution A) was prepared by dissolving pentaerythritol-tetrakis-(3-mercaptopropionate) (1 mL, 1.3 g, 2.6 mmol) in THF (50 mL). SMAMP-*co*-5%PEG copolymer (10 mg, 0.02 mmol), SMAMP-*co*-10%PEG (11 mg, 0.02 mmol), or SMAMP-*co*-25%PEG (14 mg, 0.02 mmol) were dissolved in Solution A (0.25 mL). Chloroform (0.8 mL) was added as co-solvent. The mixture was stirred for 60 s. From this solution, a polymer film was spin cast onto a 3-EBP treated silicon wafer or a LS-BP treated gold substrate at 3000 rpm for 30 s. The resulting polymer film was cross-linked at 254 nm for 30 min. It was then washed with THF to remove unbound polymer chains and dried under N_2-flow. To remove the Boc protective groups, the film was immersed in HCl (4 M in dioxane) for 12 h and washed twice with ethanol, and dried under N_2-flow.

Physical and biological characterization—All experiments were performed as reported previously for other surface-attached polymer networks [25].

3. Results and Discussion

3.1. Material Design

The two above described functionalities were combined by copolymerizing the oxanorbornene-functionalized PEG macromonomer M with the oxanorbornene monomer P via ring-opening metathesis polymerization (ROMP, Scheme 1). The P moieties carry each a propyl group and a N-Boc protected primary amine group, which would be cationic after removal of the Boc group. This would impart antimicrobial activity onto the polymer upon deprotection. The thus obtained N-Boc-protected SMAMP-*co*-PEG copolymers were simultaneously surface-attached and cross-linked by UV-activated thiol-ene reactions with pentaerythritoltetrakis(3-mercaptopropionate) and the polymer double bonds, and by C,H insertion reactions between the benzophenone groups on the substrate and polymer CH-bonds (Scheme 1). After surface-attachment and deprotection (giving SMAMP*-*co*-PEG surfaces), the cationic moieties of the SMAMP* repeat units and the PEG grafts microphase separated. Thus, by varying the PEG content of the copolymer, the domain sizes and the content of PEG at the polymer–air interface could be controlled, so that materials with simultaneous protein repellency and antimicrobial activity were obtained, as described in detail below.

3.2. Monomer Synthesis

The protected cationic monomer P and the PEG macromonomer M were each synthesized in a two-step reaction following the procedures described in the literature (Scheme 1a,b) [36,38]. To synthesize P, the oxanorbornene anhydride 1 was ring-opened by 1-propanol in the presence of catalytic amounts of a base (*N,N*-dimethylaminopyridine, DMAP), so that the propyl ester 2 was obtained. In the second step, the remaining carboxyl functionality was esterified with 2-(N-Boc)aminoethanol using standard peptide coupling conditions (DMAP and dicyclohexylcarbodiimide, DCC, Scheme 1a) [33,37]. In a similar reaction sequence, anhydride 1 was ring-opened by an ω-methoxy PEG alcohol with 16 repeat units (average molecular weight M_n = 750 g mol^{-1}, Scheme 1b) using DMAP as base; the second PEG chain was attached to the carboxyl group of intermediate 3 using DMAP/DCC, so that the symmetrical macromonomer M was obtained (Scheme 1b). The ^{1}H-NMR spectra of the monomer P and the macromonomer M obtained after monomer purification matched the literature data [36,38]. In particular, the NMR spectrum of macromonomer M, with less peaks than the spectrum of intermediate 3, indicated that a symmetrical two-armed compound was obtained (Figure S1a). Both the macromonomer M and the intermediate 3 were analyzed by gel-permeation chromatography (GPC, in THF, Figure S1b). In the GPC elugram, both M and 3 had a single peak with a low polydispersity index (M_w/M_n = 1.04). The macromonomer M eluted at shorter retention times than the intermediate 3, i.e., it had a higher molar mass. Notably, there was no peak at the elution time of the intermediate, indicating that a quite pure macromonomer was obtained. Using a calibration curve (polystyrene standards), the number average molar mass M_n of M was calculated as 1920 g mol^{-1}, and that of 3 as 960 g mol^{-1}. Both numbers are in good correlation with the expected masses (=oxanorbornene head group + PEG residues). The structure of the macromonomer was further confirmed by MALDI-TOF mass spectrometry (Figure S1c). All these findings matched literature reports [36].

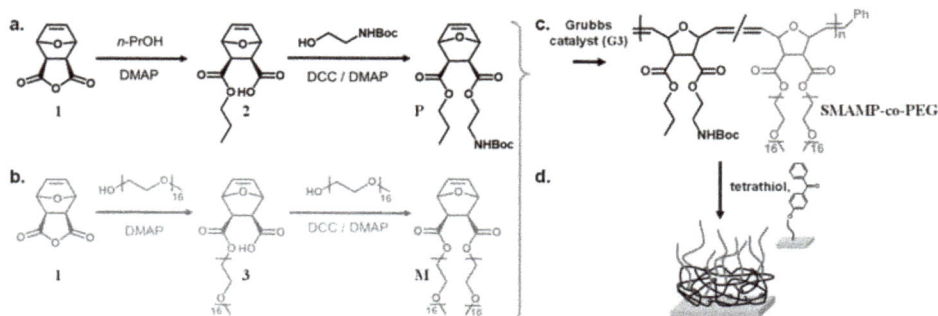

Scheme 1. Synthesis of the protected propyl monomer P (**a**) and the PEG macromonomer M (**b**). Copolymerization of P and M with different mass ratios gave the copolymer SMAMP-*co*-PEG (**c**). Using UV irradiation, the copolymers were simultaneously surface-immobilized on a substrate pre-treated with a benzophenone linker, and cross-linked using the tetrathiol cross-linker (**d**).

3.3. Polymer Synthesis

The copolymers SMAMP-*co*-PEG were obtained by ring-opening metathesis polymerization (ROMP) using Grubbs' third generation catalyst (G3). The monomer P and the macromonomer M were copolymerized with different P:M ratios in dry tetrahydrofuran, so that copolymers with 5, 10, and 25 mol % PEG content were obtained (Scheme 1c; samples were named SMAMP-*co*-5%PEG, SMAMP-*co*-10%PEG, and SMAMP-*co*-25%PEG, respectively). The isolated yield after work-up by precipitation into diethyl ether/n-hexane was about 80%. The polymers thus obtained were analyzed by ^{1}H-NMR spectroscopy (Figure 1a). Due to structural similarity, many peaks from the propyl SMAMP repeat unit and the PEG repeat unit overlapped. However, the characteristic peak at 3.64 ppm could be assigned to the methylene protons of the PEG component (OCH$_2$CH$_2$), while the peak at 1.63 ppm belonged to two propyl methylene protons of the SMAMP repeat unit (CH$_2$CH$_2$CH$_3$). Thus, the PEG content of the copolymers could be determined by integrating and comparing these two signals. The data thus obtained is summarized in Table 2. The actual PEG content, as determined by NMR, was 4.5, 8.1, and 22.1% respectively. This closely matched the initial monomer feed ratio, demonstrating the high efficiency of the polymerization. The molar masses and molar mass distributions of the polymers were determined by gel-permeation chromatography (GPC, in THF using polystyrene standards). The GPC elugrams thus obtained are shown in Figure 1b, the analytical data are summarized in Table 2. The GPC peaks of the copolymers were symmetrical, monomodal, and elute significantly earlier than macromonomer M. This confirms that high molecular masses were obtained. While the target molecular mass for all polymers were 100,000 g mol^{-1}, the calculated masses were significantly lower (66,000; 63,000; and 50,000 g mol^{-1}, respectively). This can be explained by the different hydrodynamic volumes per unit mass of the comb polymers compared to the calibration standard: first, the chemical nature of the repeat units was different, and second and more importantly, the samples were highly branched. The polydispersity indices were 1.2 to 1.6 and thus a little higher than what would be expected for RCMP, however this can be attributed to one monomer being a macromonomer: these are generally more difficult to polymerize than conventional low molar mass monomers.

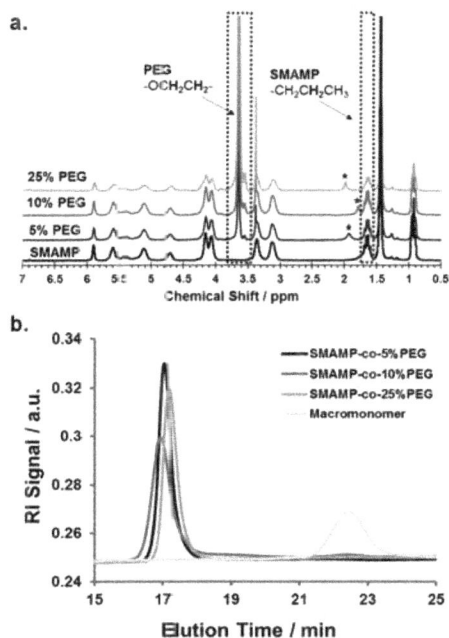

Figure 1. (a) ^{1}H-NMR spectra of the three SMAMP-*co*-PEG copolymers (in CDCl$_3$) with 5, 10, and 25 mol % PEG content. The PEG content was calculated by comparing the signal integral of the methylene protons from the SMAMP propyl chain at 1.63 ppm to the peak intensity of the ethylene glycol protons of the PEG repeat unit at 3.64 ppm. The asterisks (*) indicate water peaks. (b) GPC elugrams (in THF, calibrated with polystyrene standards) of SMAMP-*co*-PEG copolymers with 5, 10, and 25 mol % PEG content, compared to the GPC elugram of the macromonomer M.

Table 2. Characterization of the three SMAMP-*co*-PEG polymers. The PEG content determined from the ^{1}H-NMR spectra was compared to the calculated values; the number average molar mass M_n and the polydispersity index (M_w/M_n) were determined by gel permeation chromatography.

Copolymer	SMAMP to PEG Ratio	PEG Content				M_n/kg mol^{-1}	M_w/M_n
		mol %		mass %			
		calc.	NMR	calc.	NMR		
SMAMP-*co*-5%PEG	95:5	5	4.5	18.8	17.3	66	1.3
SMAMP-*co*-10%PEG	90:10	10	8.1	32.8	28.0	63	1.6
SMAMP-*co*-25%PEG	75:25	25	22.1	59.5	55.6	50	1.2

3.4. Synthesis and Physical Characterization of Surface-Attached Polymer Networks.

To obtain bifunctional polymer surfaces with antimicrobial and protein-repellent moieties, the SMAMP-*co*-PEG polymers were surface-immobilized as networks (Scheme 1d), and then activated. For this, they were each dissolved in a mixture of CHCl$_3$ and THF, to which the cross-linker pentaerythritoltetrakis(3-mercaptopropionate) (=tetrathiol) was added. This solution was spin-coated onto a solid substrate (either a silicon wafer or a gold substrate) that had been functionalized with benzophenone as reported previously [25,33]. Upon UV irradiation, the polymers were simultaneously cross-linked (by thiol-ene reaction of the polymer double bonds with the SH groups of the crosslinker) and surface-immobilized (by C,H insertion reactions between the surface-attached benzophenone groups and C-H bonds of the polymer). The samples were then washed with THF to remove unbound cross-linker and polymer chains. To remove the N-Boc protective group and thereby activate the antimicrobial function of the SMAMP repeat units, the surfaces were treated with hydrochloric acid. The resulting SMAMP*-*co*-PEG networks as well as their SMAMP-*co*-PEG precursors with the protective groups were characterized by ellipsometry, contact angle measurements, FTIR spectroscopy, atomic force microscopy (AFM), and photoelectron spectroscopy (XPS). The results of these measurements, compared to the data obtained for pure SMAMP networks (synthesized as described previously) [33], are summarized in Table 3. After deprotection, the thickness of the SMAMP*-*co*-PEG networks (determined by ellipsometry) decreased slightly (to 71 to 93 nm, Table 3) compared to the SMAMP-*co*-PEG networks, which due to the removal of the N-Boc protective groups. The hydrophilicity of the networks was investigated by static and dynamic water contact angles measurements. Overall and matching expectations, the contact angles of the protected and the deprotected networks decreased with increasing PEG content. Thus, the hydrophilicity of these surfaces increased with increasing content of hydrophilic PEG. The FTIR spectra of the protected and the deprotected networks are shown in Figure 2a. These data show that there is great structural similarity between the protected SMAMP-*co*-PEG networks (grey line) and their deprotected SMAMP*-*co*-PEG analogues (black lines). Both had the stretching vibration of C=O at 1732 cm^{-1}. The stretching vibration of C–O was found at about 1140 and 1230 cm^{-1}, and the C–H stretching vibrations of the aliphatic CH$_3$ and CH$_2$ groups were in the range of 2870 to 2950 cm^{-1}. The spectra of the protected surfaces exhibited an additional characteristic peak at about 3400 cm^{-1}, which was assigned to amide NH stretching vibration of the of N-Boc protective groups. This absorption band was weaker for the samples that had a higher PEG content and completely disappeared after acidic deprotection of the networks. Instead, a new absorption peak at about 1583 cm^{-1} was observed, which corresponds to the deformation vibration of the NH$_2$ and NH$_3$$^+$ groups. Again, the intensity of this band decreased with higher PEG content, which is plausible as this corresponds to decreasing SMAMP content. Overall, these FTIR spectra confirmed the presence of the expected functional groups in the surface-attached SMAMP-*co*-PEG and SMAMP*-*co*-PEG copolymer networks, and indicated that the activation step (removal of the N-Boc group by HCl) was successful.

Table 3. Physical characterization data of SMAMP-*co*-PEG networks (protected) and SMAMP*-*co*-PEG networks (deprotected). The dry layer thickness was determined by ellipsometry; θ_{static}, θ_{adv} and θ_{rec} = static, advancing and receding contact angles; rms roughness was determined by atomic force microscopy (AFM) from the images shown in Figure 2b.

	SMAMP	SMAMP*	SMAMP -*co*-5% PEG	SMAMP* -*co*-5% PEG	SMAMP -*co*-10% PEG	SMAMP* -*co*-10% PEG	SMAMP -*co*-25% PEG	SMAMP* -*co*-25% PEG
Thickness/nm	62 ± 2	53 ± 2	86 ± 2	79 ± 1	101 ± 2	93 ± 1	79 ± 2	71 ± 3
θ_{static}/°	82 ± 2	51 ± 1	89 ± 1	52 ± 2	85 ± 3	51 ± 2	79 ± 1	46 ± 2
θ_{adv}/°	91 ± 4	56 ± 2	90 ± 2	61 ± 2	87 ± 1	58 ± 3	85 ± 2	52 ± 1
θ_{rec}/°	43 ± 4	26 ± 2	42 ± 3	24 ± 2	41 ± 1	27 ± 1	37 ± 3	22 ± 2
Roughness/nm		2.1		1.1		1.9		0.6

The morphology of the deprotected surface-attached SMAMP*-*co*-PEG networks was characterized by atomic force microscopy (AFM) using the peak force tapping mode. The resulting height images are shown in Figure 2b, together with a height image of SMAMP* for comparison. The SMAMP* network has the typical morphology obtained for surface-attached polymer networks that are cross-linked by an additional low molecular weight cross-linker [33,39]: small pores with a diameter of about 100 nm formed (previous areas of excess cross-linker), which are surrounded by a homogeneous polymer coating with a relatively low rms roughness (2.1 nm). These pores are also observed for the SMAMP*-*co*-PEG networks. The slightly different pore sizes in the various samples reflect the different polymer compositions, with a different overall hydrophilicity. This leads to a slightly different partition of the tetrathiol cross-linker between the polymer and the pure cross-linker microphases. Interestingly, the SMAMP*-*co*-PEG networks with 5 and 10 mol % (corresponding to about 19 to 32 mass % PEG) have further structural features. For the SMAMP*-*co*-5%PEG sample, small dots of PEG-rich domains in a SMAMP matrix can be seen. When the PEG content increased to 10 mol %, the PEG domains increased further, forming irregular wormlike patterns. For surfaces with 25 mol % PEG (corresponding to about 60 mass % PEG), the morphology appeared to be more homogeneous. Apparently, the PEG arms segregate to the surface and cover a large area of the polymer–air interface. It is well known that the highly mobile and hydrophilic PEG chains can easily rearrange on the surface depending on copolymer composition and environmental influences (e.g., humidity) [40,41].

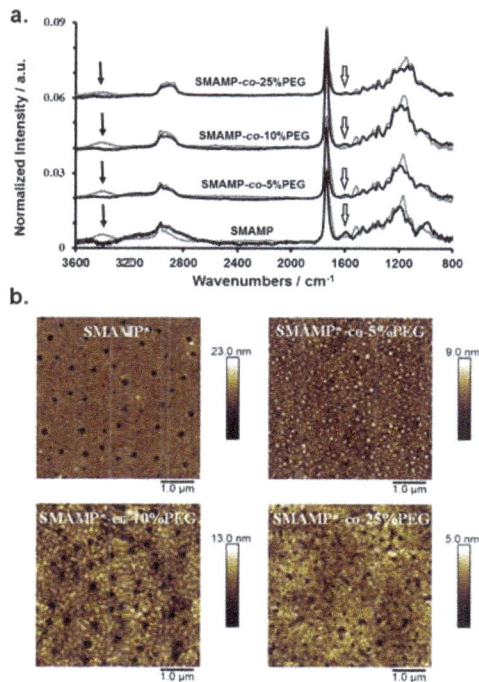

Figure 2. (**a**) FTIR spectra of the surface-attached SMAMP and SMAMP-*co*-PEG networks (grey lines), as well as the deprotected SMAMP* and SMAMP*-*co*-PEG networks (black lines). The black arrows indicate the signals of the Boc protective group of the SMAMP units, the open arrows designate the deprotected amine and ammonium groups. (**b**) AFM height images of the activated SMAMP*-*co*-PEG and SMAMP* networks (peak force tapping mode in air). The RMS roughness (Table 3) was calculated from these images using the Gwydion software package.

The SMAMP* and SMAMP*-*co*-PEG networks were also analyzed by photoelectron spectroscopy (XPS). Using this method, the elemental composition on the top of few nanometers of each sample was probed. The XPS data obtained for carbon, nitrogen, and oxygen, as well as the elemental compositions of these of these polymers calculated from their molecular formulae are presented in Table 4. For all polymers, the carbon and oxygen contents determined by XPS were a little higher than the calculated data, yet there is no trend in that data that would indicate a preferred polymer orientation at the interface with increasing PEG content. The amount of nitrogen, however, decreases disproportionately in the XPS/calc. signal ratio with increasing PEG content. Thus, with increasing PEG content, the PEG chains seem to dominate the air-polymer interface, while SMAMP gets hidden underneath. This is in line with expectations since the PEG chains are longer and more flexible than the SMAMP repeat units. Thus, the XPS data complement the FTIR, ellipsometry, contact angle, and AFM results.

Table 4. XPS data for the surface-attached SMAMP* and SMAMP*-*co*-PEG networks (deprotected), compared to the elemental composition of these polymers calculated from their respective molecular structures.

Polymer	Elemental Composition/%						Ratio XPS/calc.		
	XPS			calc.					
	C 1s	N 1s	O 1s	C	N	O	C	N	O
SMAMP*	71.2	3.2	25.6	68.4	5.3	26.3	1.04	0.61	1.02
SMAMP*-*co*-5%PEG	70.6	3.1	26.3	68.3	5.0	26.7	1.03	0.62	1.02
SMAMP*-*co*-10%PEG	70.1	2.8	27.1	68.2	4.7	27.0	1.03	0.38	1.03
SMAMP*-*co*-25% PEG	70.5	1.3	28.2	68.0	3.9	28.1	1.04	0.33	1.04

3.5. Protein Adhesion and Antimicrobial Activity of Surface-Attached Polymer Networks.

To verify that the bifunctional copolymer networks were indeed protein-repellent, their interaction with the protein fibrinogen was studied by surface plasmon resonance spectroscopy (SPR). For this, the SMAMP*-*co*-PEG networks and a pure SMAMP* reference network were surface-immobilized on gold substrates as described previously [25,31,33]. After each fabrication step, full angular reflectivity curves were recorded (see Figure S2). After that, the activated surfaces were exposed to fibrinogen (1 mg mL^{-1} in HEPES buffer). The interaction of this protein with the surfaces was monitored by SPR using the kinetics mode, where time-dependent changes in reflectivity at constant angle were measured at room temperature. Additionally, full angular reflectivity curves of the dry surfaces were recorded before and after the kinetics experiments to quantify the amount of irreversibly adhered protein.

In the kinetics experiment (Figure 3a), the surfaces were first exposed to buffer for about 10 min. The dashed line marks the time point of protein injection. If the reflectivity signal stays constant over time, no protein is adsorbed. If it increases, protein adheres to the surface and thereby changes the dielectric properties of the sample. For SMAMP*-*co*-25%PEG (light grey line), only very little protein adhesion was observed. For the other samples, protein adhesion was visible and increased with decreasing PEG content. This was confirmed by the full angular reflectivity scans recorded before and after protein adhesion (Figure 3b): For SMAMP*-*co*-25%PEG, the position of the plasmon peak was almost unchanged. For the other samples, its minimum shifted to higher angles. By simulating these curves using the Winspall software, the protein adhesion can be quantified. This data is summarized in Table 5. It shows that the PEG content on the SMAMP*-*co*-10%PEG surfaces was already too low to substantially suppress fibrinogen adhesion and to work against the adhesive forces exerted on the negatively charged protein molecules by the cationic SMAMP* groups. On the other hand, SMAMP*-*co*-25%PEG adsorbed only 0.99 ng fibrinogen per mm^2. In this case, the PEG coverage on the surface was high enough to substantially screen the electrostatic attraction of SMAMP* ammonium cations. However, quantitative protein repellency was also not observed.

The antimicrobial activity of the surface-attached SMAMP*-*co*-PEG networks was studied using the standardized airborne antimicrobial assay [35]. In this experiment, two pathogenic bacterial strains, Gram-negative *Escherichia coli* and Gram-positive *Staphylococcus aureus*, were sprayed onto the test surfaces. As reported previously [35], uncoated silicon wafers (negative control) and uncoated silicon wafers impregnated with chlorhexidine digluconate (positive control) were tested along with the polymer coated samples. Each surface was sprayed with bacterial suspensions containing 10^6 bacteria per cm^3 and incubated for four hours. After that, the surviving bacteria were transferred onto agar plates and cultivated, each surviving bacteria forming a colony. By counting these colonies, the antimicrobial activity could be quantified (Figure 3c,d). The assay was performed twice with five samples of each material type. The error bars are the standard deviation calculated from these data. The antimicrobial activity of each polymer surface was reported as bacterial growth (in percent) normalized to the negative control. Unexpected, all surface-attached SMAMP*-co-PEG networks, regardless of their PEG content, killed ≥99.9% of the adherent *E. coli* bacteria and were thus bactericidal. However, the polymer surfaces were significantly less effective against *S. aureus* bacteria—here, the killing efficiency decreased with increasing PEG content from 89% killing (11% growth) for the surfaces with 5 mol % PEG content, up to 52.6% killing (47.4% growth) for the surfaces with 25% PEG content. In contrast, the SMAMP* surface had excellent antimicrobial activity [33].

Table 5. Bioactivity data of SMAMP*-co-PEG samples and the SMAMP* reference surface: fibrinogen adhesion (in ng mm^{-2}), determined by surface plasmon resonance spectroscopy, and antimicrobial activity (% bacterial growth) against *E. coli* and *S. aureus* bacteria.

Polymer	Protein Adhesion/ng mm^{-2}	Antimicrobial Activity/% growth	
		E. coli	*S. aureus*
SMAMP*	11.3	-	-
SMAMP*-co-5%PEG	9.8	0.1	11
SMAMP*-co-10%PEG	8.4	0.1	18.2
SMAMP*-co-25% PEG	0.99	0	47.4

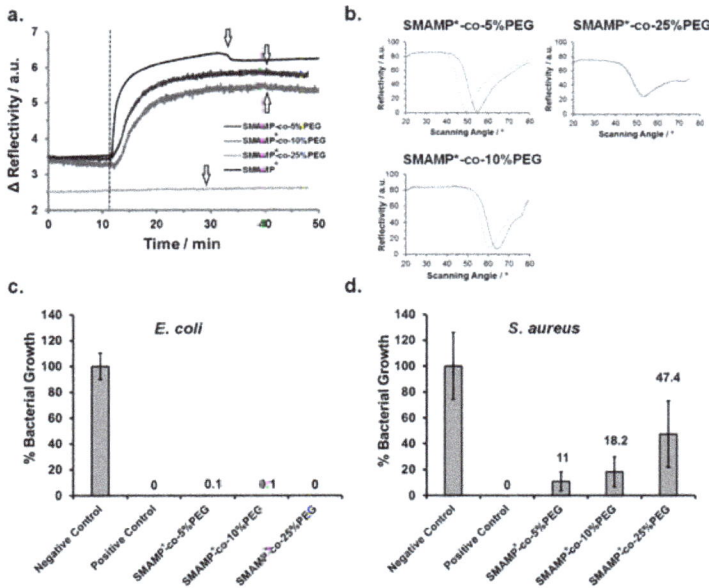

Figure 3. (**a**) Kinetics of fibrinogen adsorption on SMAMP*-co-PEG and SMAMP* monitored by surface plasmon resonance spectroscopy (SPR). (**b**) Full angular reflectivity SPR curves of the dry SMAMP*-co-PEG samples before and after protein adhesion. (**c**) and (**d**) Antimicrobial activity of the SMAMP*-co-PEG and SMAMP* surfaces against *E. coli* (**c**) and *S. aureus* (**d**). Bacterial growth (percentage of surviving colony forming units after 4 h incubation time, normalized to the negative control (=100% growth)) is shown for each material. For each data point, the assay was performed in duplicate, with five samples per repetition. The negative control was a blank silicon wafer piece, the positive control was a wafer piece to which chlorhexidine digluconate had been added.

3.6. Discussion

By copolymerization of monomer P and macromonomer M, SMAMP-*co*-PEG polymers with different amounts of the protein-repellent PEG and the (masked, N-Boc proteced) antimicrobial SMAMP functionality were obtained. These polymers were immobilized to form surface-attached networks, and activated. Physical characterization data showed that the substrates used were homogeneously covered with the SMAMP*-*co*-PEG polymer, and that the PEG moieties tended to segregate to the polymer–air interface. Thus, with increasing PEG content, the interface became increasingly hydrophilic and protein-repellent, as demonstrated by the protein adhesion assay. However, 10 mol % (more than 30 mass %) PEG was still insufficient to effectively shield the materials from fibrinogen adhesion; for this, 25 mol % PEG were required. The antimicrobial activity of the SMAMP*-*co*-PEG polymers originates from the electrostatic interaction of the activated SMAMP* groups with the negatively charged cell envelope of bacteria, as reported previously [33]. However, the SMAMP*-*co*-PEG surfaces were only fully active against *E. coli*, while their activity against *S. aureus* was already compromised by as little as 5 mol % (18.8 mass %) PEG. The long PEG chains seem to prevent the close interaction of bacteria with the surface, and also might reduce the surface charge density. Kügler et al. found that the positive charge density of grafted poly(vinylpyridine) chains necessary for killing Gram-positive *Staphylococcus epidermidis* was 10 times higher than for Gram-negative *E. coli* [21]. In the data here presented, we also see that Gram-positive *S. aureus* is significantly less affected by the SMAMP* moieties than Gram-negative *E. coli*. Previous data from our group also showed that Gram-negative bacteria were more susceptible to SMAMP*-coated surfaces than Gram-positive ones [33]. Also, bifunctional polymer surfaces with SMAMP* patches and polyzwitterionic patches were more active against Gram-negative bacteria than against Gram-positive ones. In that data, we also saw that if the SMAMP patch size was too small, antimicrobial activity against *E. coli* became compromised [32]. Thus, the emerging picture for bifunctional polymer surfaces, whether from hierarchically organized copolymers or from microstructured surfaces, is that the overall local number of cationic groups interacting with each bacterial cell is crucial for the fate of that cell. This critical number seems to depend on the global charge density of the polymer itself (functional groups per nm^2), the distance up to which these groups can be approached (i.e., whether direct surface access is blocked by PEG or polyzwitterion chains), and the electrostatic charge of the bacteria themselves. For example, it is known that *E. coli* bacteria approximately have a 5 times greater surface area per cell and an up to 15 times larger negative surface potential than *S. aureus* [42,43]. Thus, *E. coli* bacteria have more negative charges per cell available for electrostatic binding to the SMAMP moieties. In the case of the materials presented here, even though *E. coli* bacteria are larger than *S. aureus* bacteria, this negative potential enables *E. coli* to overcome the entropic barrier of the PEG groups and to sufficiently interact with the polymer surfaces to eventually get killed. On the other hand, it seems like the electrostatic interactions of the surfaces with *S. aureus* bacteria are not enough to overcome the shielding of the PEG moieties. This finding goes in line with observations reported by others: Fang et al. observed that a minimum cationic surface charge density of cationic functionality nanoparticles or polycations immobilized in a PEG brush was needed for *S. aureus* adhesion and antimicrobial activity; Cavallaro et al. showed that amine-coated surfaces had a specific threshold of surface-immobilized quaternary ammonium groups to induce significant antimicrobial effect against *E. coli*; and Gottenbos et al. reported that positively charged poly(methacrylate) surfaces showed a higher reduction of adhered viable counts for Gram-negative bacteria (*E. coli* and *P. aeruginosa*) than for Gram-positive bacteria (*S. aureus* and *S. epidermidis*), which is caused by weak electrostatic interaction with the thick bacterial cell membrane of Gram-positive bacteria [44–46].

It is always difficult to compare microbiological data from different laboratories because testing methods, controls used, and even environmental conditions may differ widely. Thus, it is not easy to say what is 'the best' bifunctional antimicrobial and protein-repellent coating currently known. However, a few trends become apparent. From the materials described in the introduction, the nisin-hyaluronic acid coating had up to 99.8% antibacterial activity against *S. epidermidis* after 3 h of contact with the bacterial suspension (at bacterial concentration about 10^8 bacteria per cm^3). However, this surface suffered from bacterial adhesion compared to the peptide-free surface [27]. The chitosan-poly(2-hydroxyethyl methacrylate) brushes had less than 10% protein adhesion, and killed up to 80% of *E. coli* bacteria after 4 h exposure to the bacterial suspension (bacterial concentration: about 10^6 bacteria per cm^3) [28]. The polymer brushes with zwitterionic groups and quaternary ammonium functionalities had an increased protein adhesion, and killed 72% of *E. coli* bacteria after 3 h of exposure the surface to the bacterial suspension (at a bacterial concentration of 10^9 bacteria per cm^3) [29]. Thus, even though the here presented materials only have significant activity against *E. coli*, they seem to be at least comparable in their bioactivity profile to these reference polymers. Further work, however, needs to be dedicated to simultaneous protein-repellency and broad spectrum antimicrobial activity, not to mention simultaneous cell compatibility.

4. Conclusion

In this report, the synthesis and characterization of bifunctional surface-attached polymer networks containing protein-repellent PEG moieties and antimicrobial SMAMP* groups were presented. Varying the PEG content of these materials from 5–25 mol % had a profound effect on the interaction of these surfaces with bacteria and proteins. An optimal dual activity was obtained for the SMAMP*-*co*-25%PEG material, which had a protein repellency >92% (compared to a pure SMAMP surface), and quantitatively killed *E. coli*, but not *S. aureus* bacteria. Apparently, a higher local charge density is necessary to also successfully eliminate the Gram-positive bacteria, although other features such as loss of hydrophobicity [33] with increasing PEG content might also play a role. Thus, this study has provided general insight into understanding of how combining antimicrobial and protein-repellent functionalities can affect the bioactivity of the resulting bifunctional surface. Additionally, the presented SMAMP*-*co*-25%PEG networks could be useful coatings for urinary catheters, which often fail due to biofilms formation involving *E. coli*. For this, however, their long term stability and sterilizability have to be evaluated.

Only very recently, we were able to combine antimicrobial activity and protein-repellency in a single polymer component using a polyzwitterionic material [25,47]. While the exact mechanism of activity and the long term performance of this material are still being investigated, it seems that such a single component material might be an easier pathway to dual antimicrobial activity and protein repellency than finding the 'sweet spot' of the perfect balance of cationic and protein-repellent components in bifunctional polymer surfaces. However, each approach has its merits, and might be of different use in different fields of application.

Supplementary Materials: The following are available online at http://www.mdpi.com/1996-1944/11/8/1411/s1, Figure S1: Characterization of the PEG macromonomer, Figure S2: Protein adhesion study.

Author Contributions: Conceptualization, K.L.; Data curation, M.K. and V.T.W.; Formal analysis, M.K.; Funding acquisition, K.L.; Investigation, M.K.; Methodology, A.A. and K.L.; Supervision, A.A. and K.L.; Validation, K.L.; Visualization, M.K.; Writing—original draft, M.K. and K.L.; Writing—review and editing, K.L.

Acknowledgments: Funding of this work by the German Research Foundation (Deutsche Forschungsgemeinschaft, DFG, Grant ID LI1714/5-1) is gratefully acknowledged. Thanks to the Institute for Macromolecular Chemistry, University of Freiburg, for measuring the MALDI-TOF mass spectra. The article processing charge was funded by the German Research Foundation (DFG) and the University of Freiburg in the funding program Open Access Publishing.

Conflicts of Interest: The authors declare no conflict of interest.

References

1. Hall-Stoodley, L.; Costerton, J.W.; Stoodley, P. Bacterial biofilms: From the natural environment to infectious diseases. *Nat. Rev. Microbiol.* **2004**, *2*, 95–108. [CrossRef] [PubMed]
2. O'Neill, J. Tackling drug-resistant infections globally: Final report and recommendations. Available online: https://amr-review.org/sites/default/files/160518_Final%20paper_with%20cover.pdf (accessed on 19 May 2016).

3.   Tiller, J.C. Antimicrobial surfaces. *Adv. Polym. Sci.* **2011**, *240*, 193–217.

4.   Timofeeva, L.; Kleshcheva, N. Antimicrobial polymers: Mechanism of action, factors of activity, and applications. *Appl. Microbiol. Biotechnol.* **2011**, *89*, 475–492. [CrossRef] [PubMed]

5.   Afacan, N.J.; Yeung, A.T.Y.; Pena, O.M.; Hancock, R.E.W. Therapeutic potential of host defense peptides in antibiotic-resistant infections. *Curr. Pharm. Des.* **2012**, *18*, 807–819. [CrossRef] [PubMed]

6.   Bazaka, K.; Jacob, M.V.; Crawford, R.J.; Ivanova, E.P. Efficient surface modification of biomaterial to prevent biofilm formation and the attachment of microorganisms. *Appl. Microbiol. Biotechnol.* **2012**, *95*, 299–311. [CrossRef] [PubMed]

7.   Engler, A.C.; Wiradharma, N.; Ong, Z.Y.; Coady, D.J.; Hedrick, J.L.; Yang, Y.Y. Emerging trends in macromolecular antimicrobials to fight multi-drug-resistant infections. *Nano Today* **2012**, *7*, 201–222. [CrossRef]

8.   Munoz-Bonilla, A.; Fernandez-Garcia, M. Polymeric materials with antimicrobial activity. *Prog. Polym. Sci.* **2012**, *37*, 281–339. [CrossRef]

9.   Siedenbiedel, F.; Tiller, J.C. Antimicrobial polymers in solution and on surfaces: Overview and functional principles. *Polymers* **2012**, *4*, 46–71. [CrossRef]

10. Wessels, S.; Ingmer, H. Modes of action of three disinfectant active substances: A review. *Regul. Toxicol. Pharmacol.* **2013**, *67*, 456–467. [CrossRef] [PubMed]

11. Armentano, I.; Fortunati, E.; Mattioli, S.; Arciola, C.R.; Ferrari, D.; Amoroso, C.F.; Rizzo, J.; Kenny, J.M.; Imbriani, M.; Visai, L. The interaction of bacteria with engineered nanostructured polymeric materials: A review. *Sci. World. J.* **2014**, *2014*, 410423. [CrossRef] [PubMed]

12. Mi, L.; Jiang, S. Integrated antimicrobial and nonfouling zwitterionic polymers. *Angew. Chem. Int. Ed.* **2014**, *53*, 1746–1754. [CrossRef] [PubMed]

13. Salwiczek, M.; Qu, Y.; Gardiner, J.; Strugnell, R.A.; Lithgow, T.; McLean, K.M.; Thissen, H. Emerging rules for effective antimicrobial coatings. *Trends Biotechnol.* **2014**, *32*, 82–90. [CrossRef] [PubMed]

14. Swartjes, J.J.T.M.; Sharma, P.K.; van Kooten, T.G.; van der Mei, H.C.; Mahmoudi, M.; Busscher, H.J.; Rochford, E.T.J. Current developments in antimicrobial surface coatings for biomedical applications. *Curr. Med. Chem.* **2015**, *22*, 2116–2129. [CrossRef] [PubMed]

15. Ganewatta, M.S.; Tang, C. Controlling macromolecular structures towards effective antimicrobial polymers. *Polymer* **2015**, *63*, A1–A29. [CrossRef]

16. Hasan, J.; Crawford, R.J.; Ivanova, E.P. Antibacterial surfaces: The quest for a new generation of biomaterials. *Trends Biotechnol.* **2013**, *31*, 295–304. [CrossRef] [PubMed]

17. Lowe, S.; O'Brien-Simpson, N.M.; Connal, L.A. Antibiofouling polymer interfaces: Poly(ethylene glycol) and other promising candidates. *Polym. Chem.* **2015**, *6*, 198–212. [CrossRef]

18. Campoccia, D.; Montanaro, L.; Arciola, C.R. A review of the biomaterials technologies for infection-resistant surfaces. *Biomaterials* **2013**, *34*, 8533–8554. [CrossRef] [PubMed]

19. Lejars, M.; Margaillan, A.; Bressy, C. Fouling release coatings: A nontoxic alternative to biocidal antifouling coatings. *Chem. Rev.* **2012**, *112*, 4347–4390. [CrossRef] [PubMed]

20. Tiller, J.C.; Liao, C.J.; Lewis, K.; Klibanov, A.M. Designing surfaces that kill bacteria on contact. *Proc. Natl. Acad. Sci. USA* **2001**, *98*, 5981–5985. [CrossRef] [PubMed]

21. Kugler, R.; Bouloussa, O.; Rondelez, F. Evidence of a charge-density threshold for optimum efficiency of biocidal cationic surfaces. *Microbiology* **2005**, *151*, 1341–1348. [CrossRef] [PubMed]

22. Asri, L.A.T.W.; Crismaru, M.; Roest, S.; Chen, Y.; Ivashenko, O.; Rudolf, P.; Tiller, J.C.; van der Mei, H.C.; Loontjens, T.J.A.; Busscher, H.J. A shape-adaptive, antibacterial-coating of immobilized quaternary-ammonium compounds tethered on hyperbranched polyurea and its mechanism of action. *Adv. Funct. Mater.* **2014**, *24*, 346–355. [CrossRef]

23. Riga, E.K.; Vöhringer, M.; Widyaya, V.T.; Lienkamp, K. Polymer-based surfaces designed to reduce biofilm formation: From antimicrobial polymers to strategies for long-term applications. *Macromol. Rapid Commun.* **2017**, *38*, 1700216. [CrossRef] [PubMed]

24. Hartleb, W.; Saar, J.S.; Zou, P.; Lienkamp, K. Just antimicrobial is not enough: Toward bifunctional polymer surfaces with dual antimicrobial and protein-repellent functionality. *Macromol. Chem. Phys.* **2016**, *217*, 225–231. [CrossRef]

25. Kurowska, M.; Eickenscheidt, A.; Guevara-Solarte, D.L.; Widyaya, V.T.; Marx, F.; Al-Ahmad, A.; Lienkamp, K. A simultaneously antimicrobial, protein-repellent and cell-compatible polyzwitterion network. *Biomacromolecules* **2017**, *18*, 1373–1386. [CrossRef] [PubMed]

26. Yu, Q.; Wu, Z.; Chen, H. Dual-function antibacterial surfaces for biomedical applications. *Acta Biomater.* **2015**, *16*, 1–13. [CrossRef] [PubMed]

27. Paris, J.B.; Seyer, D.; Jouenne, T.; Thebault, P. Elaboration of antibacterial plastic surfaces by a combination of antiadhesive and biocidal coatings of natural products. *Colloids Surf. B* **2017**, *156*, 186–193. [CrossRef] [PubMed]

28. Yang, W.J.; Cai, T.; Neoh, K.-G.; Kang, E.T.; Dickinson, G.H.; Teo, S.L.M.; Rittschof, D. Biomimetic anchors for antifouling and antibacterial polymer brushes on stainless steel. *Langmuir* **2011**, *27*, 7065–7076. [CrossRef] [PubMed]

29. Ye, G.; Lee, J.; Perreault, F.; Elimelech, M. Controlled architecture of dual-functional block copolymer brushes on thin-film composite membranes for integrated "defending" and "attacking" strategies against biofouling. *Appl. Mater. Interf.* **2015**, *7*, 23069–23079. [CrossRef] [PubMed]

30. Charnley, M.; Textor, M.; Acikgoz, C. Designed polymer structures with antifouling–antimicrobial properties. *React. Funct. Polym.* **2011**, *71*, 329–334. [CrossRef]

31. Zou, P.; Hartleb, W.; Lienkamp, K. It takes walls and knights to defend a castle—synthesis of surface coatings from antimicrobial and antibiofouling polymers. *J. Mater. Chem.* **2012**, *22*, 19579–19589. [CrossRef]

32. Vöhringer, M.; Hartleb, W.; Lienkamp, K. Surface structuring meets orthogonal chemical modifications: Toward a technology platform for site-selectively functionalized polymer surfaces and biomems. *Biomater. Sci. Eng.* **2017**, *3*, 909–921. [CrossRef]

33. Zou, P.; Laird, D.; Riga, E.K.; Deng, Z.; Perez-Hernandez, H.R.; Guevara-Solarte, D.L.; Steinberg, T.; Al-Ahmad, A.; Lienkamp, K. Antimicrobial and cell-compatible surface-attached polymer networks—how the correlation of chemical structure to physical and biological data leads to a modified mechanism of action. *J. Mater. Chem. B* **2015**, *3*, 6224–6238. [CrossRef]

34. Nurioglu, A.G.; Esteves, A.C.C.; de With, G. Non-toxic, non-biocide-release antifouling coatings based on molecular structure design for marine applications. *J. Mater. Chem.* **2015**, *3*, 6547–6570. [CrossRef]

35. Al-Ahmad, A.; Zou, P.; Guevara-Solarte, D.L.; Hellwig, E.; Steinberg, T.; Lienkamp, K. Development of a standardized and safe airborne antibacterial assay, and its evaluation on antibacterial biomimetic model surfaces. *PLoS One* **2014**, e111357. [CrossRef] [PubMed]

36. Alfred, S.F.; Al-Badri, Z.M.; Madkour, A.E.; Lienkamp, K.; Tew, G.N. Water soluble poly(ethylene oxide) functionalized norbornene polymers. *J. Polym. Sci. Part A Polym. Chem.* **2008**, *46*, 2640–2648. [CrossRef]

37. Al-Ahmad, A.; Laird, D.; Zou, P.; Tomakidi, P.; Steinberg, T.; Lienkamp, K. Nature-inspired antimicrobial polymers – assessment of their potential for biomedical applications. *PLoS ONE* **2013**, *8*, e73812. [CrossRef] [PubMed]

38. Lienkamp, K.; Madkour, A.E.; Musante, A.; Nelson, C.F.; Nüsslein, K.; Tew, G.N. Antimicrobial polymers prepared by romp with unprecedented selectivity: A molecular construction kit approach. *J. Am. Chem. Soc.* **2008**, *130*, 9836–9843. [CrossRef] [PubMed]

39. Riga, E.K.; Rühe, J.; Lienkamp, K. Non-delaminating Polymer Hydrogel Coatings via C,H Insertion Crosslinking (CHic)—A Case Study of Poly(oxanorbornenes). *Chem. Eur. J.* **2018**, submitted.

40. Sharma, S.; Johnson, R.W.; Desai, T.A. XPS and AFM analysis of antifouling peg interfaces for microfabricated silicon biosensors. *Biosens. Bioelectron.* **2004**, *20*, 227–239. [CrossRef] [PubMed]

41. Sun, W. Functionalization of surfaces with branched polymers. *RSC Adv.* **2016**, *6*, 42089–42108. [CrossRef]

42. Dickson, J.; Koohmaraie, M. Cellsurface charge characteristics and their relationship to bacterial attachment tomeatsurfaces. *Appl. Environ. Microbiol.* **1989**, *55*, 832–836. [PubMed]

43. Guo, S.; Jańczewski, D.; Zhu, X.; Quintana, R.; He, T.; Neoh, K.G. Surface charge control for zwitterionic polymer brushes: Tailoring surface properties to antifouling applications. *J. Colloid Interf. Sci.* **2015**, *452*, 43–53. [CrossRef] [PubMed]

44. Fang, B.; Jiang, Y.; Nüsslein, K.; Rotello, V.M.; Santore, M.M. Antimicrobial surfaces containing cationic nanoparticles: How immobilized, clustered, and protruding cationic charge presentation affects killing activity and kinetics. *Colloids Surf. B* **2015**, *125*, 255–263. [CrossRef] [PubMed]

45. Cavallaro, A.; Mierczynska, A.; Barton, M.; Majewski, P.; Vasilev, K. Influence of immobilized quaternary ammonium group surface density on antimicrobial efficacy and cytotoxicity. *Biofouling* **2016**, *32*, 13–24. [CrossRef] [PubMed]

46. Gottenbos, B.; Grijpma, D.W.; van der Mei, H.C.; Feijen, J.; Busscher, H.J. Antimicrobial effects of positively charged surfaces on adhering gram-positive and gram-negative bacteria. *J. Antimicrob. Chemother.* **2001**, *48*, 7–13. [CrossRef]

47. Kurowska, M.; Eickenscheidt, A.; Al-Ahmad, A.; Lienkamp, K. Simultaneously antimicrobial, protein-repellent and cell-compatible polyzwitterion networks: More insight on bioactivity and physical properties. *Appl. Bio Mater.* **2018**. [CrossRef]

materials

MDPI

Article

Effect of CNT/PDMS Nanocomposites on the Dynamics of Pioneer Bacterial Communities in the Natural Biofilms of Seawater

Yubin Ji [1,†], Yuan Sun [1,†], Yanhe Lang [2], Lei Wang [3,*], Bing Liu [1] and Zhizhou Zhang [3,4,*]

[1] School of Science, Harbin University of Commerce, Harbin 150076, China; yunbinji@sina.com (Y.J.); sunyuan.2010@163.com (Y.S.); bingliu2018@sina.com (B.L.)

[2] Key Laboratory of Saline-alkali Vegetation Ecology Restoration in Oil Field (SAVER), Ministry of Education, Alkali Soil Natural Environmental Science Center (ASNESC), Northeast Forestry University, Harbin 150040, China; langyanhe@163.com

[3] School of Chemistry and Chemical Engineering, Harbin Institute of Technology, Harbin 150001, China

[4] School of Marine Science and Technology, Harbin Institute of Technology at Weihai, Weihai 264209, China

* Correspondence: leiwang_chem@hit.edu.cn (L.W.); zhangzzbiox@outlook.com (Z.Z.); Tel.: +86-0451-8641-3708 (L.W.), +86-150-6631-7512 (Z.Z.)

† These authors contributed equally to this work.

Received: 10 May 2018; Accepted: 20 May 2018; Published: 24 May 2018

Abstract: In this study, the antifouling (AF) performance of different carbon nanotubes (CNTs)-modified polydimethylsiloxane (PDMS) nanocomposites (PCs) was examined directly in the natural seawater, and further analyzed using the Multidimensional Scale Analyses (MDS) method. The early-adherent bacterial communities in the natural biofilms adhering to different PC surfaces were investigated using the single-stranded conformation polymorphism (SSCP) technique. The PCs demonstrated differences and reinforced AF properties in the field, and they were prone to clustering according to the discrepancies within different CNT fillers. Furthermore, most PC surfaces only demonstrated weak modulating effects on the biological colonization and successional process of the early bacterial communities in natural biofilms, indicating that the presence of the early colonized prokaryotic microbes would be one of the primary causes of colonization and deterioration of the PCs. C6 coating seems to be promising for marine AF applications, since it has a strong perturbation effect on pioneer prokaryotic colonization.

Keywords: antifouling coatings; biofouling; natural biofilms; single-stranded conformation polymorphism; polydimethylsiloxane; multidimensional scale analysis

1. Introduction

The occurrence of biofouling on synthetic surfaces is a major issue for the shipping industries in marine environments [1], which has resulted in substantial economic and ecological consequences. For example, total cruise expenses are greatly increased by approximately 77% annually worldwide, primarily owing to the constantly enhanced propulsive power and fuel consumption [2]. Natural biofilms, also termed microfouling, are well-organized and complex assemblages, mainly developed by the undesirable colonization of marine microorganisms as well as their extracellular matrix materials [3–5]. Over the past few decades, the early-adherent biofilm-forming marine bacteria communities on the artificial surfaces aroused researchers' interests worldwide, since their presence was found to be closely related to the subsequent macrofouling process, which can further enhance the potential hazards of the biodeterioration and biodegradation of the selected biofouling-resistant substrata, thereby leading to a remarkable loss in antifouling (AF) performances [6–8].

So far, the most commonly used remedial strategies to retard the build-up of biofouling have taken the form of protective coatings, broadly categorized into AF and fouling-release (FR) coatings [9,10]. Traditionally, biocide-released AF coatings have been demonstrated to be environmentally damaging to non-target living marine organisms, due to the presence of a range of poisonous organic biocides, such as tributyltin (TBT). Therefore, their use in the coating industry has been globally restricted and prohibited [11]. As a consequence, the search for favorable biocide-independent coatings for biofouling management has been greatly accelerated [12], particularly in regard to FR coatings [13].

The organo–silicon polymers, typically the polydimethylsiloxane (PDMS), represent a desired non-toxic alternative and marked niche among specialty copolymers [14]. The PDMS resin possesses a superior environmentally-friendly nature, with characteristics such as high heat resistance, surface inertness, high hydrophobicity, as well as excellent fouling anti-adhesion characteristics, presenting viable options in several marine industries [15]. Furthermore, these PDMS-based nanocomposites have been systematically investigated in recent years, mainly because of their facile preparation and ecological stability [16,17]. Many research studies and testing procedures have been devoted to meeting the challenge of exploring effective, reliable and high-performance inorganic nanofillers, for the purpose of obtaining PDMS-based nanocomposites with reinforced AF and FR properties [18]. Carbon nanotubes (CNTs) are considered one of the most favorable inorganic fillers for PDMS modification [19], while the PDMS nanocomposites (PCs) seem to be the most promising candidate for marine anti-biofouling applications, although the potential impact of CNTs on the biological colonization dynamics of the early biofilm-forming bacterial communities still remains poorly understood. In addition, the culture-independent molecular fingerprinting method, i.e., the single-stranded conformation polymorphism (SSCP) technique, has been widely used to estimate the global diversity of environmental microbial communities in the field of microbial ecology in recent years [20].

Therefore, the current study aimed to investigate the effects of different CNT modified PDMS composites (PCs) on the colonization dynamics of the pioneer bacterial communities in the natural biofilms using the single-strand conformation polymorphism (SSCP) technique. The clustering patterns of the early bacterial biofilm communities adhering to various PCs were explored using Multidimensional Scale Analyses (MDS). In addition, a surface evaluation system based on the MDS method was established in order to quantify the fouling conditions among different PC surfaces examined in the field.

2. Materials and Methods

2.1. Materials

2.1.1. The Primer Coat

The primer coat, i.e., the chlorinated rubber iron-red antirust paint, was kindly supplied by the Jiamei Company (Weihai, China), and consisted primarily of chlorinated rubber resin, micaceous iron oxide, plasticizers, additives and a mixed solvent. The primer paint was cured for approximately 72 h at room temperature (RT).

2.1.2. Silicone-Based Matrix System

The silicone-based matrix used in this study was, necessarily, PDMS (P0) resin from a Sylgard 184 elastomer kit, purchased from the Dow Corning Company (Shanghai, China). This commercially available PDMS material acted as a standard resin for further preparation processes. The PDMS polymer was obtained directly by mixing the pre-polymer (Component A) to the curing agent (Component B) in a ratio of 10:1 (weight) at 105 °C within 6 h, which served as the standard coating controls.

2.1.3. Carbon Nanotubes (CNTs)

All CNTs used in the current study were purchased from the Chengdu Organic Chemicals Co., Ltd. (Chengdu, China), Chinese Academy of Sciences, including six multi-walled carbon nanotubes (MWCNTs, F1-F6), six hydroxyl-modified MWCNTs (hMWCNTs, F7-F12), and six carboxyl-modified MWNTs (cMWCNTs). Detailed information about these CNTs was summarized as presented in Table 1. The CNTs were incorporated in the PDMS matrix at concentrations of 0 (PDMS only) and 0.1% (w/w), respectively, as previously reported by Beigbeder and coworkers [21].

Table 1. The carbon nanotube (CNT) fillers and the polydimethylsiloxane (PDMS) composites (PCs) in the current study.

CNTs	Hydroxyl Content % (w/w)	Carboxyl Content % (w/w)	Diameter (nm)	Length (μm)	SSA (m²/g)	PC Sets	PC Names
F1	–	–	10–20	30–100	>165	M	M1
F2	–	–	8–15	~50	>233	M	M2
F3	–	–	10–20	10–30	>200	M	M3
F4	–	–	20–30	10–30	>110	M	M4
F5	–	–	30–50	10–20	>60	M	M5
F6	–	–	>50	10–20	>40	M	M6
F7	5.58	–	<8	10–30	>500	H	H1
F8	3.70	–	8–15	~50	>233	H	H2
F9	3.06	–	10–20	10–30	>200	H	H3
F10	1.76	–	20–30	~30	>110	H	H4
F11	1.06	–	30–50	~20	>60	H	H5
F12	0.71	–	>50	~20	>40	H	H6
F13	–	3.85	<8	~30	>500	C	C1
F14	–	2.55	8–15	~50	>233	C	C2
F15	–	2.00	10–20	10–30	>200	C	C3
F18	–	1.23	20–30	~30	>110	C	C4
F17	–	0.73	30–50	~20	>60	C	C5
F18	–	0.64	>50	~20	>40	C	C6

Note: F1–F6, F7–F12 and F13–F18 represent different types of multi-walled carbon nanotubes (MWCNTs), hydroxyl-modified MWCNTs (hMWCNTs) and carboxyl-modified MWNTs (cMWCNTs), respectively. SSA is short for specific surface area.

2.1.4. Preparation of the PDMS-Based Composites (PCs)

Eighteen kinds of PCs were freshly produced, which were largely cataloged into three sets: the M set (MPs, M1–M6), the H set (HPs, H1–H6) and the C set (CPs, C1–C6). The composition of these PCs are summarized in Table 1. These PCs were all formulated following a similar procedure [20]. Briefly, each CNT filler was blended with base elastomer (Part A) for 10 min by intense stirring at 500 rpm for 1 min. Then, the suspension was well mixed with the curing agents (Part B), and mechanically stirred for another 15 min. The air bubbles from the PDMS mixture were completely removed using a vacuum desiccator. Afterwards, these PCs were cured at 105 °C for 6 h in a constant temperature oven.

2.2. Panel Preparation

The steel panels (measuring 10 cm × 10 cm × 3 cm) for the seawater exposure assays were firstly drilled at the bottom, and then thoroughly polished with the abrasive paper of different grits in order to obtain the same surface condition in terms of roughness. Afterwards, these panels were carefully washed with sterile H_2O and rinsed with 70% (v/v) ethanol, then dried at room temperature overnight prior to use. A layer of the primer coat was coated on each panel and dried for 72 h at room temperature. Then, these pre-treated panels were coated with the PCs using a bar-coater and cured for 6 h at 105 °C in an oven. A minimum of 3 specimens of each PC was produced for further statistical evaluation.

2.3. Seawater Exposure Assays

The field exposure studies were performed at a static woody pontoon in a marina named Small Stone Island in the Weihai Western Port, China (37°31′51″ N; 121°58′19″ E, see Figure 1). Panels coated with different experimental materials were produced in triplicate throughout. These panels were randomly arranged on a wooden pontoon located in the Small Stone Island harbor waters using thin ropes, and then vertically suspended at 1.5 m below the lowest tide level over a period of 56 days (April–June, 2015). The average sea temperature during the exposures was 11 °C throughout.

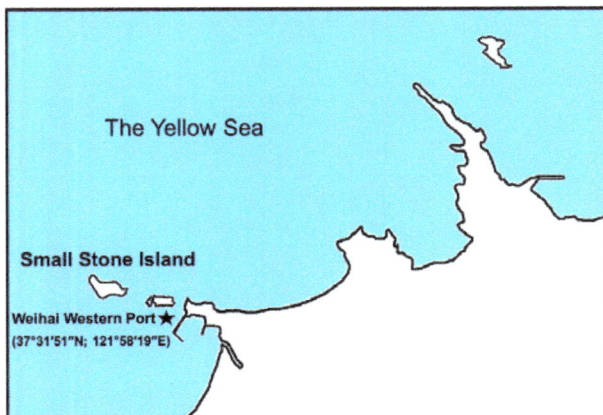

Figure 1. Location of the immersion sites for the field studies: Small Stone Island in the Western Port, Weihai, China.

The fouling conditions of each experimental material were captured using a digital camera from the fourth week after immersion at one-week intervals, namely at 28 days, 35 days, 42 days, 49 days and 56 days. After photographing, these panels were sent back to the marine realms as quickly as possible. According to the captured images, fouling conditions were further quantified according to the amount of adherence of the major fouling organisms, including barnacles (*B. Amphitrite*), mussels (*Mytilus edulis*), *Ulva pertusa*, sessile ascidian, as well as seaweeds. It is notable that the aforementioned scoring procedures were conducted at five different exposure times, i.e., each experimental material amounted to scoring fifteen times, since each experimental material was prepared in triplicate. Furthermore, owing to the edge effects, the 20 mm area from the margin of each tested panel was excluded within the scope of the assessment area. The clustering patterns of the AF properties of different PDMS-based coatings were performed by inputting the substratum and assessment outcomes as variables using the MDS method conducted by SPSS19.0 software (IBM, Armonk, NY, USA). The pure PDMS coated panels served as standards.

2.4. Sampling

The short-term in situ experiments were performed at the same field immersion sites, using the PDMS-based coatings as the artificial substrata for the biofilm recruitments. The formation of biofilm on each PDMS-based coating surface was measured throughout the two-week in situ experiment (April 2–15, 2015) at five different points in time: April 3 (2-day biofilm), April 6 (5-day biofilm), April 9 (8-day biofilm), April 12 (11-day biofilm), and April 15 (14-day biofilm). For each PDMS-based coating, a replicate of four panels (measuring 10 cm × 10 cm) was prepared throughout the investigation. For each panel, an area of approximately 80 mm × 80 mm within each PC surface was sampled.

All tested steel panels were brought back to the laboratory as quickly as possible using a cool-box. Each panel was carefully rinsed with the sterile artificial seawater prior to scrapping, in order to remove the excess sediment and temporarily attached microorganisms. The biofilm samples were gently scraped from the surfaces of each tested panel using the sterile brushes. The replicated scrapings belonging to the same PCs were collected into a sterile Eppendorf tube (2.0 mL) as a representative of all replicate biofilm samples for the subsequent microbial assays. Afterwards, the biofilm samples were suspended into 400 μL sterile deionized water, and vortexed for 60 min prior to being centrifuged, aiming to pellet the biomass at 4000 rpm for 5 min. Then, these biofilm pellets were stored at −80 °C for further analysis.

2.5. SSCP

The genomic DNA extractions were performed on all biofilm samples using the Sangon Rapid Bacterial Genomic DNA Isolation kit (Cat# B518225). The integrity of the genomic DNA was examined by 0.8% agarose gel electrophoresis and further quantified by determining the absorbance at 260 nm. Amplification of the prokaryotic 16SrRNA gene fragments was undertaken using the general primer pairs synthesized from Sangon (Shanghai, China), namely 337F (5′-GAC TCC TAC GGG AGG CWG CAG-3′) and 1100R (5′-GGG TTG CGC TCG TTG-3′), which were used to identify the early prokaryotic microbes in the pioneer natural biofilms formed on different PC surfaces, yielding a fragment of ~763 bp. The asymmetric PCR amplification of the target 16S RNA gene fragments was conducted following similar procedures as described previously [22]. A negative control was included throughout. Afterwards, the PCR products were detected by 1.5% agarose gel electrophoresis and stored at −40 °C for further analysis.

The SSCP analysis was performed on a DYCZ-24DN vertical gel electrophoresis apparatus (Liuyi, Beijing). All 16S rDNA fragments were well blended with equal volumes of the denaturation solution separately, which contained 95% formamide, 0.25% bromphenol blue and 0.25% xylene cyanol. The PCR products were denatured at 98 °C for 10 min, and then snap-frozen on ice prior to loading. Then, these denatured PCR products (6.0 μL) were loaded onto 8% (w/v) polyacrylamide (arylamide:bisacrylamide = 29:1) gel with a thickness of 1 mm, and separated at a constant voltage of 90 V for 28 h in 1× TBE buffer at 4 °C. Subsequently, the SSCP gels were silver stained. A digitized image of the SSCP gels was captured using a digital camera, and the lanes and bands in the SSCP gel images were further analyzed using the Quantity One analysis software (Bio-Rad, Hercules, CA, USA), according to the position of each nucleic acid band, thereby resulting in a matrix based on the presence/absence of bands.

2.6. Data Analysis

The SSCP presence/absence binary data matrices constructed in terms of band positions and intensities were used to identify the differences between the pioneer bacterial communities developed on the pure PDMS and CNT modified PDMS composites via the comparison of the diversity indices calculated by the Biodap software, which were able to give detailed descriptions of the dynamics of the early bacterial communities. The clustering analysis was done for the pioneer prokaryotic communities on different PDMS-based composites based on MDS using the SPSS19.0 software package (IBM, Armonk, NY, USA), primarily performed with substratum and diversity indices as variants. Furthermore, the statistical differences between the diversity indices were compared using t tests (p-value < 0.05, GraphPad Prism 6.0).

3. Results and Discussion

3.1. Fouling and Surface Evaluation

In this study, we directly examined the AF capacity of the aforementioned PDMS-based coatings in natural seawater and established a fouling evaluation system based on the MDS method in order to further quantify the fouling conditions among different PC surfaces, in the context of the adhesive number of five representative major macrofoulers, including barnacles, mussels, ascidians, *Ulva* and seaweeds (see Figures 2 and 3).

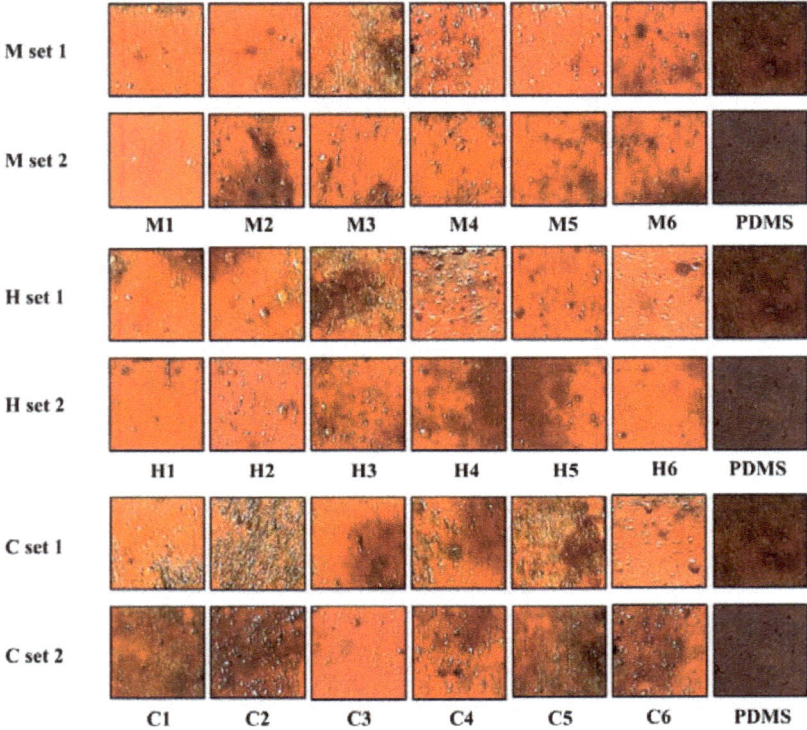

Figure 2. Appearances of various PDMS-based panels after static immersion for two months (April–June, 2015).

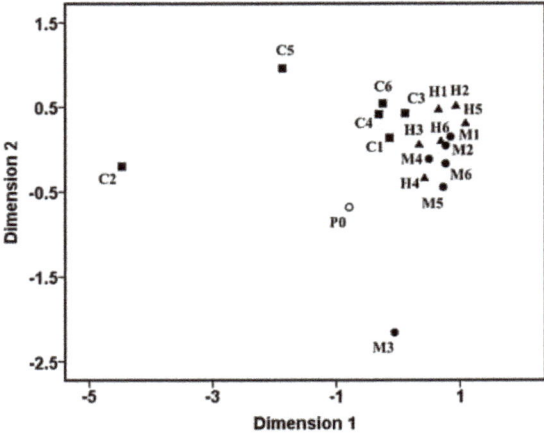

Figure 3. Clustering patterns of the antifouling (AF) capacity of the PDMS-based nanocomposites based on the MDS analysis.

From Figure 2, it is clear that the AF properties of the plain PDMS were greatly improved after the incorporation of a low amount of nanosized CNTs (0.1 wt %). Each PC displayed a differential but reinforced AF efficacy against the representative macrofoulers compared with the PDMS standard. The M1, H1 and C3 coatings performed exceptionally well in the field exposure assays, while the M3, C2 and C5 coatings were found to be heavily fouled. Furthermore, it seems that most PCs of the identical set (e.g., coatings in the M Set, except M3) tended to exhibit similar AF properties, although there were some exceptions (e.g., C2 and C5 coatings in the C Set). These differential AF behaviors may be largely owing to the differences within different CNT fillers.

In addition, in Figure 3, most PCs, including M1, M2 and M4-M6 coatings from the M set, H1–H6 coatings from the H set, and C1, C3, C4 and C6 coatings from the C set, were liable to cluster into the same group, suggesting that these PCs may have possessed similar AF performances. Nevertheless, it is noticeable that M3, C2 and C5 coatings were liable to cluster separately, and their AF properties were clearly different from those of the PDMS standards (P0) and the aforementioned PCs. This result further revealed that the physicochemical properties of the CNT filler may have differential reinforcing impacts on the AF properties of the PDMS matrix. Recently, CNTs have been applied as additives to improve the membrane properties of various polymeric matrixes worldwide, and a host of highly promising functionalized nanocomposites with excellent properties have been obtained for marine AF applications [23,24]. However, most fouling evaluation systems are still confined to laboratory assays, only involving the measurement of the adhesive number of representative hard foulings (e.g., *B. Amphitrite* and *Mytilus edulis*) [25,26] or soft foulings (e.g., *Ulva*) [27,28]. It is obvious that laboratory biological assays still remain insufficient and limited, although laboratory assays are insusceptible to environmental disturbances, unlike field exposure assays [29]. Here, we provided a feasible and effective way to solve this problem and established a novel fouling evaluation system targeting the measurement of the adhesive number of multiple natural fouling organisms in natural seawater using the MDS method, based on the data obtained from rigorous marine field assays. The advantage of this approach is obvious, since the adhesive behaviors of multiple adherent marcofoulers on different coating surfaces can be dynamically observed and recorded directly in the natural seawater, which can give a more comprehensive and objective assessment on their actual AF performance. Besides, the variations within different AF coatings can also be easily observed and captured simply using visual inspection.

3.2. SSCP Patterns of the Bacterial Biofilm Communities

Figure 4 shows the SSCP profiles of the pioneer bacterial communities in the natural biofilms developed on the PDMS-based material surfaces at different exposure times. Each band within the SSCP profiles is approximately identical to a single microbial species. As observed from SSCP patterns, eighteen kinds of PCs were generally colonized by a mixture of the early-adherent bacterial communities without exception during the two-week in situ experiment, and no significant differences were screened compared with the PDMS standards (P0) via the visual inspection. This indicated that no PCs completely resisted or deterred the colonization of pioneer prokaryotic microbes.

Early bacterial communities formed on the PDMS-based coatings belonging to the same PC set were liable to evolve similar SSCP patterns at different exposure times, while differential SSCP patterns were screened within different PC sets. For example, in the 5-day biofilm, clear differences were observed in the SSCP patterns of the pioneer biofilm communities developed on different PC sets, owing to the differences within various coating types. It is estimated that the physicochemical properties of the CNT types may be closely related to the differential and improved AF properties of the PCs. In addition, as the natural biofilm grew older (e.g., the 14-day biofilm), the early adhered bacterial communities on the PDMS-based coatings were found to be clearly increased. This result suggested that the deterrence effects of PCs against the colonization of the early bacterial communities may become increasingly weakened over time. These combined results indicated that the PCs were susceptible to microfouling when immersed in the marine environment during the short-term in situ experiment.

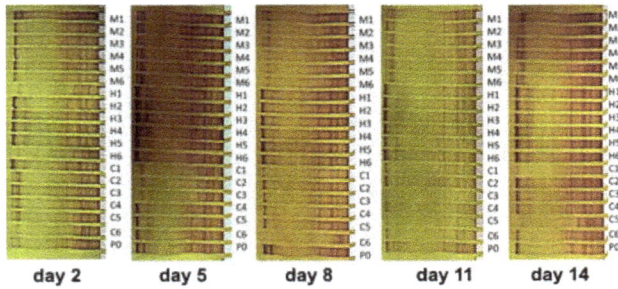

Figure 4. SSCP fingerprints of pioneer bacterial communities in the natural biofilms developed on different PDMS-based composite surfaces with different exposure times.

3.3. Clustering Patterns of the Pioneer Bacterial Communities

Figure 5 shows the clustering patterns of the pioneer bacterial communities on the PDMS-based coating surfaces using the MDS method. The pioneer bacterial communities developed on different PCs had clear differences from the PDMS standards, indicating that different PCs demonstrate differentially perturbation effects on the colonization of early bacterial communities in natural biofilms. The pioneer bacterial communities were liable to be grouped or clustered on most PC surfaces of the same PCs set (e.g., the M set), while the pioneer bacterial communities adhering to surfaces of different PCs sets were prone to show clear differences. For example, the clustering patterns of the pioneer bacterial communities attached to the MPs surfaces belonging to the M set were quite different from those attached to the CP surfaces belonging to the C set, indicating that the types of PCs may be strongly related to the differences in clustering patterns of the early biofilm communities. However, it is noticeable that the modulating effects of the PCs became gradually weaker with the growth of the natural biofilms, as evidenced by the SSCP analysis (See Figure 4). Furthermore, it seems that the AF properties of the PCs have no necessary relationships with the clustering features of the pioneer bacterial communities, as evidenced by the results shown in Figures 2 and 3, although the pioneer bacterial communities may contribute considerably to the subsequent macrofouling occurring on the surfaces of the PCs.

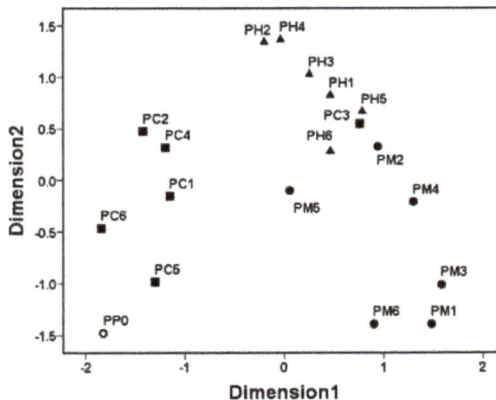

Figure 5. Clustering analysis of pioneer bacterial communities on different PDMS-based material surfaces based on the MDS method. PP0, PM, PH and PC represent the pioneer bacterial communities adhering to the surfaces of P0 coating, M1–M6 coating, coating H1–H6 coating and C1–C6 coating, respectively.

3.4. Analysis of Pioneer Bacterial Communities in the Natural Biofilms

Three diversity indices, including the Shannon diversity index (H), species richness(S), and the Simpson index (λ), were calculated and compared, as presented in Figure 6.

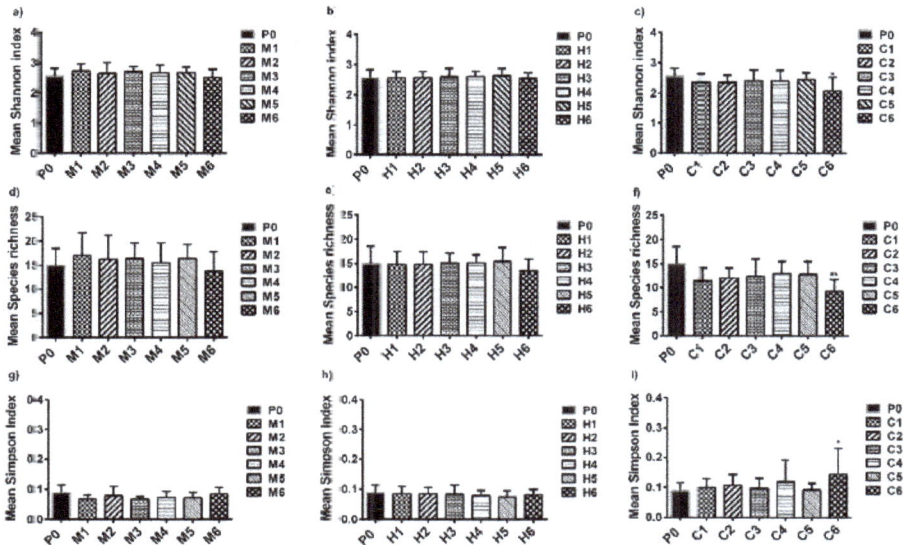

Figure 6. The comparison of the diversity indices, (a–c) Shannon diversity index, (d–f) species richness, (g–i) Simpson index of pioneer bacterial communities on different PDMS-based material surfaces.

The Shannon diversity index (H) describes the general biodiversity in environmental microbial communities, and was used to estimate the early bacterial community diversity in the natural biofilms developed on different PCs [30]. Figure 6a–c show that the H value of the bacterial communities ranged between 2.53 ± 0.27 and 2.73 ± 0.23 for all the PC surfaces, compared with the PDMS control (2.56 ± 0.26), indicating that different PC surfaces may have differential modulating effects on the colonization of pioneer bacterial communities. The highest level of early bacterial community diversity was screened on the M1 surfaces (H = 2.73 ± 0.23) among all of the PC surfaces, while the lowest level of diversity was found on the C5 surfaces (H = 2.53 ± 0.27). The pioneer prokaryotic microbial communities attached to the PCs surfaces belonging to the M set and H set (H1–H6) shared a relatively high level of diversity, with H values ranging from 2.53 ± 0.27 to 2.73 ± 0.23 and 2.54 ± 0.19 to 2.64 ± 0.23 (Figure 6a,b), respectively. However, the diversity of the early bacterial communities on the CP surfaces (C1–C6) was lower than that of the PDMS control (2.56 ± 0.26), with H values ranging from 2.09 ± 0.44 to 2.45 ± 0.22 (Figure 6c), particularly on the C6 surfaces ($p < 0.05$). No significant differences were found in the diversity level between the PCs (except C6) and the PDMS standards ($p > 0.05$). This indicated that the PC surfaces may only have weak modulating effects on colonized pioneer prokaryotic microbes, and most PC surfaces were still susceptible to the colonization and deterioration of the pioneer prokaryotic microbes.

Furthermore, species richness (S) describes the number of different species in an environmental microbial community, which was applied to give descriptions about the number of species of the early bacterial communities in the natural biofilms developed on different PCs [31]. Figure 6d–f revealed that the S value of the pioneer bacterial community (ranging from 9 ± 2 to 17 ± 5) was slightly downregulated by most PC surfaces (except C6), compared with the PDMS surface (15 ± 4). Specifically, the pioneer bacterial biofilm communities adhering to the PC surfaces belonging to the

MPs surfaces and the HPs surfaces shared a relatively high level of richness, while the early bacterial communities adhering to the PCs belonging to the C set shared a relatively low level of richness, significantly on the C6 surfaces ($p < 0.01$), compared to the PDMS control, which correlated well with the diversity level.

The Simpson index (λ) describes the number of dominant species in a particular microbial community and was used to measure the number of dominant populations in pioneer bacterial communities in the natural biofilms formed on different PCs [32]. Figure 6g–i reveal that the λ value of pioneer bacterial communities was slightly decreased on the MP surfaces (0.066 ± 0.013 to 0.083 ± 0.023), and remained almost unchanged on the HP surfaces, while it slightly increased on the CP surfaces (0.093 ± 0.021 to 0.160 ± 0.088), particularly on the C6 surfaces ($p < 0.05$), compared to the PDMS control (0.088 ± 0.027). This result indicates that the dominant bacterial population in the biofilm developed on most PCs surfaces (except C6) varied slightly, in contrast to the PDMS control. The slightly changed dominant bacterial communities in the biofilm suggest that most PC surfaces may only have a weak capacity to exert enough perturbations on the biological colonization and successional patterns of early adherent bacterial communities in natural biofilms.

Previously, a host of publications reported that AF coatings could influence and regulate the development of early-colonized bacterial communities [33–36]. However, a few publications have focused on the modulating effects of different PDMS-based nanocomposites on colonized pioneer bacterial communities in natural biofilms [37,38]. In the current study, most PCs demonstrated differential modulating effects on the colonization of pioneer prokaryotic microbes. The pioneer bacterial communities were only found to be subjected to the minor perturbations exerted by most PCs (except for coating C6). This slightly modulating effect suggested that the PCs may not exert sufficient perturbations on the biological succession patterns of the pioneer bacterial communities in the natural biofilms, which may contribute to the mechanisms causing the plain PDMS surfaces, along with most PCs surfaces (except coating C6), to be extremely susceptible heavy fouling after long-term exposure to the marine environment, since the bacterial communities in the biofilms have been found to play key roles in the biodeterioration and biodegradation of synthetic polymeric materials [39]. Data on this hypothesis still requires further study in our future work. Coating C6 seems to be promising for future marine anti-biofouling applications, owing to its strong perturbation effects on the colonization of early bacterial communities.

4. Conclusions

The present study examined the AF capacity of eighteen kinds of PDMS-based composites (PCs) via field exposure assays and provided the first example of quantifying and evaluating their AF efficacy using the MDS method. The bacterial community analysis based on the SSCP fingerprints revealed that most PCs (except C6) have weak modulating effects on the biological colonization of the early-adherent prokaryotic microbes, which may account for the mechanisms of biofouling that occurred on the PDMS-based coating surfaces. This study may lead the way to the development of a number of effective, reliable, and long-lasting ecofriendly coatings for marine anti-biofouling applications.

Author Contributions: Y.J. and Y.S. conceived and designed the experiments, Y.S. and Y.L. performed the experiments; L.W. and B.L. analyzed the data; Z.Z. contributed some valuable advice; Y.S. wrote the paper and Y.L. edited the whole manuscript.

Funding: This work was funded by the Chinese Postdoctoral Science Foundation (grant number 2017M621294), China Postdoctoral Science Foundation for international exchange programme (grant number 20170045), the Supporting Program of the Postdoctoral Research at the Harbin University of Commerce (grant number 2017BSH002), the Training Program of the Young Creative Talents at the Harbin University of Commerce (grant number 17XN011) and National Natural Science Foundation of China (grant number 31071170). The Key research and development plan of Shandong Province (grant number 2016GSF115022); the Natural Science Foundation of Shandong Province (grant number ZR2018MC002).

Acknowledgments: The corresponding author L.W. (currently working at Institute for Bioengineering of Catalonia (IBEC, Barcelona, Spain) would thank the European Commission under Horizon 2020's Marie Skłodowska-Curie Actions cofund scheme and the Severo Ochoa programme of the Spanish Ministry of Science and Competitiveness.

Conflicts of Interest: The authors declare no conflict of interest.

References

1. Fitridge, I.; Dempster, T.; Guenther, J.; de Nys, R. The impact and control of biofouling in marine aquaculture: A review. *Biofouling* **2012**, *28*, 649–669. [CrossRef] [PubMed]
2. Schultz, M.; Bendick, J.; Holm, E.; Hertel, W. Economic impact of biofouling on a naval surface ship. *Biofouling* **2011**, *27*, 87–98. [CrossRef] [PubMed]
3. Salta, M.; Wharton, J.A.; Blache, Y.; Stokes, K.R.; Briand, J.F. Marine biofilms on artificial surfaces: Structure and dynamics. *Environ. Microbiol.* **2013**, *15*, 2879–2893. [CrossRef] [PubMed]
4. Hoiby, N. A short history of microbial biofilms and biofilm infections. *APMIS* **2017**, *125*, 272–275. [CrossRef] [PubMed]
5. Meyer, B. Approaches to prevention, removal and killing of biofilms. *Int. Biodeterior. Biodegrad.* **2003**, *51*, 249–253. [CrossRef]
6. Dang, H.Y.; Lovell, C.R. Microbial surface colonization and biofilm development in marine environments. *Microbiol. Mol. Biol. Rev.* **2016**, *80*, 91–138. [CrossRef] [PubMed]
7. Xu, D.; Jia, R.; Li, Y.; Gu, T. Advances in the treatment of problematic industrial biofilms. *World J. Microbiol. Biotechnol.* **2017**, *33*, 97. [CrossRef] [PubMed]
8. Riga, E.K.; Vohringer, M.; Widyaya, V.T.; Lienkamp, K. Polymer-based surfaces designed to reduce biofilm formation: From antimicrobial polymers to strategies for long-term applications. *Macromol. Rapid Commun.* **2017**, *38*, 1700216. [CrossRef] [PubMed]
9. Gule, N.P.; Begum, N.M.; Klumperman, B. Advances in biofouling mitigation: A review. *Crit. Rev. Environ. Sci. Technol.* **2016**, *46*, 535–555. [CrossRef]
10. Selim, M.S.; Shenashen, M.A.; El-Safty, S.A.; Higazy, S.A.; Selim, M.M.; Isago, H.; Elmarakbi, A. Recent progress in marine foul-release polymeric nanocomposite coatings. *Prog. Mater. Sci.* **2017**, *87*, 1–32. [CrossRef]
11. Nir, S.; Reches, M. Bio-inspired antifouling approaches: The quest towards non-toxic and non-biocidal materials. *Curr. Opin. Biotechnol.* **2016**, *39*, 48–55. [CrossRef] [PubMed]
12. Stupak, M.E.; García, M.T.; Pérez, M.C. Non-toxic alternative compounds for marine antifouling paints. *Int. Biodeterior. Biodegrad.* **2003**, *52*, 49–52. [CrossRef]
13. Lejars, M.; Margaillan, A.; Bressy, C. Fouling release coatings: A nontoxic alternative to biocidal antifouling coatings. *Chem. Rev.* **2012**, *112*, 4347–4390. [CrossRef] [PubMed]
14. Eduok, U.; Faye, O.; Szpunar, J. Recent developments and applications of protective silicone coatings: A review of PDMS functional materials *Prog. Org. Coat.* **2017**, *111*, 124–163. [CrossRef]
15. Zhang, H.B.; Chiao, M. Anti-fouling coatings of poly(dimethylsiloxane) devices for biological and biomedical applications. *J. Med. Biol. Eng.* **2015**, *35*, 143–155. [CrossRef] [PubMed]
16. Eduok, U.; Suleiman, R.; Gittens, J.; Khaled, M.; Smith, T.J.; Akid, R.; El Ali, B.; Khalil, A. Anticorrosion/antifouling properties of bacterial spore-loaded sol-gel type coating for mild steel in saline marine condition: A case of thermophilic strain of bacillus licheniformis. *RSC Adv.* **2015**, *5*, 93818–93830. [CrossRef]
17. Eduok, U.; Khaled, M.; Khalil, A.; Suleiman, R.; El Ali, B. Probing the corrosion inhibiting role of a thermophilic bacillus licheniformis biofilm on steel in a saline axenic culture. *RSC Adv.* **2016**, *6*, 18246–18256. [CrossRef]
18. Kochkodan, V.; Hilal, N. A comprehensive review on surface modified polymer membranes for biofouling mitigation. *Desalination* **2015**, *356*, 187–207. [CrossRef]
19. Jhaveri, J.H.; Murthy, Z.V.P. Nanocomposite membranes. *Desalin. Water Treat.* **2016**, *57*, 26803–26819. [CrossRef]
20. Lyszcz, M.; Galazka, A. Genetic differentiation methods of microorganisms in the soil—Plant system. *Postepy Mikrobiol.* **2017**, *56*, 341–352.
21. Beigbeder, A.; Mincheva, R.; Pettitt, M.E.; Callow, M.E.; Callow, J.A.; Claes, M.; Dubois, P. Marine fouling release silicone/carbon nanotube nanocomposite coatings: On the importance of the nanotube dispersion state. *J. Nanosci. Nanotechnol.* **2010**, *10*, 2972–2978. [CrossRef] [PubMed]
22. Sun, Y.; Zhang, Z. New anti-biofouling carbon nanotubes-filled polydimethylsiloxane composites against colonization by pioneer eukaryotic microbes. *Int. Biodeterior. Biodegrad.* **2016**, *110*, 147–154. [CrossRef]

23. Cavas, L.; Yildiz, P.G.; Mimigianni, P.; Sapalidis, A.; Nitodas, S. Reinforcement effects of multiwall carbon nanotubes and graphene oxide on PDMS marine coatings. *J. Coat. Technol. Res.* **2018**, *15*, 105–120. [CrossRef]

24. Mahdavi, M.R.; Delnavaz, M.; Vatanpour, V.; Farahbakhsh, J. Effect of blending polypyrrole coated multi-walled carbon nanotube on desalination performance and antifouling property of thin film nanocomposite nanofiltration membranes. *Sep. Purif. Technol.* **2017**, *184*, 119–127. [CrossRef]

25. Yang, J.-L.; Shen, P.-J.; Liang, X.; Li, Y.-F.; Bao, W.-Y.; Li, J.-L. Larval settlement and metamorphosis of the mussel *Mytilus coruscus* in response to monospecific bacterial biofilms. *Biofouling* **2013**, *29*, 247–259. [CrossRef] [PubMed]

26. Oliva, M.; Martinelli, E.; Galli, G.; Pretti, C. PDMS-based films containing surface-active amphiphilic block copolymers to combat fouling from barnacles *B. Amphitrite* and *B. Improvisus*. *Polymer* **2017**, *108*, 476–482. [CrossRef]

27. Beigbeder, A.; Labruyere, C.; Viville, P.; Pettitt, M.E.; Callow, M.E.; Callow, J.A.; Bonnaud, L.; Lazzaroni, R.; Dubois, P. Surface and fouling-release properties of silicone/organomodified montmorillonite coatings. *J. Adhes. Sci. Technol.* **2011**, *25*, 1689–1700. [CrossRef]

28. Sathya, S.; Murthy, P.S.; Das, A.; Sankar, G.G.; Venkatnarayanan, S.; Pandian, R.; Sathyaseelan, V.S.; Pandiyan, V.; Doble, M.; Venugopalan, V.P. Marine antifouling property of PMMA nanocomposite films: Results of laboratory and field assessment. *Int. Biodeterior. Biodegrad.* **2016**, *114*, 57–66. [CrossRef]

29. Noguer, A.C.; Olsen, S.M.; Hvilsted, S.; Kiil, S. Field study of the long-term release of block copolymers from fouling-release coatings. *Prog. Org. Coat.* **2017**, *112*, 101–108. [CrossRef]

30. Eichner, C.A.; Erb, R.W.; Timmis, K.N.; Wagner-Dobler, I. Thermal gradient gel electrophoresis analysis of bioprotection from pollutant shocks in the activated sludge microbial community. *Appl. Environ. Microbiol.* **1999**, *65*, 102–109. [PubMed]

31. Wiens, J.J.; Donoghue, M.J. Historical biogeography, ecology and species richness. *Trends Ecol. Evol.* **2004**, *19*, 639–644. [CrossRef] [PubMed]

32. He, F.L.; Hu, X.S. Hubbell's fundamental biodiversity parameter and the simpson diversity index. *Ecol. Lett.* **2005**, *8*, 386–390. [CrossRef]

33. Camps, M.; Barani, A.; Gregori, G.; Bouchez, A.; Le Berre, B.; Bressy, C.; Blache, Y.; Briand, J.F. Antifouling coatings influence both abundance and community structure of colonizing biofilms: A case study in the northwestern mediterranean sea. *Appl. Environ. Microbiol.* **2014**, *80*, 4821–4831. [CrossRef] [PubMed]

34. Casse, F.; Swain, G.W. The development of microfouling on four commercial antifouling coatings under static and dynamic immersion. *Int. Biodeterior. Biodegrad.* **2006**, *57*, 179–185. [CrossRef]

35. Briand, J.F.; Djeridi, I.; Jamet, D.; Coupe, S.; Bressy, C.; Molmeret, M.; Le Berre, B.; Rimet, F.; Bouchez, A.; Blache, Y. Pioneer marine biofilms on artificial surfaces including antifouling coatings immersed in two contrasting french mediterranean coast sites. *Biofouling* **2012**, *28*, 453–463. [CrossRef] [PubMed]

36. Chen, C.L.; Maki, J.S.; Rittschof, D.; Teo, S.L.M. Early marine bacterial biofilm on a copper-based antifouling paint. *Int. Biodeterior. Biodegrad.* **2013**, *83*, 71–76. [CrossRef]

37. Yang, J.L.; Li, Y.F.; Guo, X.P.; Liang, X.; Xu, Y.F.; Ding, D.W.; Bao, W.Y.; Dobretsov, S. The effect of carbon nanotubes and titanium dioxide incorporated in PDMS on biofilm community composition and subsequent mussel plantigrade settlement. *Biofouling* **2016**, *32*, 763–777. [CrossRef] [PubMed]

38. Ling, G.C.; Low, M.H.; Erken, M.; Longford, S.; Nielsen, S.; Poole, A.J.; Steinberg, P.; McDougald, D.; Kjelleberg, S. Micro-fabricated polydimethyl siloxane (PDMS) surfaces regulate the development of marine microbial biofilm communities. *Biofouling* **2014**, *30*, 323–335. [CrossRef] [PubMed]

39. Gu, J.-D. Microbiological deterioration and degradation of synthetic polymeric materials: Recent research advances. *Int. Biodeterior. Biodegrad.* **2003**, *52*, 69–91. [CrossRef]

materials

MDPI

Review

Electroactive Smart Polymers for Biomedical Applications

Humberto Palza [1,2,*], Paula Andrea Zapata [3] and Carolina Angulo-Pineda [1]

[1] Departamento de Ingeniería Química, Biotecnología y Materiales, Facultad de Ciencias Físicas y Matemáticas, Universidad de Chile, 8370456 Santiago, Chile; cangulo@u.uchile.cl
[2] Millenium Nuclei in Soft Smart Mechanical Metamaterials, Universidad de Chile, 8370456 Santiago, Chile
[3] Grupo de Polímeros, Facultad de Química y Biología, Universidad de Santiago de Chile, 8350709 Santiago, Chile; paula.zapata@usach.cl
* Correspondence: hpalza@ing.uchile.cl; Tel.: +56-229-784-085

Received: 6 December 2018; Accepted: 9 January 2019; Published: 16 January 2019

Abstract: The flexibility in polymer properties has allowed the development of a broad range of materials with electroactivity, such as intrinsically conductive conjugated polymers, percolated conductive composites, and ionic conductive hydrogels. These smart electroactive polymers can be designed to respond rationally under an electric stimulus, triggering outstanding properties suitable for biomedical applications. This review presents a general overview of the potential applications of these electroactive smart polymers in the field of tissue engineering and biomaterials. In particular, details about the ability of these electroactive polymers to: (1) stimulate cells in the context of tissue engineering by providing electrical current; (2) mimic muscles by converting electric energy into mechanical energy through an electromechanical response; (3) deliver drugs by changing their internal configuration under an electrical stimulus; and (4) have antimicrobial behavior due to the conduction of electricity, are discussed.

Keywords: Electrically conductive polymers; Electroactive biomaterials; Electrical stimulation; Smart composites; Bioelectric effect; Drug delivery; Artificial muscle

---

1. Introduction

Polymers have emerged in recent decades as one of the most promising materials in biomedical applications due to their high biocompatibility and degradation/absorption in physiological media [1]. Another key characteristic of polymers is their flexibility in terms of properties and functionalities, allowing their development from bioactive hydrogels to biodegradable thermoplastic polymers [2,3]. The polymer flexibility also includes a broad range of processing techniques, such as: extrusion [4], electro-spinning [5,6], 3D printing [7–9], microfluidity [10], and casting [11], among others [5]. Remarkably, by adding/embedding nanoparticles into a polymer matrix, novel nanocomposites can be developed further extending the range of properties and functionalities of polymers. For these reasons, polymers are extensively studied today for tissue engineering [12,13], wound healing [14], artificial muscles [15], and drug delivery [16], among other bio-applications [17].

Of recent interest in polymer science is the development of smart materials with a rationally designed stimulus/response behavior. In this context, electroactive smart polymer materials are stressed because of their ability to transfer electrons/ions under a specific electric field, having multiple applications in several engineering areas, such as soft robots and sensors [18,19]. The advantages of an electric field as external stimulus, compared to others, is related to the availability of equipment that allows precise control in terms of the current magnitude, the duration of electric pulses, intervals between pulses, etc. However, compared to other functional/smart polymer systems, electroactive smart polymers have been less studied for biomedical applications, despite their multiple applications in tissue engineering [20–22]. For instance, these electroactive biomaterials can be applied to obtain adhesion and proliferation of human cells, accelerating the process of regeneration in muscles, organs and bones [23–26]. They can also be used for smart drug delivery or as artificial muscle systems, both triggered by electric stimuli. Even less studied is the development of biocidal materials based on their electric conductivity despite that today any biomaterial used in tissue engineering must not only be biocompatible in the response of the host (patient) but also active in avoiding the adhesion of bacteria or the formation of biofilms on its surface. Based on the bactericidal effect of electrical stimulation (ES), novel electroactive materials can be produced with the ability to prevent the formation of biofilms and future bacterial infections in the host. Therefore, a polymer able to deliver ES can merge the requirements needed for any biomaterial designed for tissue engineering purposes: to promote cellular adhesion and proliferation while avoiding biofilm formation through a bactericidal effect (see Figure 1).

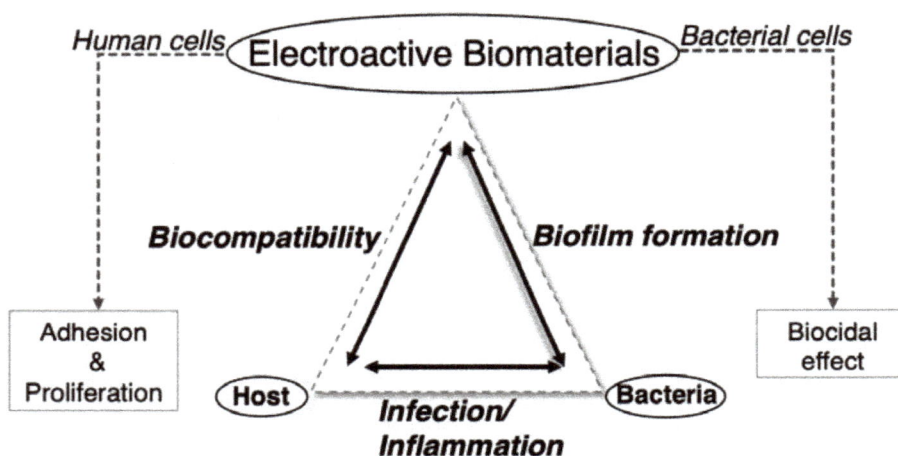

Figure 1. Relationship between electroactive biomaterials and human and bacterial cells in the context of tissue engineering.

In this review, we provide a general overview of the potential of electroactive polymer biomaterials considered as a new generation of smart systems able to respond specifically to an electric field in the context of biomedical applications. These smart systems range from polymers delivering an electric signal to polymers changing some properties under an electric stimulus [27]. The review focuses on the capacity of these electroactive polymers to stimulate: (1) cells in the context of tissue engineering; (2) an electromechanical response for artificial muscles; (3) drug delivery; and (4) antimicrobial mechanisms. From a material point of view, the electroactive polymers include intrinsically conductive polymers, percolated conductive polymer nanocomposites, and ionic conductive hydrogels. A general overview of this review is summarized in Figure 2. For further details about one or more of the above-mentioned electroactive properties or polymers, there are several excellent reviews (for instance, see references [17,21,27–33]) in which specific information can be obtained for a deeper understanding of an application.

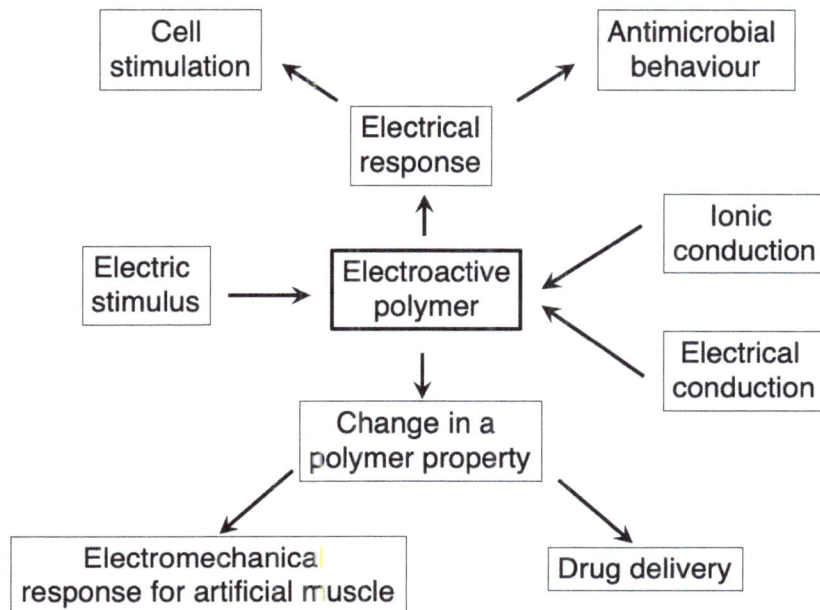

Figure 2. A general overview of electroactive polymers. The mechanism for the specific response to an electric stimulus can be through ionic or electric conduction. These mechanisms can trigger either a direct electric current to the material and the medium producing cell stimulation, or antimicrobial behavior or a change in some polymer properties, producing an electromechanical behavior and specific drug delivery.

2. Electroactive Conductive Polymers

Electroactive polymers can be classified according to the mechanism of conduction in ionic conductive polymers and electric conductive polymers. The latter are further classified as intrinsic and extrinsic based on their mechanism of electron conduction. While ionic conductive polymers present conductivities due to the presence of both ionic groups in their main chain and electrolytes in the medium, electric conductive polymers are conductive due to the high electron mobility arising from either the constitutive bonds between atoms or the presence of conductive particles, as summarized in Figure 3. Regarding electric conductive polymers, different mechanisms of electron conduction produce changes in the achieved conductivity, as summarized in Figure 4.

These conductive materials retain the good properties and flexibility of polymers, so they can be further functionalized for specific applications by optimizing properties such as roughness, porosity, hydrophobicity, conductivity, and degradability [17]. One route for increasing the functionality is to add monomers covalently bonded to functional molecules, although the conductivity is reduced [34]. For biomedical applications, the biocompatibility and biodegradability of electroactive polymers should be further considered. For instance, the application of intrinsically conductive polymers in tissue engineering is limited by the doping concentrations used to obtain electrical conduction, as high concentrations can produce inflammatory responses in tissues [28]. To increase the biocompatibility of conductive polymers, they can be doped with biomolecules or ions, taking advantages of their chemical, electric, and physical structures [17,30].

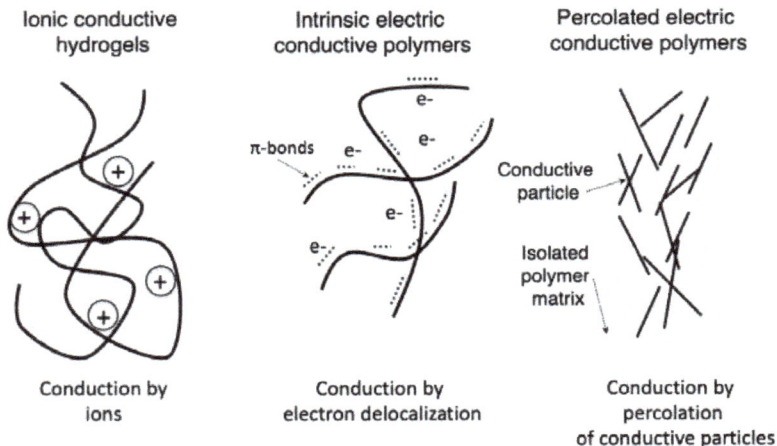

Figure 3. Simplified schematic diagrams showing the different conduction mechanisms of electroactive smart polymers. Ionic polymers present conductivities associated with the presence of polyelectrolytes (left side), while electric conductive polymers can transfer electrons by either an intrinsic mechanism associated with their chemical bonds (middle) or conductive particles percolated into the isolated matrix (right side). See text for details.

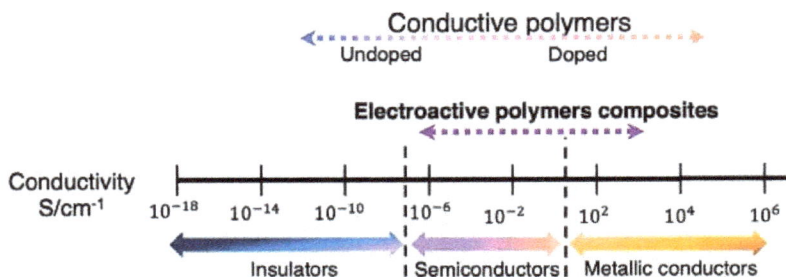

Figure 4. Conductivity range of intrinsically conductive polymers and electroactive conductive composites. Based on reference [29].

2.1. Intrinsically Conductive Polymers

Intrinsically conductive polymers present a conductivity mechanism arising from the polymer molecule itself having a conjugated chain that contains localized carbon–carbon single bonds (σ) and less localized carbon–carbon double bonds (π) (see Figure 5). The p-orbitals overlap in the π bonds and give greater electron mobility between atoms, allowing the electrons to move along the polymer chain [27,35]. The conductivity of intrinsic polymers is further based on the incorporation of dopant ions balancing the charge introduced through oxidation (p-doping) or reduction (n-doping) [27]. The dopant introduces a charge carrier by removing/adding electrons from/to the polymer chain and relocalizing them as polarons or bipolarons. The dopants are able to move in or out of the polymer (depending on the polarity) when an electrical potential is applied, disrupting the stable backbone and allowing charge to be passed through the polymer [27]. Intrinsically conductive polymers have attractive properties for use in drug delivery, sensors, electrochemistry, etc. [36–38].

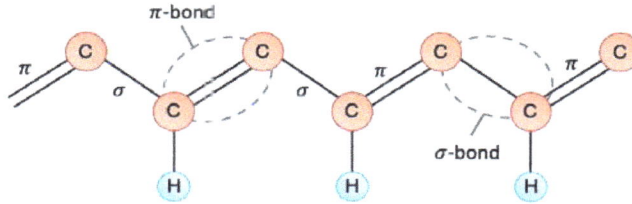

Figure 5. Diagram of the conjugated chain of an intrinsically conductive polymer. Based on reference [35].

Poly[3,4-(ethylenedioxy)thiophene] (PEDOT) [39], polypyrrole (PPy) [40], and polyaniline (PANi) [41] are some of the most widely used intrinsically conductive polymers in tissue engineering scaffolds and biomaterials [17]. However, for biomedical applications, their use is limited mainly because of their poor processability and mechanical properties [36]. The doping of these polymers with long chains can overcome these limitations although it can affect the conductivity of the resulting materials [34,42]. Another solution is to blend the intrinsic conductive polymer with another polymer possessing easier processability in order to obtain a composite with improved mechanical and biocompatibility properties [30]. Such is the case of a 3D coating made of PPy doped with dodecylbenzenesulfonic acid (DBSA) used for electrodes promoting neuronal induction [43]. As discussed above, conductive polymers can be further functionalized with bio-dopants to improve their biocompatibility in medical applications [39]. This method adapts the polymer chains for several applications, improving, for instance, the selectivity/sensitivity of biosensors or the cell-surface interaction [38,40]. Commonly used bio-dopants include glycosaminoglycans such as chondroitin sulfate, hyaluronic acid, and dextran sulfate [30].

2.2. Percolated Polymer Composites

By embedding electric conductive particles into a polymer matrix a percolation transition can occur associated with the formation of a continuum network of fillers throughout the polymer. Below the percolation threshold, the conductivity change is negligible, and the conductivity of the composite is equal to that of the polymer. However, the percolation produces a drastic increase of several orders of magnitude in the electric conductivity of the resulting composites. In this case, the polymer matrix is an insulator and the filler is responsible for the electric conduction. In the classical theory of percolation, the conductivity of the composite depends on the filler conductivity, its volume fraction, a critical filler volume fraction at which percolation takes place, and the critical index of conductivity that relates with the dimensionality of the filler [44]. This theory predicts a power-law correlation between these parameters by assuming physical contact between particles. However, electrically conductive polymer composites are more complex systems, as the electric conductivity cannot be fully predicted by this theory [28,45,46]. In polymer composites, the conductive particles are separated by energy barriers (polymer molecules) and the tunnel effect becomes relevant, modifying the percolation model by introducing a tunnel parameter that varies according to the dimensionality of the particle [44]. In this modified percolation model, the composite conductivity depends on the filler conductivity and its volume fraction, but also on the tunnel parameter. Under this model, the effect of the filler on the percolation threshold is rather explained considering the average interparticle distance related to the probability of contact between conductive particles [47], which depends on both the aspect ratio and the particle sizes [48,49]. The introduction of these parameters can explain the different electric behavior found in these polymer composites from a sharp increase in the conductivity reaching a plateau to a broad percolation curve with a growing conductivity [44]. Indeed, although the percolation theory is able to predict some experimental results, currently a different and complementary approach based on the excluded volume theory for percolation transition can explain most of the experimental findings [50]. This theory is based on the evidence that the percolation threshold is not linked to the true volume of the object itself, but rather to its excluded volume [50].

Electrically conductive polymeric composites are currently being developed in order to have light materials that combine the inherent processability of the organic matrix with the electric conductivity of the fillers [50]. Since the first polymer with silver filler was developed for electrically conductive adhesives in 1956, conductive polymer composites have been studied extensively by using gold, palladium, silver, nickel, copper, graphite, and carbon fiber [51]. Among the different fillers, those based on nanoparticles, such as carbon nanotubes (CNT), have emerged as some of the most interesting due to their outstanding properties [52]. In percolated composites, the particle aspect ratio can be considered as one of the most relevant variable, explaining, for instance, that CNT-based composites presented percolation thresholds lower than composites containing metallic particles, carbon black, or carbon fibers, or even some graphite derivatives [53,54]. Actually, the percolation threshold in polymer composites is inversely proportional to their aspect ratio [55]. This is explained by changes in the average inter-particle distance in the composite, with more somewhat spherical structures presenting longer distances. Although nanoparticles render lower percolation transitions than microfillers, their high surface energy produces composites with agglomerated rather than isolated structures, affecting negativelly the electric percolation threshold. It is well known that improving the dispersion state of nanoparticles produces a reduction in the percolation threshold [53,55,56]. For high-aspect-ratio fillers, their alignment is another variable affecting the electric conductivity of polymer/CNT composites [57]. Monte Carlo simulations have confirmed that the conductivity decreases with applied strain, because inter-particle distance increases due to CNT alignment [57]. In general, the percolation transition depends on all these variables in a complex way, and, for instance, an optimal agglomeration and aligment level can be found [57].

The advantage of electric conductive polymer composites is the flexibility of the kind of filler that can be used and other properties emerging from the electric conductivity through the filler. For instance, the current passing through the polymer composite can induce a Joule heating, raising the internal temperature of the polymer composite to above the transition temperature [58–61].

2.3. Conductive Polyelectrolite Hydrogels

Hydrogels are three-dimensional polymeric networks possessing hydrophilic characteristics and high water absorbtion capacities. Due to their high water content, porosity and soft consistency, they can mimmick natural living tissue better than any other class of synthetic biomaterials [62]. Hydrogels can be reversible when the network is formed by molecular entanglements and/or secondary forces such as ionic, H-bonding or hydrophobic forces. If the network is based on covalent bonds joining the macromolecular chains or cross-linking polymers, the hydrogels are permanent. Due to their porous networks and high water content allowing transport of water and small solutes, hydrogels present ionic conductivity, especially in the case of polyelectrolytes, as recently studied by comparing different hydrogels [63]. This ionic conductivity depends on several variables such as polymer polarity, water content, salt/ions, and hydrogel structure. Higher water content increases the ionic conductivity of the hydrogel and leads to a high ion transfer rate [63]. The conductivity of the hydrogel is further controlled by two parameters: the mobility and concentration of ions. In low-concentration electrolyte solutions, the concentration of total ions plays a dominant role in conductivity. In high-salt solutions, the fraction of counterions to the total ions is significantly reduced, so the mobility of the ions becomes the dominant parameter. Ionic hydrogels show higher conductivity than nonionic hydrogels, because cationic and anionic hydrogels have higher concentrations of counterions functioning as charge carriers, leading to high conductivity [63]. Besides the cations and anions of the electrolyte itself, the mobile counterions of the ionic polymers also function as charge carriers, and electrolytes and polymer counterions together contribute to higher ionic conductivity [63].

3. Polymers for Tissue Engineering through Electrostimulation of Cells

3.1. Electrostimulation

Living cells use electric fields for several activities associated with: the generation of electromotive force, the control of a specific potential difference, the capacity to control and switch current on and off, and the stored charge. Indeed, an electric voltage exists across the plasma membrane, with the inside of the cell remaining more negative than the outside. Bioelectricity present in the human body plays an integral role in maintaining normal biological functions, such as signaling of the nervous system, muscle contraction and wound healing. During major cellular events like cell division, development, and migration, there is the generation of electric fields [33]. Therefore, a large variety of cell types respond to electrical stimulation, including fibroblasts, osteoblasts, myoblasts, chick embryo dorsal root ganglia, and neural crest cells [33].

The inherent bioelectricity present in different cellular events explains the use of electrical stimulation (ES) for tissue repair through either direct current (DC) or alternating current (AC) [22,25]. By applying electric fields, the cell behavior can be modified, including orientation, proliferation, and rate and direction of cell migration, as tested in corneal, epithelial, and vascular cells, among others [17,20,21,31,63,64]. For instance, ES produces electrotaxis or galvanotaxis, the phenomenon by which there is a directional migration of cells in response to the electric field [65–67]. There is further evidence showing the great influence of an ES on growth and development of nerve cells, wound healing, and angiogenesis, among other cellular properties, the former being one of the most relevant in this field [68]. In addition, by means of controlled ES, a greater cellular differentiation is achieved; for instance, stem cell differentiation to neurons [33,68,69]. One of the main effect of ES is the opening of ion channels, triggering the production of ions that can be deposited on tissues [70]. This change results in the alteration of ionic fluxes like calcium ions, contributing to cellular locomotion or electrophoretic/electroosmotic effects that cause a redistribution of membrane components [65,67]. The effect of an electrical field is not only valid for cells but also for tissues [71].

3.2. Polymers for Electrostimulation of Cells

The construction of scaffolds based on electrically conductive polymers for nerve tissue engineering to enhance the nerve regeneration process have been one of the most studied applications of electroactive polymers [33]. For instance, PC12 cells were seeded on electrochemically synthesized PPy films, producing a ~91% increase in median neurite length when a positive potential of 100 mV was passed through the PPy for 2 h [72]. Applying electric stimuli to nerve cells through conductive nanofibrous scaffolds of PANi/gelatin enhanced cell proliferation and neurite outgrowth compared with non-stimulated scaffolds can also be achieved [73]. Poly(D,L-lactide-co-ε-caprolactone) membrane coated with PPy and the composite scaffolds increased the proliferation and differentiation of PC12 into neuronal phenotypes as well as sciatic nerve regeneration in rats, showing that they can be used for ES enhancing the neurite outgrowth [74]. These studies demonstrate that cell growth and function can be drastically enhanced at the interface of PPy undergoing ES.

ES has also been used in fibroblast cells. For instance, conductive biodegradable PPy-Polylactide (PLA) membranes and poly(D,L-lactide)/PPy nanocomposites are able to upregulate the mitochondrial activity of human skin fibroblasts [75]. Under a constant electrical field strength of 100 mV/mm, a greater cell viability was observed than that shown by the non-stimulated cells cultured on the same substrate of identical surface morphology and chemistry. Moreover, electrical field seems to play a more substantial role than does electrical current in modulating the activity of cells cultured on conductive polymeric scaffold. DC applied to nanofibrous scaffolds of PANi and poly(L-lactide-co-ε-caprolactone) enhanced the growth of NIH-3T3 fibroblasts [76]. Electric stimulus in conductive polymers may offer a novel engineering technique to regulate cell adhesion and orientation of bone marrow-derived mesenchymal stem cells (MSCs) and fibroblasts [77].

Regarding the mechanisms behind the effects of an electric potential and/or an electric field on cell activity through an intrinsic conductive polymer, it is speculated that reduction of the polymer (for instance PPy) and electric conduction itself can both affect cells in several ways [34]. For example, the process of neutralization of PPy, under a reducing potential, causes the expulsion of negative ions or the uptake of positive ions from the medium. An uptake of positive ions such as Na^+ from the medium is speculated to affect several processes, including protein adsorption and the cell cycle. For instance, human Adipose-Derived Mesenchymal Stem Cells (AD-MSCs) attached to PPy/chitosan composite scaffolds and stimulated under DC for 7 days presented a calcium deposition 346% higher than non-stimulated scaffolds [78]. The adsorption of serum proteins, specifically fibronectin, on the electrically conducting polymer can further explain the improved cell behavior under ES as reported in PC12 cells [34,79].

Electroactive polymer composites can also be used for tissue regenerating scaffolds, biosensors, and bioapplications, leaving in evidence several potential applications in tissue engineering [29]. For instance, a polymeric composite scaffold of polyacrylonitrile/carbon nanofibers was developed, yielding promising results under ES for applications in nerve tissue regeneration. Electrostimulated cells attached on this conductive scaffold improve neuronal differentiation, and maturation of neural stem cell under 5 V (AC) for 4 h during 7 days [80]. The intracellular and extracellular fluids, which possessed different potentials under ES, produced an extra depolarization, generating these improvements and cell extension [78,79]. Poly(lactic-co-glycolic acid) (PLGA)/CNT electroactive scaffolds were also tested under an electric current (AC) using similar cells with better results than the non-stimulated cell/samples [81]. In particular, an increase in proliferation, differentiation, and growth of long neurites attached to the scaffolds were found under ES in these composite scaffolds [81].

The use of graphene in biomaterials is well known due to its excellent mechanical and electric properties, as well as its biocompatibility with human cells [49,82]. Graphene particles are used as a mechanical support strengthening hydrogels and as electric fillers for percolated conductivity polymers [83]. For instance, electrically conductive graphene hydrogels based on Reduced Graphene Oxide (rGO) and polyacrylamide (PAAm) can be considered as a composite useful for the development of skeletal muscle in soft tissue engineering scaffolds and bioelectrodes. Moreover, ES of myoblasts by the soft electroactive composite can upregulate myogenic gene expressions [83]. Polymer/graphene composite scaffolds can further be designed for cardiac tissue engineering [84]. For instance, Polycaprolactone (PCL) and Graphene composite scaffolds were obtained by an electrospinning technique, producing changes in cardiomyocyte functions and significantly increasing the flux and concentration of Ca^{2+} after ES [84].

Despite the several advantages of electroactive polymers for tissue engineering, some relevant challenges should be addressed in order to continuously improve their behavior in this field. For intrinsic conductive polymers, one the most relevant drawback is the lack of a proper biodegradation among other issues such as poor polymer–cell interactions, the absence of cell interaction sites, hydrophobicity, processability, and mechanical properties [85]. The most common strategy to overcome these issues is to mix the electroactive polymer with another polymer possessing the desired property, such as: PLA, PCL, PLGA, polyurethane (PU), chitosan, gelatin, and collagen, among others, for biodegradation improvements. However, even minimizing the amounts of electroactive polymers in these blends, they are expected to stay in the body. Another route to overcome this limitation is by synthesizing erodible conducting polymers able to have a gradual dissolution [86] or by preparing degradable conductive polymers containing conducting oligomers [85]. For electroactive polymer composites, the potential toxicity of the carbon nanostructures is one the main drawbacks [87]. Moreover, carbon nanomaterials are not biodegradable in general, adding another limitation, although they can be excreted in vivo and cleared from the body once it is no longer needed. The increment in the polymer resistivity after applying an electrical current can further add limitations together with the likely cytotoxicity effect of long-term electrical exposure of cells [27].

4. Electroactive Polymers for Drug Delivery

After the discovery more than 50 years ago that hydrophobic and low-molecular-weight drug molecules are able to diffuse through silicone materials at a controlled rate, polymers have been extensively studied for drug delivery systems [88]. The flexibility of polymeric materials can be used to modulate the properties of the materials such as biodegradability and biocompatibility, because of their diversity in chemistry, topology, and dimension. Indeed, polymers show usually improved pharmacokinetics compared to pure small molecule drugs. Polymers are not drugs themselves, and therefore they are designed to provide a passive function as drug carriers, reducing immunogenicity, toxicity, or degradation, while improving circulation time [88]. Relevant in drug delivery is the study of stimuli-responsive polymers mimicking biological systems in the capacity to change under external stimulation [89,90]. These smart polymer biomaterials should present their response within biological conditions. Typical stimuli are temperature [89], pH [91], light [92], electric field [93], and electrolytes [94], among others [95–98]. The responses triggering the drug release can be: dissolution/precipitation, degradation, change in hydration state, swelling/collapsing, hydrophilic/hydrophobic surface, change in shape, conformational change and micellization. The most important stimuli are pH, temperature, ionic strength, light, and redox potential. However, electric fields can also be a stimuli for drug delivery and today electro-responsive polymers can be considered smart drug carriers [21].

In drug delivery, hydrogels are highlighted because their highly porous structure permits loading of drugs into the gel matrix, subsequently allowing drug release at a rate dependent on the diffusion coefficient of the active molecule through the gel [89,93]. In stimulus/response electroactive hydrogels the final effect of ES on drug release depends to a large extent on how the gel responds to the stimulus, how the drug is released from the gel, and the interactions between the gel network and the drug [99]. The main mechanisms of drug release in these electroactive hydrogels are: (1) forced convection of the drug out of the gel along with syneresed/expelled water due to the electric field [98]; (2) diffusion [100]; (3) electrophoresis of charged drugs [101]; and (4) drug release upon erosion of electro-erodible gels [102]. For charged drugs, the migration of the charged entities towards the electrode bearing an opposite charge should be further considered [103]. The first mechanism is, however, the most important mechanism of drug release in these systems, since under the influence of an electric field, hydrogels generally deswell, causing the movement of solutes out of the gel. In particular, when an electric field is applied, water is syneresed/expelled from the gel, causing the ejection of the drug [93]. When the electric field is removed, the gel absorbs fluid and swells. Thus, upon sequential switching "on" and "off" of the electric field, the gel deswells and swells, following the electric field program [104–106]. Three main mechanisms of the electro-induced gel deswelling process exist: (1) the establishment of a stress gradient in the gel; (2) changes in local pH around the electrodes; and (3) electro-osmosis of water coupled with electrophoresis [102]. When diffusion is the major drug release mechanism, electro-induced gel shrinking may inhibit drug release from the gel as the "pores" in the polymer network of the gel become smaller and the pathway for drug movement out of the gel becomes more tortuous. In this case, the application of an electric field stops/reduces drug release from gels. This is especially significant for large drug molecules whose movement out of the gels can be more effectively hindered by a "shrunken" polymeric network [102]. Electro-induced anisotropic gel swelling can also occur when the gel is placed in a fixed position away from the electrodes. In this case, gel expansion occurs when the mobile cations in the aqueous medium migrate towards the cathode, penetrating into the gel network inducing ionization of the carboxyl groups on the gel network that causes the gel on the anode side to swell as the ionized groups become hydrated [107]. These kinds of gels, which swell in response to an electric field (and thus allow drug diffusion out of the gel) may be more appropriate vehicles for electro-controlled release of such large molecules. Finally, pH changes can lead to disruption of the ionic bonds responsible for the gel complex, and for instance the gel surface facing the cathode can dissolve and erode under some conditions. This process triggers drug release.

Intrinsic conductive polymers can also be used for electroactive drug delivery devices, as they can undergo controllable and reversible redox reactions. These reactions alter their redox state, causing simultaneous changes in polymer charge, conductivity, and volume that result in the uptake or expulsion of charged molecules from the bulk of the polymer [107]. By exploiting these changes, the rate of drug release from these conductive polymers can be modified. Anionic drugs can be loaded into the polymers during the oxidative polymerization process or via ion exchange through redox cycling after polymerization. By an electrochemical reduction, anionic molecules can be released [108]. For instance, glutamate anions can be released more than 14 times better from PPy during the application of a reducing voltage, compared to the system without ES. In this case, PPy was prepared with mobile anions that would be released on electric reduction accompanied by polymer contraction (anion-driven actuation), therefore releasing the anionic drug [108]. So the drug release is triggered by reduction and the reincorporation of drug by oxidation. For cationic drug release, when the neutral intrinsic conductive polymer is oxidized, the resulting net positive charge in the polymer repels the drug out of the film. PPy prepared with immobilized anions will incorporate cations on reduction accompanied by swelling (cation driven actuation), and cations can then be released on oxidation. Of interest is mixing intrinsic conductive polymers with hydrogels for the development of electro-conductive hydrogels. In particular, a poly(ethyleneimine) (PEI) and 1-vinylimidazol(VI) polymer blend containing polyacrylic acid (PAA) and poly(vinyl alcohol) (PVA) semi-interpenetrating networks (semi-IPNs) was recently produced for therapeutic electro-responsive drugs [109]. Another electrically active hydrogel was prepared by mixing chitosan-graft-polyaniline copolymer with oxidized dextran (OD) as a cross-linking agent. The copolymer acted as a drug carrier with electrically driven release at a release rate that dramatically increased when an increase in voltage was applied [93]. The electrically driven release of drug molecules from conductive hydrogels has been directly associated with (1) electric field-driven migration of the charged molecules [93] and (2) change in the overall net charge within the polymer upon reduction or oxidation [110].

More complex structures can also be produced using intrinsic conductive polymers for instance those based on nanotubes and microcups. In the former case, biodegradable polymer fibers having the drug were produced by electrospinning, and then the conductive polymer was added on the surface by electrochemical deposition [111]. A local dilation of the tube by the ES promotes mass transport, accounting for the drug release in a desired fashion by ES of the nanotubes. Microcups made of PPy were produced using PLGA polymer as template, with the capacity to control the drug loading/release characteristics [112]. PPy nanoparticles can also be used for drug delivery externally stimulated through a weak and external DC electric field having excellent spatial, temporal, and dosage control [113]. In this case, the conductive polymer was coupled with a temperature-sensitive hydrogel, and the mechanism involved a synergistic process of electrochemical reduction/oxidation and electric-field-driven movement of charged molecules. Recently, electrically responsive micro-reservoirs made of arrays of vertical microtubes were used as support for PPy polymers sealed with PLGA were produced as microcontainers for anti-inflammatory drugs. This system was able to accelerate the cells' osteogenic differentiation via electrically controlled release of dexamethasone [114].

The electric conductivity of many electroactive polymeric materials used is not high enough to achieve an effective modulation of drug release, leading to the use of more conducting materials (e.g., carbon-based nanomaterials) in polymeric networks as a strategy to enhance the electro-sensitivity of hydrogels. The addition of conductive particles such as CNT can improve the electrically stimulated drug delivery behavior of the intrinsic conductive polymers [115,116]. For instance, a semi-interpenetrating polymer network based on polyethylene oxide and pentaerythritol triacrylate polymers was prepared by electrospinning, and CNT was used to increase the electric sensitivity. The amount of released drug increased under the presence of the conductive particles due to the polymer dissolution under the effects of carbon nanotubes, thereby releasing the drug. A similar tendency was found using an aligned CNT array membrane electrode as a platform for the production of PPy films, showing significant improvement in the controlled release of neurotrophin [117]. Electrospinning was used to prepare poly(vinyl alcohol)/poly(acrylic acid)/multi-walled carbon nanotubes (MWCNTs) nanocomposites where the drug release of nanofibers depended on the electric voltage applied due to the variation of the ionization of functional groups in the polymer matrices [118]. In this context, spherical hybrid hydrogels composed of gelatin with CNT were produced as drug delivery systems for the electro-responsive release of diclofenac sodium salt, where the electrical stimulation increased the drug release associated with a reduction of swelling behavior by built-in osmotic pressure [119]. Electro-responsive hybrid hydrogels can also be produced by this route such as gelatin-coated CNT mixed with acrylamide and polyethylene glycol dimethacrylate as plasticizing and crosslinking monomer, respectively [120]. These composites were highly versatile in modulating the drug delivery of neutral drugs as a function of both nanotube content and voltage magnitude, with drug release being dependent on the balance between electrostatic attractive and repulsive forces and the degree of hydrogel swelling. Another electro-responsive poly(methylacrylic acid)/CNT composite was also reported, presenting controlled drug release upon the On/Off application of an electric field as tested both in vitro and in vivo [121].

The above-mentioned drawbacks of electroactive polymers in tissue engineering are still valid for drug delivery, in particular lack of a proper biodegradation, high hydrophobicity, and poor mechanical behavior [103]. In the particular case of non-biodegradable drug delivery implanted devices, after an initial procedure to administer the device, a second procedure will be required for removal [103]. In addition to these issues, these electroactive polymers will require attachment to an electrode and some electronic circuitry for operating, limiting their use. Another limitation is related to the low levels of drug than can be incorporated and released [103].

5. Artificial Muscle Based on Polymer Composites

Artificial muscles can be defined as electromechanical actuators, meaning that they can directly convert electric energy into mechanical energy. They are relevant for a broad range of applications, especially in biomedical engineering, as they can be used in applications such as: microsurgical devices, artificial limbs, or even, in the future, implants like artificial ocular muscles, or hearts [122]. Specific examples are blood vessel (microanastomosis) connectors, tubes that hold open the ear drum (myringotomy tubes), and microvalves for prevention of urinary incontinence [123]. Artificial muscles based on conductive polymer actuators have many advantages for biomedical applications as they (1) can be electrically controlled; (2) have a large strain which is favorable for linear, volumetric, or bending actuators; (3) possess great strength; (4) require low voltage for actuation (1 V or less); (5) can be positioned continuously between minimum and maximum values; (6) work at room/body temperature; (7) can be readily microfabricated and have light weight; and (8) can operate in body fluids [123]. Although different materials are used as artificial muscles, most of them are polymers based on electroactive PPy, ionic metal–polymer composites (IMPCs), hydrogels, or liquid crystal elastomers (LCEs). Today, polymer actuators can even exceed the performance of natural muscle in many respects, making them particularly attractive for use anywhere a muscle-like response is desirable [124]. Each polymer system presents a specific mechanism for the electromechanical actuation and for instance, electronically intrinsic conducting polymers such as PANi and PPy provide one type of high-strain actuator based on dimensional changes produced by electrochemically inserting solvated dopant ions into a conducting-polymer electrode [124]. Dielectric elastomers present actuation through "Maxwell stress" due to the attraction between charges on opposite capacitor electrodes and the repulsion between like charges [125]. The volume change of an electrolyte and electrostatic repulsion can be further used as a mechanism such as in ionic polymer/metal composite actuator. Depending on the conductive mechanisms, these polymers can be divided into two major groups: (1) electroactive polymers (EAPs) such as intrinsic conductive polymers, dielectric elastomer actuators (DEAs), ferroelectric polymers, and liquid crystal elastomers; and (2) ionic EAPs characterized by the presence and movement of ions triggering the actuation [124].

In ionic conducting polymers, an ion is mobile within the matrix and when a positive voltage is applied to a conducting polymer electrode, electrons leave the polymer electrode and anions are attracted to and inserted into the polymer to balance the electric charge, resulting in an expansion. To complete the circuit, a second electrode is used which acts in the opposite direction, expelling ions when it is negatively biased. This inclusion and exclusion of ions can create expansion and contraction on opposite sides of a structure such as a catheter, producing bending [126]. For instance, during oxidation of PPy films, electrons are extracted from the polymer chains, double bonds are rearranged, and positive charges (polarons or bipolarons) are stored along the chains. To maintain the electroneutrality, conformational movements of the chains stimulated by the electrochemical process generate free volume, which is occupied by the counterions and water molecules, producing the film swells. Otherwise, during reduction of the polymer, electrons are injected into the chains and positive charges are compensated. The original structure of the double bonds is restored and counterions and water molecules are expelled towards the solution by the electrochemically stimulated conformational relaxation, promoting a shrinking [127]. Artificial muscles from these conducting polymers are fully reliable Faradaic motors and the movement rate is under linear control of the flowing current and the consumed charge [128]. Design of PPy electroactuators can use a monolithic, bilayered, or trilayered structure, and while monolithic and bilayered implementations are primarily used in applications involving a supporting liquid electrolyte (either aqueous or organic), trilayered ones are employed with an ionic gel electrolyte sandwiched between two PPy films for operation in air. Bending bilayers are one of the most efficient structures transducing reaction that drive from small volume variations in the conductive polymer film to large bending movements [129]. In this case, the second layer is required to translate the volume variation from the polymer film into mechanical stress gradient across the bilayer, producing the macroscopic bending movement. Thus, the second layer is a passive layer that must be bent, although it consumes a fraction of the applied electric energy for bending it. As a result, the muscular energetic efficiency and the angular displacement, for the same consumed charge, decreases [130]. Two layers of the same conducting polymer constituting an asymmetric bilayer muscle can overcome this limitation as one PPy is expected to swell during oxidation by entrance of anions pushing the bending movement while the second layer must shrink during oxidation (simultaneously) by expulsion of cations pulling the bending movement. The improvement arising from the asymmetric bilayer can be seven and four times that of the two layer artificial muscles [128]. A cooperative electro-chemo-mechanical actuation of each of the individual layers occurs in each asymmetric bilayer.

A different kind of material broadly used for actuators in artificial muscles is the family of ion-exchange polymer–metal composites (IPMCs) showing large deformation in the presence of a low applied voltage. IPMCs consist of a solvent swollen ion-exchange polymer membrane laminated between two thin flexible metal (typically percolated Pt nanoparticles or Au) or carbon-based electrodes [131]. The mechanism in IPMCs is based on the characteristic of polyelectrolytes to possess ionizable groups on their molecular backbone that can be dissociated to obtain a net charge in a variety of solvent media. Therefore, the capacity of these polymers to interact with externally applied fields as well as their own internal field triggers the electromechanical deformation of such polyelectrolyte. For instance, polyelectrolytes filled with liquid containing ions can also deform under an external electric field due to the electrophoretic migration of such ions inside the structure [132]. An IPMC bends toward the anode if it is cationic under the influence of an imposed electric potential, and can oscillate in response to an alternating input voltage. Furthermore, the appearance of water at the surface of the expansion side and the disappearance of water on the contraction side occur near the electrodes, meaning that charged particles drag water molecules parasitically with them when they are electrophoretically transported within the IPMC. Therefore, the imposition of an electric field produces an electrophoretic dynamic migration of the mobile cations that are conjugated with the polymeric anions that can result in a local deformation of the material [132]. These composites are produced in two steps: (1) a compositing process to metallize the inner surface of the polymer by a chemical-reduction where the metallic particles are concentrated predominantly near the interface boundaries; and (2) a surface electroding process in which multiple reducing agents are introduced, and where the original roughened surface disappears [133]. The particles improve the conductivity between the polymer and the electrodes.

Another approach for artificial muscle is based on dielectric elastomer actuators that are essentially compliant variable capacitors consisting of a thin elastomeric film coated on both sides with compliant electrodes [131]. When an electric field is applied across the electrodes, the electrostatic attraction between the opposite charges on opposing electrode and the repulsion of the like charges on each electrode generate stress on the film causing it to contract in thickness and expand in area. This concept extends toward the construction of flexible dielectric elastomers by the production of a soft dielectric sandwiched between two soft conductors that are subject to a voltage producing electric charges of the opposite polarities accumulate on the faces of the dielectric, causing the dielectric to reduce thickness and expand area [134]. This approach can be further extended to soft robots, where an encapsulated hydrogel serves as an ionically conductive electrode and surrounding tap water can be used as the other electrode [135]. When a voltage is applied the positive and negative charges accumulate on both sides of the dielectric elastomer, inducing Maxwell stress that deforms the membranes. The net effect is a reduction of the body's curvature, corresponding to the actuated state. The resulting strain in these systems is proportional to the quadratic of the applied voltage and the material electrical strength [136]. Indeed, by increasing the electrical breakdown strength, lager range of input operating voltages and reduced probability of material degradation can be obtained. Notably, the electrical breakdown and the dielectric losses can be changed by controlling processing parameters of the polymer synthesis and fabrication procedure as recently shown for Poly(vinylidenefluoride–trifluoroethylene–chlorotrifluoroethylene) terpolymer [136].

A much less studied material for electro-actuators are percolated electric conductive polymer composites, where the mechanism is based on heating the polymer by the joule effect due to the current passing through the conductive paths. This heating produces an observable expansion of the composites and the buckling of the device when the boundaries are restricted [137]. Although these composites can present low volume changes at high voltages, it depends on the materials used and a chitosan/CNT composite can present larger electromechanical actuations [138]. Recently, environmentally friendly electrothermal bimetallic actuators based on waterborne polyurethane and a silicone rubber matrix filler with CNT presented an improved behavior. Under 7 V AC, the actuator achieved a bending displacement up to 28 mm, which is greater than most of other electrothermal actuators reported [137].

Despite the potentiality of artificial muscles based on electroactive polymers and hydrogels, they present a major drawback related with the small electrochemical stability window of aqueous electrolytes ($\approx$1.23 V) [139]. Beyond this window, electrolysis of water can lead to catastrophe due to hydrogen and oxygen evolution reactions at the electrodes. Indeed, although these systems can be stable in air, they exhibit slow response time. Moreover, it can be a drift in the bending amplitude which may require correction by a feedback-loop control system. In hydrogels, the main drawback relates with the relatively slow response time as well as chemical stability and performance degradation over time [139].

6. Antimicrobial and Antifouling Polymers Based on Electrical Stimulation

6.1. Microbial Infections and Biofouling

Microorganisms are present at all time in different environments, so it is necessary for the design of any kind of biomaterials to consider their antimicrobial properties [140]. Biofouling is the formation of a microbial consortium which contributes to the development of biofilms capable of adhering to the surface of materials, facilitating the adhesion of other microorganisms on wet surfaces. The development of biofilms on different surfaces is a problem that affects several materials in applications such as food, drinking water quality, and medicines, among others [141]. The bacterial colonies on the surface of a biomaterial, which are highly resistant to antibiotic treatments, are difficult to be eliminated by conventional methods [142], leading to a chronic inflammatory response. Once formed, biofilms cause serious and even fatal clinical complications. In biomedical applications, bacterial infections can cause tissue destruction, premature device failure, and the spread of infection to other areas [143,144]. For instance, bone implants are always associated with risks of bacterial infection that leads to implant failure or, in critical cases, amputation or death of the patient [144,145]. In contact with the eye lens this process causes serious eye infections [146]. Other examples of fouling formation are in catheter-associated urinary tract infections [147,148], and dental implants cause periodontal diseases and gingivitis [144]. Therefore, it is of great importance to eradicate biofilm formation avoiding the reversible anchoring of bacterial colonies [149]. The formation of bacterial films on a surface can be classified as follows: State 1—reversible anchoring of bacterial colonies; State 2—bacterial colonies irreversibly anchored to the surface, losing the flagella that give spatial mobility; State 3—beginning of the first maturation stage; State 4—completion of the maturation phase; and State 5—movement of bacterial colonies and dispersal into microcolonies [140]. These states are summarized in Figure 6. The different strategies for the control of biofilms are still under discussion, although some of them are: inhibit microbial adhesion on the surface, interfere with the surface by molecules that modulate the development of the biofilm, and the dissociation of the biofilm matrix [140].

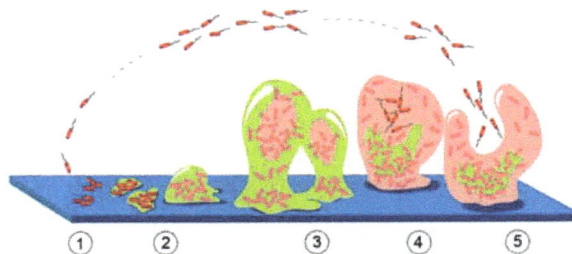

Figure 6. States of biofilm (or biofouling) formation on a material surface (based on reference [140]) State 1: reversible anchoring of bacterial colonies; State 2: bacterial colonies are irreversibly anchored to the surface; State 3: maturation; State 4: the maturation phase is completed and State 5: the bacterial colonies begin to move again, dispersing in microcolonies.

6.2. Electrical Stimulation as an Antimicrobial Method

ES has been applied to promote the inactivation of different biofilms and bacterial strains, such as S. aureus, Pseudomonas, and E. coli on different metals and amorphous carbon substrates, among other types of electroactive materials [150,151]. Because all naturally occurring surfaces, including those of bacterial cells, are generally negatively charged, the electrostatic force between bacteria and a biomaterial surface is repulsive. These repulsive forces can be enhanced by application of an electric current, thereby increasing the negative charge and consequently the repulsive force [152]. Therefore, this electrostatic repulsion between the resulting electrically charged material surface and biofoulants such as soluble microbial product molecules and extracellular polymeric substances which are negatively charged, and microbial cells can facilitate their removal [153]. An electric current can further enhance the activities of antimicrobial agents such as aminoglycosides, quinolones, and oxytetracycline against Pseudomonas aeruginosa, Klebsiella pneumoniae, Staphylococcus epidermidis, Escherichia coli, and Streptococcus gordonii biofilms, a phenomenon referred to as the bioelectric effect [154]. This effect can be related to pH modification, the production and transportation of antimicrobial agents into the biofilm by an electrophoretic process, the genesis of additional biocidal ions, or hyperoxygenation [155]. The bioelectric effect has been studied mainly in infections associated with metal prostheses, although studies have also been conducted to the treatment of infections in the auditory canals, through non-invasive transcutaneous or minimally invasive applications such as subcutaneous [154].

In addition to the mechanism based on electrostatic repulsion, there are others not yet fully understood [148,149]. For instance, the electric field causes an increase in the permeability of the cell membrane, causing electropermeabilization or irreversible electroporation and the production of reactive oxygen species (ROS) [156–164]. This electrolytic damage of the internal cell membranes generates an irreversible loss of the semipermeable barrier function, the release of intracellular content, a loss of motility, and synthesis of some enzymes such as lactate dehydrogenase and trypsin [160]. The effects of electroporation produced by low electric fields (1.5–20 V/cm) promotes biocidal action in the different existing biofilms [160–162]. Free radicals and ROS are generated as hydrogen peroxide and reactive nitrogen species (RNS) at low electric field and low current [155,164]. Moreover, electric current, even at a low intensity, can cause an increase of hydrogen ion concentration inside the cytoplasm and disorganization of membrane functionality, causing the alteration of cells. It has been demonstrated that the use of AC causes the inhibition of yeast cell metabolism because it induces the migration of electrons from the cell to the graphite electrode and the accumulation of H^+ ions in the cell, thereby modifying the membrane potential [160].

6.3. Electroactive Polymers as Antimicrobial and Antifouling Materials

Despite the relevance of electric field to avoid biofouling, its use in electroactive antimicrobial polymers for biomedical applications has been barely reported. For instance, modified PPy membranes coated with graphene derivatives were produced to enhance their electric conductivity and improve biofouling suppression because of higher electrostatic repulsions [153]. In general, most of the research has focused on antifouling membranes for bioreactors. The mechanisms of fouling prevention and cleaning with conductive membranes are also mainly based on electrostatic interactions or electrochemical redox reactions on the membrane surface [165]. For instance, during filtration of charged macromolecules and particles, the charged conducting membrane pushes back the foulants due to the electrostatic effect, and this reduces membrane fouling. In electrochemical fouling, the membrane acts either as the electrode where direct or indirect oxidation of foulants takes place on the membrane surface or at the electrode, where foulants are removed via bubble generation on the surface [165]. In this context, intrinsic conductive polymers are able to show antimicrobial behavior without any external electric stimulus due to the oxidative stress that these polymers can generate on the bacterial cells, suppressing the formation of the bacterial cell wall [165]. Nanocomposites of PANi with zinc oxide (ZnO) nanorods, and epoxy resins with PANi, showed excellent antifouling properties [165,166] By developing an electrically conductive membranes through a graphene (Gr) and PANi coating doped with phytic acid (PA) on polyester filter cloth, a membrane with good conductivity was obtained, presenting excellent antifouling properties. The membrane with a higher conductivity had better antifouling property [166].

One of the first reports about polymer composites for electric antimicrobial effect in biomedical applications used carbon particles where two modified catheters were placed vertically in a nutrient agar plate and connected to an electric device with one catheter acting as a cathode and the other as an anode [167]. The bactericidal activity possessed by negatively charged electroconducting polymers was explained by the establishment of electrostatic repulsions between the negatively charged bacterial cell wall and the polymer [168]. Recently, Arriagada et al., 2018 [169] achieved 100% antimicrobial activity by applying 9V by means of an electroactive composite based on Poly(lactic acid) (PLA) with Thermally Reduced Graphene Oxide particles (TrGO). The results are attributed to the electrostatic effect and the transfer of electrons in conductive materials under an electric current, which causes the death bacteria attached to the electroactive materials [169]. Future research should focus on polymeric compounds capable of eradicating in the early states of microorganisms attaching to surfaces through new smart electroactive biomaterials [170]. In Zhang et al., 2014 [170] Polypyrrol (PPy)/chitosan films with a synergic effect of DC current and gentamycin treatment against biofilm bacterial were fabricated, and they were able to produce biofilm disruption by compromising the integrity of the cell wall by an autolysis-induced cell disruption, i.e., through the action of enzymes produced under the applied DC [170].

In hydrogels the effect of an electric current on the bacterial growth has also barely been reported. For instance, a DC electric field was used as a practical nonthermal procedure to reduce or modify the microbial distribution in alginate and agarose gel beads. The viability of bacteria entrapped in the beads decreases as the field intensity and duration of electric field increase [171].

7. Conclusions

The flexibility of polymers makes possible the development not only of highly compatible and degradable biomaterials, but also a broad set of conductive materials such as: intrinsically electric conductive polymers, percolated electric conductive composites, and ionic conductive hydrogels. This unique flexibility of polymers can be used for the design of electroactive materials for specific biomedical applications such as ES of cells; drug delivery; artificial muscles; and antimicrobial materials. While the use of ES in conductive polymers has been well documented for drug delivery and artificial muscles, more research should take place regarding the potential use of these smart polymeric materials for cell proliferation and antimicrobial scaffolds. For instance, additive manufacturing can extend the range of possibilities for designing electroactive scaffolds that would certainly impact the applications of electroactive polymers.

Author Contributions: Conceptualization, H.P. and P.Z.; Validation, H.P. and P.Z.; Formal Analysis, H.P. and C.A.-P.; Writing-Original Draft Preparation H.P. and C.A.-P.; Writing-Review & Editing, H.P. and P.Z.

Funding: This research was funded by CONICYT under FONDECYT grant number 1150130 and under CONICYT-PCHA Doctorado Nacional grant number 2015-21150921, and the Millennium Science Initiative of the Ministry of Economy, Development and Tourism, grant "Nuclei for Soft Smart Mechanical Metamaterials".

Acknowledgments: The authors gratefully acknowledge the financial support of CONICYT under FONDECYT Project 1150130, and funding from Millennium Science Initiative of the Ministry of Economy, Development and Tourism, grant "Nuclei for Soft Smart Mechanical Metamaterials". C. Angulo-Pineda acknowledges the funding by CONICYT-PCHA under Doctorado Nacional/2015-21150921.

Conflicts of Interest: The authors declare no conflict of interest.

References

1. Woodard, L.N.; Grunlan, M.A. Hydrolytic Degradation and Erosion of Polyester Biomaterials. *ACS Macro Lett.* **2018**, *7*, 976–982. [CrossRef]
2. Zheng, Y.; Li, Y.; Hu, X.; Shen, J.; Guo, S. Biocompatible Shape Memory Blend for Self-Expandable Stents with Potential Biomedical Applications. *ACS Appl. Mater. Interfaces* **2017**, *9*, 13988–13998. [CrossRef] [PubMed]
3. Peterson, G.I.; Dobrynin, A.V.; Becker, M.L. Biodegradable Shape Memory Polymers in Medicine. *Adv. Healthc. Mater.* **2017**, *6*, 1700694. [CrossRef] [PubMed]
4. Dutta, R.C.; Dey, M.; Dutta, A.K.; Basu, B. Competent processing techniques for scaffolds in tissue engineering. *Biotechnol. Adv.* **2017**, *35*, 240–250. [CrossRef] [PubMed]
5. Preethi Soundarya, S.; Haritha Menon, A.; Viji Chandran, S.; Selvamurugan, N. Bone tissue engineering: Scaffold preparation using chitosan and other biomaterials with different design and fabrication techniques. *Int. J. Biol. Macromol.* **2018**, *119*, 1228–1239. [CrossRef] [PubMed]
6. Cheng, J.; Jun, Y.; Qin, J.; Lee, S.-H. Electrospinning versus microfluidic spinning of functional fibers for biomedical applications. *Biomaterials* **2017**, *114*, 121–143. [CrossRef]
7. Ngo, T.D.; Kashani, A.; Imbalzano, G.; Nguyen, K.T.Q.; Hui, D. Additive manufacturing (3D printing): A review of materials, methods, applications and challenges. *Compos. Part B Eng.* **2018**, *143*, 172–196. [CrossRef]
8. Wang, X.; Jiang, M.; Zhou, Z.; Gou, J.; Hui, D. 3D printing of polymer matrix composites: A review and prospective. *Compos. Part B Eng.* **2017**, *110*, 442–458. [CrossRef]

9.   Wu, L.; Virdee, J.; Maughan, E.; Darbyshire, A.; Jell, G.; Loizidou, M.; Emberton, M.; Butler, P.; Howkins, A.; Reynolds, A.; et al. Stiffness memory nanohybrid scaffolds generated by indirect 3D printing for biologically responsive soft implants. *Acta Biomater.* **2018**, *80*, 188–202. [CrossRef]

10. Bolaños Quiñones, V.A.; Zhu, H.; Solovev, A.A.; Mei, Y.; Gracias, D.H. Origami Biosystems: 3D Assembly Methods for Biomedical Applications. *Adv. Biosyst.* **2018**, 1800230. [CrossRef]

11. Yin, K.; Divakar, P.; Wegst, U.G.K. Freeze-Casting Porous Chitosan Ureteral Stents for Improved Drainage. *Acta Biomater.* **2018**. [CrossRef]

12. Ye, H.; Zhang, K.; Kai, D.; Li, Z.; Loh, X.J. Polyester elastomers for soft tissue engineering. *Chem. Soc. Rev.* **2018**, *47*, 4545–4580. [CrossRef] [PubMed]

13. Kolosnjaj-Tabi, J.; Gibot, L.; Fourquaux, I.; Golzio, M.; Rols, M.-P. Electric field-responsive nanoparticles and electric fields: physical, chemical, biological mechanisms and therapeutic prospects. *Adv. Drug Deliv. Rev.* **2018**. [CrossRef]

14. Saghazadeh, S.; Rinoldi, C.; Schot, M.; Kashaf, S.S.; Sharifi, F.; Jalilian, E.; Nuutila, K.; Giatsidis, G.; Mostafalu, P.; Derakhshandeh, H.; et al. Drug delivery systems and materials for wound healing applications. *Adv. Drug Deliv. Rev.* **2018**, *127*, 138–166. [CrossRef] [PubMed]

15. Yang, C.; Suo, Z. Hydrogel ionotronics. *Nat. Rev. Mater.* **2018**, *3*, 125–142. [CrossRef]

16. Shah, A.; Malik, M.S.; Khan, G.S.; Nosheen, E.; Iftikhar, F.J.; Khan, F.A.; Shukla, S.S.; Akhter, M.S.; Kraatz, H.-B.; Aminabhavi, T.M. Stimuli-responsive peptide-based biomaterials as drug delivery systems. *Chem. Eng. J.* **2018**, *353*, 559–583. [CrossRef]

17. Nezakati, T.; Seifalian, A.; Tan, A.; Seifalian, A.M. Conductive Polymers: Opportunities and Challenges in Biomedical Applications. *Chem. Rev.* **2018**, *118*, 6766–6843. [CrossRef] [PubMed]

18. Zhang, F.; Xia, Y.; Wang, L.; Liu, L.; Liu, Y.; Leng, J. Conductive Shape Memory Microfiber Membranes with Core–Shell Structures and Electroactive Performance. *ACS Appl. Mater. Interfaces* **2018**, *10*, 35526–35532. [CrossRef]

19. Han, D.; Farino, C.; Yang, C.; Scott, T.; Browe, D.; Choi, W.; Freeman, J.W.; Lee, H. Soft Robotic Manipulation and Locomotion with a 3D Printed Electroactive Hydrogel. *ACS Appl. Mater. Interfaces* **2018**, *10*, 17512–17518. [CrossRef]

20. Thrivikraman, G.; Boda, S.K.; Basu, B. Unraveling the mechanistic effects of electric field stimulation towards directing stem cell fate and function: A tissue engineering perspective. *Biomaterials* **2018**, *150*, 60–86. [CrossRef]

21. Tandon, B.; Magaz, A.; Balint, R.; Blaker, J.J.; Cartmell, S.H. Electroactive biomaterials: Vehicles for controlled delivery of therapeutic agents for drug delivery and tissue regeneration. *Adv. Drug Deliv. Rev.* **2018**, *129*, 148–168. [CrossRef]

22. Rotman, S.G.; Guo, Z.; Grijpma, D.W.; Poot, A.A. Preparation and characterization of poly(trimethylene carbonate) and reduced graphene oxide composites for nerve regeneration. *Polym. Adv. Technol.* **2017**, *28*, 1233–1238. [CrossRef]

23. Supronowicz, P.R.; Ajayan, P.M.; Ullmann, K.R.; Arulanandam, B.P.; Metzger, D.W.; Bizios, R. Novel current-conducting composite substrates for exposing osteoblasts to alternating current stimulation. *J. Biomed. Mater. Res.* **2002**, *59*, 499–506. [CrossRef]

24. Rivers, T.J.; Hudson, T.W.; Schmidt, C.E. Synthesis of a Novel, Biodegradable Electrically Conducting Polymer for Biomedical Applications. *Adv. Funct. Mater.* **2002**, *12*, 33. [CrossRef]

25. McLeod, K.J.; Rubin, C.T. The effect of low-frequency electrical fields on osteogenesis. *J. Bone Joint Surg. Am.* **1992**, *74*, 920–929. [CrossRef] [PubMed]

26. Shao, S.; Zhou, S.; Li, L.; Li, J.; Luo, C.; Wang, J.; Li, X.; Weng, J. Osteoblast function on electrically conductive electrospun PLA/MWCNTs nanofibers. *Biomaterials* **2011**, *32*, 2821–2833. [CrossRef] [PubMed]

27. Balint, R.; Cassidy, N.J.; Cartmell, S.H. Conductive polymers: Towards a smart biomaterial for tissue engineering. *Acta Biomater.* **2014**, *10*, 2341–2353. [CrossRef]

28. Qazi, T.H.; Rai, R.; Boccaccini, A.R. Tissue engineering of electrically responsive tissues using polyaniline based polymers: A review. *Biomaterials* **2014**, *35*, 9068–9086. [CrossRef]

29. Kaur, G.; Adhikari, R.; Cass, P.; Bown, M.; Gunatillake, P. Electrically conductive polymers and composites for biomedical applications. *RSC Adv.* **2015**, *5*, 37553–37567. [CrossRef]

30. Hackett, A.J.; Malmström, J.; Travas-Sejdic, J. Functionalization of conducting polymers for biointerface applications. *Prog. Polym. Sci.* **2017**, *70*, 18–33. [CrossRef]

31. Khan, S.; Narula, A.K. Bioactive Materials Based on Biopolymers Grafted on Conducting Polymers: Recent Trends in Biomedical Field and Sensing. *Biopolym. Grafting* **2018**, 441–467. [CrossRef]

32. Li, X.; Zhao, T.; Sun, L.; Aifantis, K.E.; Fan, Y.; Feng, Q.; Cui, F.; Watari, F. The applications of conductive nanomaterials in the biomedical field. *J. Biomed. Mater. Res. Part A* **2016**, *104*, 322–339. [CrossRef]

33. Ghasemi-Mobarakeh, L.; Prabhakaran, M.P.; Morshed, M.; Nasr-Esfahani, M.H.; Baharvand, H.; Kiani, S.; Al-Deyab, S.S.; Ramakrishna, S. Application of conductive polymers, scaffolds and electrical stimulation for nerve tissue engineering. *J. Tissue Eng. Regen. Med.* **2011**, *5*, e17–e35. [CrossRef] [PubMed]

34. Guimard, N.K.; Gomez, N.; Schmidt, C.E. Conducting polymers in biomedical engineering. *Prog. Polym. Sci.* **2007**, *32*, 876–921. [CrossRef]

35. Le, T.-H.; Kim, Y.; Yoon, H.; Le, T.-H.; Kim, Y.; Yoon, H. Electrical and Electrochemical Properties of Conducting Polymers. *Polymers* **2017**, *9*, 150. [CrossRef]

36. Chan, E.W.C.; Bennet, D.; Baek, P.; Barker, D.; Kim, S.; Travas-Sejdic, J. Electrospun Polythiophene Phenylenes for Tissue Engineering. *Biomacromolecules* **2018**, *19*, 1456–1468. [CrossRef]

37. Adhikari, S.; Richter, B.; Mace, Z.S.; Sclabassi, R.J.; Cheng, B.; Whiting, D.M.; Averick, S.; Nelson, T.L. Organic Conductive Fibers as Nonmetallic Electrodes and Neural Interconnects. *Ind. Eng. Chem. Res.* **2018**, *57*, 7866–7871. [CrossRef]

38. Feldman, D. Polymer nanocomposites in medicine. *J. Macromol. Sci. Part A* **2016**, *53*, 55–62. [CrossRef]

39. Stříteský, S.; Marková, A.; Víteček, J.; Šafaříková, E.; Hrabal, M.; Kubáč, L.; Kubala, L.; Weiter, M.; Vala, M. Printing inks of electroactive polymer PEDOT:PSS: The study of biocompatibility, stability, and electrical properties. *J. Biomed. Mater. Res. Part A* **2018**, *106*, 1121–1128. [CrossRef]

40. Mao, J.; Zhang, Z. *Polypyrrole as Electrically Conductive Biomaterials: Synthesis, Biofunctionalization, Potential Applications and Challenges*; Springer: Singapore, 2018; pp. 347–370.

41. Humpolíček, P.; Radaszkiewicz, K.A.; Capáková, Z.; Pacherník, J.; Bober, P.; Kašpárková, V.; Rejmontová, P.; Lehocký, M.; Ponížil, P.; Stejskal, J. Polyaniline cryogels: Biocompatibility of novel conducting macroporous material. *Sci. Rep.* **2018**, *8*, 135. [CrossRef]

42. Çetin, M.Z.; Camurlu, P. An amperometric glucose biosensor based on PEDOT nanofibers. *RSC Adv.* **2018**, *8*, 19724–19731. [CrossRef]

43. Zhang, Q.; Beirne, S.; Shu, K.; Esrafilzadeh, D.; Huang, X.-F.; Wallace, G.G. Electrical Stimulation with a Conductive Polymer Promotes Neurite Outgrowth and Synaptogenesis in Primary Cortical Neurons in 3D. *Sci. Rep.* **2018**, *8*, 9855. [CrossRef]

44. Garzon, C.; Palza, H. Electrical behavior of polypropylene composites melt mixed with carbon-based particles: Effect of the kind of particle and annealing process. *Compos. Sci. Technol.* **2014**. [CrossRef]

45. Wang, P.; Chong, H.; Zhang, J.; Yang, Y.; Lu, H. Ultralow electrical percolation in melt-compounded polymer composites based on chemically expanded graphite. *Compos. Sci. Technol.* **2018**, *158*, 147–155. [CrossRef]

46. Araby, S.; Meng, Q.; Zhang, L.; Zaman, I.; Majewski, P.; Ma, J. Elastomeric composites based on carbon nanomaterials. *Nanotechnology* **2015**, *26*, 112001. [CrossRef] [PubMed]

47. Zare, Y.; Rhee, K.Y. A power model to predict the electrical conductivity of CNT reinforced nanocomposites by considering interphase, networks and tunneling condition. *Compos. Part B Eng.* **2018**, *155*, 11–18. [CrossRef]

48. Lovett, J.R.; Derry, M.J.; Yang, P.; Hatton, F.L.; Warren, N.J.; Fowler, P.W.; Armes, S.P. Can percolation theory explain the gelation behavior of diblock copolymer worms? *Chem. Sci.* **2018**, *9*, 7138–7144. [CrossRef] [PubMed]

49. Palza, H.; Zapata, P.; Sagredo, C. Shape memory composites based on a thermoplastic elastomer polyethylene with carbon nanostructures stimulated by heat and solar radiation having piezoresistive behavior. *Polym. Int.* **2018**, *67*, 1046–1053. [CrossRef]

50. Palza, H.; Garzon, C.; Rojas, M. Elastomeric ethylene copolymers with carbon nanostructures having tailored strain sensor behavior and their interpretation based on the excluded volume theory. *Polym. Int.* **2016**, *65*, 1441–1448. [CrossRef]

51. Ezquerra, T.; Connor, M.; Roy, S.; Kulescza, M.; Fernandes-Nascimento, J.; Baltá-Calleja, F. Alternating-current electrical properties of graphite, carbon-black and carbon-fiber polymeric composites. *Compos. Sci. Technol.* **2001**, *61*, 903–909. [CrossRef]

52. Li, C.; Thostenson, E.T.; Chou, T.-W. Effect of nanotube waviness on the electrical conductivity of carbon nanotube-based composites. *Compos. Sci. Technol.* **2008**, *68*, 1445–1452. [CrossRef]

53. Li, J.; Ma, P.C.; Chow, W.S.; To, C.K.; Tang, B.Z.; Kim, J.-K. Correlations between Percolation Threshold, Dispersion State, and Aspect Ratio of Carbon Nanotubes. *Adv. Funct. Mater.* **2007**, *17*, 3207–3215. [CrossRef]

54. Ahir, S.V.; Huang, Y.Y.; Terentjev, E.M. Polymers with aligned carbon nanotubes: Active composite materials. *Polymer* **2008**, *49*, 3841–3854. [CrossRef]

55. Nan, C.-W.; Shen, Y.; Ma, J. Physical Properties of Composites Near Percolation. *Annu. Rev. Mater. Res.* **2010**, *40*, 131–151. [CrossRef]

56. Bauhofer, W.; Kovacs, J.Z. A review and analysis of electrical percolation in carbon nanotube polymer composites. *Compos. Sci. Technol.* **2009**, *69*, 1486–1498. [CrossRef]

57. Román, S.; Lund, F.; Bustos, J.; Palza, H. About the relevance of waviness, agglomeration, and strain on the electrical behavior of polymer composites filled with carbon nanotubes evaluated by a Monte-Carlo simulation. *Mater. Res. Express* **2018**, *5*, 015044. [CrossRef]

58. Meng, Q.; Hu, J. A review of shape memory polymer composites and blends. *Compos. Part A Appl. Sci. Manuf.* **2009**, *40*, 1661–1672. [CrossRef]

59. Leng, J.; Lan, X.; Liu, Y.; Du, S. Shape-memory polymers and their composites: Stimulus methods and applications. *Prog. Mater. Sci.* **2011**, *56*, 1077–1135. [CrossRef]

60. Yu, K.; Zhang, Z.; Liu, Y.; Leng, J. Carbon nanotube chains in a shape memory polymer/carbon black composite: To significantly reduce the electrical resistivity. *Appl. Phys. Lett.* **2011**, *98*, 074102. [CrossRef]

61. Le, H.H.; Kolesov, I.; Ali, Z.; Uhard, M.; Osazuwa, O.; Ilisch, S.; Radusch, H.-J. Effect of filler dispersion degree on the Joule heating stimulated recovery behaviour of nanocomposites. *J. Mater. Sci.* **2010**, *45*, 5851–5859. [CrossRef]

62. Caló, E.; Khutoryanskiy, V.V. Biomedical applications of hydrogels: A review of patents and commercial products. *Eur. Polym. J.* **2015**, *65*, 252–267. [CrossRef]

63. Lee, C.-J.; Wu, H.; Hu, Y.; Young, M.; Wang, H.; Lynch, D.; Xu, F.; Cong, H.; Cheng, G. Ionic Conductivity of Polyelectrolyte Hydrogels. *ACS Appl. Mater. Interfaces* **2018**, *10*, 5845–5852. [CrossRef]

64. Gamboa, O.L.; Pu, J.; Townend, J.; Forrester, J.V.; Zhao, M.; McCaig, C.; Lois, N. Electrical stimulation of retinal pigment epithelial cells. *Exp. Eye Res.* **2010**, *91*, 195–204. [CrossRef]

65. Pullar, C.E.; Baier, B.S.; Kariya, Y.; Russell, A.J.; Horst, B.A.J.; Marinkovich, M.P.; Isseroff, R.R. β4 Integrin and Epidermal Growth Factor Coordinately Regulate Electric Field-mediated Directional Migration via Rac1. *Mol. Biol. Cell* **2006**, *17*, 4925–4935. [CrossRef]

66. Cho, Y.; Son, M.; Jeong, H.; Shin, J.H. Electric field–induced migration and intercellular stress alignment in a collective epithelial monolayer. *Mol. Biol. Cell* **2018**, *29*, 2292–2302. [CrossRef]

67. Cortese, B.; Palamà, I.E.; D'Amone, S.; Gigli, G. Influence of electrotaxis on cell behaviour. *Integr. Biol.* **2014**, *6*, 817–830. [CrossRef]

68. Park, S.Y.; Park, J.; Sim, S.H.; Sung, M.G.; Kim, K.S.; Hong, B.H.; Hong, S. Enhanced Differentiation of Human Neural Stem Cells into Neurons on Graphene. *Adv. Mater.* **2011**, *23*, H263–H267. [CrossRef]

69. Zhang, L.G.; Kaplan, D. *Neural Engineering: From Advanced Biomaterials to 3D Fabrication Techniques*; Springer: Berlin, Germany, 2016; pp. 145–158. [CrossRef]

70. Zhang, J.; Li, M.; Kang, E.-T.; Neoh, K.G. Electrical stimulation of adipose-derived mesenchymal stem cells in conductive scaffolds and the roles of voltage-gated ion channels. *Acta Biomater.* **2016**, *32*, 46–56. [CrossRef]

71. Basser, P.J.; Roth, B.J. New Currents in Electrical Stimulation of Excitable Tissues. *Annu. Rev. Biomed. Eng.* **2000**, *2*, 377–397. [CrossRef]

72. Schmidt, C.E.; Shastri, V.R.; Vacanti, J.P.; Langer, R. Stimulation of neurite outgrowth using an electrically conducting polymer. *Proc. Natl. Acad. Sci. USA* **1997**, *94*, 8948–8953. [CrossRef]

73. Ghasemi-Mobarakeh, L.; Prabhakaran, M.P.; Morshed, M.; Nasr-Esfahani, M.H.; Ramakrishna, S. Electrical stimulation of nerve cells using conductive nanofibrous scaffolds for nerve tissue engineering. *Tissue Eng. Part A* **2009**, *15*, 3605–3619. [CrossRef]

74. Zhang, Z.; Rouabhia, M.; Wang, Z.; Roberge, C.; Shi, G.; Roche, P.; Li, J.; Dao, L.H. Electrically Conductive Biodegradable Polymer Composite for Nerve Regeneration: Electricity-Stimulated Neurite Outgrowth and Axon Regeneration. *Artif. Organs* **2007**, *31*, 13–22. [CrossRef]

75. Shi, G.; Rouabhia, M.; Meng, S.; Zhang, Z. Electrical stimulation enhances viability of human cutaneous fibroblasts on conductive biodegradable substrates. *J. Biomed. Mater. Res. Part A* **2008**, *84A*, 1026–1037. [CrossRef]

76. Jeong, S.I.; Jun, I.D.; Choi, M.J.; Nho, Y.C.; Lee, Y.M.; Shin, H. Development of Electroactive and Elastic Nanofibers that contain Polyaniline and Poly(L-lactide-*co*-ε-caprolactone) for the Control of Cell Adhesion. *Macromol. Biosci.* **2008**, *8*, 627–637. [CrossRef]

77. Sun, S.; Titushkin, I.; Cho, M. Regulation of mesenchymal stem cell adhesion and orientation in 3D collagen scaffold by electrical stimulus. *Bioelectrochemistry* **2006**, *69*, 133–141. [CrossRef]

78. Zhang, J.; Neoh, K.G.; Kang, E.-T. Electrical stimulation of adipose-derived mesenchymal stem cells and endothelial cells co-cultured in a conductive scaffold for potential orthopaedic applications. *J. Tissue Eng. Regen. Med.* **2018**, *12*, 878–889. [CrossRef]

79. Kotwal, A.; Schmidt, C.E. Electrical stimulation alters protein adsorption and nerve cell interactions with electrically conducting biomaterials. *Biomaterials* **2001**, *22*, 1055–1064. [CrossRef]

80. Zhu, W.; Ye, T.; Lee, S.-J.; Cui, H.; Miao, S.; Zhou, X.; Shuai, D.; Zhang, L.G. Enhanced neural stem cell functions in conductive annealed carbon nanofibrous scaffolds with electrical stimulation. *Nanomedicine Nanotechnol. Biol. Med.* **2018**, *14*, 2485–2494. [CrossRef]

81. Wang, J.; Tian, L.; Chen, N.; Ramakrishna, S.; Mo, X. The cellular response of nerve cells on poly-L-lysine coated PLGA-MWCNTs aligned nanofibers under electrical stimulation. *Mater. Sci. Eng. C* **2018**, *91*, 715–726. [CrossRef]

82. Mohan, V.B.; Lau, K.; Hui, D.; Bhattacharyya, D. Graphene-based materials and their composites: A review on production, applications and product limitations. *Compos. Part B Eng.* **2018**, *142*, 200–220. [CrossRef]

83. Jo, H.; Sim, M.; Kim, S.; Yang, S.; Yoo, Y.; Park, J.-H.; Yoon, T.H.; Kim, M.-G.; Lee, J.Y. Electrically conductive graphene/polyacrylamide hydrogels produced by mild chemical reduction for enhanced myoblast growth and differentiation. *Acta Biomater.* **2017**, *48*, 100–109. [CrossRef]

84. Hitscherich, P.; Aphale, A.; Gordan, R.; Whitaker, R.; Singh, P.; Xie, L.; Patra, P.; Lee, E.J. Electroactive graphene composite scaffolds for cardiac tissue engineering. *J. Biomed. Mater. Res. Part A* **2018**. [CrossRef]

85. Guo, B.; Glavas, L.; Albertsson, A.-C. Biodegradable and electrically conducting polymers for biomedical applications. *Prog. Polym. Sci.* **2013**, *38*, 1263–1286. [CrossRef]

86. Zelikin, A.N.; Lynn, D.M.; Farhadi, J.; Martin, I.; Shastri, V.; Langer, R. Erodible Conducting Polymers for Potential Biomedical Applications. *Angew. Chem. Int. Ed.* **2002**, *41*, 141–144. [CrossRef]

87. Wang, W.; Zhu, L.; Shan, B.; Xie, C.; Liu, C.; Cui, F.; Li, G. Preparation and characterization of SLS-CNT/PES ultrafiltration membrane with antifouling and antibacterial properties. *J. Membr. Sci.* **2018**, *548*, 459–469. [CrossRef]

88. Qiu, L.Y.; Bae, Y.H. Polymer Architecture and Drug Delivery. *Pharm. Res.* **2006**, *23*, 1–30. [CrossRef]

89. Schmaljohann, D. Thermo- and pH-responsive polymers in drug delivery. *Adv. Drug Deliv. Rev.* **2006**, *58*, 1655–1670. [CrossRef]

90. Hu, X.; Zhang, Y.; Xie, Z.; Jing, X.; Bellotti, A.; Gu, Z. Stimuli-Responsive Polymersomes for Biomedical Applications. *Biomacromolecules* **2017**, *18*, 649–673. [CrossRef]

91. Li, Y.; Bui, Q.N.; Duy, L.T.M.; Yang, H.Y.; Lee, D.S. One-Step Preparation of pH-Responsive Polymeric Nanogels as Intelligent Drug Delivery Systems for Tumor Therapy. *Biomacromolecules* **2018**, *19*, 2062–2070. [CrossRef]

92. Jia, S.; Fong, W.-K.; Graham, B.; Boyd, B.J. Photoswitchable Molecules in Long-Wavelength Light-Responsive Drug Delivery: From Molecular Design to Applications. *Chem. Mater.* **2018**, *30*, 2873–2887. [CrossRef]

93. Qu, J.; Zhao, X.; Ma, P.X.; Guo, B. Injectable antibacterial conductive hydrogels with dual response to an electric field and ph for localized "smart" drug release. *acta biomater.* **2018**, *72*, 55–69. [CrossRef]

94. Ramasamy, T.; Ruttala, H.B.; Gupta, B.; Poudel, B.K.; Choi, H.-G.; Yong, C.S.; Kim, J.O. Smart chemistry-based nanosized drug delivery systems for systemic applications: A comprehensive review. *J. Control. Release* **2017**, *258*, 226–253. [CrossRef]

95. Jin, Z.; Wu, K.; Hou, J.; Yu, K.; Shen, Y.; Guo, S. A PTX/nitinol stent combination with temperature-responsive phase-change 1-hexadecanol for magnetocaloric drug delivery: Magnetocaloric drug release and esophagus tissue penetration. *Biomaterials* **2018**, *153*, 49–58. [CrossRef]

96. Kang, T.; Li, F.; Baik, S.; Shao, W.; Ling, D.; Hyeon, T. Surface design of magnetic nanoparticles for stimuli-responsive cancer imaging and therapy. *Biomaterials* **2017**, *136*, 98–114. [CrossRef]

97. Yilmaz, N.D. Multicomponent, Semi-interpenetrating-Polymer-Network and Interpenetrating-Polymer-Network Hydrogels: Smart Materials for Biomedical Applications. In *Functional Biopolymers*; Springer Series on Polymer and Composite Materials; Thakur, V., Thakur, M., Eds.; Springer: Cham, Switzerland, 2018; pp. 281–342. [CrossRef]

98. Kennedy, S.; Bencherif, S.; Norton, D.; Weinstock, L.; Mehta, M.; Mooney, D. Rapid and extensive collapse from electrically responsive macroporous hydrogels. *Adv. Healthc. Mater.* **2014**, *3*, 500–507. [CrossRef]

99. Priya James, H.; John, R.; Alex, A.; Anoop, K.R. Smart polymers for the controlled delivery of drugs—A concise overview. *Acta Pharm. Sin. B* **2014**, *4*, 120–127. [CrossRef]

100. Lee, H.; Song, C.; Baik, S.; Kim, D.; Hyeon, T.; Kim, D.-H. Device-assisted transdermal drug delivery. *Adv. Drug Deliv. Rev.* **2018**, *127*, 35–45. [CrossRef]

101. Wang, Y.; Kohane, D.S. External triggering and triggered targeting strategies for drug delivery. *Nat. Rev. Mater.* **2017**, *2*, 17020. [CrossRef]

102. Murdan, S. Electro-responsive drug delivery from hydrogels. *J. Control. Release* **2003**, *92*, 1–17. [CrossRef]

103. Fenton, O.S.; Olafson, K.N.; Pillai, P.S.; Mitchell, M.J.; Langer, R. Advances in Biomaterials for Drug Delivery. *Adv. Mater.* **2018**, *30*, 1705328. [CrossRef]

104. Merino, S.; Martín, C.; Kostarelos, K.; Prato, M.; Vázquez, E. Nanocomposite Hydrogels: 3D Polymer–Nanoparticle Synergies for On-Demand Drug Delivery. *ACS Nano* **2015**, *9*, 4686–4697. [CrossRef]

105. Guiseppi-Elie, A. Electroconductive hydrogels: Synthesis, characterization and biomedical applications. *Biomaterials* **2010**, *31*, 2701–2716. [CrossRef]

106. Tanaka, T.; Nishio, I.; Sun, S.-T.; Ueno-Nishio, S. Collapse of Gels in an Electric Field. *Science* **1982**, *218*, 467–469. [CrossRef]

107. Svirskis, D.; Travas-Sejdic, J.; Rodgers, A.; Garg, S. Electrochemically controlled drug delivery based on intrinsically conducting polymers. *J. Control. Release* **2010**, *146*, 6–15. [CrossRef]

108. Guo, J.; Fan, D. Electrically Controlled Biochemical Release from Micro/Nanostructures for in vitro and in vivo Applications: A Review. *ChemNanoMat* **2018**, *4*, 1023–1038. [CrossRef]

109. Indermun, S.; Choonara, Y.E.; Kumar, P.; du Toit, L.C.; Modi, G.; Luttge, R.; Pillay, V. An interfacially plasticized electro-responsive hydrogel for transdermal electro-activated and modulated (TEAM) drug delivery. *Int. J. Pharm.* **2014**, *462*, 52–65. [CrossRef]

110. Ge, J.; Neofytou, E.; Cahill, T.J.; Beygui, R.E.; Zare, R.N.; Zare, R.N. Drug release from electric-field-responsive nanoparticles. *ACS Nano* **2012**, *6*, 227–233. [CrossRef]

111. Abidian, M.R.; Kim, D.-H.; Martin, D.C. Conducting-Polymer Nanotubes for Controlled Drug Release. *Adv. Mater.* **2006**, *18*, 405–409. [CrossRef]

112. Antensteiner, M.; Khorrami, M.; Fallahianbijan, F.; Borhan, A.; Abidian, M.R. Conducting Polymer Microcups for Organic Bioelectronics and Drug Delivery Applications. *Adv. Mater.* **2017**, *29*, 1702576. [CrossRef]

113. Amjadi, M.; Sheykhansari, S.; Nelson, B.J.; Sitti, M. Recent Advances in Wearable Transdermal Delivery Systems. *Adv. Mater.* **2013**, *30*, 1704530. [CrossRef]

114. Paun, I.A.; Zamfirescu, M.; Luculescu, C.R.; Acasandrei, A.M.; Mustaciosu, C.C.; Mihailescu, M.; Dinescu, M. Electrically responsive microreservoires for controllable delivery of dexamethasone in bone tissue engineering. *Appl. Surf. Sci.* **2017**, *392*, 321–331. [CrossRef]

115. Im, J.S.; Bai, B.C.; Lee, Y.-S. The effect of carbon nanotubes on drug delivery in an electro-sensitive transdermal drug delivery system. *Biomaterials* **2010**, *31*, 1414–1419. [CrossRef]

116. Beg, S.; Rahman, M.; Jain, A.; Saini, S.; Hasnain, M.S.; Swain, S.; Imam, S.; Kazmi, I.; Akhter, S. Emergence in the functionalized carbon nanotubes as smart nanocarriers for drug delivery applications. *Fuller. Graphenes Nanotub.* **2018**, 105–133. [CrossRef]

117. Thompson, B.C.; Chen, J.; Moulton, S.E.; Wallace, G.G. Nanostructured aligned CNT platforms enhance the controlled release of a neurotrophic protein from polypyrrole. *Nanoscale* **2010**, *2*, 499. [CrossRef]

118. Yun, J.; Im, J.S.; Lee, Y.-S.; Kim, H.-I. Electro-responsive transdermal drug delivery behavior of PVA/PAA/MWCNT nanofibers. *Eur. Polym. J.* **2011**, *47*, 1893–1902. [CrossRef]

119. Spizzirri, U.G.; Hampel, S.; Cirillo, G.; Nicoletta, F.P.; Hassan, A.; Vittorio, O.; Picci, N.; Iemma, F. Spherical gelatin/CNTs hybrid microgels as electro-responsive drug delivery systems. *Int. J. Pharm.* **2013**, *448*, 115–122. [CrossRef]

120. Cirillo, G.; Curcio, M.; Spizzirri, U.G.; Vittorio, O.; Tucci, P.; Picci, N.; Iemma, F.; Hampel, S.; Nicoletta, F.P. Carbon nanotubes hybrid hydrogels for electrically tunable release of Curcumin. *Eur. Polym. J.* **2017**, *90*, 1–12. [CrossRef]

121. Servant, A.; Methven, L.; Williams, R.P.; Kostarelos, K. Electroresponsive Polymer-Carbon Nanotube Hydrogel Hybrids for Pulsatile Drug Delivery In Vivo. *Adv. Healthc. Mater.* **2013**, *2*, 806–811. [CrossRef]

122. Vohrer, U.; Kolaric, I.; Haque, M.; Roth, S.; Detlaff-Weglikowska, U. Carbon nanotube sheets for the use as artificial muscles. *Carbon* **2004**, *42*, 1159–1164. [CrossRef]

123. Smela, E. Conjugated Polymer Actuators for Biomedical Applications. *Adv. Mater.* **2003**, *15*, 481–494. [CrossRef]

124. Mirfakhrai, T.; Madden, J.D.W.; Baughman, R.H. Polymer artificial muscles. *Mater. Today* **2007**, *10*, 30–38. [CrossRef]

125. Baughman, R.H. MATERIALS SCIENCE: Playing Nature's Game with Artificial Muscles. *Science* **2005**, *308*, 63–65. [CrossRef]

126. Farajollahi, M.; Woehling, V.; Plesse, C.; Nguyen, G.T.M.; Vidal, F.; Sassani, F.; Yang, V.X.D.; Madden, J.D.W. Self-contained tubular bending actuator driven by conducting polymers. *Sens. Actuators A Phys.* **2016**, *249*, 45–56. [CrossRef]

127. Otero, T.F.; Sansieña, J.M. Soft and Wet Conducting Polymers for Artificial Muscles. *Adv. Mater.* **1998**, *10*, 491–494. [CrossRef]

128. Fuchiwaki, M.; Martinez, J.G.; Otero, T.F. Polypyrrole Asymmetric Bilayer Artificial Muscle: Driven Reactions, Cooperative Actuation, and Osmotic Effects. *Adv. Funct. Mater.* **2015**, *25*, 1535–1541. [CrossRef]

129. Yan, B.; Wu, Y.; Guo, L.; Yan, B.; Wu, Y.; Guo, L. Recent Advances on Polypyrrole Electroactuators. *Polymers* **2017**, *9*, 446. [CrossRef]

130. Fuchiwaki, M.; Martinez, J.G.; Otero, T.F. Asymmetric Bilayer Muscles. Cooperative and Antagonist Actuation. *Electrochim. Acta* **2016**, *195*, 9–18. [CrossRef]

131. Brochu, P.; Pei, Q. Advances in Dielectric Elastomers for Actuators and Artificial Muscles. *Macromol. Rapid Commun.* **2010**, *31*, 10–36. [CrossRef]

132. Shahinpoor, M.; Bar-Cohen, Y.; Simpson, J.O.; Smith, J. Ionic polymer-metal composites (IPMCs) as biomimetic sensors, actuators and artificial muscles—A review. *Smart Mater. Struct.* **1998**, *7*, R15–R30. [CrossRef]

133. Shahinpoor, M.; Kim, K.J. Ionic polymer-metal composites: I. Fundamentals. *Smart Mater. Struct.* **2001**, *10*, 819–833. [CrossRef]

134. Chen, B.; Bai, Y.; Xiang, F.; Sun, J.-Y.; Chen, Y.M.; Wang, H.; Zhou, J.; Suo, Z. Stretchable and Transparent Hydrogels as Soft Conductors for Dielectric Elastomer Actuators. *J. Polym. Sci. Part B Polym. Phys.* **2014**, *52*, 1055–1060. [CrossRef]

135. Li, T.; Li, G.; Liang, Y.; Cheng, T.; Dai, J.; Yang, X.; Liu, B.; Zeng, Z.; Huang, Z.; Luo, Y.; et al. Fast-moving soft electronic fish. *Sci. Adv.* **2017**, *3*. [CrossRef]

136. Pedroli, F.; Marrani, A.; Le, M.-Q.; Froidefond, C.; Cottinet, P.-J.; Capsal, J.-F. Processing optimization: A way to improve the ionic conductivity and dielectric loss of electroactive polymers. *J. Polym. Sci. Part B Polym. Phys.* **2018**, *56*, 1164–1173. [CrossRef]

137. Zeng, Z.; Jin, H.; Zhang, L.; Zhang, H.; Chen, Z.; Gao, F.; Zhang, Z. Low-voltage and high-performance electrothermal actuator based on multi-walled carbon nanotube/polymer composites. *Carbon* **2015**, *84*, 327–334. [CrossRef]

138. Hu, Y.; Chen, W.; Lu, L.; Liu, J.; Chang, C. Electromechanical Actuation with Controllable Motion Based on a Single-Walled Carbon Nanotube and Natural Biopolymer Composite. *ACS Nano* **2010**, *4*, 3498–3502. [CrossRef]

139. Mirvakili, S.M.; Hunter, I.W. Artificial Muscles: Mechanisms, Applications, and Challenges. *Adv. Mater.* **2018**, *30*, 1704407. [CrossRef]

140. Srivastava, S.; Bhargava, A. Biofilms and human health. *Biotechnol. Lett.* **2016**, *38*, 1–22. [CrossRef] [PubMed]

141. Callow, J.A.; Callow, M.E. Trends in the development of environmentally friendly fouling-resistant marine coatings. *Nat. Commun.* **2011**, *2*, 244. [CrossRef]

142. Debiemme-Chouvy, C.; Cachet, H. Electrochemical (pre)treatments to prevent biofouling. *Curr. Opin. Electrochem.* **2018**. [CrossRef]

143. Kavanagh, N.; Ryan, E.J.; Widaa, A.; Sexton, G.; Fennell, J.; O'Rourke, S.; Cahill, K.C.; Kearney, C.J.; O'Brien, F.J.; Kerrigan, S.W. Staphylococcal Osteomyelitis: Disease Progression, Treatment Challenges, and Future Directions. *Clin. Microbiol. Rev.* **2018**, *31*, e00084-17. [CrossRef]

144. Benčina, M.; Mavrič, T.; Junkar, I.; Bajt, A.; Krajnović, A.; Lakota, K.; Žigon, P.; Sodin-Šemrl, S.; Kralj-Iglič, V. The Importance of Antibacterial Surfaces in Biomedical Applications. *Adv. Biomembr. Lipid Self-Assembly* **2018**, *28*, 115–165. [CrossRef]

145. Bixler, G.D.; Bhushan, B. Biofouling: Lessons from nature. *Philos. Trans. A. Math. Phys. Eng. Sci.* **2012**, *370*, 2381–2417. [CrossRef]

146. Palioura, S.; Gibbons, A.; Miller, D.; O'Brien, T.P.; Alfonso, E.C.; Spierer, O. Clinical Features, Antibiotic Susceptibility Profile, and Outcomes of Infectious Keratitis Caused by Stenotrophomonas maltophilia. *Cornea* **2018**, *37*, 326–330. [CrossRef] [PubMed]

147. Fernández, J.; Ribeiro, I.A.C.; Martin, V.; Martija, O.L.; Zuza, E.; Bettencourt, A.F.; Sarasua, J.-R. Release mechanisms of urinary tract antibiotics when mixed with bioabsorbable polyesters. *Mater. Sci. Eng. C* **2018**, *93*, 529–538. [CrossRef] [PubMed]

148. Voegele, P.; Badiola, J.; Schmidt-Malan, S.M.; Karau, M.J.; Greenwood-Quaintance, K.E.; Mandrekar, J.N.; Patel, R. Antibiofilm Activity of Electrical Current in a Catheter Model. *Antimicrob. Agents Chemother.* **2015**, *60*, 1476–1480. [CrossRef] [PubMed]

149. Bhushan, B. Bio-and Inorganic Fouling. In *Biomimetics*; Springer Series in Materials Science; Springer: Cham, Switzerland, 2018; pp. 621–664.

150. Singh, A.; Dubey, A.K. Various Biomaterials and Techniques for Improving Antibacterial Response. *ACS Appl. Bio Mater.* **2018**, *1*, 3–20. [CrossRef]

151. Pandit, S.; Shanbhag, S.; Mauter, M.; Oren, Y.; Herzberg, M. Influence of Electric Fields on Biofouling of Carbonaceous Electrodes. *Environ. Sci. Technol.* **2017**, *51*, 10022–10030. [CrossRef]

152. Van der Borden, A.J.; Maathuis, P.G.M.; Engels, E.; Rakhorst, G.; van der Mei, H.C.; Busscher, H.J.; Sharma, P.K. Prevention of pin tract infection in external stainless steel fixator frames using electric current in a goat model. *Biomaterials* **2007**, *28*, 2122–2126. [CrossRef]

153. Asam, M.; Ahmad, R.; Kim, J. Recent developments in biofouling control in membrane bioreactors for domestic wastewater treatment. *Sep. Purif. Technol.* **2018**, *206*, 297–315. [CrossRef]

154. De Pozo, J.L.; Rouse, M.S.; Mandrekar, J.N.; Sampedro, M.F.; Steckelberg, J.M.; Patel, R. Effect of Electrical Current on the Activities of Antimicrobial Agents against Pseudomonas aeruginosa, Staphylococcus aureus, and Staphylococcus epidermidis Biofilms. *Antimicrob. Agents Chemother.* **2009**, *53*, 35–40. [CrossRef] [PubMed]

155. Del Pozo, J.L.; Rouse, M.S.; Patel, R. Bioelectric Effect and Bacterial Biofilms. a Systematic Review. *Int. J. Artif. Organs* **2008**, *31*, 786–795. [CrossRef] [PubMed]

156. Khan, S.I.; Blumrosen, G.; Vecchio, D.; Golberg, A.; McCormack, M.C.; Yarmush, M.L.; Hamblin, M.R.; Austen, W.G. Eradication of multidrug-resistant pseudomonas biofilm with pulsed electric fields. *Biotechnol. Bioeng.* **2016**, *113*, 643–650. [CrossRef]

157. Wolfmeier, H.; Pletzer, D.; Mansour, S.C.; Hancock, R.E.W. New Perspectives in Biofilm Eradication. *ACS Infect. Dis.* **2018**, *4*, 93–106. [CrossRef]

158. Korem, M.; Goldberg, N.S.; Cahan, A.; Cohen, M.J.; Nissenbaum, I.; Moses, A.E. Clinically applicable irreversible electroporation for eradication of micro-organisms. *Lett. Appl. Microbiol.* **2018**, *67*, 15–21. [CrossRef] [PubMed]

159. Gibot, L.; Golberg, A. Electroporation in Scars/Wound Healing and Skin Response. In *Handbook of Electroporation*; Springer International Publishing: Cham, Swizerland, 2017; pp. 531–548. [CrossRef]

160. Ranalli, G.; Iorizzo, M.; Lustrato, G.; Zanardini, E.; Grazia, L. Effects of low electric treatment on yeast microflora. *J. Appl. Microbiol.* **2002**, *93*, 877–883. [CrossRef] [PubMed]

161. Costerton, J.W.; Ellis, B.; Lam, K.; Johnson, F.; Khoury, A.E. Mechanism of electrical enhancement of efficacy of antibiotics in killing biofilm bacteria. *Antimicrob. Agents Chemother.* **1994**, *38*, 2803–2809. [CrossRef] [PubMed]

162. Ronen, A.; Duan, W.; Wheeldon, I.; Walker, S.; Jassby, D. Microbial Attachment Inhibition through Low-Voltage Electrochemical Reactions on Electrically Conducting Membranes. *Environ. Sci. Technol.* **2015**, *49*, 12741–12750. [CrossRef]

163. Eynard, N.; Rodriguez, F.; Trotard, J.; Teissié, J. Electrooptics Studies of Escherichia coli Electropulsation: Orientation, Permeabilization, and Gene Transfer. *Biophys. J.* **1998**, *75*, 2587–2596. [CrossRef]

164. Wang, C.; Yue, L.; Wang, S.; Pu, Y.; Zhang, X.; Hao, X.; Wang, W.; Chen, S. Role of Electric Field and Reactive Oxygen Species in Enhancing Antibacterial Activity: A Case Study of 3D Cu Foam Electrode with Branched CuO–ZnO NWs. *J. Phys. Chem. C* **2018**, *122*, 26454–26463. [CrossRef]

165. Ahmed, F.; Lalia, B.S.; Kochkodan, V.; Hashaikeh, R. Electrically conductive polymeric membranes for fouling prevention and detection: A review. *Desalination* **2016**, *391*, 1–15. [CrossRef]

166. Li, N.; Liu, L.; Yang, F. Highly conductive graphene/PANi-phytic acid modified cathodic filter membrane and its antifouling property in EMBR in neutral conditions. *Desalination* **2014**, *338*, 10–16. [CrossRef]

167. Liu, W.-K.; Tebbs, S.E.; Byrne, P.O.; Elliott, T.S.J. The effects of electric current on bacteria colonising intravenous catheters. *J. Infect.* **1993**, *27*, 261–269. [CrossRef]

168. Francolini, I.; Donelli, G.; Stoodley, P. Polymer Designs to Control Biofilm Growth on Medical Devices. *Rev. Environ. Sci. Bio/Technol.* **2003**, *2*, 307–319. [CrossRef]

169. Arriagada, P.; Palza, H.; Palma, P.; Flores, M.; Caviedes, P. Poly(lactic acid) composites based on graphene oxide particles with antibacterial behavior enhanced by electrical stimulus and biocompatibility. *J. Biomed. Mater. Res. Part A* **2018**, *106*, 1051–1060. [CrossRef] [PubMed]

170. Zhang, J.; Neoh, K.G.; Hu, X.; Kang, E.-T. Mechanistic insights into response of Staphylococcus aureus to bioelectric effect on polypyrrole/chitosan film. *Biomaterials* **2014**, *35*, 7690–7698. [CrossRef]

171. Zvitov, R.; Zohar-Perez, C.; Nussinovitch, A. Short-duration low-direct-current electrical field treatment is a practical tool for considerably reducing counts of gram-negative bacteria entrapped in gel beads. *Appl. Environ. Microbiol.* **2004**, *70*, 3781–3784. [CrossRef] [PubMed]

![materials logo] *materials*

MDPI

Article

Antimicrobial Porous Surfaces Prepared by Breath Figures Approach

Alexandra Muñoz-Bonilla [1,*], Rocío Cuervo-Rodríguez [2] , Fátima López-Fabal [3] ,
José L. Gómez-Garcés [3] and Marta Fernández-García [1,*]

[1] Instituto de Ciencia y Tecnología de Polímeros (ICTP-CSIC), C/Juan de la Cierva 3, 28006 Madrid, Spain
[2] Facultad de Ciencias Químicas, Universidad Complutense de Madrid, Avenida Complutense s/n,
    Ciudad Universitaria, 28040 Madrid, Spain; rociocr@quim.ucm.es
[3] Hospital Universitario de Móstoles, C/Río Júcar, s/n, Móstoles, 28935 Madrid, Spain;
    flopezf@salud.madrid.org (F.L.-F.); jlgarces@microb.net (J.L.G.-G.)
*   Correspondence: sbonilla@ictp.csic.es (A.M.-B.); martafg@ictp.csic.es (M.F.-G.)

Received: 2 July 2018; Accepted: 20 July 2018; Published: 24 July 2018

Abstract: Herein, efficient antimicrobial porous surfaces were prepared by breath figures approach
from polymer solutions containing low content of block copolymers with high positive charge
density. In brief, those block copolymers, which were used as additives, are composed of a
polystyrene segment and a large antimicrobial block bearing flexible side chain with 1,3-thiazolium
and 1,2,3-triazolium groups, PS_{54}-b-PTTBM-M$_{44}$, PS_{54}-b-PTTBM-B$_{44}$, having different alkyl groups,
methyl or butyl, respectively. The antimicrobial block copolymers were blended with commercial
polystyrene in very low proportions, from 3 to 9 wt %, and solubilized in THF. From these solutions,
ordered porous films functionalized with antimicrobial cationic copolymers were fabricated, and the
influence of alkylating agent and the amount of copolymer in the blend was investigated. Narrow pore
size distribution was obtained for all the samples with pore diameters between 5 and 11 μm. The
size of the pore decreased as the hydrophilicity of the system increased; thus, either as the
content of copolymer was augmented in the blend or as the copolymers were quaternized with
methyl iodide. The resulting porous polystyrene surfaces functionalized with low content of
antimicrobial copolymers exhibited remarkable antibacterial efficiencies against Gram positive
bacteria *Staphylococcus aureus*, and *Candida parapsilosis* fungi as microbial models.

Keywords: antimicrobial coatings; porous surfaces; breath figures

1. Introduction

Healthcare-associated infections are a major problem nowadays, causing high morbidity and
mortality rates and substantial increase in health care costs. These infections are mainly associated
with surgery procedures and medical devices such as ventilators or catheters. Prescription of
antibiotics is typically used as a prevention method and/or treatment to avoid such transmissions of
nosocomial pathogens; however, antibiotic consumption is a primary cause of antibiotic resistance [1],
and consequently, there is an urgent need for alternatives, as well as strategies to prevent
healthcare-acquired infections [2]. Inhibition of bacterial growth on the surfaces of medical equipment
and devices is necessary to prevent the transmission of diseases by contact, and one promising
approach is the development of antimicrobial coatings. Many of these self-disinfecting coatings are
based on impregnating the surfaces with antimicrobial agents including antibiotics, silver compounds,
light active species, and antimicrobial polymers such as polycations [3–7]. Other strategies limit the
bacterial colonization of surfaces by inducing micro and nano-roughness, which modifies the surface
area of contact with the microorganisms [8–10]. Although in the past the effect of surface topography
on bacterial inhibition has received less attention, it is gaining popularity nowadays [11,12]. Many of

these works study antifouling effects [13,14] based on superhydrophobic surfaces with reduced surface contact [10,15]. Most of the antifouling surfaces are able to effectively reduce the initial bacterial attachment, but only within a relatively short period. If adhesion occurs, the bacteria rapidly proliferate, leading to the formation of the biofilm.

Alternatively, the introduction of roughness can produce the opposite effect, dramatically increasing the contact adhesion area. Thus, combinations of surface roughness with chemical biocidal functionalities can create more effective bactericidal properties than flat surfaces [16–18]. Nowadays, many techniques are available to create surfaces with finely controlled topography, including lithography approaches and colloidal templates [19,20]. Most of these techniques usually require multiple stages, expensive equipment and prefabricated masks. One of the most versatile and simple methodologies to create polymeric porous surfaces with controlled pore size, and thus tailored roughness, is the so-called breath figures approach [21–24]. This method prepares ordered porous films with water droplets as the template. Basically, polymer solution is cast onto a substrate under humid atmosphere. The solvent evaporation induces the condensation of water droplets, which self-organize into a hexagonal array and after solvent evaporation, honeycomb-patterned films are obtained. Although this technique can be used with a diversity of polymers and functionalities [25], it is limited to polymers soluble in non-polar organic solvents such as CS_2 or chloroform, which are mostly hydrophobic polymers, polymers with polar end groups or some amphiphilic copolymers. An alternative for obtaining porous surfaces functionalized with highly hydrophilic polymers or polyelectrolytes is the use of polymer blends, consisting of incorporating low amount of hydrophilic polymers into a hydrophobic polymeric matrix such as polystyrene [26]. Due to the formation mechanism of the breath figures, the hydrophilic polymers tend to migrate towards the condensed water droplets, which imply their localization at the surface, on the wall of the pores [26–28]. In this context, cationic antimicrobial polymers based on quaternary ammonium groups have been incorporated into breath figures films by using blends [29]. However, only systems based on copolymers with low positive charge density have been prepared due to the difficulty to dissolve them in organic solvents [29,30]. In this work, we prepared porous films by a breath figures approach functionalized with high charge density by the incorporation of antimicrobial cationic polymers bearing two quaternary ammonium groups per monomeric unit. In fact, these structures are based on methacrylic monomers with 1,3-thiazolium and 1,2,3-triazolium side-chain groups, and have demonstrated a broad spectrum of antimicrobial activity in solution [31,32], and also when immobilized onto a surface [16,33]. It is well known that surface positive charge density is an important parameter for defining antimicrobial efficiency [33,34], and the incorporation of polymers with high charge density as blend component will enhance the biocidal activity of the microstructured surfaces, maintaining the physicochemical properties of the resulting coating.

2. Experimental Section

2.1. Materials

Film preparation: high molecular weight polystyrene (PS, Aldrich, Schnelldorf, Germany, weight-average molecular weight, $M_w = 2.50 \times 10^5$ g mol^{-1}) was employed as polymeric matrix and used as received. The block copolymers polystyrene-*b*-poly(4-(1-(2-(4-methylthiazol -5-yl)ethyl)-1H-1,2,3-triazol-4-yl)butyl methacrylate) (M_n = 22,000 g/mol, M_w/M_n = 1.53) quaternized with either butyl iodide (PS$_{54}$-*b*-PTTBM-B$_{44}$) or methyl iodide (PS$_{54}$-*b*-PTTBM-M$_{44}$) were synthesized as previously reported [16]. Briefly, the first block of polystyrene was prepared by atom transfer radical polymerization (ATRP) and then was used as macroinitiator for the synthesis of the PTTBM block. This second block was obtained by combination of ATRP and copper-catalyzed azide-alkyne cycloaddition (CuAAC) click reaction, using the same catalyst (CuCl/PMDETA). The TTBM monomer was synthesized in situ during its ATRP polymerization by click chemistry between 2-(4-methylthiazol-5-yl)ethanol azide and hex-5-yn-1-yl methacrylate. The cationic copolymers

PS$_{54}$-*b*-PTTBM-R$_{44}$ (R = butyl or methyl) were finally obtained by quaternization of their thiazole and triazole groups. ^{1}H-NMR spectroscopy confirms that quaternization was achieved quantitatively [16]. Tetrahydrofuran (THF, Aldrich, St Quentin Fallavier, France, ACS reagent) was employed as organic solvent for the preparation of the porous films without further purification. Round glass coverslips of 12 mm diameter were obtained from Ted Pella Inc (Redding, CA, USA).

Microbiological assays: Sodium chloride (NaCl, 0.9%, BioXtra, St Quentin Fallavier, France, suitable for cell cultures) and phosphate-buffered saline (PBS, pH 7.4) were obtained from Aldrich. Sheep blood (5%) Columbia Agar plates were purchased from bioMérieux (Madrid, Spain). American Type Culture Collection (ATCC): Gram-positive *Staphylococcus aureus* (*S. aureus*, ATCC 29213) bacteria and *Candida parapsilosis* (*C. parapsilosis*, ATCC 22019) fungi were obtained from Oxoid™ (Madrid, Spain).

2.2. Antimicrobial Film Formation

Porous films with antimicrobial activity were fabricated by a breath figures approach. For this purpose, blends of polymers composed of commercial PS as the main component, and cationic copolymers PS$_{54}$-*b*-PTTBM-R$_{44}$ (PS$_{54}$-*b*-PTTBM-B$_{44}$ or PS$_{54}$-*b*-PTTBM-M$_{44}$) as minor components were prepared at different compositions: 3, 6 and 9 wt % of copolymer. The polymer mixtures were dissolved in THF at a concentration of 30 mg mL^{-1}. THF was selected as organic solvent because is compatible with all the components. Then, polymer films of 1 cm diameter were obtained by drop casting of 30 µL of each solution onto glass substrate at room temperature in a closed chamber under controlled humidity. The humidity conditions of the chamber were set to 60, 70 and 90% for each solution.

2.3. Film Characterization

The surface structures of the films were observed by a scanning electron microscope (Philips XL30, Eindhoven, The Netherlands) with an acceleration voltage of 25 kV. The films were coated with gold-palladium (80/20) prior to imaging. The pore diameters and the quantitative order of porous patterns were analyzed by the image analysis software Image-J (NIHimage, National Institutes of Health, Bethesda, MD, USA). Water contact angle measurements of the prepared films were carried out in a KSV Theta goniometer (KSV Instruments Ltd., Helsinki, Finland) from digital images of 3.0 µL water droplets on the surface. The measurements were made in at least quintuplicate.

2.4. Evaluation of Antimicrobial Activity in Films

Antimicrobial activities of the prepared films were evaluated following the E2149-01 standard method from the American Society for Testing and Materials (ASTM) [35], which is a quantitative and standardized method, typically used for the analysis of material surfaces in antimicrobial materials. Firstly, the microorganisms *S. aureus* and *C. parapsilosis* were incubated on 5% sheep blood Columbia agar plates for 24 h for bacteria and 48 h for yeast at 37 °C in a Jouan IQ050 incubator (Winchester, VA, USA). Subsequently, the microorganism concentration was adjusted with saline solution to a turbidity equivalent to ca. 0.5 McFarland turbidity standard, about 10^8 colony-forming units (CFU) mL^{-1}. The optical density of the microorganism suspensions was measured in a DensiCHEKTM Plus (VITEK, bioMérieux, Madrid, Spain). Then, these suspensions were further diluted (1:200) with PBS to obtain 10^6 CFU mL^{-1}. Each film was introduced in a sterile falcon tube containing 1 mL of the tested inoculum and 9 mL of PBS to reach a working solution of ca. 10^5 CFU mL^{-1}. Control experiments were also carried out on films made exclusively with commercial PS, and also blank experiments with only the tested inoculum in the absence of films. The suspensions were shaken for 24 h at 120 rpm. After this period, 1 mL of each solution was taken and serially diluted. The dilutions, 1 mL, were placed on 5% sheep blood Columbia agar plates and incubated for 24 h for bacteria and 48 h for fungi at 37 °C. Then, the number of bacteria in each sample was determined by the plate counting method [36]. The measurements were made at least in triplicate.

3. Results and Discussion

Porous films were prepared by the breath figures method from blends containing antimicrobial copolymers with high charge density as a minor component or additive. By this approach, bactericidal films on contact with antimicrobial chemical functionalities at the surface and controlled microstructure can be fabricated in a very simple and effective way. In addition, structural parameters, such as pore size and pore density, can be easily controlled by selecting the experimental conditions of humidity, concentration of the solution or the type of polymer [22]. In fact, the use of polymeric structures with polar moieties, i.e., amphiphilic polymers, favors the formation of ordered porous arrays, because these structures help the stabilization of the condensed water droplets. Herein, amphiphilic copolymers based on a hydrophilic block with two quaternary ammonium moieties per monomeric unit (Figure 1) were added to commercial polystyrene as a modifier and antimicrobial component. THF solutions of these blends were cast onto glass substrates under controlled humidity. It has to be mentioned that the cationic copolymers PS_{54}-*b*-PTTBM-R_{44} are not soluble in the common organic solvents typically used in the breath figures approach, such as CS_2 and chloroform. Thus, THF was selected to be more compatible with both components of the blend, although it is known that THF is not an ideal solvent and typically leads to irregular arrays due to its miscibility with water. Nevertheless, THF only allows the solubility of low content of copolymers. For this reason, THF solution was prepared with a polymeric concentration of 30 mg/mL, with PS/PS_{54}-*b*-PTTBM-R_{44} ratios of 97/3, 94/6 and 91/9 wt %.

Films were prepared from these THF solutions under different humidities: 60%, 70% and 90%. As mentioned above, the relative humidity is a fundamental parameter in the preparation of porous films by the breath figures approach and has a large impact on the morphology of the films. As shown in Figure 1, when the humidity was set at 60%, only flat surfaces were obtained; thus, in this case, higher humidity was necessary to fabricate porous films. In effect, at higher humidity values, such as 70% and 90%, porous films were found; however, films prepared at 90% show irregular patterns containing large and heterogeneous pores mixed with smaller pores, resulting from the coagulation of the rapidly condensing water droplets during the breath figure process, which leads to a dramatic increase in the droplet size (Figure 1c,d) [37]. On the other hand, when the humidity was set at 70%, more homogeneous patterns were obtained, as shown in Figure 1b. Thus, the following experiments for the fabrication of antimicrobial surfaces were carried out under this relative humidity.

Films were prepared at 70% relative humidity from blend solutions composed of commercial PS and a low amount of the antimicrobial copolymer quaternized with butyl or methyl iodide, PS_{54}-*b*-PTTBM-B_{44} or PS_{54}-*b*-PTTBM-M_{44}, respectively. It is well known that the alkylating agents affect the efficacy of the antimicrobial polymers based on quaternary ammonium groups, because they modify the hydrophobic/hydrophilic balance [31,38]. Additionally, the alkylating agents would also influence the microstructure of the breath figure films. Figure 2 shows SEM images of the films containing different contents of copolymer prepared at 70% humidity, in which porous films are observed in all cases. Additionally, it is observed from the cross-section image that there is only a single layer with pores. In these SEM images, the influence of the type of copolymer and its concentration on the pore structure and morphology of the porous breath figure films can also be seen.

Previous works indicate that, in general, copolymers with large hydrophilic blocks produce poorly ordered structures because the interfacial tension tends to decrease and, consequently, the coalescence of the water droplets increases [29,37,39]. However, in this case, relatively ordered porous arrays are obtained for all the blends, using both types of antimicrobial copolymers with large cationic segments as additives. Table 1 shows the quantitative evaluation of the order obtained by using Voronoi polygon construction on low-magnification SEM images. The images were processed and analyzed by the software ImageJ to calculate the conformational entropy, which is compared with the entropy for an ideal hexagonal array (S = 0) and a randomly organized array (S = 1.71) [40]. The large entropies obtained between 1.17 and 0.86 indicate relatively poorly ordered arrays, although these values are also substantially less than S = 1.71 for random packing. It has to be mentioned that they are typical values for breath figures made from water-miscible solvent such as THF [41–43]. Concerning the pore size, highly homogeneous pore diameters can clearly be seen for all the samples in SEM images. When the different films are compared, in general, higher diameters are obtained in films containing the copolymer quaternized with butyl, PS_{54}-*b*-PTTBM-B_{44} (Figure 2a–c), which is more hydrophobic than PS_{54}-*b*-PTTBM-M_{44} quaternized with methyl groups. Nevertheless, the differences are slight, as seen in Table 1, which summarizes the mean pore sizes of all prepared porous films, determined by measuring at least 100 pores from the SEM images. Additionally, when the concentration of both copolymers incorporated into the films is varied from 3 to 9 wt %, a slight influence is noted in the pore size, which decreases when the content of copolymer increases.

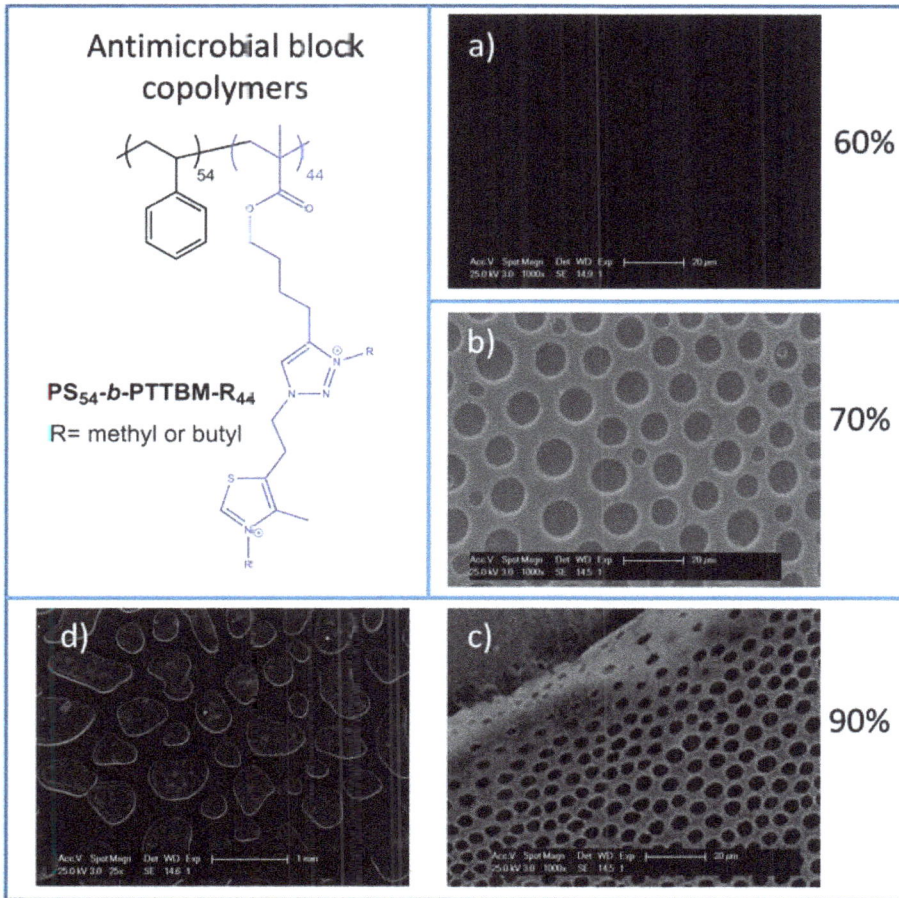

Figure 1. Chemical structure of the antimicrobial block copolymer quaternized with butyl (PS_{54}-*b*-PTTBM-B_{44}) or methyl (PS_{54}-*b*-PTTBM-M_{44}) iodides, and SEM images of the films of PS/PS_{54}-*b*-PTTBM-B_{44} blends, 97/3 w: % obtained from THF solutions at (**a**) 60%; (**b**) 70%; and (**c**,**d**) 90% relative humidity.

PS$_{54}$-*b*-PTTBM-B$_{44}$

PS$_{54}$-*b*-PTTBM-M$_{44}$

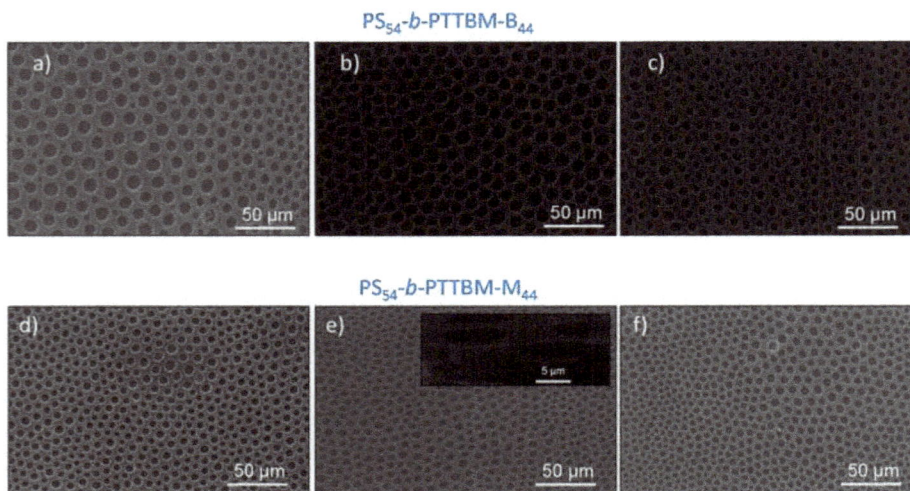

Figure 2. SEM micrographs of films obtained at 70% relative humidity composed of PS/PS$_{54}$-*b*-PTTBM-B$_{44}$ blends: (**a**) 97/3 wt %; (**b**) 94/6 wt %; (**c**) 91/9 wt %; and PS/ PS$_{54}$-*b*-PTTBM-M$_{44}$ blends: (**d**) 97/3 wt %; (**e**) 94/6 wt % (inset: cross-section); (**f**) 91/9 wt %.

Table 1. Average pore diameter $\pm$ SD (standard deviation), conformational entropy (S), and roughness factor (r_f) of porous films containing variable content of the cationic copolymers PS$_{54}$-*b*-PTTBM-B$_{44}$ and PS$_{54}$-*b*-PTTBM-M$_{44}$.

Cationic Copolymer	Concentration (wt. %)	S	Pore Size (µm)	r_f
PS$_{54}$-*b*-PTTBM-B$_{44}$	3	1.16	11 $\pm$ 1	1.33
	6	0.91	10 $\pm$ 1	1.42
	9	1.17	7 $\pm$ 1	1.41
PS$_{54}$-*b*-PTTBM-M$_{44}$	3	0.86	7 $\pm$ 1	1.47
	6	0.98	6 $\pm$ 2	1.48
	9	0.94	5 $\pm$ 1	1.48

In general, we can conclude that as the hydrophilicity of the system is augmented, either by the use of more hydrophilic copolymer or by the use of more percentage of the cationic copolymer, the size of the pores decreases. It is well known in the breath figures approach that amphiphilic structures help to stabilize the water droplets condensed at the surface of the polymeric solution; thus, in polymeric blends, the content of amphiphilic copolymers significantly influences the porous structures [22]. As the content of copolymer increases in the blend, more droplets can be stabilized and, therefore, more and smaller pores can finally be formed at the surface [22,44].

The surface wettability of the films and, then, their contact with culture media mainly depends on both the chemical functionality and the roughness of the surface. The water contact angle values of obtained films were found to be ~120° for all the samples measured, independent of the copolymer content. However, as the cationic copolymer content increases in the sample, so does the hydrophilicity, and the contact angle should decrease. Therefore, the roughness of the sample, as expected, also contributes to the wettability of the films [45,46]. Table 1 summarizes the Wenzel roughness factor, r $_f$, defined as the ratio between the actual and the projected areas of the surface [47]. This factor is equal to one for flat surfaces and is greater than one for rough surfaces. It is observed that the roughness of the films increases with the content of the copolymers, and in films containing the copolymer quaternized with methyl iodide. Pore diameter slightly decreases with the content of the cationic copolymer, but at the same time, pore density also increases, which contributes to the augmentation of the roughness. These contrary contributions to wettability, chemical functionality and roughness could be the reason for the similar contact angles values found in the films. Therefore, in principle, microbial contact with the surface would be rather similar for all the samples.

The antimicrobial activity of the prepared breath figure films was evaluated against *S. aureus* Gram-positive bacteria and the fungi *C. parapsilosis* as model microbes, since they are common pathogens responsible of many nosocomial infections. The shake flask method [36] was employed to quantify the antimicrobial activity of the films under dynamic contact conditions. Table 2 summarizes the cell killing percentage in microbial medium in contact with the films for 24 h, and then the growth in agar plates for 24 h and 48 h for bacteria and fungi, respectively. The cell killing percentages were expressed with respect to control experiments in which the microbial reduction was null (experiments performed with films prepared from commercial PS, 0 wt % of copolymers, and without any films).

Table 2. Cell killing percentage of the breath figure films for *S. aureus* and *C. parapsilosis* microorganisms.

Cationic Copolymer	Concentration (wt %)	Cell Killing (%)	
		S. aureus	*C. parapsilosis*
PS$_{54}$-*b*-PTTBM-B$_{44}$	3	99.99	50
	6	99.99	90
	9	99.99	90
PS$_{54}$-*b*-PTTBM-M$_{44}$	3	99.99	90
	6	99.99	90
	9	99.99	90

It can be seen that all films exhibit high killing efficiency against *S. aureus* bacteria, with a reduction of more than 99.99% in the culture medium. On the other hand, moderate activity was found against *C. parapsilosis* fungi, with reduction of up to 90% for contents of copolymer higher than 6%. It is worth mentioning that these films present relatively high antimicrobial activity even with very low content of cationic copolymer; films containing only 6 wt % copolymers can reduce 99.99% of *S. aureus* and 90% of *C. parapsilosis* exposure to the films. Thus, these results reveal that the preparation method provides films with enough accessible active groups at the surfaces to kill the microorganisms by surface contact, even when low amounts of copolymer are incorporated in the film. Remarkably, these breath figure films provide better efficiencies than flat films prepared directly from the copolymer solution; that is, 100 wt % of PS$_{54}$-*b*-PTTBM-B$_{44}$, PS$_{54}$-*b*-PTTBM-M$_{44}$ [16]. These findings demonstrate the importance of the surface roughness on the antimicrobial activity of contact-active films, which allows the use of very low amounts of antimicrobial component in the coating while maintaining excellent biocidal activity.

4. Conclusions

In summary, efficient antimicrobial porous coatings were fabricated by the breath figures approach from blends containing very low contents of antimicrobial polymers. Highly active amphiphilic copolymers with a large cationic block bearing a flexible side chain with 1,3-thiazolium and 1,2,3-triazolium groups were used as antimicrobial polymers with high charge density. Due to the high biocidal effectiveness of the copolymers and the controlled roughness of the porous surfaces, the resulting films exhibit high killing efficiency against the studied microorganisms. Thus, we can conclude that this breath figures approach, using only a low content of cationic polymers, allows the formation of surfaces with accessible polycationic chains for killing the microorganisms *S. aureus* and *C. parapsilosis* by surface contact.

Author Contributions: Conceptualization, A.M.-B. and M.F.-G.; Methodology, R.C.-R.; Software, A.M.-B.; Validation, A.M.-B., R.C.-R. and M.F.-G.; Formal Analysis, F.L.-F. and J.L.G.-G.; Investigation, A.M.-B.; Resources, F.L.-F. and J.L.G.-G.; Data Curation, A.M.-B., R.C.-R. and M.F.-G.; Writing-Original Draft Preparation, A.M.-B.;

Writing-Review & Editing, M.F.-G.; Visualization, X.X.; Supervision, A.M.-B., M.F.-G.; Project Administration, M.F.-G.; Funding Acquisition, M.F.-G.

Funding: This work was supported financially by the MINECO (Project MAT2016-78437-R), the Agencia Estatal de Investigación (AEI, Spain) and Fondo Europeo de Desarrollo Regional (FEDER, EU).

Conflicts of Interest: The authors declare no conflict of interest.

References

1. Klein, E.Y.; Van Boeckel, T.P.; Martinez, E.M.; Pant, S.; Gandra, S.; Levin, S.A.; Goossens, H.; Laxminarayan, R. Global increase and geographic convergence in antibiotic consumption between 2000 and 2015. *Proc. Natl. Acad. Sci. USA* **2018**, *115*, E3463–E3470. [CrossRef] [PubMed]
2. Sugden, R.; Kelly, R.; Davies, S. Combatting antimicrobial resistance globally. *Nat. Microbiol.* **2016**, *1*, 16187. [CrossRef] [PubMed]
3. Noimark, S.; Dunnill, C.W.; Wilson, M.; Parkin, I.P. The role of surfaces in catheter-associated infections. *Chem. Soc. Rev.* **2009**, *38*, 3435–3448. [CrossRef] [PubMed]
4. Yang, C.; Ding, X.; Ono, R.J.; Lee, H.; Hsu, L.Y.; Tong, Y.W.; Hedrick, J.; Yang, Y.Y. Brush-like polycarbonates containing dopamine, cations, and PEG providing a broad-spectrum, antibacterial, and antifouling surface via one-step coating. *Adv. Mater.* **2014**, *26*, 7346–7351. [CrossRef] [PubMed]
5. Vaterrodt, A.; Thallinger, B.; Daumann, K.; Koch, D.; Guebitz, G.M.; Ulbricht, M. Antifouling and Antibacterial Multifunctional Polyzwitterion/Enzyme Coating on Silicone Catheter Material Prepared by Electrostatic Layer-by-Layer Assembly. *Langmuir* **2016**, *32*, 1347–1359. [CrossRef] [PubMed]
6. Wang, G.; Zreiqat, H. Functional Coatings or Films for Hard-Tissue Applications. *Materials* **2010**, *3*, 3994–4050. [CrossRef] [PubMed]
7. Alvarez-Paino, M.; Juan-Rodriguez, R.; Cuervo-Rodriguez, R.; Tejero, R.; Lopez, D.; Lopez-Fabal, F.; Gomez-Garces, J.L.; Munoz-Bonilla, A.; Fernandez-Garcia, M. Antimicrobial films obtained from latex particles functionalized with quaternized block copolymers. *Colloids Surf. B Biointerfaces* **2016**, *140*, 94–103. [CrossRef] [PubMed]
8. Manabe, K.; Nishizawa, S.; Shiratori, S. Porous surface structure fabricated by breath figures that suppresses Pseudomonas aeruginosa biofilm formation. *ACS Appl. Mater. Interfaces* **2013**, *5*, 11900–11905. [CrossRef] [PubMed]
9. Hasan, J.; Jain, S.; Padmarajan, R.; Purighalla, S.; Sambandamurthy, V.K.; Chatterjee, K. Multi-scale surface topography to minimize adherence and viability of nosocomial drug-resistant bacteria. *Mater. Des.* **2018**, *140*, 332–344. [CrossRef] [PubMed]
10. Hasan, J.; Jain, S.; Chatterjee, K. Nanoscale Topography on Black Titanium Imparts Multi-biofunctional Properties for Orthopedic Applications. *Sci. Rep.* **2017**, *7*, 41118. [CrossRef] [PubMed]
11. Hasan, J.; Chatterjee, K. Recent advances in engineering topography mediated antibacterial surfaces. *Nanoscale* **2015**, *7*, 15568–15575. [CrossRef] [PubMed]
12. Muñoz-Bonilla, A.; Fernández-García, M. The roadmap of antimicrobial polymeric materials in macromolecular nanotechnology. *Eur. Polym. J.* **2015**, *65*, 46–62. [CrossRef]
13. May, R.M.; Magin, C.M.; Mann, E.E.; Drinker, M.C.; Fraser, J.C.; Siedlecki, C.A.; Brennan, A.B.; Reddy, S.T. An engineered micropattern to reduce bacterial colonization, platelet adhesion and fibrin sheath formation for improved biocompatibility of central venous catheters. *Clin. Transl. Med.* **2015**, *4*, 9. [CrossRef] [PubMed]
14. Truong, V.K.; Webb, H.K.; Fadeeva, E.; Chichkov, B.N.; Wu, A.H.; Lamb, R.; Wang, J.Y.; Crawford, R.J.; Ivanova, E.P. Air-directed attachment of coccoid bacteria to the surface of superhydrophobic lotus-like titanium. *Biofouling* **2012**, *28*, 539–550. [CrossRef] [PubMed]
15. Ivanova, E.P.; Hasan, J.; Webb, H.K.; Gervinskas, G.; Juodkazis, S.; Truong, V.K.; Wu, A.H.; Lamb, R.N.; Baulin, V.A.; Watson, G.S.; et al. Bactericidal activity of black silicon. *Nat. Commun.* **2013**, *4*, 2838. [CrossRef] [PubMed]
16. Cuervo-Rodriguez, R.; Lopez-Fabal, F.; Gomez-Garces, J.L.; Munoz-Bonilla, A.; Fernandez-Garcia, M. Contact Active Antimicrobial Coatings Prepared by Polymer Blending. *Macromol. Biosci.* **2017**, *17*, 1700258. [CrossRef] [PubMed]
17. Su, L.; Yu, Y.; Zhao, Y.; Liang, F.; Zhang, X. Strong Antibacterial Polydopamine Coatings Prepared by a Shaking-assisted Method. *Sci. Rep.* **2016**, *6*, 24420. [CrossRef] [PubMed]

18. Wei, T.; Yu, Q.; Zhan, W.; Chen, H. A Smart Antibacterial Surface for the On-Demand Killing and Releasing of Bacteria. *Adv. Healthc. Mater* **2015**, *5*, 449–456. [CrossRef] [PubMed]

19. Acikgoz, C.; Hempenius, M.A.; Huskens, J.; Vancso, G.J. Polymers in conventional and alternative Lithography for the fabrication of nanostructures. *Eur. Polym. J.* **2011**, *47*, 2033–2052. [CrossRef]

20. Gi-Ra Yi, G.R.; Moon, J.H.; Yang, S.M. Ordered Macroporous Particles by Colloidal Templating. *Chem. Mater.* **2001**, *13*, 2613–2618.

21. Zhang, A.; Bai, H.; Li, L. Breath Figure: A Nature-Inspired Preparation Method for Ordered Porous Films. *Chem. Rev.* **2015**, *115*, 9801–9868. [CrossRef] [PubMed]

22. Muñoz-Bonilla, A.; Fernández-García, M.; Rodríguez-Hernández, J. Towards hierarchically ordered functional porous polymeric surfaces prepared by the breath figures approach. *Prog. Polym. Sci.* **2014**, *39*, 510–554. [CrossRef]

23. Widawski, G.; Rawiso, M.; François, B. Self-organized honeycomb morphology of star-polymer polystyrene films. *Nature* **1994**, *369*, 387–389. [CrossRef]

24. Escalé, P.; Rubatat, L.; Billon, L.; Save, M. Recent advances in honeycomb-structured porous polymer films prepared via breath figures. *Eur. Polym. J.* **2012**, *48*, 1001–1025. [CrossRef]

25. Hernández-Guerrero, M.; Stenzel, M.H. Honeycomb structured polymer films via breath figures. *Polym. Chem.* **2012**, *3*, 563–577. [CrossRef]

26. de Leon, A.S.; del Campo, A.; Fernandez-Garcia, M.; Rodriguez-Hernandez, J.; Munoz-Bonilla, A. Tuning the pore composition by two simultaneous interfacial self-assembly processes: Breath figures and coffee stain. *Langmuir* **2014**, *30*, 6134–6141. [CrossRef] [PubMed]

27. Bolognesi, A.; Galeotti, F.; Giovanella, U.; Bertini, F.; Yunus, S. Nanophase separation in polystyrene-polyfluorene block copolymers thin films prepared through the breath figure procedure. *Langmuir* **2009**, *25*, 5333–5338. [CrossRef] [PubMed]

28. Böker, A.; Lin, Y.; Chiapperini, K.; Horowitz, R.; Thompson, M.; Carreon, V.; Xu, T.; Abetz, C.; Skaff, H.; Dinsmore, A.D.; et al. Hierarchical nanoparticle assemblies formed by decorating breath figures. *Nat. Mater.* **2004**, *3*, 302–306. [CrossRef] [PubMed]

29. Vargas-Alfredo, N.; Santos-Coquillat, A.; Martinez-Campos, E.; Dorronsoro, A.; Cortajarena, A.L.; Del Campo, A.; Rodriguez-Hernandez, J. Highly Efficient Antibacterial Surfaces Based on Bacterial/Cell Size Selective Microporous Supports. *ACS Appl. Mater. Interfaces* **2017**, *9*, 44270–44280. [CrossRef] [PubMed]

30. Vargas-Alfredo, N.; Dorronsoro, A.; Cortajarena, A.L.; Rodriguez-Hernandez, J. Antimicrobial 3D Porous Scaffolds Prepared by Additive Manufacturing and Breath Figures. *ACS Appl Mater. Interfaces* **2017**, *9*, 37454–37462. [CrossRef] [PubMed]

31. Tejero, R.; López, D.; López-Fabal, F.; Gómez-Garcés, J.L.; Fernández-García, M. Antimicrobial polymethacrylates based on quaternized 1,3-thiazole and 1,2,3-triazole side-chain groups. *Polym. Chem.* **2015**, *6*, 3449–3459. [CrossRef]

32. Tejero, R.; Lopez, D.; Lopez-Fabal, F.; Gomez-Garces, J.L.; Fernandez-Garcia, M. High Efficiency Antimicrobial Thiazolium and Triazolium Side-Chain Polymethacrylates Obtained by Controlled Alkylation of the Corresponding Azole Derivatives. *Biomacromolecules* **2015**, *16*, 1844–1854. [CrossRef] [PubMed]

33. Tejero, R.; Gutiérrez, B.; López, D.; López-Fabal, F.; Gómez-Garcés, J.; Muñoz-Bonilla, A.; Fernández-García, M. Tailoring Macromolecular Structure of Cationic Polymers towards Efficient Contact Active Antimicrobial Surfaces. *Polymers* **2018**, *10*, 241. [CrossRef]

34. Kaur, R.; Liu, S. Antibacterial surface design—Contact kill. *Prog. Surf. Sci.* **2016**, *91*, 136–153. [CrossRef]

35. ASTM E2149-01. *Standard Test Method for Determining the Antimicrobial Activity of Immobilized Antimicrobial Agents under Dynamic Contact Conditions (Withdrawn 2010)*; ASTM International: West Conshohocken, PA, USA, 2001. Available online: www.astm.org (accessed on 14 April 2010).

36. Clinical and Laboratory Standards Institute. *Methods for Dilution Antimicrobial Susceptibility Tests for Bacteria That Grow Aerobically*, 10th ed.; CLSI Document M07-A10; Clinical and Laboratory Standards Institute: Wayne, PA, USA, 2015.

37. Wong, K.H.; Davis, T.P.; Barner-Kowollik, C.; Stenzel, M.H. Honeycomb structured porous films from amphiphilic block copolymers prepared via RAFT polymerization. *Polymer* **2007**, *48*, 4950–4965. [CrossRef]

38. Muñoz-Bonilla, A.; Fernández-García, M. Polymeric materials with antimicrobial activity. *Prog. Polym. Sci.* **2012**, *37*, 281–339. [CrossRef]

39. Bunz, U.H.F. Breath Figures as a Dynamic Templating Method for Polymers and Nanomaterials. *Adv. Mater.* **2006**, *18*, 973–989. [CrossRef]

40. Song, L.; Sharma, V.; Park, J.O.; Srinivasarao, M. Characterization of ordered array of micropores in a polymer film. *Soft Matter* **2011**, *7*, 1890–1896. [CrossRef]

41. de Leon, A.S.; del Campo, A.; Fernandez-Garcia, M.; Rodriguez-Hernandez, J.; Munoz-Bonilla, A. Hierarchically structured multifunctional porous interfaces through water templated self-assembly of ternary systems. *Langmuir* **2012**, *28*, 9778–9787. [CrossRef] [PubMed]

42. Ferrari, E.; Fabbri, P.; Pilati, F. Solvent and substrate contributions to the formation of breath figure patterns in polystyrene films. *Langmuir* **2011**, *27*, 1874–1881. [CrossRef] [PubMed]

43. Soo Park, M.; Kon Kim, J. Breath Figure Patterns Prepared by Spin Coating in a Dry Environment. *Langmuir* **2004**, *20*, 5347–5352. [CrossRef]

44. de León, A.S.; del Campo, A.; Rodríguez-Hernández, J.; Muñoz-Bonilla, A. Switchable and pH responsive porous surfaces based on polypeptide-based block copolymers. *Mater. Des.* **2017**, *131*, 121–126. [CrossRef]

45. Yabu, H.; Takebayashi, M.; Tanaka, M.; Shimomura, M. Superhydrophobic and Lipophobic Properties of Self-Organized Honeycomb and Pincushion Structures. *Langmuir* **2005**, *21*, 3235–3237. [CrossRef] [PubMed]

46. Feng, X.J.; Jiang, L. Design and Creation of Superwetting/Antiwetting Surfaces. *Adv. Mater.* **2006**, *18*, 3063–3078. [CrossRef]

47. De León, A.S.; Campo, A.D.; Labrugère, C.; Fernández-García, M.; Muñoz-Bonilla, A.; Rodríguez-Hernández, J. Control of the chemistry outside the pores in honeycomb patterned films. *Polym. Chem.* **2013**, *4*, 4024. [CrossRef]

materials

MDPI

Article

The Effect of the Isomeric Chlorine Substitutions on the Honeycomb-Patterned Films of Poly(x-chlorostyrene)s/Polystyrene Blends and Copolymers via Static Breath Figure Technique

Leire Ruiz-Rubio [1,2,*] , Leyre Pérez-Álvarez [1,2], Julia Sanchez-Bodón [1], Valeria Arrighi [3] and José Luis Vilas-Vilela [1,2]

[1]   Grupo de Química Macromolecular (LABQUIMAC) Dpto. Química-Física, Facultad de Ciencia y Tecnología, Universidad del País Vasco (UPV/EHU), 48940 Leioa, Bizkaia, Spain; leyre.perez@ehu.eus (L.P.-Á.); jsanchez903@ikasle.ehu.eus (J.S.-B.); joseluis.vilas@ehu.eus (J.L.V.-V.)
[2]   BCMaterials, Basque Center for Materials, Applications and Nanostructures, UPV/EHU Science Park, 48940 Leioa, Spain
[3]   Chemical Sciences, School of Engineering & Physical Sciences, Heriot-Watt University, Edinburgh EH14 4AS, UK; V.Arrighi@hw.ac.uk
*   Correspondence: leire.ruiz@ehu.eus; Tel.: +34-946017972

Received: 30 November 2018; Accepted: 31 December 2018; Published: 7 January 2019

Abstract: Polymeric thin films patterned with honeycomb structures were prepared from poly(x-chlorostyrene) and statistical poly(x-chlorostyrene-co-styrene) copolymers by static breath figure method. Each polymeric sample was synthesized by free radical polymerization and its solution in tetrahydrofuran cast on glass wafers under 90% relative humidity (RH). The effect of the chorine substitution in the topography and conformational entropy was evaluated. The entropy of each sample was calculated by using Voronoi tessellation. The obtained results revealed that these materials could be a suitable toolbox to develop a honeycomb patterns with a wide range of pore sizes for a potential use in contact guidance induced culture.

Keywords: poly(x-chlorostyrene); honeycomb; breath figures; conformational entropy

1. Introduction

The control over water condensing phenomenon is a useful approach to form highly ordered honeycomb structures from polymer by condensation of water droplets onto a drying polymer solution. These honeycomb polymer thin films have attracted much attention due to the increasing range of applications, such as energy storage [1,2], membranes [3–7], catalytic surfaces [8,9], and sensing materials [10], among others. The effectiveness and low cost of these methods in comparison to other techniques, such as photolithography or templating methods, have stimulated interest on breath figures to fabricate substrates for biological applications [11–13].

In brief, the breath figures were formed when a polymer solution in a high volatile solvent was cast onto a substrate under adequate humidity (Figure 1). The evaporation of the solvent induced a cooling at the solvent/air interface. This process favored the condensation of water from the humidity in the solution. Condensed water droplets formed a honeycomb pattern on the surfaces [14].

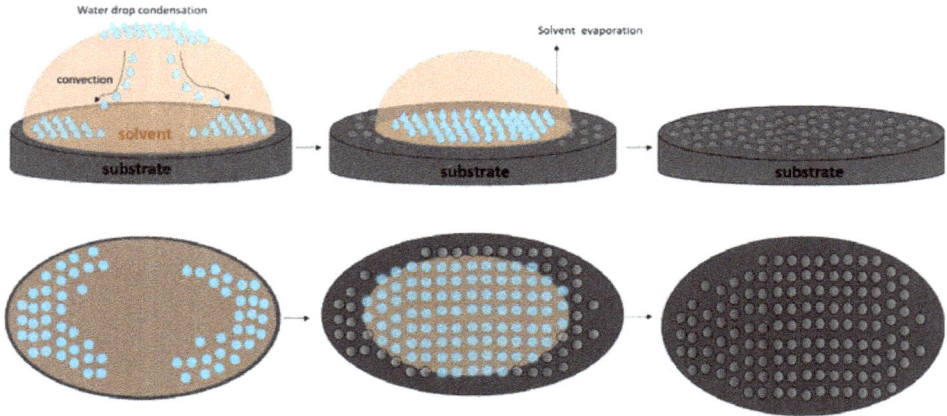

Figure 1. Mechanism of breath figure formation.

The formation of the breath figures could be performed by different approaches, such as dip coating, spin coating, air-flow or dynamic technique, and solvent cast or static method, with the last two techniques more extended (Figure 2). On the one hand, in the static breath figure technique, a polymer solution is cast dropwise on a solid substrate under a controlled relative humidity, where the experiment is placed in a close chamber that allows the control over temperature and RH. On the other hand, in the dynamic method, the polymer casting is carried out under airflow, with controlled flow and humidity. This flow forces the rapid evaporation of the solvent due to the formed temperature gradient between the solution and the bulk [15,16]. In this study, a static breath figure method was used.

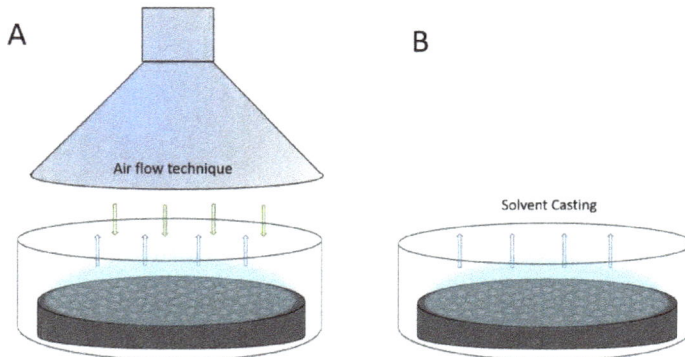

Figure 2. (**A**) Air flow or dynamic technique, and (**B**) solvent cast or static technique.

Several factors could affect the breath figure method, such as humidity, solvent, concentration, polymer, or substrate, among others. For example, an increase on the RH usually induces an increase on the pore size, whereas a solvent with a higher volatility produces a lower pore size. These effects were summarized in several reviews devoted to breath figures [17,18].

Cell-material interaction is considered one of the fundamental fields in biomaterials [19–21]. Apart from the chemical composition of the surfaces, physical factors, such as pore size and stiffness, are important in cell adhesion, spreading, and proliferation [22]. Studies indicate that a reduction of the cell adhesive sites on a substrate could be a key factor in the design of the surface structure of cell culture substrates [11]. In this context, recently 2D and 3D patterns have been fabricated, from submicrometer to less than one hundred of micrometer, capable of modeling in vivo topographic microenvironments of cells inducing the contact guidance [20,23–25].

Several researchers have demonstrated the successful adhesion of the most common cells to honeycomb structures [26–30]. It has been demonstrated as a promotion of the adhesion for most common cells when using honeycomb patterns, including hepatocytes [28,31] or endothelial cells [27,32], among others. As an example, Arai et al. [33] have reported the influence of the pore size on the incubation of cardiac myocytes. They fabricated poly(ε-caprolactone)–based honeycomb structures with diameters ranging from 4 to 13 µm. This range could be divided in subcellular (around 4.00 µm), cellular (8 µm), and overcellular (12.5 µm), by comparison with the cellular size (7–10 µm for cardiac myocytes). These authors observed that the pore size could crucially affect both cellular adhesion and morphologies. Similar results were reported by Tsuruma and co-workers for neural cell stems [12,34]. Considering these honeycomb patterns could be used to successfully grow several types of cells. However, some studies also reported an excellent growth when unidirectional/anisotropic patterns were used as scaffolds [35]. Besides, the cost of this technique, added to it ease to perform, could arise interest in its industrial fabrication.

The present study reports the fabrication and characterization of highly ordered honeycomb films of polystyrene and poly(x-chlorostyrene)s, their copolymers, and polystyrene/poly(x-chlorostyrene) blends by using the static breath figure method. The polymer solutions were cast on glass wafers under 90% relative humidity (RH). We show that the use of different chlorine substitutions allowed varying the pore size of the surface, in order to obtain topographies with subcellular, cellular, and overcellular diameter (compared to an average cellular diameter of 7 µm), as a potential toolbox in the development of surfaces for contact guidance induced culture. The polystyrene was chosen as a reference material since it is a very commonly used polymer and not expensive, so its use as a template could be highly interesting from an industrial point of view. On the other hand, the synthetic process of these chloro-substituted polymers is quite similar to the polystyrene, and their synthesis could be considered cost-effective when compared to the polymers that have been used for breath figure applications obtained from time–consuming reactions, such as atom transfer radical polymerization or reversible addition–fragmentation chain-transfer polymerization.

2. Materials and Methods

2.1. Materials

Styrene (S), 2-chlorostyrene (2ClS), 3-chlorostyrene (3ClS), 4-chlorostyrene (4ClS), and the initiator, α,α'-azobisisobutyronitrile (AIBN), were obtained from Sigma Aldrich. Polystyrene (PS, weight-average molecular weight (M_w) = 3 × 10^6 g·mol^{-1}) was purchased from Polysciences (Warrington, PA, USA). Tetrahydrofuran HPLC grade, (Scharlab, Sentmenat, Spain) was used without further purification. The polymer solutions were cast in round glass coverslips of 20 mm diameter purchased from Marienfeld (Lauda-Königshofen, Germany).

2.2. Synthesis of the Homopolymers and Random Copolymers

Phenolic inhibitors from the 2, 3 or 4-chlorostyrene monomers were first removed by washing with sodium hydroxide (0.1 M) solution, then with hydrochloric acid solution, and finally, with distilled water until neutral pH. The monomers were dried over anhydrous $MgSO_4$ and filtered. The initiator, AIBN, was purified by crystallization from methanol.

Homopolymers and statistical copolymers of styrene and x-chlorostyrenes were prepared by free radical bulk polymerization, at 60 °C, under nitrogen atmosphere, with 0.5 mol% AIBN with respect to the total amount of monomer. Copolymers were synthesized at low conversions to prevent composition drifts.

The polymers were precipitated into excess methanol, re-dissolved in chloroform, and re-precipitated in methanol. Finally, the purified samples were vacuum dried at 70 °C for two days.

The copolymer compositions were determined by Fourier Transform Infrared spectroscopy (FTIR) and Elemental Analysis. FTIR spectra were recorded using a Nicolet Nexus FTIR spectrophotometer (Thermo Fisher Scientific, Loughborough, UK), in KBr pellets with a resolution of 4 cm^{-1}, and a total of 32 scans were averaged in all cases. The copolymer compositions were obtained from the FTIR data collected for the copolymers, using a calibration curve obtained by measuring the absorbance ratio of two characteristic peaks for each homopolymer, for known homopolymer composition mixtures. Elemental Analysis was carried out using the Eurovector EA3000 (Eurovector, Milan, Italy) analyzer. Copolymer compositions were obtained from the chlorine content.

Molecular weights of all polymer samples were determined by gel permeation chromatography (GPC), using a Waters chromatograph (Mildford, MA, USA) and THF as the eluent. M_w values reported in Tables 1 and 2 are relative to polystyrene standards.

Table 1. Molecular weight of the studied polymer samples.

Sample	M_w (10^4) (g mol^{-1})	M_w/M_n
Polystyrene	300	1.8
Poly(2-chlrostyrene)	23.3	1.3
Poly(3-chlorostyrene)	17.1	1.5
Poly(4-chlorostyrene)	11.7	1.2

Table 2. Molecular weight and fraction of styrene in the copolymer for the different samples.

Sample	M_w (10^4) (g mol^{-1})	M_w/M_n	F_S
Polystyrene	300	1.80	1
Poly(2-chlrostyrene-co-styrene)	11.3	1.9	0.526
Poly(3-chlorostyrene-co-styrene)	11.2	1.6	0.608
Poly(4-chlorostyrene-co-styrene)	13.9	1.5	0.619

2.3. Preparation of the Films

The polymer solutions were prepared by dissolution of the homopolymers or copolymers in THF. The polymer concentration used in this study was 30 mg mL^{-1}. The films were prepared from these solutions by casting (50 µL) onto glass wafers under controlled humidity in a closed chamber. The relative humidity (RH) was controlled by a saturated salt solution of KNO$_3$ in water to obtain 90% RH. Round films of 150 $\pm$ 2 mm diameter and 0.04–0.05 mm thickness were obtained.

2.4. Characterization

The morphology of the honeycomb patterned films on a glass substrate was studied by a scanning electron microscopy (SEM). The samples were coated with gold, prior to the SEM measurements, using a Fine Coat Ion Sputter JFC-1100 (JEOL, Tokyo, Japan). SEM micrographs were taken using a Hitachi S-4800 (Hitachi, Tokyo, Japan). The images were processed and analyzed using image analysis freeware ImageJ to obtain pore size (mean diameter) and size distribution of the patterned surfaces. The regularity of the obtained honeycomb-like patterns shown in the SEM images was evaluated by Voronoi tessellation of the images, also performed by ImageJ.

3. Results

In this study, the influence of the chlorine substitution and the geometrical positioning of the Cl substituent on the topography and conformational entropy of the honeycomb surfaces were studied. Highly ordered patterns were successfully fabricated from polystyrene and its derivatives by the static breath figure method, using THF as the solvent at 90% RH, at room temperature. The regularity of the patterned films was quantified by measuring the conformational entropy calculated by Voronoi polygons. This method has not only been used to assess the regularity of the surface patterns, but also to describe defects in the membranes [36–38].

A Voronoi polygon is defined as the smallest convex polygon surrounding a point whose sides are perpendicular bisectors of the lines between a point and its neighbors [39]. The conformational entropy, which can be related to the degree of order of the arrays, is defined as (Equation (1)):

$$S = -\sum_n P_n \ln P_n \qquad (1)$$

where n is the coordination number of each Voronoi, i.e., the number of sides of the polygon, and P_n is the fraction of the polygons having the coordination number n. That is, Voronoi tessellation method calculates the probability of the occurrence of four (P_4), five ((P_5), six—which is a perfectly hexagonal lattice—(P_6), seven (P_7), or eight (P_8) As a reference, for a perfectly ordered hexagonal array, i.e., an ideal hexagonal lattice, the conformational entropy is 0, whereas the entropy of a completely random pattern is 1.71 [18,39–43].

The formation of the patterns was carried out by exploiting the static breath figure method from different samples at 30 mg mL^{-1} and 90% RH. First, the formation of honeycomb arrays in poly(x-chlorostyrene) and polystyrene homopolymers were studied. After, the formation in the patterns obtained for random copolymers of poly(x-chlorostyrene-co-styrene) and, finally, their blends of the polystyrene and poly(x-chlorostyrene) were analyzed.

3.1. Formation of Breath Figures in Poly(x-chlorostyrene) Isomers and Polystyrene

The poly(x-chlorostyrene)s were synthesized by free radical polymerization (0.5 mol% AIBN, 60 °C). The weight average molecular weights (M_w) and the polydispersity index (M_w/M_n) of poly(x-chlorostyrene) and polystyrene used in this study were determined by GPC and are summarized in Table 1.

The homopolymer samples were cast on a glass substrate, and the obtained arrays were analyzed by SEM. As shown in Figure 3, all samples present cavities at this condition. However, honeycomb patterns based on poly(x-chlorostyrene) samples seem to be less homogenous than polystyrene. These changes could be induced by the change on the chlorine substitution, so different parameters, such as pore size or surface entropy, have been analyzed to evaluate these variations.

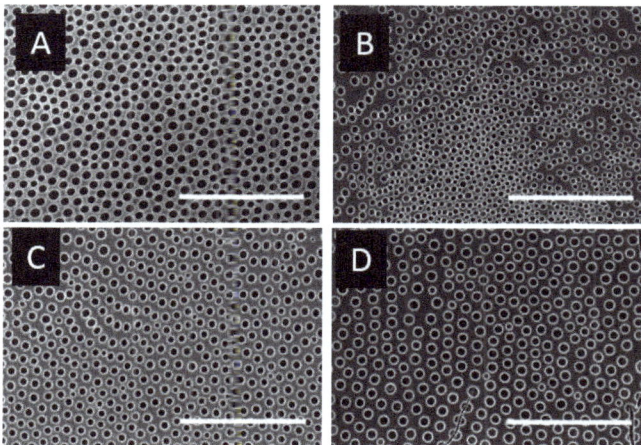

Figure 3. SEM images for (**A**) poly(styrene), (**B**) poly(2-chlorostyrene) (**C**) poly(3-chlorostyrene), and (**D**) poly(4-chlorostyrene). (Scale bar = 100 μm).

The pore diameter and their standard deviation (SD) for x-chloro-substituted samples were 4.1 ± 0.45, 5.25 ± 0.35 and 6.26 ± 0.20 μm for poly(2-chlorostyrene) (P2ClS), poly(3-chlorostyrene) (P3ClS) and poly(4-chlorostyrene) (P4ClS), respectively. Thus, pore size increases from ortho to para substitution, with the cavity diameter of the 4-chloro-substitued polymer similar to that of polystyrene, 6.15 ± 0.38 μm. Size distribution diagrams are presented in Figure S1 of the Supporting Information (SI).

Several authors reported that an increase of the polymer molecular weight usually leads to larger pore sizes [44,45]. However, in our poly(x-chlorostyrene) systems there was no significant variation between the diameters of the polystyrene and the chlorine substituted polymers. That is, the presence of the chloro-substitution could have increase the pore size, due to the influence of the side groups, avoiding the use of high molecular weight polystyrene samples to obtain similar pore size.

The conformational entropy based on the Voronoi tessellation (Figure 4) indicated that the maximum order was reached for poly(4-chlorostyrene), with a value of entropy equal to 1.035, similar to that obtained for polystyrene (S = 1.033). The other two isomers showed higher entropy values, being equal to 1.294 and 1.112 for P2ClS and P3ClS, respectively. As expected, the irregular chemical structure of the polymer leads to highly disordered structures. The chlorine substituent in poly(4-chlorostyrene) samples are in the less hindered position, and this makes P4ClS similar to polystyrene, considering both pore size and conformational entropy, even if the molecular weight of the 4-chloro-substituted homopolymer is one magnitude order lower than molecular weight of polystyrene.

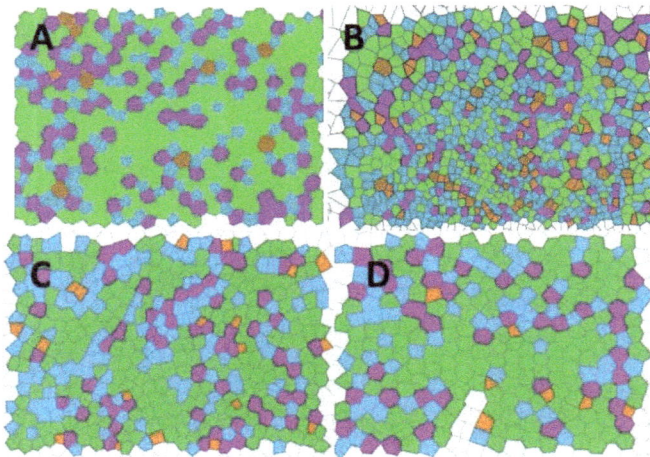

Figure 4. Voronoi tessellation for (**A**) poly(styrene), (**B**) poly(2-chlorostyrene) (**C**) poly(3-chlorostyrene), and (**D**) poly(4-chlorostyrene). (Orange = P_4, Blue = P_5, Green = P_6, Purple = P_7, Brown = P_8).

3.2. Formation of Breath Figures with Poly(x-chlorostyrene-co-styrene) Copolymers

Similar to the homopolymers, statistical copolymers of x-chlorostyrene and styrene, i.e., poly(x-chlorostyrene-co-styrene) were synthesized by free radical polymerization (0.5 mol% AIBN, 60 °C), at the average molecular weight of poly(x-chlorostyrene)s and the fraction of styrene (F_s) in the copolymer, summarized in the Table 2.

The SEM images of the breath figure arrays for the statistical copolymers are compared to polystyrene in Figure 5. As shown in this Figure, the patterns obtained from the copolymers were more homogeneous compared to those of the corresponding homopolymers. As for the homopolymers, polystyrene and poly(4-chlorostyrene-co-styrene) (P(4Cl-co-S)) present more similarities between them.

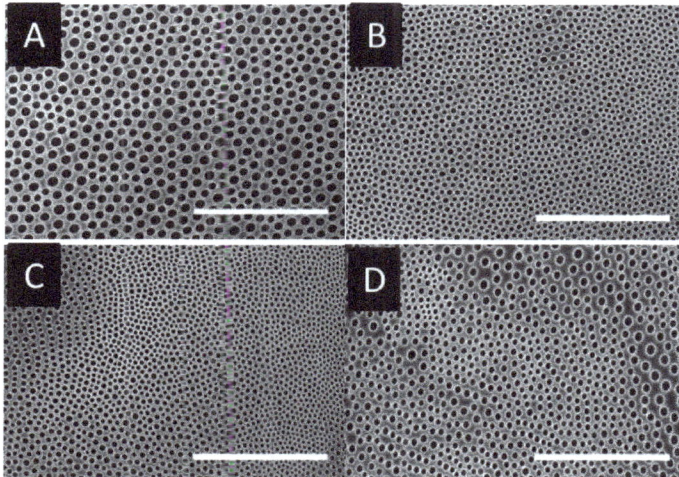

Figure 5. SEM images for (**A**) poly(styrene), (**B**) poly(2-chlorostyrene-co-styrene), (**C**) poly(3-chlorostyrene-co-styrene), and (**D**) poly(4-chlorostyrene-co-styrene). (Scale bar = 100 μm).

All of the patterns obtained from the copolymers showed a significant decrease in pore diameter when compared to the homopolymers (Table 3). It is important to notice the improvement in the homogeneity of the honeycomb arrays of the copolymers based on 2-chlorostyrene and 3-chlorostyrene (Figure 5B,D) when compared with those formed from the homopolymers (Figure 3). In addition, these two copolymers with greater steric hindrance groups (with ortho and metha substitutions) presented similar molecular weight, slightly below of the 4-chloro-substituted copolymer, which in this case could have enhanced the slight reduction on pore size. The size distribution diagrams (Figure S2 in the Supplementary) showed a broad distribution that also affected the standard deviation of the pore size diameter that could be attributed to the aforementioned decrease in the size. Homopolymer and copolymer samples presented a good pore distribution in the film surface lower than several cell sizes (commonly, around 10 μm), and these kinds of patterns have been successfully used by several authors for a contact guidance in a cellular growth [12,23,33,34].

Table 3. Pore diameter of poly(x-chlorostyrene-co-styrene) samples compared to the homopolymers.

Copolymer Sample	Pore Diameter (μm)		Homopolymer Sample
Polystyrene	6.15 ± 0.38		Polystyrene
P(2ClS-co-S)	3.47 ± 0.44	4.1 ± 0.45	P2ClS
P(3ClS-co-S)	2.9 ± 0.46	5.25 ± 0.35	P3ClS
P(4ClS-co-S)	4.56 ± 0.55	6.26 ± 0.20	P4ClS

The reduction of the pore size causes a drastic increase in the number of holes present in the patterns. This fact could induce an increase in the disorder of the system. In order to evaluate the influence of the chloro-substitution on the disorder of obtained structures, the conformational entropy was calculated from the Voronoi tessellation (Figure 6). Values are tabulated in Table 4 and compared to those of the homopolymers.

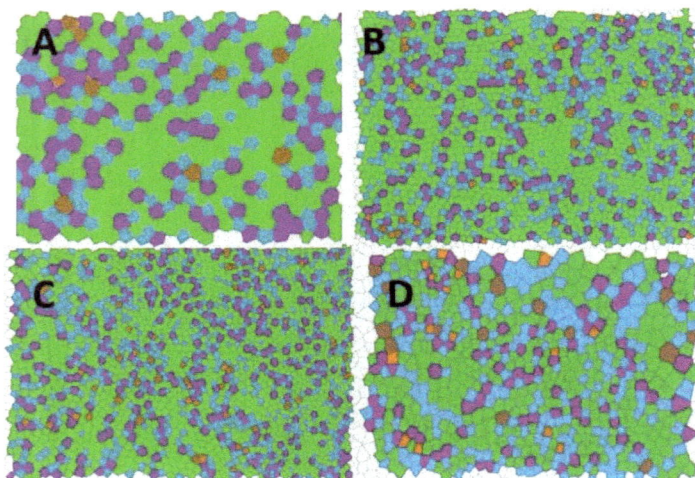

Figure 6. Voronoi tessellation for (**A**) poly(styrene), (**B**) poly(2-chlorostyrene-co-styrene), (**C**) poly(3-chlorostyrene-co-styrene), and (**D**) poly(4-chlorostyrene-co-styrene). (Orange = P_4, Blue = P_5, Green = P_6, Purple = P_7, Brown = P_8).

Table 4. Pore diameter of poly(x-chlorostyrene-co-styrene) samples compared with the homopolymers.

Copolymer Sample	Conformational Entropy		Homopolymer Sample
Polystyrene	1.033		Polystyrene
P(2ClS-co-S)	1.061	1.291	P2ClS
P(3ClS-co-S)	1.089	1.112	P3ClS
P(4ClS-co-S)	1.133	1.035	P4ClS

The reduction in pore size positively affected the conformational entropy when P(2ClS-co-S) and P(3ClS-co-S) was studied. For these copolymers, the order of the obtained patterns increased significantly when compared to the homopolymer samples. However, the variation on the 4-chlorostyrene–based samples presented an opposite behavior, and a considerable increase on the entropy was obtained, so in these samples the reduction in the pore size induced an increase on the surface disorder.

The octanol–water partition coefficient (P_{ow}) is a well-known indicator of the hydrophobicity of an organic compound that could be used for comparing the benzene and chlorobenzene substitution. These compounds present a P_{ow} of 2.16 and 2.84, respectively [46,47]. The higher hydrophobicity of the x-chlorosubstituted samples could increase the interface tension, inducing a reduction in the drop size and, therefore, a reduction in the pore size when compared to the polystyrene. In addition, this effect is more relevant to compounds where substitution results in greater steric hindrance.

3.3. Formation of Breath Figures in Poly(x-chlorostyrene) Isomers and Polystyrene Blends

Several authors have reported the use of blends to develop breath figures [48–50]. In view of this, blends of polystyrene with poly(x-chlorostyrene) homopolymers were used to fabricate breath figure patterns. A 90/10 PS/poly(x-chlorostyrene) composition was chosen, considering that most commonly used approach for breath figure used a polymer matrix (polystyrene) and an additive polymer in order to vary the formed patterns [9,49,51,52]; in this case, poly(x-chlorostyrenes)s were used as additives. Both polymers are miscible [53,54] and their miscibility was analyzed by Differential Scanning Calorimetry (DSC) and the obtained glass transition temperatures were summarized in the

Supporting Information (Tables S1–S3). However, as shown in Figure 7, even if the polymers were miscible at this composition, the obtained patterns were less regular when compared to the polystyrene matrix. Therefore, this approach, traditionally used for several systems to improve the honeycomb patterns, was not adequate for these materials.

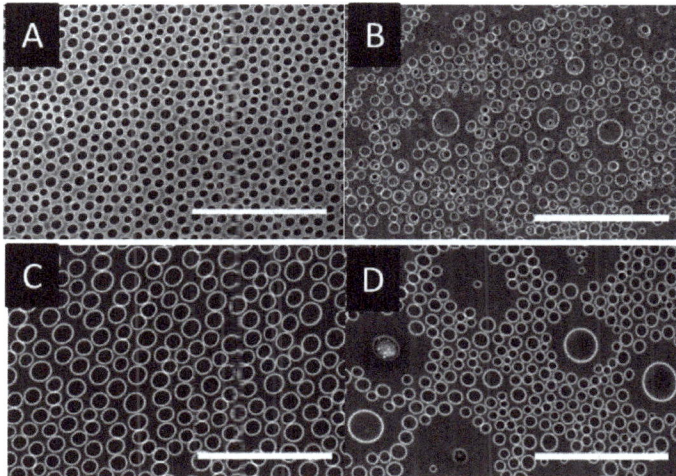

Figure 7. SEM images for (**A**) poly(styrene) and blends (90/10) (**B**) polystyrene/poly(2-chlorostyrene, (**C**) polystyrene/poly(3-chlorostyrene), and (**D**) polystyrene/poly(4-chlorostyrene). (Scale bar = 100 μm).

However, a regular pattern was obtained for polystyrene/poly(3-chlorostyrene) blends, with a pore size of 10.62 ± 1.53 μm and conformational entropy equal to 1.18 (Figure S3). The entropy value of this sample was similar to that reported for the homopolymers, but the pore size was found to be even higher than for polystyrene. This discrepancy on the behavior of P3ClS has been also observed when properties of P2ClS, P3ClS and P4ClS were studied. In this line, investigations devoted to the study of local dynamic of poly(x-chlorostyrene)s reported by Casalini et al. [55] described that segmental relaxations depend on the chloro-substitution in the phenyl ring, leading to unexpected results for 3-chlorostyrene. Indeed, this work asserts that the reorientation of pendant of P4ClS was independent of the movement in the main chain of the polymer, and for P2ClS, the rotation of the side group was restricted by the coupling to rotation of the backbone. However, when P3ClS was evaluated, clear results could not be observed, given the constraints of the pendant group intermediate, between P2ClS and P4ClS. In our case, it was observed (Figure 7) that this side group induced a change in the water/polymer interaction. In the case of PS/P2ClS, the lack of mobility of the 2-chloro substitution prevented a uniform rearrangement of the chains forming irregular patterns (Figure 7B). On the other hand, the absence of hindrance for PS/P4ClS blends induced the coalescence of the water droplets forming domains with higher pore size and other domains with smaller pores (Figure 7D). However, in the case of PS/P3ClS, similar to the behavior observed for dielectric relaxation, an intermediate characteristic of 3-chloro substitution allowed the formation of regular arrays (Figure 7C). Moreover, the patterns obtained for polystyrene/poly(3-chlorostyrene) (Figure 7C) presented an average pore size, higher than 10 μm, that could be useful for the culture of larger cells [27].

In order to apply this procedure to larger areas, it is important to notice that the use of raw materials, such as homopolymers or statistical copolymers, leads to homogeneous surfaces of adequate diameter range to be used for future cell cultures. However, in these systems, contrary to many other studies, the use of a polymer blend was not successful. Thus, x-chlorosubstitution and random copolymerization are preferred routes rather than traditional approaches, such as blending.

4. Conclusions

In the present study, a series of honeycomb films presenting different pore sizes and conformational entropies were developed using the static breath figure method. This simple film fabrication technique added to the variety in pores' sizes, making these systems highly suitable for cell culture uses.

Supplementary Materials: The following are available online at http://www.mdpi.com/1996-1944/12/1/167/s1, Figure S1: Pore size distribution in THF at 90% RH for: (A) polystyrene, (B) poly(2-chlorostryrene), (C) poly(3-chlorostyrene) and (D) poly(4-chlorostyrene), Figure S2: Pore size distribution in THF at 90% RH for: (A) polystyrene, (B) poly(2-chlorostryrene-co-styrene), (C) poly(3-chlorostryrene-co-styrene) and (D) poly(4-chlorostyrene-co-styrene), Figure S3: Voronoi tessellation for polystyrene/poly(3-chlorostyrene blend. (Orange = P4, Blue = P5, Green = P6, Purple = P7, Brown = P8). Table S1: Glass transition temperatures of PS and P(xClS)s, Table S2: Glass transition temperatures of P(S-co-xClS)s, Table S3: Glass transition temperatures of PS/P(xClS) blends 90/10.

Author Contributions: Conceptualization, L.R.-R.; validation, L.R.-R. and L.P.-Á.; formal analysis, L.R.-R. and J.S.-B; investigation, L.R.-R., J.S.-B.; writing—original draft preparation, L.R.-R. and J.S.-B.; writing—review and editing, L.R.-R. L.P.-Á. and V.A.; supervision, V.A. and J.L.V.-V.; project administration, L.R.-R., J.L.V.-V.; funding acquisition, J.L.V.-V.

Funding: This research was funded by the Government of Basque Country, grant ELKARTEK FRONTIERS KK-2017/0096 and grant Grupos de Investigación IT718-13.

Acknowledgments: Authors are grateful to technical and human support provided by SGIKER (UPV/EHU, MICINN, GV/EJ, ERDF, and ESF).

Conflicts of Interest: The authors declare no conflict of interest.

References

1. Zhang, N.; Li, J.; Ni, D.; Sun, K. Preparation of honeycomb porous La0.6Sr0.4Co0.2Fe0.8O3−δ–Gd0.2Ce0.8O2−δ composite cathodes by breath figures method for solid oxide fuel cells. *Appl. Surf. Sci.* **2011**, *258*, 50–57. [CrossRef]

2. Li, J.; Zhang, N.; Ni, D.; Sun, K. Preparation of honeycomb porous solid oxide fuel cell cathodes by breath figures method. *Int. J. Hydrogen Energy* **2011**, *36*, 7641–7648. [CrossRef]

3. Lu, Y.; Zhao, B.; Ren, Y.; Xiao, G.; Wang, X.; Li, C. Water-assisted formation of novel molecularly imprinted polymer membranes with ordered porous structure. *Polymer (Guildf)* **2007**, *48*, 6205–6209. [CrossRef]

4. Mansouri, J.; Yapit, E.; Chen, V. Polysulfone filtration membranes with isoporous structures prepared by a combination of dip-coating and breath figure approach. *J. Memb. Sci.* **2013**, *444*, 237–251. [CrossRef]

5. Sakatani, Y.; Boissière, C.; Grosso, D.; Nicole, L.; Soler-Illia, G.J.A.A.; Sanchez, C. Coupling Nanobuilding Block and Breath Figures Approaches for the Designed Construction of Hierarchically Templated Porous Materials and Membranes. *Chem. Mater.* **2008**, *20*, 1049–1056. [CrossRef]

6. Wan, L.; Li, J.; Ke, B.; Xu, Z. Ordered Microporous Membranes Templated by Breath Figures for Size-Selective Separation. *J. Am. Chem. Soc.* **2012**, *134*, 95–98. [CrossRef] [PubMed]

7. Ou, Y.; Lv, C.; Yu, W.; Mao, Z.; Wan, L.; Xu, Z. Fabrication of Perforated Isoporous Membranes via a Transfer-Free Strategy: Enabling High-Resolution Separation of Cells. *ACS Appl. Mater. Interfaces* **2014**, *6*, 22400–22407. [CrossRef]

8. Kon, K.; Brauer, C.N.; Hidaka, K.; Löhmannsröben, H.-G.; Karthaus, O. Preparation of patterned zinc oxide films by breath figure templating. *Langmuir* **2010**, *26*, 12173–12176. [CrossRef]

9. De León, A.S.; Garnier, T.; Jierry, L.; Boulmedais, F.; Muñoz-Bonilla, A.; Rodríguez-Hernández, J. Enzymatic Catalysis Combining the Breath Figures and Layer-by-Layer Techniques: Toward the Design of Microreactors. *ACS Appl. Mater. Interfaces* **2015**, *7*, 12210–12219. [CrossRef]

10. Chen, P.-C.; Wan, L.-S.; Ke, B.-B.; Xu, Z.-K. Honeycomb-patterned film segregated with phenylboronic acid for glucose sensing. *Langmuir* **2011**, *27*, 12597–12605. [CrossRef]

11. Nishikawa, T.; Nishida, J.; Ookura, R.; Nishimura, S.I.; Wada, S.; Karino, T.; Shimomura, M. Honeycomb-patterned thin films of amphiphilic polymers as cell culture substrates. *Mater. Sci. Eng. C* **1999**, *8–9*, 495–500. [CrossRef]

12. Tsuruma, A.; Tanaka, M.; Yamamoto, S.; Shimomura, M. Control of neural stem cell differentiation on honeycomb films. *Colloids Surf. A Physicochem. Eng. Asp.* **2008**, *313–314*, 536–540. [CrossRef]

13. Chen, S.; Gao, S.; Jing, J.; Lu, Q. Designing 3D Biological Surfaces via the Breath-Figure Method. *Adv. Healthc. Mater.* **2018**, *7*, 1701043. [CrossRef] [PubMed]

14. Bunz, U.H.F. Breath figures as a dynamic templating method for polymers and nanomaterials. *Adv. Mater.* **2006**, *18*, 973–989. [CrossRef]

15. Song, L.; Bly, R.K.; Wilson, J.N.; Bakbak, S.; Park, J.O.; Srinivasarao, M.; Bunz, U.H.F. Facile Microstructuring of Organic Semiconducting Polymers by the Breath Figure Method: Hexagonally Ordered Bubble Arrays in Rigid Rod-Polymers. *Adv. Mater.* **2004**, *16*, 115–118. [CrossRef]

16. Nishikawa, T.; Nonomura, M.; Arai, K.; Hayashi, J.; Sawadaishi, T.; Nishiura, Y.; Hara, M.; Shimomura, M. Micropatterns Based on Deformation of a Viscoelastic Honeycomb Mesh. *Langmuir* **2003**, *19*, 6193–6201. [CrossRef]

17. Escalé, P.; Rubatat, L.; Billon, L.; Save, M. Recent advances in honeycomb-structured porous polymer films prepared via breath figures. *Eur. Polym. J.* **2012**, *48*, 1001–1025. [CrossRef]

18. Hernández-Guerrero, M.; Stenzel, M.H. Honeycomb structured polymer films via breath figures. *Polym. Chem.* **2012**, *3*, 563–577. [CrossRef]

19. Chen, C.S. Geometric Control of Cell Life and Death. *Science (80-.)* **1997**, *276*, 1425–1428. [CrossRef]

20. Flemming, R.G.; Murphy, C.J.; Abrams, G.A.; Goodman, S.L.; Nealey, P.F. Effects of synthetic micro- and nano-structured surfaces on cell behavior. *Biomaterials* **1999**, *20*, 573–588. [CrossRef]

21. Walboomers, X.F.; Croes, H.J.E.; Ginsel, L.A.; Jansen, J.A. Growth behavior of fibroblasts on microgrooved polystyrene. *Biomaterials* **1998**, *19*, 1861–1868. [CrossRef]

22. Liu, X.; Liu, R.; Cao, B.; Ye, K.; Li, S.; Gu, Y.; Pan, Z.; Ding, J. Subcellular cell geometry on micropillars regulates stem cell differentiation. *Biomaterials* **2016**, *111*, 27–39. [CrossRef] [PubMed]

23. Kawano, T.; Sato, M.; Yabu, H.; Shimomura, M. Honeycomb-shaped surface topography induces differentiation of human mesenchymal stem cells (hMSCs): Uniform porous polymer scaffolds prepared by the breath figure technique. *Biomater. Sci.* **2014**, *2*, 52–56. [CrossRef] [PubMed]

24. Clark, P.; Connolly, P.; Curtis, A.S.; Dow, J.A.; Wilkinson, C.D. Cell guidance by ultrafine topography in vitro. *J. Cell Sci.* **1991**, *99 Pt 1*, 73–77.

25. Abagnale, G.; Sechi, A.; Steger, M.; Zhou, Q.; Kuo, C.C.; Aydin, G.; Schalla, C.; Müller-Newen, G.; Zenke, M.; Costa, I.G.; et al. Surface Topography Guides Morphology and Spatial Patterning of Induced Pluripotent Stem Cell Colonies. *Stem Cell Rep.* **2017**, *9*, 654–666. [CrossRef] [PubMed]

26. Wu, X.; Wang, S. Regulating MC3T3-E1 Cells on Deformable Poly(ε-caprolactone) Honeycomb Films Prepared Using a Surfactant-Free Breath Figure Method in a Water-Miscible Solvent. *ACS Appl. Mater. Interfaces* **2012**, *4*, 4966–4975. [CrossRef]

27. Yamamoto, S.; Tanaka, M.; Sunami, H.; Ito, E.; Yamashita, S.; Morita, Y.; Shimomura, M. Effect of honeycomb-patterned surface topography on the adhesion and signal transduction of porcine aortic endothelial cells. *Langmuir* **2007**, *23*, 8114–8120. [CrossRef]

28. Fukuda, J.; Sakai, Y.; Nakazawa, K. Novel hepatocyte culture system developed using microfabrication and collagen/polyethylene glycol microcontact printing. *Biomaterials* **2006**, *27*, 1061–1070. [CrossRef]

29. Cristallini, C.; Cibrario Rocchietti, E.; Accomasso, L.; Folino, A.; Gallina, C.; Muratori, L.; Pagliaro, P.; Rastaldo, R.; Raimondo, S.; Saviozzi, S.; et al. The effect of bioartificial constructs that mimic myocardial structure and biomechanical properties on stem cell commitment towards cardiac lineage. *Biomaterials* **2014**, *35*, 92–104. [CrossRef]

30. Choi, H.; Tanaka, M.; Hiragun, T.; Hide, M.; Sugimoto, K. Non-tumor mast cells cultured in vitro on a honeycomb-like structured film proliferate with multinucleated formation. *Nanomed. Nanotechnol. Biol. Med.* **2014**, *10*, 313–319. [CrossRef]

31. Tanaka, M.; Nishikawa, K.; Okubo, H.; Kamachi, H.; Kawai, T.; Matsushita, M.; Todo, S.; Shimomura, M. Control of hepatocyte adhesion and function on self-organized honeycomb-patterned polymer film. *Colloids Surf. A Physicochem. Eng. Asp.* **2006**, *284–285*, 464–469. [CrossRef]

32. Sunami, H.; Ito, E.; Tanaka, M.; Yamamoto, S.; Shimomura, M. Effect of honeycomb film on protein adsorption, cell adhesion and proliferation. *Colloids Surf. A Physicochem. Eng. Asp.* **2006**, *284–285*, 548–551. [CrossRef]

33. Arai, K.; Tanaka, M.; Yamamoto, S.; Shimomura, M. Effect of pore size of honeycomb films on the morphology, adhesion and cytoskeletal organization of cardiac myocytes. *Colloids Surf. A Physicochem. Eng. Asp.* **2008**, *313–314*, 530–535. [CrossRef]

34. Tsuruma, A.; Tanaka, M.; Fukushima, N.; Shimomura, M. Morphological changes of neurons on self-organized honeycomb patterned films. *Kobunshi Ronbunshu* **2004**, *61*, 628–633. [CrossRef]

35. Annabi, N.; Tsang, K.; Mithieux, S.M.; Nikkhah, M.; Ameri, A.; Khademhosseini, A.; Weiss, A.S. Highly Elastic Micropatterned Hydrogel for Engineering Functional Cardiac Tissue. *Adv. Funct. Mater.* **2013**, *23*, 4950–4959. [CrossRef]

36. Rogers, J.D.; Long, R.L. Modeling hollow fiber membrane contactors using film theory, Voronoi tessellations, and facilitation factors for systems with interface reactions. *J. Memb. Sci.* **1997**, *134*, 1–17. [CrossRef]

37. Broughton, J.; Davies, G.A. Porous cellular ceramic membranes: A stochastic model to describe the structure of an anodic oxide membrane. *J. Memb. Sci.* **1995**, *106*, 89–101. [CrossRef]

38. Alinchenko, M.G.; Anikeenko, A.V.; Medvedev, N.N.; Voloshin, V.P.; Mezei, M.; Jedlovszky, P. Morphology of Voids in Molecular Systems. A Voronoi–Delaunay Analysis of a Simulated DMPC Membrane. *J. Phys. Chem. B* **2004**, *108*, 19056–19067. [CrossRef]

39. Limaye, A.; Narhe, R.; Dhote, A.; Ogale, S. Evidence for Convective Effects in Breath Figure Formation on Volatile Fluid Surfaces. *Phys. Rev. Lett.* **1996**, *76*, 3762–3765. [CrossRef]

40. Steyer, A.; Guenoun, P.; Beysens, D.; Knobler, C.M. Two-dimensional ordering during droplet growth on a liquid surface. *Phys. Rev. B* **1990**, *42*, 1086–1089. [CrossRef]

41. Steyer, A.; Guenoun, P.; Beysens, D.; Review, P.; Steyer, A.; Guenoun, P.; Beysens, D. Hexatic and fat-fractal structures for water droplets condensing on oil. *Phys. Rev. E* **1993**, *48*, 428–431. [CrossRef]

42. Choi, Y.W.; Lee, H.; Song, Y.; Sohn, D. Colloidal stability of iron oxide nanoparticles with multivalent polymer surfactants. *J. Colloid Interface Sci.* **2015**, *443*, 8–12. [CrossRef] [PubMed]

43. Song, L.; Sharma, V.; Park, J.O.; Srinivasarao, M. Characterization of ordered array of micropores in a polymer film. *Soft Matter* **2011**, *7*, 1890. [CrossRef]

44. Lin, C.-L.; Tung, P.-H.; Chang, F.-C. Synthesis of rod-coil diblock copolymers by ATRP and their honeycomb morphologies formed by the 'breath figures' method. *Polymer (Guildf)* **2005**, *46*, 9304–9313. [CrossRef]

45. Muñoz-Bonilla, A.; Fernández-García, M.; Rodríguez-Hernández, J. Towards hierarchically ordered functional porous polymeric surfaces prepared by the breath figures approach. *Prog. Polym. Sci.* **2014**, *39*, 510–554. [CrossRef]

46. Khaledi, M.G.; Breyer, E.D. Quantitation of hydrophobicity with micellar liquid chromatography. *Anal. Chem.* **1989**, *61*, 1040–1047. [CrossRef]

47. Sasaki, T.; Tanaka, S. Adsorption behavior of some aromatic compounds on hydrophobic magnetite for magnetic separation. *J. Hazard. Mater.* **2011**, *196*, 327–334. [CrossRef]

48. De León, A.S.; Muñoz-Bonilla, A.; Fernández-García, M.; Rodríguez-Hernández, J. Breath figures method to control the topography and the functionality of polymeric surfaces in porous films and microspheres. *J. Polym. Sci. Part A Polym. Chem.* **2012**, *50*, 851–859. [CrossRef]

49. S. de León, A.; del Campo, A.; Fernández-García, M.; Rodríguez-Hernández, J.; Muñoz-Bonilla, A.; Leo, A.S. De Fabrication of Structured Porous Films by Breath Figures and Phase Separation Processes: Tuning the Chemistry and Morphology Inside the Pores Using Click Chemistry. *ACS Appl. Mater. Interfaces* **2013**, *5*, 3943–3951. [CrossRef]

50. Muñoz-Bonilla, A.; Ibarboure, E.; Papon, E.; Rodriguez-Hernandez, J. Self-Organized Hierarchical Structures in Polymer Surfaces: Self-Assembled Nanostructures within Breath Figures. *Langmuir* **2009**, *25*, 6493–6499. [CrossRef]

51. De León, A.S.; Rodríguez-Hernández, J.; Cortajarena, A.L. Honeycomb patterned surfaces functionalized with polypeptide sequences for recognition and selective bacterial adhesion. *Biomaterials* **2013**, *34*, 1453–1460. [CrossRef] [PubMed]

52. Farbod, F.; Pourabbas, B.; Sharif, M. Direct breath figure formation on PMMA and superhydrophobic surface using in situ perfluoro-modified silica nanoparticles. *J. Polym. Sci. Part B Polym. Phys.* **2013**, *51*, 441–451. [CrossRef]

53. Alexandrovich, P.S.; Karasz, F.E.; Macknight, W.J. Dielectric study of polymer compatibility: Blends of polystyrene/poly-2-chlorostyrene. *J. Macromol. Sci. Part B* **1980**, *17*, 501–516. [CrossRef]

54. Leffingwell, J.; Bueche, F. Molecular Motion in 2-Chlorostyrene-Styrene Copolymers from Dielectric Measurements. *J. Appl. Phys.* **1968**, *39*, 5910–5912. [CrossRef]

55. Casalini, R.; Roland, C.M. Effect of Regioisomerism on the Local Dynamics of Polychlorostyrene. *Macromolecules* **2014**, *47*, 4087–4093. [CrossRef]

![materials logo] *materials*

Article

Creation of Superhydrophobic and Superhydrophilic Surfaces on ABS Employing a Nanosecond Laser

Cristian Lavieja [1], Luis Oriol [2] and José-Ignacio Peña [1],*

[1] Instituto de Ciencia de Materiales de Aragón, Universidad de Zaragoza-CSIC Dpto. Ciencia y Tecnología de Materiales y Fluidos Maria de Luna 3, 50018 Zaragoza, Spain; clavieja@unizar.es

[2] Instituto de Ciencia de Materiales de Aragón, Universidad de Zaragoza-CSIC Dpto. Química Orgánica-Facultad de Ciencias Pedro Cerbuna 12, 50009 Zaragoza, Spain; loriol@unizar.es

* Correspondence: jipena@unizar.es

Received: 27 November 2018; Accepted: 12 December 2018; Published: 14 December 2018

Abstract: A nanosecond green laser was employed to obtain both superhydrophobic and superhydrophilic surfaces on a white commercial acrylonitrile-butadiene-styrene copolymer (ABS). These wetting behaviors were directly related to a laser-induced superficial modification. A predefined pattern was not produced by the laser, rather, the entire surface was covered with laser pulses at 1200 DPI by placing the sample at different positions along the focal axis. The changes were related to the laser fluence used in each case. The highest fluence, on the focal position, induced a drastic heating of the material surface, and this enabled the melted material to flow, thus leading to an almost flat superhydrophilic surface. By contrast, the use of a lower fluence by placing the sample 0.8 µm out of the focal position led to a poor material flow and a fast cooling that froze in a rugged superhydrophobic surface. Contact angles higher than 150° and roll angles of less than 10° were obtained. These wetting behaviors were stable over time.

Keywords: nanosecond laser surface modification; ABS (Acrylonitrile-Butadiene-Styrene); surface wettability; superhydrophobic; superhydrophilic

---

1. Introduction

The wettability of solid surfaces is of particular interest because of the wide range of applications derived from the control of this property, which is directly related to the surface structure or roughness, and the chemical interactions with a liquid. Nature has provided some examples of structures that affect the wettability properties, such as self-cleaning superhydrophobic surfaces on the Lotus leaf [1], hierarchical structures on plants [2], butterflies that exhibit directional wettability properties [3], or the adhesion properties of the Gecko toe [4]. The surface interaction of a water drop deposited on a surface is characterised by a contact angle (CA). This angle, θ is measured between the solid/liquid and liquid/air interfaces. A surface is considered to be hydrophilic if its static CA is below 90°, and superhydrophilic if this CA is lower than 10°. By contrast, surfaces with a static CA higher than 90° are hydrophobic, and if the CA is higher than 150°, the surface can be considered as superhydrophobic [5]. Additionally, a non-equilibrium situation is measured by defining the advancing and receding angle of a drop sliding on a vertical surface. The contact angle hysteresis (CAH) measures the difference between the two angles. A low CAH is exhibited by a surface on which the drop descends easily. In contrast, a surface with a relatively high CAH has a strong interaction with the drop (i.e., the drop tends to adhere to the surface) [6,7]. The roughness of the surface influences the wettability according to either the Wenzel model [8], where the liquid fills all of the space, or the Cassie-Baxter model, which considers that some air bubbles are trapped between the drop and the surface [9].

Control of the wettability properties of materials is widely studied and applied in numerous fields [10]. For example, there is special interest in surfaces that have self-cleaning properties [11], anti-frosting and ice-resistance behavior [12,13], anti-fouling [14,15], anti-corrosion [16,17], or transparent and anti-reflective properties [18,19]. Different chemical coatings or physical patterning processes have been used to mimic surfaces with these properties, especially on polymeric materials. The proposed methods usually work properly at a laboratory scale, but are challenging when implemented for mass production. However, in the last two decades, an increasing number of techniques have demonstrated the potential of scale-up material processing for the control of surface properties [20]. One interesting industrial strategy to change the surface topography is the use of a laser beam as a micromachining tool [21].

Lasers have been widely used to tune the wettability properties of materials like metal [22–27], silicon [28,29], or different polymeric materials [30–32]. Various approaches have been employed to change polymer wetting behavior using lasers, in particular, the chemical activation of the surface [33], controlled isomerisation of photoresponsive moieties [34], or nano- and micro-patterning [35–38]. Different strategies can be studied to modify the material surface in a controlled way using a patterning approach [39]. For example, ultra-short lasers are used to produce very precise ablation, thus reducing the thermal impact and enabling highly defined patterns [40,41] or the creation of ripples [42,43]. However, the use of these lasers does not extend to the manufacturing industry, because of their high cost. Other proposed techniques involve processes like beam shaping control [44] or micro-machining of the injection mold [45,46]. The use of nanosecond pulse-width lasers to modify the wettability of polymers, particularly ABS, has not been widely reported. These lasers have the advantage of a low cost compared with the ultra-short lasers. In the work described here, a nanosecond green laser was used to modify the roughness of a commercial ABS—and thus its wetting properties—in order to obtain both superhydrophobic and superhydrophilic behaviour.

2. Materials and Methods

The laser used was a Nd:YVO$_4$ nanosecond laser (Trumark 6230 from Trumpf, Ditzingen, Germany). The laser generates a gaussing beam with no defined polarization at a wavelength of 532 nm. A 100 mm lens was used to provide a beam spot size with a focus of 30 μm. The pulse width was 8 ns, operating at a pulse frequency of 15 kHz. The power of the laser was 3.5 W, providing a fluence in focus of 34.4 J/cm^2, that is, far above the ablation threshold of the ABS, which is 0.3 J/cm^2 [47]. The laser source was attached to a motorised arm that was able to move the system in the vertical axis with micro-scale precision. Furthermore, the system had a beam scanner controlled by CAD software (TruTops Mark 2.6).

The material selected was a white commercial ABS employed in the automotive and home appliance industries (Elix P2H-AT, ELIX Polymers, La Pineda, Spain); in particular, in this last industrial sector, the control of the wettability properties can be of interest for different applications. The ABS samples were cleaned with isopropanol prior to irradiation. The material surface was analyzed using a confocal microscope (Sensofar Plu 2300, Barcelona, Spain), which allows for 3D images to be obtained as well as for the measurement of the physical dimensions of the superficial structures. Environmental scanning electron microscopes (ESEM, QUANTA FEG 250, Thermo Fisher Scientific, Waltham, MA, USA) was also used to provide information about the topography. Information of the chemical modification after laser treatment was obtained by employing X-ray photoelectron spectroscopy (XPS). The equipment was a Kratos AXIS Ultra DLD (Manchester, UK, Mono Al Kα, Power: 120 W (10 mA, 15 kV)). The binding energies were calibrated using the C 1s signal (284.6 eV) of adventitious carbon. The wettability of the material and the stability of the wetting properties were measured using a contact angle goniometer (Theta Lite Optical Tensiometer from Biolin Scientific, Gothenburg, Sweden). The samples were stored under controlled environmental conditions for at least six months (temperature ~22 °C and relative humidity ~40%, UV light (below ~400 nm) was avoided, and special protection against dust was not taken).

3. Results and Discussion

The topography and wetting properties of the material can be tuned by irradiation and by controlling the laser parameters such as the overlap between the pulses, the number of spots per area (DPI), and/or the laser fluence over the material [38,48]. The study described here was carried out at a constant value of 1200 DPI (consequently, the total energy deposited on the surface was constant), and the fluence of each laser pulse was modified by placing the sample in different positions along the focal axis. Specifically, a distance of 2 mm up and down out of the focal plane in successive steps of 0.1 mm was employed. Thus, the fluence ranged from 5 to 35 J/cm^2. As a consequence of this approach, the overlap of the laser pulses changed depending on the sample position.

3.1. Topographical Characterization

The material surface was studied by confocal microscopy before and after the laser treatment. Two types of topography were created, depending on the deposited fluence. Although a clear transition was not detected, a change was observed when the samples were irradiated at around 0.6 mm out of focus. The two general types of generated surfaces corresponding to the irradiation in focus and ±0.8 mm out of the focal position (the results were similar in both directions) are shown in Figure 1, together with the unmarked material. The surface prior to marking was flat (Figure 1a,b). After green laser treatment in focus (Figure 1c,d), the topography that was detected was characterised as being almost flat, although small ripples could be observed (changes from green to pale blue in Figure 1c,d). The presence of some small holes was also observed. Thus, the total surface area had increased slightly in comparison to the unmarked material. By contrast, the ABS treated ±0.8 mm out of the focal position (Figure 1e,f) had a rugged topography characterized by the presence of peaks and valleys in the irradiated area. In this case, the increase in the surface area was significantly higher than in the previous case.

A more detailed image was obtained using ESEM. The images taken by this technique on the unmarked material surface and of the materials marked in focus and 0.8 mm out of focus are presented in Figure 2 (at three different magnifications). Once again, clear differences between the analyzed surfaces were observed. The unmarked material (Figure 2a–c) exhibited a smooth surface. White dots were visible in the image taken at 24,000 × magnification, and these mainly correspond to the TiO$_2$ particles (used as a whitening additive), according to the EDX (Energy Dispersive X-ray spectroscopy) analysis (the intensity of the Ti signal was around three times higher than in the rest of the material). When the surface was observed at the highest magnification, a wavy area was evident, probably due to the injection process. The ABS marked in focus (Figure 2d–f) also had a smooth surface. However, as observed by confocal microscopy, this surface had some holes distributed throughout the area. The presence of white dots corresponding to the TiO$_2$ particles was also observed. In addition, the appearance of a sub-micron dotted structure was detected using the highest magnification, probably associated with the laser treatment. The diameter of these points was less than 100 nm. Finally, the surface of the material marked 0.8 mm out of focus (Figure 2g–i) had a rugged topography. The surface was characterised by the presence of peaks in all directions. A slight tilting of the samples clearly showed that the surface was chaotic and that holes and channels had formed in some of the areas. These areas may be associated with the material filaments that were melted during the laser treatment, and then rapidly solidified. The sub-micron structure of the points was also visible in this sample. Finally, the presence of white points associated with the TiO$_2$ particles was also detected, although these were difficult to observe because of the high roughness of the surface.

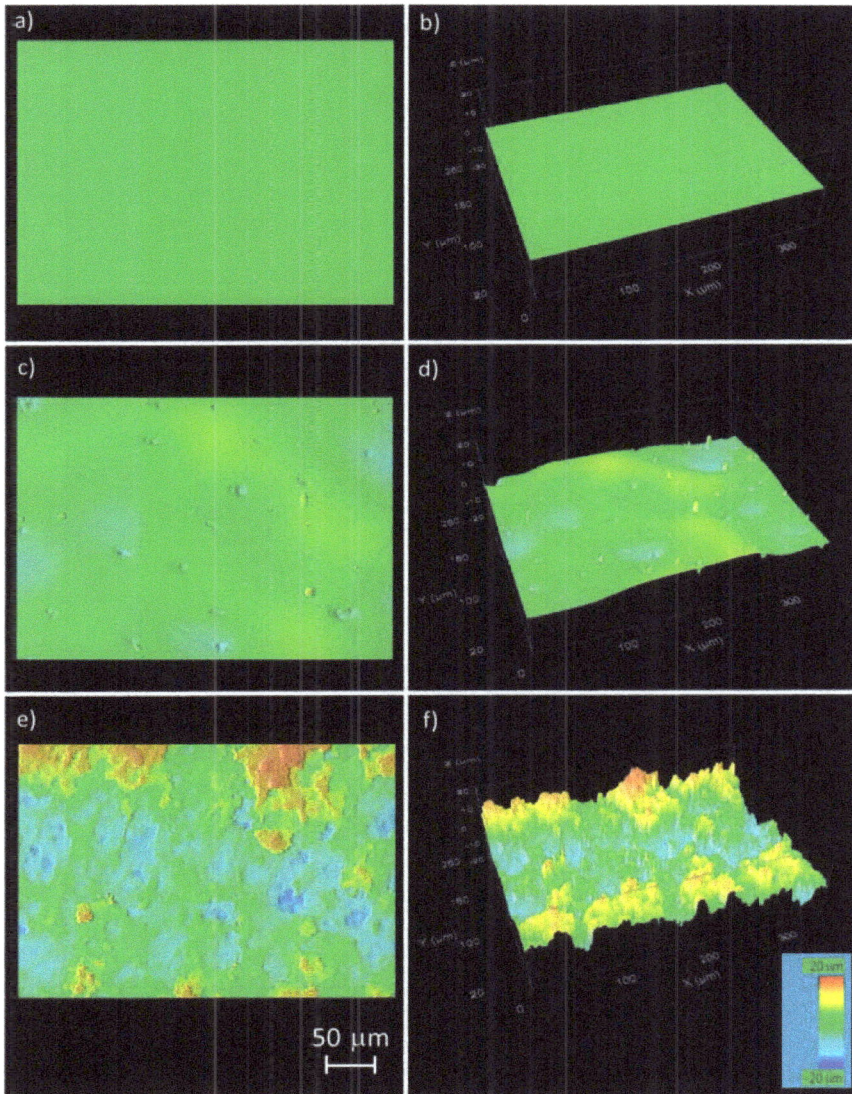

Figure 1. Topographical images of the ABS surface roughness observed by confocal microscopy: (a,b) before laser irradiation; (c,d) after green laser irradiation in focus; (e,f) after green laser irradiation 0 8 mm out of focus.

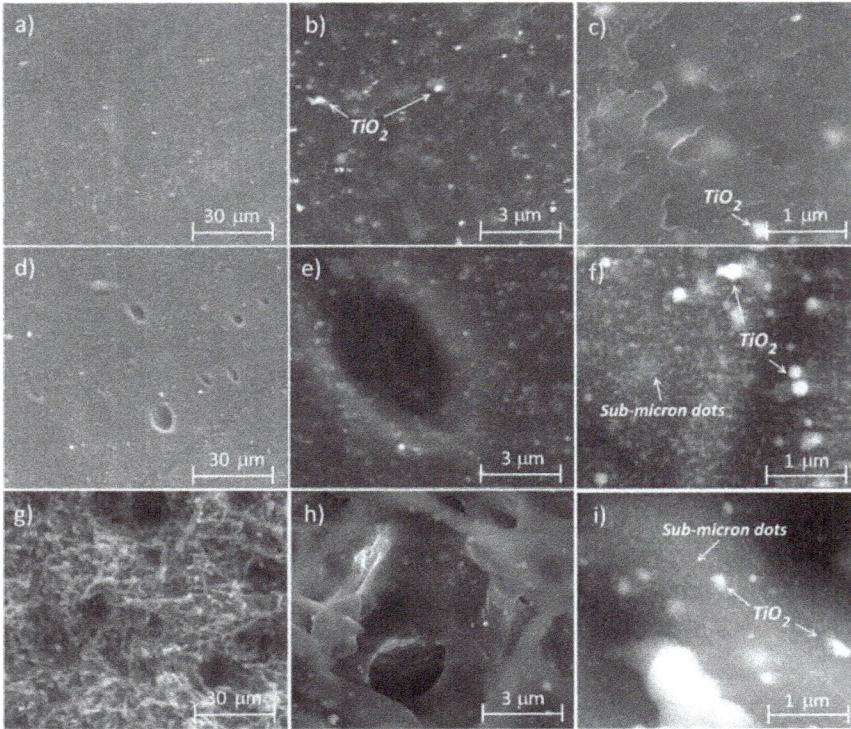

Figure 2. ESEM images of the ABS surface roughness: (**a–c**) before laser irradiation; (**d–f**) after green laser irradiation in focus; (**g–i**) after green laser irradiation 0.8 mm out of focus. Magnification: (**a**) 2800× scale bar 30 μm; (**b**) 24,000×, scale bar 3 μm; (**c**) 80,000×, scale bar 1 μm.

These changes can be explained by the fluence differences on the surface. The highest fluence used (i.e., in focus) causes the heating of the material, which leads to a quasi-pure ablation effect together with the formation of some melted surface. The material was heated up very rapidly and it then began to flow before cooling down, which led to bubbles being trapped within the material, giving the appearance of small holes. By contrast, on positioning the sample out of focus, the temperature reached at the surface was lower and the temperature of the material was not sufficient to achieve flow before it dropped again. Furthermore, the laser generated additional pressure and this led to bubble formation and the ejection of the material. The cooling of the surface caused the freezing of this rugged structure.

In addition to the topographical properties, an XPS analysis was carried out to study the chemical changes on the ABS surface after the laser irradiation. No significant modification on the total percentages of the total atomic percentages was detected after the marking process. However, the signal intensity corresponding to the Ti peak, detected at 455 eV, was higher (around five times) compared with the other peaks after the laser treatment. High-resolution experiments of the region of the binding energies correspond to Ti 2p (Figure 3) showed changes after the laser treatment, while negligible differences were observed between the treated surfaces on and out of focus. Two main peaks were detected, at 459 eV and at 464 eV (although this one is barely visible in the unmarked sample) corresponding to Ti^{4+} $2p^{3/2}$. After the laser irradiation (Figure 3b), another peak at lower binding energies was detected, around 454 eV. According to the literature, this peak can be ascribed to TiC [49–51]. No significant evidence of other peaks after the laser treatment were detected on the analysed samples.

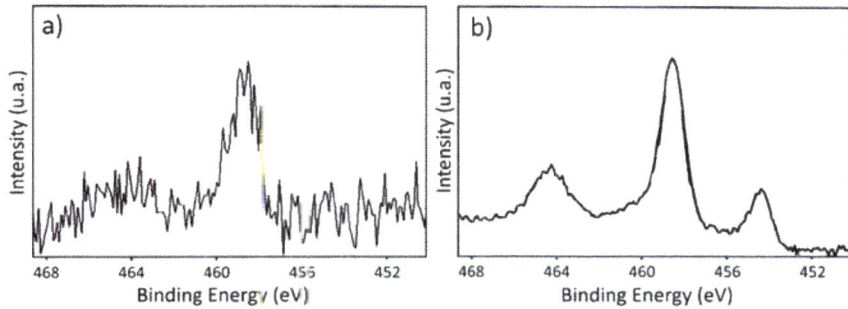

Figure 3. XPS high-resolution spectrum of the Ti region of the ABS: (**a**) before laser irradiation; (**b**) after green laser irradiation.

3.2. Wetting Properties

The wetting properties of the irradiated materials were studied by employing a contact angle goniometer. The contact angles cannot be simulated because of the chaotic topography of the structures created. As a result of the high CA obtained in some cases, the CAH was also measured. For this purpose, the plane containing the droplet was tilted, and the advancing and receding contact angles (θ_A and θ_R) were measured when the droplet started to slide down [7]. The CA of the unmarked material was $62.8 \pm 1.7°$.

The most significant results were obtained after the surface had been processed with the laser in terms of the position along the focal axis, and the corresponding fluence is shown in Table 1. Although only two differentiated structures were observed using confocal microscopy, three different wetting behaviors were detected. The samples marked close to the focus (less than ± 0.5 mm from the focal position) exhibited superhydrophilic behaviour [52]. The water drop wetted the entire surface, and, consequently, the CA was considered to be lower than 10°. By contrast, when the material was treated slightly further away from the focus (± 0.8 mm or ± 0.9 mm), dramatic changes in the wetting properties of the surface were observed. The CA increased to values higher than 150°, and the surface was considered to show superhydrophobic behavior [52]. Furthermore, the measured CAH was less than 10°; the drop did not adhere to the surface and tended to slide when the plane was slightly tilted. Finally, the samples marked out of the focal position by more than ± 1.2 mm also exhibited a high CA value (i.e., close to superhydrophobic behaviour), although in this case, the drop was pinned to the surface and did not slide down upon tilting, even to the vertical position. The temporal evolution of the CA was studied for a period of six months, but significant changes were not detected in the samples.

Table 1. The CA and CAH measured on some of the structures created with the green laser.

Position Respect to Focus (mm)	Fluence (J/cm^2)	Overlap between Pulses	Topography Observed	CA	CAH
0	34.4	29.4	smooth	<10°	-
± 0.4	30.1	34.1	smooth	<10°	-
± 0.6	26.0	38.7	slightly rugged	unstable	-
± 0.8	21.8	43.9	rugged	$168 \pm 3°$	<10°
± 0.9	19.9	46.4	rugged	$170 \pm 3°$	<10°
± 1.2	14.9	53.5	rugged	$153 \pm 5°$	pinned
± 1.3	13.6	55.7	rugged	$143 \pm 5°$	pinned

The images of the water drops on the ABS surfaces irradiated with the green laser and observed by the contact angle goniometer camera are shown in Figure 4. The behavior of a water drop deposited on the untreated material is shown in Figure 4a as a reference. Figure 4b corresponds to a surface treated with the green laser 0.9 mm out of the focal position, and this image represents superhydrophobic behavior. The behavior of a water drop on the surface marked 1.3 mm out of focus is shown in

Figure 4c, and the images were taken at different tilt angles (0°, 45°, and 90°). The water drop exhibited a high adhesion and was pinned to the surface, as can be seen for a tilt angle of 90°. Finally, time-lapse images of a drop on the surface marked by the laser in focus are shown in Figure 4d. The water was dispersed on the surface within one second, thus revealing superhydrophilic behavior. A slow-motion video sequence of water drops falling on the superhydrophilic and on the superhydrophobic surfaces is provided in the Supplementary Materials.

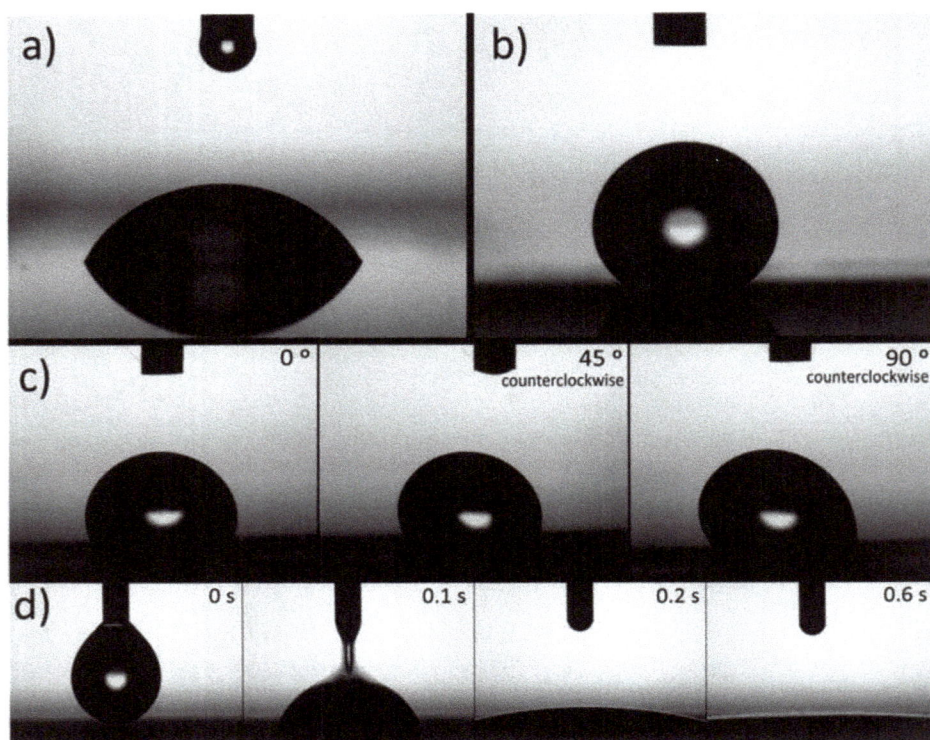

Figure 4. Images of a water drop deposited on ABS: (**a**) untreated surface; (**b**) surface irradiated by laser 0.9 mm out of focus; (**c**) surface irradiated by laser 1.3 mm out of focus and taken at different tilt angles (the camera rotated simultaneously with the surface); (**d**) time-lapse images of a drop on the surface treated by the laser in focus.

It is clear that the wettability of the ABS can be changed by irradiation with a green laser, and three different wetting behaviors can be obtained depending on the irradiation conditions, as follows: a region of superhydrophilic behavior and two regions characterised by high CAs, one associated with low CAH and the other one exhibiting a high adhesion to the surface. These different behaviors (Figure 5) can be achieved depending on the sample position along the focal axis and the corresponding spot width and fluence of the green laser. The superhydrophilic region was associated with the focal position. The topography observed (Figure 1c,d) was characterised as being almost flat. This region is characterised by the highest fluence values (i.e., above 25 J/cm^2). The surfaces treated in the transition zone to the adjacent region showed unstable behaviour; initially, the surface seemed to be hydrophobic, but under a small perturbation (or after several seconds), the drop began to wet the surface in a manner consistent with superhydrophilic behavior, probably because of the inhomogeneous topography observed. The adjacent region showed a high CA together with a low CAH. Also, the fluence associated with this region was less than in the case of the superhydrophilic region, and the laser fluences in the approximate range between 15 and 25 J/cm^2 seemed to produce this behavior. The transition samples to the next region displayed heterogeneous behavior, in which adhered and non-adhered drops were observed on the same marked surface. Finally, the last region showed hydrophobic behavior and a high CAH. The lowest fluence studied gave rise to this behavior, namely values below 15 J/cm^2.

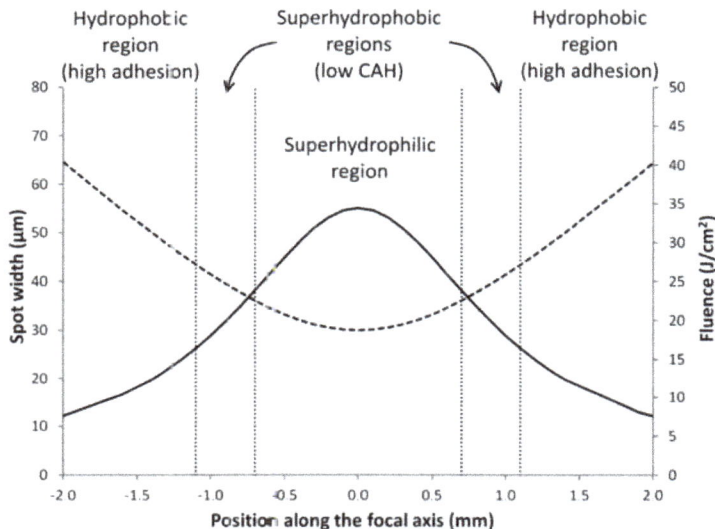

Figure 5. Different wetting behaviors in terms of the sample position along the focal axis compared with the spot width (dashed line, units on the left axis) and fluence (continuous line, units on the right axis) of the green laser.

It is possible to correlate the topography of the surfaces analysed with their wetting properties. The superhydrophobic surface (Figure 2g) had a rugged topography with numerous peaks and holes, which caused a reduction in the top material surface area. Thus, following the Cassie-Baxter approach, this reduction leads to an increase in the CA, and superhydrophobic behaviour is observed. Furthermore, the water did not adhere to the material and it slid easily upon the surface. By contrast, the material that presented a high CA but with a high hysteresis behaviour also had a rough surface, but with a lower peak density. In this case, the water was able to penetrate slightly into the topography, thus leading to a high adhesion to the surface. Finally, the superhydrophilic surface (Figure 2d) had a structure characterised by smooth hills, which increased the ratio of the surface area when compared to the unmarked material. According to the Wenzel approach, the CA, in this case, should be lower than that for the untreated surface.

4. Conclusions

The use of a green laser in the range of nanosecond pulses provided an effective method to control the final wettability of ABS, with both superhydrophilic and superhydrophobic behaviors being observed. These wetting properties were reliant on the surface topography created, and this was directly related to the laser parameters such as the fluence. Three types of topographies were created—one smooth surface that exhibited superhydrophilic behavior and two surfaces with a rugged topography that showed superhydrophobic behavior, one with a low CAH and the other with a high adhesion effect. No significant chemical differences were detected between the marked samples that could affect to the wetting properties. All of the structures were stable over time.

Supplementary Materials: The following are available online at http://www.mdpi.com/1996-1944/11/12/2547/s1, Video: A slow-motion video sequence of water drops falling on the superhydrophilic and on the superhydrophobic surfaces.

Author Contributions: Conceptualization, methodology, and investigation, C.L.; review, and editing and supervision, L.O. and J.-I.P.

Acknowledgments: The authors wish to thank the Advanced Microscopy Laboratory (LMA) for their technical support with sample characterization, and BSH Electrodomésticos España S.A. for their financial support for this research. The authors declare no conflict of interest.

Conflicts of Interest: The funders had no role in the design of the study; in the collection, analyses, or interpretation of data; in the writing of the manuscript, or in the decision to publish the results.

References

1. Barthlott, W.; Neinhuis, C. Purity of the sacred lotus, or escape from contamination in biological surfaces. *Planta* **1997**, *202*, 1–8. [CrossRef]
2. Koch, K.; Bohn, H.F.; Barthlott, W. Hierarchically sculptured plant surfaces and superhydrophobicity. *Langmuir* **2009**, *25*, 14116–14120. [CrossRef] [PubMed]
3. Zheng, Y.; Gao, X.; Jiang, L. Directional adhesion of superhydrophobic butterfly wings. *Soft Matter* **2007**, *3*, 178–182. [CrossRef]
4. Hansen, W.R.; Autumn, K. Evidence for self-cleaning in gecko setae. *Proc. Natl. Acad. Sci. USA* **2005**, *102*, 385–389. [CrossRef] [PubMed]
5. Wang, S.; Jiang, L. Definition of superhydrophobic states. *Adv. Mater.* **2007**, *19*, 3423–3424. [CrossRef]
6. Gao, L.; McCarthy, T.J. Contact angle hysteresis explained. *Langmuir* **2006**, *22*, 6234–6237. [CrossRef] [PubMed]
7. Eral, H.B.; 't Mannetje, D.J.C.M.; Oh, J.M. Contact angle hysteresis: a review of fundamentals and applications. *Colloid Polym. Sci.* **2013**, *291*, 247–260. [CrossRef]
8. Wenzel, R.N. Resistance of solid surfaces to wetting by water. *Ind. Eng. Chem.* **1936**, *28*, 988–994. [CrossRef]
9. Cassie, A.B.D.; Baxter, S. Wettability of porous surfaces. *Trans. Faraday Soc.* **1944**, *40*, 546–551. [CrossRef]
10. Bhushan, B.; Jung, Y.C. Natural and biomimetic artificial surfaces for superhydrophobicity, self-cleaning, low adhesion, and drag reduction. *Prog. Mater. Sci.* **2011**, *56*, 1–108. [CrossRef]
11. Nosonovsky, M.; Bhushan, B. *Multiscale Dissipative Mechanisms and Hierarchical Surfaces: Friction, Superhydrophobicity, and Biomimetics*, 1st ed.; Springer Science & Business Media: Berlin, Germany, 2008.
12. Zhang, P.; Lv, F.Y. A review of the recent advances in superhydrophobic surfaces and the emerging energy-related applications. *Energy* **2015**, *82*, 1068–1087. [CrossRef]
13. Sarkar, D.K.; Farzaneh, M. Superhydrophobic coatings with reduced ice adhesion. *J. Adhes. Sci. Technol.* **2009**, *23*, 1215–1237. [CrossRef]
14. Marmur, A. Super-hydrophobicity fundamentals: implications to biofouling prevention. *Biofouling* **2006**, *22*, 107–115. [CrossRef] [PubMed]
15. Callow, J.A.; Callow, M.E. Trends in the development of environmentally friendly fouling-resistant marine coatings. *Nat. Commun.* **2011**, *2*, 244. [CrossRef]
16. Xu, W.; Song, J.; Sun, J.; Lu, Y.; Yu, Z. Rapid fabrication of large-area, corrosion-resistant superhydrophobic Mg alloy surfaces. *ACS Appl. Mater. Interfaces* **2011**, *3*, 4404–4414. [CrossRef] [PubMed]
17. Barkhudarov, P.M.; Shah, P.B.; Watkins, E.B.; Doshi, D.A.; Brinker, C.J.; Majewski, J. Corrosion inhibition using superhydrophobic films. *Corros. Sci.* **2008**, *50*, 897–902. [CrossRef]
18. Deng, X.; Mammen, L.; Zhao, Y.; Lellig, P.; Müllen, K.; Li, C.; Butt, H.; Vollmer, D. Transparent, thermally stable and mechanically robust superhydrophobic surfaces made from porous silica capsules. *Adv. Mater.* **2011**, *23*, 2962–2965. [CrossRef]
19. Bravo, J.; Zhai, L.; Wu, Z.; Cohen, R.E.; Rubner, M.F. Transparent superhydrophobic films based on silica nanoparticles. *Langmuir* **2007**, *23*, 7293–7298. [CrossRef]
20. Daoud, W.A. *Self-Cleaning Materials and Surfaces: A Nanotechnology Approach*, 1st ed.; John Wiley & Sons: New Delhi, India, 2013.
21. Mishra, S.; Yadava, V. Laser beam micromachining (LBMM)–a review. *Opt. Lasers Eng.* **2015**, *73*, 89–122. [CrossRef]

22. Bizi-Bandoki, P.; Benayoun, S.; Valette, S.; Beaugiraud, B.; Audouard, E. Modifications of roughness and wettability properties of metals induced by femtosecond laser treatment. *Appl. Surf. Sci.* **2011**, *257*, 5213–5218. [CrossRef]

23. Luo, B.H.; Shum, P.W.; Zhou, Z.F.; Li, K.Y. Preparation of hydrophobic surface on steel by patterning using laser ablation process. *Surf. Coat. Technol.* **2010**, *204*, 1180–1185. [CrossRef]

24. Wu, B.; Zhou, M.; Li, J.; Ye, X.; Li, G.; Cai, L. Superhydrophobic surfaces fabricated by microstructuring of stainless steel using a femtosecond laser. *Appl. Surf. Sci.* **2009**, *256*, 61–66. [CrossRef]

25. Cunha, A.; Serro, A.P.; Oliveira, V.; Almeida, A.; Vilar, R.; Durrieu, M.C. Wetting behaviour of femtosecond laser textured Ti–6Al–4V surfaces. *Appl. Surf. Sci.* **2013**, *265*, 688–696. [CrossRef]

26. Vorobyev, A.Y.; Guo, C. Multifunctional surfaces produced by femtosecond laser pulses. *J. Appl. Phys.* **2015**, *117*, 33103. [CrossRef]

27. Ta, V.D.; Dunn, A.; Wasley, T.J.; Li, J.; Kay, R.W.; Stringer, J.; Smith, P.J.; Esenturk, E.; Connaughton, C.; Shephard, J.D. Laser textured surface gradients. *Appl. Surf. Sci.* **2016**, *371*, 583–589. [CrossRef]

28. Skantzakis, E.; Zorba, V.; Papazoglou, D.G.; Zergioti, I.; Fotakis, C. Ultraviolet laser microstructuring of silicon and the effect of laser pulse duration on the surface morphology. *Appl. Surf. Sci.* **2006**, *252*, 4462–4466. [CrossRef]

29. Zorba, V.; Persano, L.; Pisignano, D.; Athanassiou, A.; Stratakis, E.; Cingolani, R.; Tzanetakis, P.; Fotakis, C. Making silicon hydrophobic: wettability control by two-lengthscale simultaneous patterning with femtosecond laser irradiation. *Nanotechnology* **2006**, *17*, 3234. [CrossRef]

30. Cardoso, M.R.; Tribuzi, V.; Balogh, D.T.; Misoguti, L.; Mendonça, C.R. Laser microstructuring for fabricating superhydrophobic polymeric surfaces. *Appl. Surf. Sci.* **2011**, *257*, 3281–3284. [CrossRef]

31. Malinauskas, M.; Farsari, M.; Piskarskas, A.; Juodkazis, S. Ultrafast laser nanostructuring of photopolymers: A decade of advances. *Phys. Rep.* **2013**, *533*, 1–31. [CrossRef]

32. Primo, G.A.; Igarzabal, C.I.A.; Pino, G.A.; Ferrero, J.C.; Rossa, M. Surface morphological modification of crosslinked hydrophilic co-polymers by nanosecond pulsed laser irradiation. *Appl. Surf. Sci.* **2016**, *369*, 422–429. [CrossRef]

33. Gotoh, K.; Kikuchi, S. Improvement of wettability and detergency of polymeric materials by excimer UV treatment. *Colloid Polym. Sci.* **2005**, *283*, 1356–1360. [CrossRef]

34. Athanassiou, A.; Lygeraki, M.I.; Pisignano, D.; Lakiotaki, K.; Varda, M.; Mele, E.; Fotakis, C.; Cingolani, R.; Anastasiadis, S.H. Photocontrolled variations in the wetting capability of photochromic polymers enhanced by surface nanostructuring. *Langmuir* **2006**, *22*, 2329–2333. [CrossRef] [PubMed]

35. van Pelt, S.; Frijns, A.; Mandamparambil, R.; den Toonder, J. Local wettability tuning with laser ablation redeposits on PDMS. *Appl. Surf. Sci.* **2014**, *303*, 456–464. [CrossRef]

36. Waugh, D.G.; Lawrence, J.; Walton, C.D.; Zakaria, R.B. On the effects of using CO$_2$ and F$_2$ lasers to modify the wettability of a polymeric biomaterial. *Opt. Laser Technol.* **2010**, *42*, 347–356. [CrossRef]

37. Mirzadeh, H.; Dadsetan, M. Influence of laser surface modifying of polyethylene terephthalate on fibroblast cell adhesion. *Radiat. Phys. Chem.* **2003**, *67*, 381–385. [CrossRef]

38. Riveiro, A.; Soto, R.; del Val, J.; Comesaña, R.; Boutinguiza, M.; Quintero, F.; Lusquiños, F.; Pou, J. Texturing of polypropylene (PP) with nanosecond lasers. *Appl. Surf. Sci.* **2016**, *374*, 379–386. [CrossRef]

39. Li, L.; Hong, M.; Schmidt, M.; Zhong, M.; Malshe, A.; Huis, B.; Kovalenko, V. Laser nano-manufacturing–state of the art and challenges. *CIRP Ann. Manuf. Technol.* **2011**, *60*, 735–755. [CrossRef]

40. McCann, R.; Bagga, K.; Groarke, R.; Stalcup, A.; Vázquez, M.; Brabazon, D. Microchannel fabrication on cyclic olefin polymer substrates via 1064 nm Nd:YAG laser ablation. *Appl. Surf. Sci.* **2016**, *387*, 603–608. [CrossRef]

41. Suriano, R.; Kuznetsov, A.; Eaton, S.M.; Kiyan, R.; Cerullo, G.; Osellame, R.; Chichkov, B.N.; Levi, M.; Turri, S. Femtosecond laser ablation of polymeric substrates for the fabrication of microfluidic channels. *Appl. Surf. Sci.* **2011**, *257*, 6243–6250. [CrossRef]

42. Berta, M.; Biver, É.; Maria, S.; Phan, T.N.T.; D'Aleo, A.; Delaporte, P.; Fages, F.; Gigmes, D. Nanosecond laser-induced periodic surface structuring of cross-linked azo-polymer films. *Appl. Surf. Sci.* **2013**, *282*, 880–886. [CrossRef]

43. Rebollar, E.; Castillejo, M.; Ezquerra, T.A. Laser induced periodic surface structures on polymer films: From fundamentals to applications. *Eur. Polym. J.* **2015**, *73*, 162–174. [CrossRef]

44. Sanner, N.; Huot, N.; Audouard, E.; Larat, C.; Huignard, J.P. Direct ultrafast laser micro-structuring of materials using programmable beam shaping. *Opt. Lasers Eng.* **2007**, *45*, 737–741. [CrossRef]

45. Conrad, D.; Richter, L. Ultra-short pulse laser structuring of molding tools. *Phys. Procedia* **2014**, *56*, 1041–1046. [CrossRef]

46. Sarbada, S.; Shin, Y.C. Superhydrophobic contoured surfaces created on metal and polymer using a femtosecond laser. *Appl. Surf. Sci.* **2017**, *405*, 465–475. [CrossRef]

47. Kreutz, E.W.; Frerichs, H.; Stricker, J.; Wesner, D.A. Processing of polymer surfaces by laser radiation. *Nucl. Instrum. Methods Phys. Res. Sect. B Beam Interact. Mater. Atoms* **1995**, *105*, 245–249. [CrossRef]

48. Farshchian, B.; Gatabi, J.R.; Bernick, S.M.; Park, S.; Lee, G.H.; Droopad, R.; Kim, N. Laser-induced superhydrophobic grid patterns on PDMS for droplet arrays formation. *Appl. Surf. Sci.* **2017**, *396*, 359–365. [CrossRef]

49. Biesinger, M. X-ray Photoelectron Spectroscopy (XPS) Reference Pages. Available online: http://www.xpsfitting.com/search/label/Titanium (accessed on 20 November 2018).

50. Do, H.; Yen, T.C.; Tian, C.S.; Wu, Y.H.; Chang, L. Epitaxial growth of titanium oxycarbide on MgO (001) substrates by pulsed laser deposition. *Appl. Surf. Sci.* **2011**, *257*, 2990–2994. [CrossRef]

51. Huang, Z.; Xue, K.; Zhang, Y.; Cheng, X.; Dong, P.; Zhang, X. TiO_2 hybrid material film with high CO_2 adsorption for CO_2 photoreduction. *J. Alloys Compd.* **2017**, *729*, 884–889. [CrossRef]

52. Xiu, Y.; Zhu, L.; Hess, D.W.; Wong, C.P. Hierarchical silicon etched structures for controlled hydrophobicity/superhydrophobicity. *Nano Lett.* **2007**, *7*, 3388–3393. [CrossRef]

Article

Ring Wrinkle Patterns with Continuously Changing Wavelength Produced Using a Controlled-Gradient Light Field

Hongye Li, Bin Sheng *, He Wu, Yuanshen Huang, Dawei Zhang and Songlin Zhuang

Engineering Research Center of Optical Instruments and Systems, Ministry of Education and Shanghai Key Laboratory of Modern Optical Systems, University of Shanghai for Science and Technology, Shanghai 200093, China; 167710556@st.usst.edu.cn (H.L.); 18318226501@163.com (H.W.); hyshyq@sina.com (Y.H.); dwzhang@usst.edu.cn (D.Z.); slzhuang@yahoo.com (S.Z.)

* Correspondence: bsheng@usst.edu.cn; Tel.: +86-18964196659

Received: 7 August 2018; Accepted: 27 August 2018; Published: 1 September 2018

Abstract: We report a facile method to prepare gradient wrinkles using a controlled-gradient light field. Because of the gradient distance between the ultraviolet (UV) lamp and polydimethylsiloxane (PDMS) substrate during UV/ozone treatment, the irradiance reaching the substrate continuously changed, which was transferred into the resulting SiO_x film with a varying thickness. Therefore, wrinkles with continuously changing wavelength were fabricated using this approach. It was found that the wrinkle wavelength decreased as the distance increased. We fabricated 1-D wrinkle patterns and ring wrinkles with a gradient wavelength. The ring wrinkles were prepared using radial stresses, which were achieved by pulling the center of a freely hanging PDMS film. The resulting wrinkles with changing wavelength can be used in fluid handling systems, biological templates, and optical devices.

Keywords: gradient wrinkles; UV/ozone; irradiance

---

1. Introduction

Wrinkling and buckling are common phenomena in nature, e.g., surface patterns of plants, skin wrinkles, and mountain ranges [1–3]. Surface wrinkling has recently attracted attention as a useful method to fabricate micro- and nanostructures because it is facile, cost effective, and does not use conventional lithography [4–7]. The most commonly used model for surface wrinkling is a bilayer membrane composed of a rigid, thin elastic surface layer on top of a soft, thick elastic substrate. When depositing a metal film, or conducting oxygen or ultraviolet (UV)/ozone (O_3) treatment of a pre-stretched polydimethylsiloxane (PDMS) substrate, sinusoidal wrinkles are formed after release of the pre-strain because of the strain mismatch between layers with different elastic moduli [8–10].

Many applications of wrinkles have been reported, such as tunable diffraction gratings, microlenses, mechanical property measurement of thin films, and microfluidic channels based on anisotropic wetting [4,11–15]. In these applications, wrinkle patterns with different feature sizes exhibit different surface properties such as transmittance and anisotropic wetting characteristics. Therefore, there is a demand for various wrinkle groove structures with different feature sizes. There have been many attempts to fabricate gradient wrinkle patterns [16–18]. Claussen et al. [19] presented a method to continuously change the elastic modulus of PDMS. Yu and co-workers achieved different symmetry breaking of the wrinkle patterns caused by the tunable thickness gradient in metal films deposited on soft elastic substrates [20]. Lee et al. [21] generated a stepwise gradient wrinkle pattern on a PDMS substrate in which the wavelength could be spatially controlled in each distinct region with a clear boundary.

In this paper, we report a new facile method to fabricate wrinkles with continuously changing wavelength based on a controlled gradient light field. We inclined the PDMS substrate to fabricate gradient 1-D wrinkle patterns. The period of wrinkles is indirectly controlled by simply changing the distance between the UV lamp and PDMS substrate. Based on the conclusion of this experiment, we exposed a PDMS substrate to UV/O$_3$ using a cone to provide radial stretch, which in turn enables the thickness of the SiO$_x$ thin film along the radial distribution to be controlled. Then we obtained variable-period ring wrinkles.

2. Experimental Section

2.1. Materials

The PDMS we used was Sylgard 184 of DOW Corning (Dow Corning, Midland, TX, USA), which consisted of two parts: one was the silicone elastomer; the other was the curing agent. Silicone elastomer was mixed with the curing agent at a 10:1 weight ratio [22], cured at 100 °C for 2 h, and then naturally cooled in air. The cured PDMS with a thickness of 0.5 mm was cut into a circular substrate with a diameter of 20 mm. The PDMS substrate was exposed to UV/O$_3$ using a UV lamp (low-pressure mercury lamp, BHK, Claremont, NH, USA) that emitted 185- and 254-nm radiation in atmospheric oxygen (O$_2$).

2.2. Film Fabrication

The film samples were prepared by UV/O$_3$ treatment at room temperature in the presence of atmosphere oxygen [23]. The 185-nm radiation from the lamp produced O$_3$, while the 254-nm radiation decomposed the O$_3$ into O$_2$ and atomic oxygen (O). The atomic oxygen was the chief reactant with PDMS. The oxidation was initiated at the PDMS surface and gradually penetrated into the PDMS. Oxidation converted the organic portion of PDMS to carbon dioxide, water, and some volatile organic compounds that escaped from the PDMS. In contrast, the silicon components did not form volatile compounds under these conditions, thus forming a residual hard layer of SiO$_x$. The exposure time was 80 min. The film thickness was easily tuned by changing the distance between the lamp and PDMS substrate. To fabricate SiO$_x$ thin films with a thickness gradient, the PDMS substrate was held on an incline or stretched by cones with different angles.

2.3. Characterization

The surface morphologies of the samples were observed by an optical microscope (10XB-PC, Shanghai Optical Instrument Factory, Shanghai, China) equipped with a charge-coupled device camera. The profiles of the wrinkle structures were scanned by a white light interferometer (Contour GT-KO, Bruker, Tucson, AZ, USA).

3. Results and Discussion

3.1. Fabrication of Gradient 1-D Wrinkle Patterns and Sample Characterization

The wrinkle wavelength λ of a buckled bilayer system can be calculated for high strains according to Equation (1) [24]:

$$\lambda = \frac{2\pi h_f}{(1 + \varepsilon_{pre})(1 + \delta)^{\frac{1}{3}}} \left[\frac{E_f\left(1 - \nu_s^2\right)}{3E_s\left(1 - \nu_f^2\right)} \right]^{\frac{1}{3}}, \tag{1}$$

where $\delta = \frac{5}{32}\left[\varepsilon_{pre}\left(1 + \varepsilon_{pre}\right)\right]$ represents the large deformation and geometrical nonlinearity in the substrate, h_f is the thickness of the SiO$_x$ film, ε_{pre} is the prestrain, and E and ν are the Young's modulus and Poisson's ratio of the substrate (s) and film (f), respectively. Therefore, λ can be controlled by changing h_f, ε_{pre}, or E_s. In our experiment, λ was tuned by changing the thickness of the SiO$_x$ film exposed to the UV/O$_3$ treatment. It is well known that in UV/O$_3$ treatment, O$_3$ is not the primary reactant; instead atomic oxygen produced by the photodissociation of O$_3$, reacts with PDMS to form the SiO$_x$ film. Therefore, the thickness of the SiO$_x$ film is related to the concentration of atomic oxygen, which is influenced by the irradiance (E) In our experiments, the UV lamp was a line source. Therefore, E can be expressed as follows:

$$E = \frac{d\Phi}{dA} = \frac{I}{d}\cos\theta, \tag{2}$$

where $d\Phi$ is the radiation flux on facet dA, and I is the radiation intensity of the radiation source. θ is the angle between the facet and radiation source, and d represents the distance between the illuminated surface and radiation source. The thickness of the SiO_x film and the concentration of the atomic oxygen have a logarithmic relationship [25]. Therefore, the thickness of the SiO_x film can be tuned by E. According to Equation (1), $\lambda \propto h_f$. Thus, at a fixed radiation intensity and the same exposure time, the continuously changing λ can be easily tuned by controlling d.

The SiO_x film with a thickness gradient was prepared by inclining the PDMS substrate during UV/O$_3$ exposure, as shown in Figure 1. The UV lamp was uniformly scanned by the stepper motor to illuminate the substrate. The sinusoidal wrinkling of the sample with a continuously changing λ was fabricated by exposing an inclined PDMS substrate subjected to 20% pre-strain to UV/O$_3$ treatment for 80 min. At a fixed exposure time, when d was smaller, the irradiance was greater and the film was thicker. As a result, a film with a thickness gradient formed on the inclined PDMS substrate. The thickness gradient reflected the slope of the film and strongly depended on the dip angle θ.

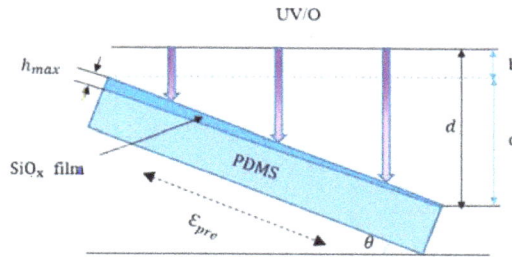

Figure 1. Schematic view of the cross section of a film with a thickness gradient prepared by inclining the PDMS substrate during exposure.

To investigate the wrinkle surface of the thickness-gradient film in more detail, corresponding optical images of the wrinkles at different positions were collected, as presented in Figure 2. Sample position (a) corresponds to h_{max} It was found that λ decreased significantly from (a) to (c) as d increased. Because of the uniform λ at each measuring spot, these optical images were representative of the topography at the specific sample positions.

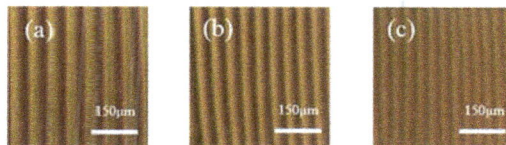

Figure 2. Optical images showing the profiles of wrinkle patterns with different wavelengths at sample positions of (a) 0.6 mm, (b) 3.3 mm and (c) 4.5 mm. Corresponding distances between the lamp and PDMS substrate d were (a) 7.83 mm, (b) 9.15 mm and (c) 9.98 mm.

Figure 3 plots λ as a function of the sample position. λ changed continuously from 75 to 39 µm as the sample position changed. λ at different positions was obtained by a white light interferometer.

The experimental values showed that the gradient λ could be attributed to d. We proposed the following correlation between the wavelength and some variables.

$$k = \frac{2\pi}{(1+\varepsilon_{pre})(1+\delta)^{\frac{1}{3}}}\left[\frac{E_f(1-\nu_s{}^2)}{3E_s(1-\nu_f{}^2)}\right]^{\frac{1}{3}}, \tag{3}$$

$$\lambda \sim k{\cdot}\lg\left(\frac{I}{d}\cos\theta\right) = k{\cdot}\lg\left(\frac{I}{x\sin\theta+b}\cos\theta\right), \tag{4}$$

where k is related to ε_{pre}, x is the sample position, I is the radiation intensity of radiation source, and d, θ, and b are shown in Figure 1. The data for λ showed excellent agreement with the prediction of Equations (3) and (4). Therefore, we realized the modulation of λ by controlling the irradiance.

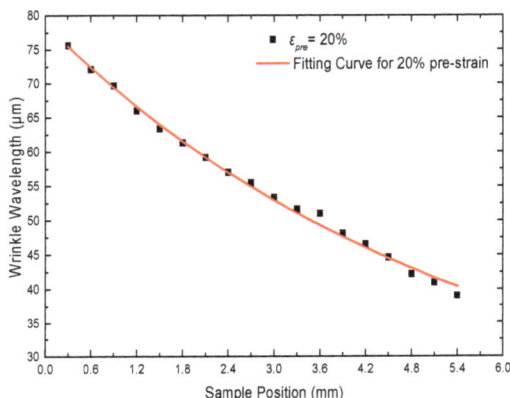

Figure 3. Wrinkle wavelength λ as a function of sample position at 20% pre-strain. The red curve represents the fit according to the relation between λ and the distance d using Equations (3) and (4).

3.2. Variable-Period Ring Wrinkles Fabrication and Characterization

From the 1-D stretching experiment, we know that the wrinkles were perpendicular to the direction of compressive stress. Based on this principle, we produced ring wrinkles using radial stresses [26,27]. In our experiment, ring wrinkles were achieved by pulling the center of a freely hanging PDMS film, as shown in Figure 4. In the pulling device, a Teflon cone with a semi-spheroidal tip pushed the center of the PDMS film from below to stretch the film. We moved the cone by rotating a screw underneath it, so that the cone itself did not rotate to avoid friction with the PDMS film and decrease unwanted strain in the azimuthal direction. The cone was made of Teflon to lower the friction between the cone and PDMS film.

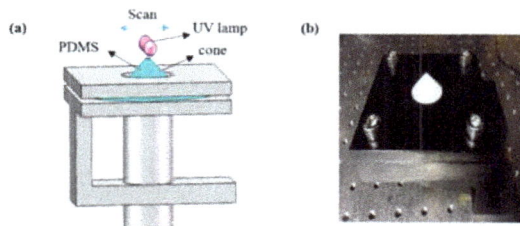

Figure 4. (a) Schematic diagram of the experimental device used to prepare ring wrinkles with a continuously changing wavelength. The PDMS membrane was fixed between two plates with round holes and then stretched by a Teflon cone. The UV lamp evenly scanned the substrate. (b) Photograph of the experimental setup.

The ring wrinkle fabrication procedure is shown in Figure 5a. First, the PDMS membrane was fixed between two plates with round holes. The PDMS membrane was 0.5 mm thick. Then the Teflon cone pushed the center of the PDMS membrane. The radius of the stretched membrane was 10 mm. The stretched PDMS substrate was then evenly exposed to UV/O$_3$. Ring wrinkles were generated after releasing the pre-strain. According to the theory behind the gradient 1-D wrinkle pattern formation, a SiO$_x$ film with a thickness gradient could be generated in this experiment (Figure 5b). Therefore, variable-period ring wrinkles could be fabricated by this process. The approximate stretching ratio (*SR*) can be calculated using Equation (5),

$$SR = \frac{\sqrt{OA^2 + R^2}}{R},$$

(5)

where the *OA* is the height of the cone and *R* is the effective radius of the PDMS membrane (Figure 5b).

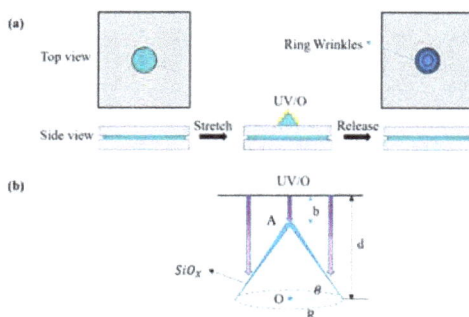

Figure 5. Schematic illustration of the production of concentric ring wrinkles with continuously changing wavelength. (a) The center of the substrate was extended upward using a cone and simultaneously treated with UV/O$_3$. Concentric wrinkles with continuously changing wavelength were generated after the removal of the cone. (b) The thickness gradient of the film was controlled by the distance between the lamp and PDMS substrate *d*.

The ring wrinkles with continuously changing λ were fabricated by exposing UV/O$_3$ for 80 min on a PDMS substrate subjected to 20% pre-strain generated by the cone. As shown in the optical micrograph in Figure 6b, nonuniformly oriented patterns in the central region were produced because of the friction between the membrane and semi-spheroidal tip of the cone. The size of this region is related to the radian of the cone. Ideally, if the top of a cone does not have a radian, it should avoid the formation of disordered wrinkles in the central region. In the other regions of the membrane, there was no stretching in the circumferential direction, and the strain was directly determined by the percentage of radial stretching. Similar to the 1-D stretching experiment presented above, the distance from the center of the membrane pushed by the cone to the edge d gradually increased. Figure 6c–f reveals that λ decreased from 102 to 69 µm with increasing d.

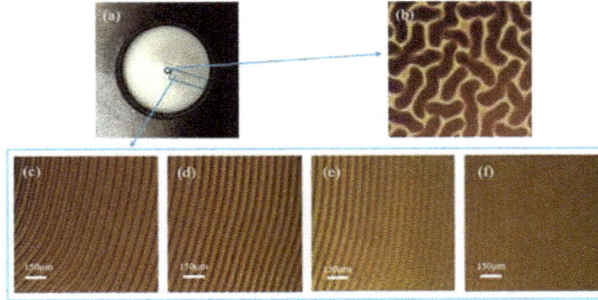

Figure 6. After releasing the pre-strain, the resulting ring wrinkles were observed by an optical microscope. (**a**) The whole membrane in ambient light. (**b**) Optical image of the center of the membrane. (**c–f**) Concentric ring wrinkles with continuously changing wavelength from the center of the membrane to the edge in the blue rectangle of Figure 6a corresponding to the gradual increase of d.

To explore the effect of different pre-strains on λ of the resulting wrinkles, we stretched the PDMS membrane by the cone with different SR. The UV/O_3 exposure time was fixed at 80 min. The distance between the lamp and the top of the PDMS membrane was the same in all samples. The λ values of the samples were measured by a white light interferometer starting from the boundary of the disordered center of each sample. The measured results are presented in Figure 7. The data for λ showed excellent agreement with the prediction of Equations (3) and (4). At each pre-strain, the experimental values showed that λ was influenced by the distance d between the lamp and the PDMS substrate; that is, λ gradually decreased as d increased. Figure 7 reveals that for different SR, λ decreased as SR increased at the same sample position. The larger θ corresponded to the larger pre-strain shown in Figure 5b. According to Equations (3) and (4), λ decreased as the pre-strain increased. In the same way, λ decreased as θ increased. Thus, at the same sample position, larger pre-strain resulted in smaller λ.

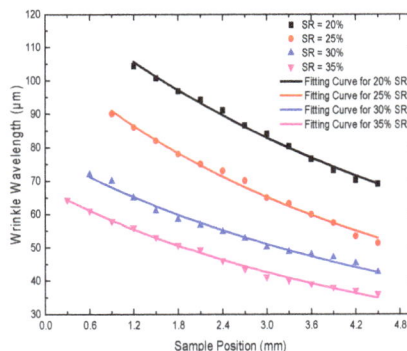

Figure 7. Continuously changing wrinkle wavelengths obtained at different SR are schematically depicted as a function of the sample position. The solid lines correspond to the fits according to Equations (3) and (4).

4. Conclusions

In summary, we developed a facile method to produce gradient wrinkles using a gradient light field. The continuously changing distance between the lamp and PDMS substrate influenced the irradiance during UV/O$_3$ treatment, which then resulted in SiO$_x$ films with varying thicknesses. Therefore, wrinkles with continuously changing λ can be fabricated using this approach. We fabricated gradient 1-D wrinkle patterns using a gradient light field. In addition, we also prepared variable-period ring wrinkles by using a cone to stretch the PDMS membrane. Our results demonstrated that the gradient wavelength of wrinkles can be easily controlled by modulating the distance between the lamp and PDMS substrate. The experimental technique presented in this paper can be further developed to effectively control pattern formation, which may be beneficial to provide patterned surfaces for use in optical devices, diffraction gratings, hydrophobicity, fluid handling systems, and biological templates.

Author Contributions: Writing–original draft, H.L.; Writing–review & editing, B.S., Y.H. and H.W.; Supervision, D.Z. and S.Z.

Funding: This research was funded by the National Natural Science Foundation of China (61775140, 61775141 and 11105149) and the National Key Research and Development Program of China (2016YFB1102303).

Acknowledgments: We would like to express our gratitude to the editors and the reviewers for their constructive and helpful review comments.

Conflicts of Interest: The authors declare no conflict of interest.

References

1. Efimenko, K.; Rackaitis, M.; Marias, E.; Vaziri, A.; Mahadevan, L.; Genzer, J. Nested self-similar wrinkling patterns in skins. *Nat. Mater.* **2005**, *4*, 293–297. [CrossRef] [PubMed]
2. Genzer, J.; Groenewold, J. Soft matter with hard skin: From skin wrinkles to templating and material characterization. *Soft Matter* **2006**, *2*, 310–323. [CrossRef]
3. Li B.; Cao, Y.P.; Feng, X.Q.; Gao, H. Mechanics of morphological instabilities and surface wrinkling in soft materials: A review. *Soft Matter* **2012**, *8*, 5728–5745. [CrossRef]
4. Miao, L.; Cheng, X.; Chen, H.; Song, Y.; Guo, H.; Zhang, J.; Chen, X.; Zhang, H. Fabrication of controlled hierarchical wrinkle structures on polydimethylsiloxane via one-step C$_4$F$_8$ plasma treatment. *J. Micromech. Microeng.* **2018**, *28*, 015007. [CrossRef]
5. Park, S.K.; Kwark, Y.J.; Nam, S.; Moon, J.; Dong, W.K.; Park, S.; Park, B.; Yun, S.; Lee, J.I.; Yu, B. A variation in wrinkle structures of UV-cured films with chemical structures of prepolymers. *Mater. Lett.* **2017**, *199*, 105–109. [CrossRef]
6. Pretzl, M.; Schweikart, A.; Hanske, C.; Chiche, A.; Zettl, U.; Horn, A.; Böker, A.; Fery, A. A lithography-free pathway for chemical microstructuring of macromolecules from aqueous solution based on wrinkling. *Langmuir* **2008**, *24*, 12748–12753. [CrossRef] [PubMed]
7. Yang, S.; Khare, K.; Lin, P.C. Harnessing surface wrinkle patterns in soft matter. *Adv. Funct. Mater.* **2010**, *20*, 2550–2564. [CrossRef]
8. Bowden, N.; Brittain, S.; Evans, A.G.; Hutchinson, J.W.; Whitesides, G.M. Spontaneous formation of ordered structures in thin films of metals supported on an elastomeric polymer. *Nature* **1998**, *393*, 146–149.
9. Bowden, N.; Huck, W.T.S.; Paul, K.E.; Whitesides, G.M. The controlled formation of ordered, sinusoidal structures by plasma oxidation of an elastomeric polymer. *Appl. Phys. Lett.* **1999**, *75*, 2557–2559. [CrossRef]
10. Efimenko, K.; Wallace, W.E.; Genzer, J. Surface modification of Sylgard-184 poly(dimethyl siloxane) networks by ultraviolet and ultraviolet/ozone treatment. *J. Colloid Interface Sci.* **2002**, *254*, 306–315. [CrossRef] [PubMed]
11. Chakraborty, A.; Xiang, M.; Luo. C. Fabrication of super-hydrophobic microchannels via strain-recovery deformations of polystyrene and oxygen reactive ion etch. *Materials* **2013**, *6*, 3610–3623. [CrossRef] [PubMed]
12. Chan, E.P.; Crosby, A.J. Fabricating microlens arrays by surface wrinkling. *Adv. Mater.* **2010**, *18*, 3238–3242. [CrossRef]
13. Chan, E.P.; Smith, E.J.; Hayward, R.C.; Crosby, A.J. Surface wrinkles for smart adhesion. *Adv. Mater.* **2010**, *20*, 711–716. [CrossRef]

14. Khare, K.; Zhou, J.; Yang, S. Tunable open-channel microfluidics on soft poly(dimethylsiloxane) (PDMS) substrates with sinusoidal grooves. *Langmuir* **2009**, *25*, 12794–12799. [CrossRef] [PubMed]
15. Chung, J.Y.; Youngblood, J.P.; Stafford, C.M. Anisotropic wetting on tunable micro-wrinkled surfaces. *Soft Matter* **2007**, *3*, 1163–1169. [CrossRef]
16. Zhao, J.; Guo, X.; Lu, L. Controlled wrinkling analysis of thin films on gradient substrates. *Appl. Math. Mech. -Engl. Ed.* **2017**, *38*, 1–8. [CrossRef]
17. Li, K.; Wang, J.; Shao, B.; Xiao, J.; Zhou, H.; Yu, S. Wrinkling patterns of tantalum films on modulus-gradient compliant substrates. *Thin Solid Films* **2018**, *654*, 100–106. [CrossRef]
18. Yu, S.; Sun, Y.; Ni, Y.; Zhang, X.; Zhou, H. Controlled formation of surface patterns in metal films deposited on elasticity-gradient PDMS substrates. *ACS Appl. Mater. Interfaces* **2016**, *8*, 5706–5714. [CrossRef] [PubMed]
19. Kai, U.C.; Tebbe, M.; Giesa, R.; Schweikart, A.; Fery, A.; Schmidt, H.W. Towards tailored topography: Facile preparation of surface-wrinkled gradient poly(dimethyl siloxane) with continuously changing wavelength. *RSC Adv.* **2012**, *2*, 10185–10188.
20. Filiatrault, H.L.; Carmichael, R.S.; Boutette, R.A.; Carmichael, T.B. A self-assembled, low-cost, microstructured layer for extremely stretchable gold films. *ACS Appl. Mater. Interfaces* **2015**, *37*, 20745–20752. [CrossRef] [PubMed]
21. Lee, J.S.; Hong, H.; Park, S.J.; Lee, S.J.; Dong, S.K. A simple fabrication process for stepwise gradient wrinkle pattern with spatially-controlled wavelength based on sequential oxygen plasma treatment. *Microelectron. Eng.* **2017**, *176*, 101–105. [CrossRef]
22. Dow Consumer Solutions. Available online: https://consumer.dow.com/en-us/pdp.sylgardTM184siliconeelastomerkit.01064291z.html (accessed on 21 January 2018).
23. Ouyang, M.; Muisener, R.J.; Boulares, A.; Koberstein, J.T. UV–ozone induced growth of a SiO_x surface layer on a cross-linked polysiloxane film: Characterization and gas separation properties. *J. Membr. Sci. Technol.* **2000**, *177*, 177–187. [CrossRef]
24. Jiang, H.; Khang, D.Y.; Song, J.; Sun, Y.; Huang, Y.; Rogers, J.A. Finite deformation mechanics in buckled thin films on compliant supports. *Proc. Natl. Acad. Sci. USA* **2007**, *104*, 15607–15612. [CrossRef] [PubMed]
25. Bayley, F.A.; Liao, J.L.; Stavrinou, P.N.; Chiche, A.; Cabral, J.T. Wavefront kinetics of plasma oxidation of polydimethylsiloxane: Limits for sub-μm wrinkling. *Soft Matter* **2014**, *10*, 1155–1166. [CrossRef] [PubMed]
26. Li, R.; Yi, H.; Hu, X.; Chen, L.; Shi, G.; Wang, W.; Yang, T. Generation of diffraction free optical beams using wrinkled membranes. *Sci. Rep.* **2013**, *3*, 2775. [CrossRef] [PubMed]
27. Saito, A.C.; Matsui, T.S.; Sato, M.; Deguchi, S. Aligning cells in arbitrary directions on a membrane sheet using locally formed microwrinkles. *Biotechnol. Lett.* **2014**, *36*, 391–396. [CrossRef] [PubMed]